P9-CBD-815

Basic Electromagnetic Fields

BASIC ELECTROMAGNETIC FIELDS

HERBERT P. NEFF, JR.

University of Tennessee, Knoxville

HARPER & ROW, PUBLISHERS, New York

Cambridge, Philadelphia, San Francisco,
London, Mexico City, São Paulo, Sydney

To Barbara and the kids

Cover: On the cover is reproduced Figure 5.10, page 141.

Sponsoring Editor: Charlie Dresser
Project Editor: Karla Billups Philip
Production Manager: Marion A. Palen
Compositor: The Composition House Limited
Printer and Binder: The Murray Printing Company
Art Studio: Danmark & Michaels Inc.

BASIC ELECTROMAGNETIC FIELDS

Library of Congress Cataloging in Publication Data

Neff, Herbert P 1930–
Basic electromagnetic fields.

Includes index.
1. Electromagnetic fields. I. Title.
QC665.E4N43 537 80-20147
ISBN 0-06-044785-0

Contents

Preface

It is assumed in *Basic Electromagnetic Fields* that the student is familiar with vector addition and the scalar and vector products. Other, more advanced, vector operations are introduced and explained as they are needed. This approach is entirely consistent with the desire for a modern textbook, even though it may be a break with tradition. A rather complete summary of vector relations is included in Appendix A for reference.

Although it is not necessary, it is very helpful if the student has also had a basic course in electric circuit theory in the sophomore year. This is quite likely the case for most engineering colleges. The concept and use of the Dirac delta function is now being taught in circuit theory at this level in many institutions. The Dirac delta function proves convenient for use in supplying the proofs of many integral relations encountered in field theory. These proofs are not supplied in the body of this text, but, rather, are included in Appendix G for the interested reader.

Since one can find the response of a linear system to arbitrary excitation if its response to impulse (or Dirac delta function) excitation is known, and since the field theory in this text is almost exclusively "linear," impulse responses will be identified and used wherever possible. This proves to be not only conceptually effective, but quite often avoids the expenditure of much time and labor in pursuing classical methods of deriving superposition type integrals of field theory.

This text has been designed primarily for use at the junior or senior level in electrical engineering. From an applications viewpoint, it is by no means complete, but is heavily slanted toward guiding systems. For this reason, some mention of the cylindrical waveguide and cavity and the spherical cavity is included in Appendix H. A detailed study of these systems requires the introduction of certain special mathematical functions with which the undergraduate student may not be familiar. In order to obtain a qualitative and quantitative understanding of the behavior of these systems at this level of exposure, it is not necessary for the student to have a detailed knowledge of these functions. It is very easy, however, for the instructor to simply omit this material, if so desired, with no loss in continuity.

One of the most frustrating experiences the beginner can encounter in problems involving integration with several variables is the lack of a consistent notation. Even if the student understands the statement of a fundamental equation or law, he or she may have considerable difficulty in solving a problem involving the use of such a law. Many times this difficulty can be overcome if a precise and consistent notation is followed. This does not, of course, guarantee that the student will be able to solve the problem, but at least the solution can be started properly. One of the goals of this text has been that of following a precise and consistent notation.

Even experienced engineers occasionally find themselves in an apparent dilemma when working in an area as mathematically oriented as is field theory. Quite often the difficulty arises because of naive attitudes developed as an undergraduate. For example, in certain mathematical formulations, we sometimes overlook built-in constraints in plunging headlong toward our final result. It is worthwhile to stop along the way occasionally and ask if the last step was valid, and if so, under what conditions? Are the results reasonable on physical grounds? An attempt has been made in this textbook to point out explicitly these pitfalls when they arise.

The procedure followed in *Basic Electromagnetic Fields* is that of adopting experimental laws as the need arises, maintaining a perhaps slow, but steady, pace toward Maxwell's equations. We could start with Maxwell's equations, if desired, and show that the experimental laws are correct, but that they are merely special cases. The latter procedure is often followed at the graduate level. The former procedure has the distinct advantage of allowing the student to gradually accept the fundamental laws and the mathematics, instead of being engulfed by them.

There are many example problems that are worked out in the text when a particularly important, or perhaps difficult, point is being made. A modest number of problems has been included at the end of each chapter. These problems refer, with few exceptions, explicitly to the material just covered, and should, therefore, be considered an intrinsic part of the chapter.

A reference list is included at the end of each chapter and is referred to occasionally. The books that are mentioned explicitly are those that have had the most influence on this author, but it is certainly not implied that these are the only useful books. Many discussions of historical interest can be found in the literature and the student is invited to look into these. This material, often called "bedtime reading," is usually very interesting and informative, but, at the same time, represents one more time-consuming task in the already crowded schedule of the undergraduate engineering student. For this reason, outside reading is recommended but not demanded.

Listed below are three possible outlines assuming that three hours per week are allowed for the course.

TWO QUARTERS

FIRST QUARTER	SECOND QUARTER
(STATICS)	(DYNAMICS)
Chapter 1	Chapter 9[a]
Chapter 2	Chapter 10[b]
Chapter 3	Chapter 11
Chapter 4	Chapter 12
Chapter 5[a]	Chapter 13[b]
Chapter 6[a]	Chapter 14[a]
Chapter 7	Chapter 15[b]
Chapter 8[a]	Chapter 16[b,c]

THREE QUARTERS

FIRST QUARTER	SECOND QUARTER	THIRD QUARTER
(ELECTROSTATICS)	(MAGNETOSTATICS AND DYNAMICS)	(GUIDING SYSTEMS)
Chapter 1	Chapter 7	Chapter 11
Chapter 2	Chapter 8	Chapter 12
Chapter 3	Chapter 9	Chapter 13[b]
Chapter 4	Chapter 10[a]	Chapter 14
Chapter 5		Chapter 15[b]
Chapter 6		Chapter 16[a]

[a] Selected topics only.
[b] Can be omitted entirely with no loss in continuity.
[c] If time allows.

TWO SEMESTERS

FIRST SEMESTER	SECOND SEMESTER
Chapter 1	Chapter 9
Chapter 2	Chapter 10
Chapter 3	Chapter 11
Chapter 4	Chapter 12
Chapter 5	Chapter 13[b]
Chapter 6	Chapter 14
Chapter 7	Chapter 15[b]
Chapter 8	Chapter 16[b,c]

[a] Selected topics only.
[b] Can be omitted entirely with no loss in continuity.
[c] If time allows.

The author would like to acknowledge his appreciation to Professor V. F. Schultz, Purdue University (retired), Professor C. H. Weaver, University of Tennessee, and Professor E. R. Graf, Auburn University, for providing the opportunities which led to the development of this book. Particular thanks go to Professor J. D. Tillman, Jr., University of Tennessee, for the many fruitful discussions we have shared on electromagnetics. The author is also indebted to the typists and reviewers who labored on the manuscript for this text.

Herbert P. Neff, Jr.

Symbols and Units

The International System of Units is an expanded form of the rationalized meter-kilogram-second-ampere (MKSA) system of units, and is used throughout this textbook. It is based on six basic units: the meter (length), the kilogram (mass), the second (time), the ampere (electric current), the kelvin degree (temperature), and the candela (luminous intensity). All other units may be derived from these six. Definitions of these fundamental units may be found in the references. See Hayt in the references at the end of Chapter 2, for example.

The symbols used in this text are, more or less, conventional. That is, they are usually the same as those for the same quantities found in other textbooks. Occasionally, it is convenient, because of inference, to use the same symbol for two different quantities. For example, ω_c is used to represent the angular cutoff frequency of a waveguide, and it may also be used to represent the cyclotron frequency. In a situation such as this, these quantities will not appear together in the discussion, and certainly not in the same equations. There will never be any confusion about what is meant. The following table lists the quantities that are encountered in the text. It also includes the symbol used for the quantity, the unit of the quantity, and the abbreviation of the unit.

NAMES AND UNITS OF THE QUANTITIES FOUND IN THE TEXT IN THE INTERNATIONAL SYSTEM

QUANTITY	SYMBOL	UNIT	ABBREVIATION
ac capacitivity	$\varepsilon'(\omega)$	farads/meter	F/m
ac inductivity	$\mu'(\omega)$	henrys/meter	H/m
Acceleration	$\mathbf{a}$	meters/second^2	m/s^2
Admittance	Y	mhos	℧
Admittivity	$\hat{y}$	mhos/meter	℧/m
Angular frequency	ω	radians/second	s^{-1}
Angular velocity	$\boldsymbol{\omega}$	radians/second	s^{-1}
Antenna aperture	A	meters^2	m^2
Antenna gain	G	—	—
Associated Legendre function, first kind	$P_n^m(x)$	—	—
Associated Legendre function, second kind	$Q_n^m(x)$	—	—
Attenuation constant	α	nepers/meter	m^{-1}
Bessel function, first kind, order n	$J_n(x)$	—	—
Bessel function, second kind, order n	$N_n(x)$	—	—
Bound charge density	ρ_v^b	coulombs/meter^3	C/m^3
Bound current density	$\mathbf{J}^b$	amperes/meter^2	A/m^2
Capacitance	C	farads	F
Characteristic impedance	Z_0	ohms	Ω
Characteristic resistance	R_0	ohms	Ω
Charge	Q, q	coulombs	C
Coefficient of magnetic coupling	k_{mc}	—	—

QUANTITY	SYMBOL	UNIT	ABBREVIATION
Coefficient of reflection	Γ	—	—
Coefficient of transmission	T	—	—
Complex propagation constant	γ	meter^{-1}	m^{-1}
Conductance	G	mhos	℧
Conductivity	σ	mhos/meter	℧/m
Current	$i(t)$, I	amperes	A
Current density	$\mathbf{J}$	amperes/meter^2	A/m^2
Cutoff angular frequency	ω_c	radians/second	s^{-1}
Cutoff frequency	f_c	hertz	Hz
Del operator	∇	—	—
Dirac delta function	$\delta(x)$ (x, meters)	meter^{-1}	m^{-1}
Dielectric loss angle	δ_d	—	—
Dielectric loss factor	$\varepsilon''(\omega)$	farads/meter	F/m
Drift velocity	$\mathbf{u}_d$	meters/second	m/s
Effective length	l_E	meters	m
Electric dipole moment	$\mathbf{p}$	coulomb-meter	C·m
Electric energy	W_E	joules	J
Electric energy density	w_E	joules/meter^3	J/m^3
Electric field intensity	$\mathbf{E}$	volts/meter	V/m
Electric flux	Ψ_E	coulombs	C
Electric flux density	$\mathbf{D}$	coulombs/meter^2	C/m^2
Electric potential difference	Φ_{ab}	volts	V
Electric scalar potential	Φ	volts	V
Electric susceptibility	χ_E	—	—

QUANTITY	SYMBOL	UNIT	ABBREVIATION
Electric vector potential	$\mathbf{F}$	volts	V
Electromotive force	emf	volts	V
Electron charge	$-e$	coulombs	C
Electron mobility	μ_e	meter²/volt-second	$m^2/V\cdot s$
Electron rest mass	m_e	kilograms	kg
Energy (work)	W	joules	J
Force	$\mathbf{F}$	newtons	N
Frequency	f	hertz	Hz
Group velocity	u_g	meters/second	m/s
Hankel function, first kind, order n	$H_n^{(1)}(x)$	—	—
Hankel function, second kind, order n	$H_n^{(2)}(x)$	—	—
Heat flux rector	$\mathbf{s}$	watts/meter²	W/m^2
Hole mobility	μ_h	meter²/volt-second	$m^2/V\cdot s$
Impedance	Z	ohms	Ω
Inductance	L	henrys	H
Intrinsic impedance	η	ohms	Ω
Length	R, l, d, etc.	meters	m
Line charge density	ρ_l	coulombs/meter	C/m
Mass	M, m	kilograms	kg
Magnetic dipole moment	$\mathbf{m}$	ampere-meter²	$A\cdot m^2$
Magnetic energy	W_H	joules	J
Magnetic energy density	w_H	joules/meter³	J/m^3
Magnetic field intensity	$\mathbf{H}$	amperes/meter	A/m
Magnetic flux	ψ_m	webers	Wb
Magnetic flux density	$\mathbf{B}$	webers/meter² (tesla)	Wb/m^2
Magnetic loss angle	δ_m	—	—

QUANTITY	SYMBOL	UNIT	ABBREVIATION
Magnetic loss factor	$\mu''(\omega)$	henrys/meter	H/m
Magnetic scalar potential	Φ_m, Φ_f	amperes	A
Magnetic susceptibility	χ_m	—	—
Magnetic vector potential	$\mathbf{A}$	webers/meter	Wb/m
Magnetization	$\mathbf{M}$	amperes/meter	A/m
Magnetomotive force	mmf	ampere-turns	A·t
Modified Bessel function, first kind, order n	$I_n(x)$	—	—
Permeability	μ	henrys/meter	H/m
Permeability of vacuua	μ_0	henrys/meter	H/m
Permittivity	ε	farads/meter	F/m
Permittivity of vacuua	ε_0	farads/meter	F/m
Phase constant	β	radians/meter	m^{-1}
Phase velocity	u_p	meters/second	m/s
Polarization	$\mathbf{P}$	coulombs/meter^2	C/m^2
Power	P	watts	W
Power per unit solid angle	P_s	watts/steradian	W
Power radiated	P_r	watts	W
Poynting vector (power density)	$\mathbf{S}$	watts/meter^2	W/m^2
Proton rest mass	m_p	kilograms	kg
Quality factor	Q	—	—
Radiation impedance	Z_r	ohms	Ω
Radiation resistance	R_r	ohms	Ω
Relative permeability	μ_R	—	—

QUANTITY	SYMBOL	UNIT	ABBREVIATION
Relative permittivity	ε_R	—	—
Reluctance	$\mathscr{R}$	henrys^{-1}	H^{-1}
Resistance	R	ohms	Ω
Resonant frequency	f_r	hertz	Hz
Schelkunoff's Bessel function, first kind	$\hat{J}_n(x)$	—	—
Schelkunoff's Bessel function, second kind	$\hat{N}_n(x)$	—	—
Spherical Bessel function, first kind	$j_n(x)$	—	—
Spherical Bessel function, second kind	$n_n(x)$	—	—
Skin depth (depth of penetration)	δ	meters	m
Standing wave ratio	SWR	—	—
Surface area	s	meter^2	m^2
Surface charge density	ρ_s	coulombs/meter^2	C/m^2
Surface current density	$\mathbf{J}_s$	amperes/meter	A/m
Surface impedance	η	ohms	Ω
System operator	$H\{-\}$	—	—
Temperature	T	degrees kelvin	°K
Time delay	τ_D	seconds	s
Torque	$\mathbf{T}$	newton-meters	N·m
Transfer function	H, $\mathbf{H}$, H(ω)	varies	varies
Transverse wave impedance	$\eta^{\pm}$	ohms	Ω
Unit impulse response	h, $\mathbf{h}$	varies	varies
Unit step function	$u(x)$	—	—

QUANTITY	SYMBOL	UNIT	ABBREVIATION
Vectors	$\mathbf{v}$	—	—
Velocity	$\mathbf{u}$	meters/second	m/s
Velocity of energy flow	u_e	meters/second	m/s
Velocity of light	c	meters/second	m/s
Voltage	$v(t)$, V_0, V_{21}, etc.	volts	V
Voltage standing wave ratio	VSWR, s	—	—
Volume	v, vol	meter^3	m^3
Volume charge density	ρ_v	coulombs/meter^3	C/m^3
Waveguide wavelength	λ_g	meters	m
Wavelength	λ	meters	m
Wave number (intrinsic phase constant)	k	radians/meter	m^{-1}

Basic Electromagnetic Fields

Chapter 1
Coulomb's Law and Electric Field Intensity

Electromagnetic theory is intimately associated with the properties of electric charges. Starting with Maxwell's equations one can obtain all of the desired results as special cases. This rather abrupt approach may be acceptable at the graduate level, but would certainly leave the undergraduate student in a dazed condition. This occurs because most students at the junior or senior level have very little mathematical experience in the use of the required vector operations. The simplest approach for the beginning student is to consider first the forces present in a system of *fixed* charges. This is the *electrostatic* case. The forces between these charges are given by a famous experimental law, called Coulomb's law. After the properties of fixed charges are well understood, charges in motion can be considered. In this way the more difficult aspects of vector analysis can be mastered gradually as they are needed to describe Coulomb's law and its consequences. It is this course that will be followed here.

As another possibility, charges in relative motion can be considered in an introductory section. This procedure provides a more general result in that electric fields and magnetic fields can be simultaneously introduced. Results for this approach are given in Appendix F for the interested reader.

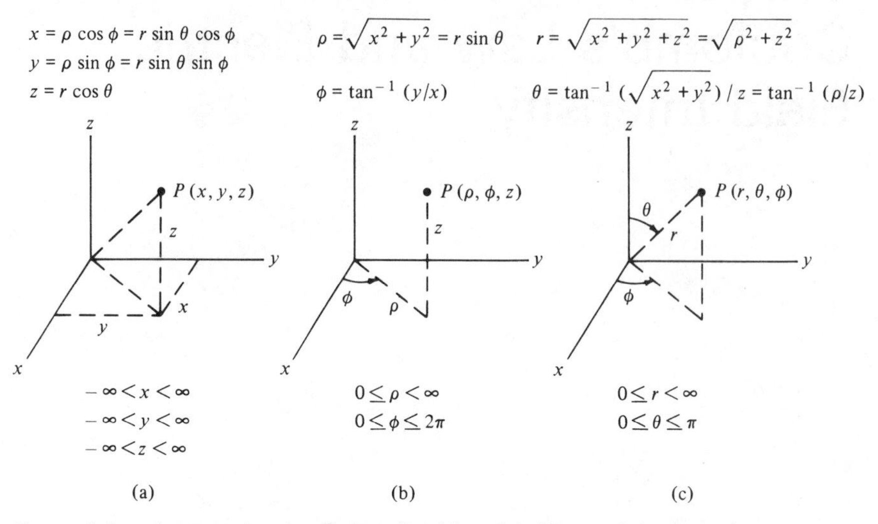

Figure 1.1. A general point P described by: (a) $P(x, y, z)$ in Cartesian coordinates, (b) $P(\rho, \phi, z)$ in circular cylindrical coordinates, and (c) $P(r, \theta, \phi)$ in spherical coordinates.

In this chapter we introduce Coulomb's experimental law and with it define electric field intensity. The unique relation between the electric field and its source, the electric charge, is then examined. Completely general results are derived in mathematical form, but only those examples where a high degree of symmetry is present in the distribution of charge are investigated explicitly.

1.1 COORDINATE SYSTEMS

Before introducing Coulomb's law, the reader should become familiarized with the coordinate systems we will be using in this textbook. They are the Cartesian (rectangular), circular cylindrical, and spherical coordinate systems. Figure 1.1 shows a point P described by the three coordinates associated with these systems. Also shown are coordinate transformations between the systems. These are easy to derive with Figure 1.1 and simple trigonometry.

The point P is also the locus of the intersection of the three orthogonal surfaces on which each of the coordinates is constant. This is shown in Figure 1.2. Notice that in Cartesian coordinates $z =$ constant is the same infinite plane as in cylindrical[1] coordinates, and in cylindrical coordinates ϕ constant is the same half-plane as in spherical coordinates.

Figure 1.3 shows the three orthogonal unit vectors for each system. These systems are all right-handed. That is, if the fingers of the right hand

[1] When we refer to cylindrical coordinates in what follows, we mean circular cylindrical coordinates. There are other types of cylindrical coordinates which we will not use.

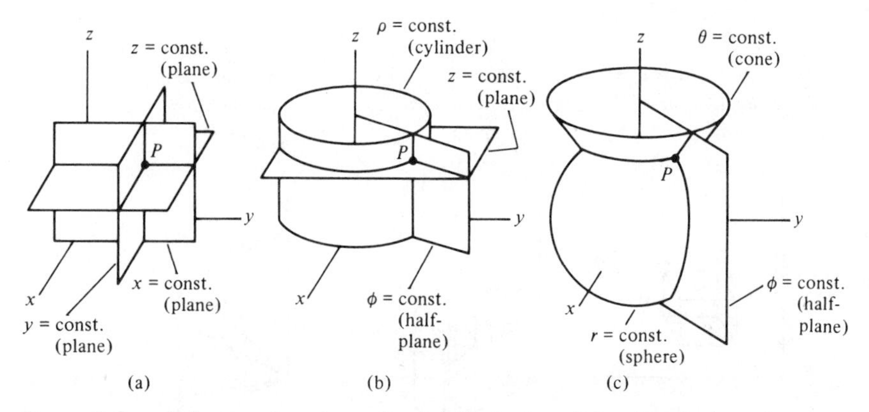

Figure 1.2. Orthogonal surfaces for (a) Cartesian, (b) cylindrical, and (c) spherical coordinates.

are rotated from one unit vector toward the next unit vector in the correct order ($\mathbf{a}_x$ to $\mathbf{a}_y$, $\mathbf{a}_y$ to $\mathbf{a}_z$, $\mathbf{a}_z$ to $\mathbf{a}_x$; $\mathbf{a}_\rho$ to $\mathbf{a}_\phi$, $\mathbf{a}_\phi$ to $\mathbf{a}_z$, $\mathbf{a}_z$ to $\mathbf{a}_\rho$; $\mathbf{a}_r$ to $\mathbf{a}_\theta$, $\mathbf{a}_\theta$ to $\mathbf{a}_\phi$, $\mathbf{a}_\phi$ to $\mathbf{a}_r$), then the correct direction for the third unit vector is obtained from the direction of the thumb. This is also demonstrated by the cross products listed in Figure 1.3. Notice that each unit vector is normal to its respective coordinate surface, and the direction of the unit vector is that for which its coordinate is *increasing*. It is important to recognize that the Cartesian coordinate system is the only one for which *all three unit vectors* are constant. All other coordinate systems possess at least one unit vector whose direction is not constant. When it becomes necessary (and it will) to integrate an expression containing a nonconstant unit vector with respect to a variable upon which the unit vector depends, the unit vector should be resolved into Cartesian components. For example,

$$\mathbf{a}_\rho \, d\phi = (\mathbf{a}_x \cos \phi + \mathbf{a}_y \sin \phi) \, d\phi$$

Figure 1.3. The unit vectors for (a) Cartesian, (b) cylindrical, and (c) spherical coordinates.

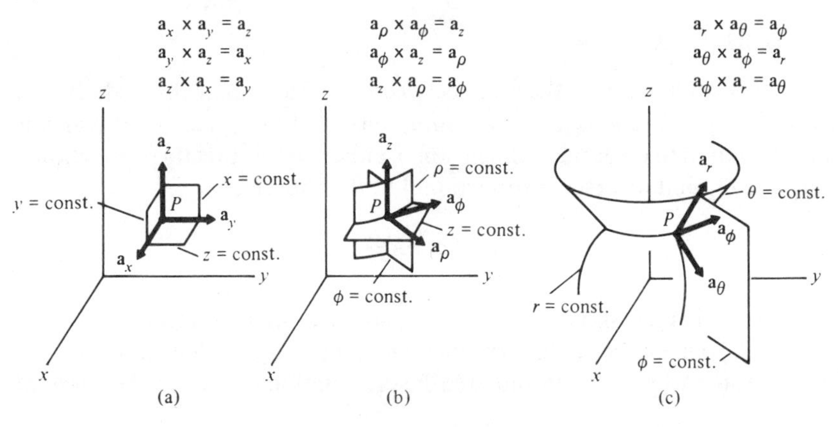

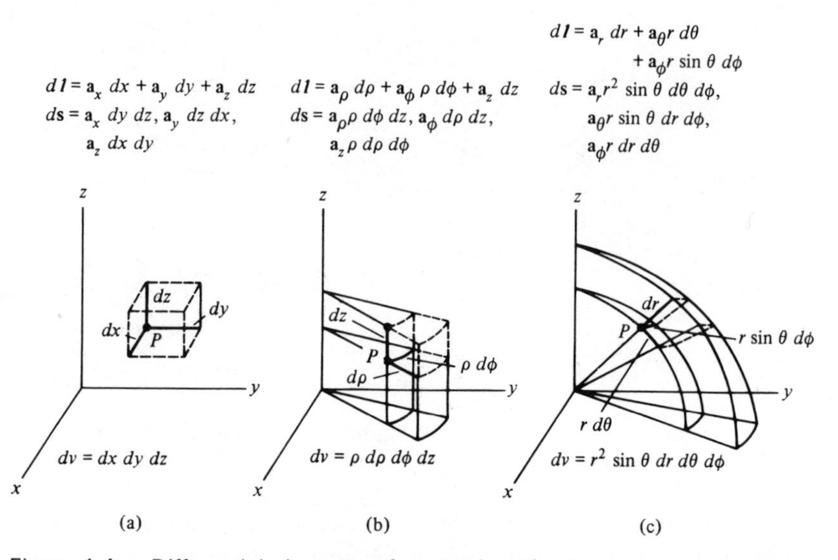

Figure 1.4. Differential elements of vector length, vector area, and volume for (a) Cartesian, (b) cylindrical, and (c) spherical coordinates.

since

$$\mathbf{a}_\rho = \frac{\boldsymbol{\rho}}{\rho} = \frac{1}{\rho}(\mathbf{a}_x x + \mathbf{a}_y y) = \frac{1}{\rho}(\mathbf{a}_x \rho \cos \phi + \mathbf{a}_y \rho \sin \phi)$$

Differential elements of vector length, vector area, and scalar volume can be found using Figure 1.4. Notice that each component of vector area is normal to its coordinate surface. In the case of cylindrical and spherical coordinates, the differential area and volume are obtained from the first order approximation that the differential volumes are rectangular boxes. The use of these quantities is explained in more detail as the need arises. Vector relations are summarized in Appendix A.

1.2 COULOMB'S LAW

Coulomb's law states that the force between two fixed small charged objects, having charges Q_1 and Q_2, in a vacuum, separated by a distance R which is large compared to the largest dimension of either one, is directly proportional to Q_1 and Q_2 and inversely proportional to the square of R, or

$$F = K\frac{Q_1 Q_2}{R^2} \tag{1.1}$$

Ideally the charges should have dimensions so small that they may be considered *point charges*. The hypothetical point charge is defined in Section 1.3. Newton had already formulated his gravitational law for the force of

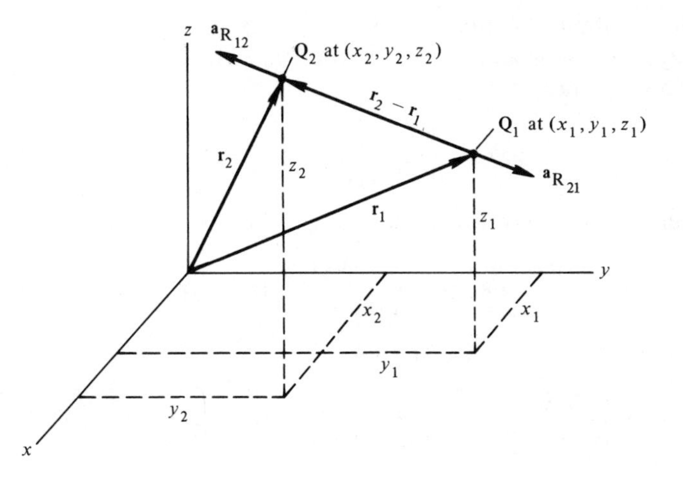

Figure 1.5. The location of two point charges.

attraction between two masses, M_1 and M_2,

$$F = C\frac{M_1 M_2}{R^2}$$

where C is a constant of proportionality and R is the distance between their centers. This equation is in the same form as equation 1.1, so it is quite possible that Coulomb suspected an inverse distance square type behavior before he verified equation 1.1 experimentally. Coulomb's law further asserts that the force (a vector quantity) is directed along the line joining the charge centers. The force is repulsive if the charges are like in sign and attractive otherwise.

In order to formulate Coulomb's law in a systematic manner such that its vector nature may be clearly seen, let us consider two point charges in the coordinate system shown in Figure 1.5. Call the charges Q_1 and Q_2, and locate them by means of the radius vectors $\mathbf{r}_1$ and $\mathbf{r}_2$, where

$$\mathbf{r}_1 = \mathbf{a}_x x_1 + \mathbf{a}_y y_1 + \mathbf{a}_z z_1 \tag{1.2}$$

and

$$\mathbf{r}_2 = \mathbf{a}_x x_2 + \mathbf{a}_y y_2 + \mathbf{a}_z z_2 \tag{1.3}$$

in Cartesian coordinates. The *constant* unit vectors $\mathbf{a}_x$, $\mathbf{a}_y$, and $\mathbf{a}_z$ lie along the x, y, and z axes, respectively. Thus, the end point of $\mathbf{r}_1$ is located at (x_1, y_1, z_1) or (ρ_1, ϕ_1, z_1) or (r_1, θ_1, ϕ_1) for rectangular, cylindrical, and spherical coordinates, respectively. Then the desired separation, R, of equation 1.1 is $|\mathbf{r}_2 - \mathbf{r}_1|$, or in Cartesian coordinates

$$R = |\mathbf{r}_2 - \mathbf{r}_1| = [(x_2 - x_1)^2 + (y_2 - y_1)^2 + (z_2 - z_1)^2]^{1/2} \tag{1.4}$$

We must next decide about the direction of the force. Suppose we want the vector force on Q_2. A vector along the line from Q_1 *to* Q_2 is $\mathbf{r}_2 - \mathbf{r}_1$, and a *unit* vector in this direction is

$$\mathbf{a}_{R_{12}} = \frac{\mathbf{r}_2 - \mathbf{r}_1}{|\mathbf{r}_2 - \mathbf{r}_1|} \tag{1.5}$$

as shown in Figure 1.5. We may now write

$$\mathbf{F}_2 = K\frac{Q_1Q_2}{|\mathbf{r}_2 - \mathbf{r}_1|^2}\mathbf{a}_{R_{12}} = K\frac{Q_1Q_2}{|\mathbf{r}_2 - \mathbf{r}_1|^3}(\mathbf{r}_2 - \mathbf{r}_1) \tag{1.6}$$

It should be obvious that $\mathbf{F}_1 = -\mathbf{F}_2$, and

$$\mathbf{F}_1 = K\frac{Q_1Q_2}{|\mathbf{r}_1 - \mathbf{r}_2|^2}\mathbf{a}_{R_{21}} = K\frac{Q_1Q_2}{|\mathbf{r}_1 - \mathbf{r}_2|^3}(\mathbf{r}_1 - \mathbf{r}_2) \tag{1.7}$$

Remember that $|\mathbf{r}_1 - \mathbf{r}_2| = |\mathbf{r}_2 - \mathbf{r}_1|$, but $\mathbf{a}_{R_{12}} = -\mathbf{a}_{R_{21}}$, since the unit vectors are oppositely directed.

The constant of proportionality, K, in Coulomb's law may be eliminated in favor of another constant if we define

$$K \equiv \frac{1}{4\pi\varepsilon_0} \quad \text{(vacuum)} \tag{1.8}$$

The units of this new constant for a MKS system must be newton (meter)^2 per (coulomb)^2 with Q_1 and Q_2 in coulombs and $|\mathbf{r}_1 - \mathbf{r}_2|$ in meters. We can show that the units of ε_0 are the same as farads per meter. (Units of quantities will be abbreviated according to the list of symbols and units following the Preface.) Thus, equation 1.7 becomes

$$\boxed{\mathbf{F}_1 = \frac{Q_1Q_2(\mathbf{r}_1 - \mathbf{r}_2)}{4\pi\varepsilon_0|\mathbf{r}_1 - \mathbf{r}_2|^3}} \quad \text{(Newtons, N)*} \tag{1.9}$$

The constant 4π in equation 1.9 looks unattractive, but will disappear in many of the equations in the following chapters. ε_0 is called the permittivity of a vacuum and is given numerically by

$$\varepsilon_0 = 8.854 \times 10^{-12} \approx \frac{10^{-9}}{36\pi} \quad \text{(F/m)} \tag{1.10}$$

It was stated initially that the charges are fixed. This requires some external force to hold the charges in place, otherwise they would accelerate away from each other (like in sign) or toward each other (unlike in sign), giving a nonstatic or dynamic system.

Coulomb's law asserts that the force depends *linearly* on the charge on either particle. In simple terms the principle of superposition states that the

* The answer should be expressed in the unit with the parentheses.

response ($\mathbf{F}_1$ in the present case) of a linear system due to several inputs (point charges Q_2, Q_3, Q_4, etc. in the present case) acting simultaneously is equal to the sum of the responses of each input (charge) acting alone. Thus, the force on one charged particle due to the presence of several other charged particles can be found as the superposition (sum) of the forces on the one charged particle due to the others acting separately. An example is now in order.

EXAMPLE 1

Find the force on a 10^{-5} C point charge at (0, 0, −1) due to a point charge (2×10^{-4} C) located at (1, 1, 1) and a (3×10^{-4} C) point charge located at (2, −1, 3). The given charges are labeled in Figure 1.6.

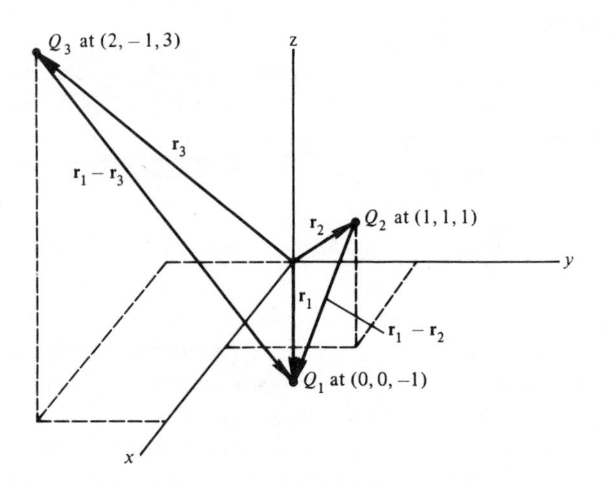

Figure 1.6. The location of three point charges.

Using equation 1.9 (with superposition),

$$\mathbf{F}_1 = \frac{1}{4\pi\varepsilon_0}\left[\frac{Q_1Q_2}{|\mathbf{r}_1 - \mathbf{r}_2|^3}(\mathbf{r}_1 - \mathbf{r}_2) + \frac{Q_1Q_3}{|\mathbf{r}_1 - \mathbf{r}_3|^3}(\mathbf{r}_1 - \mathbf{r}_3)\right]$$

Inspection of Figure 1.6 reveals that

$$\mathbf{r}_1 - \mathbf{r}_2 = \mathbf{a}_x(0 - 1) + \mathbf{a}_y(0 - 1) + \mathbf{a}_z(-1 - 1) = -\mathbf{a}_x - \mathbf{a}_y - 2\mathbf{a}_z$$

$$\mathbf{r}_1 - \mathbf{r}_3 = \mathbf{a}_x(0 - 2) + \mathbf{a}_y(0 + 1) + \mathbf{a}_z(-1 - 3) = -2\mathbf{a}_x + \mathbf{a}_y - 4\mathbf{a}_z$$

$$|\mathbf{r}_1 - \mathbf{r}_2|^3 = [1^2 + 1^2 + 2^2]^{3/2} = 6^{3/2}$$

and

$$|\mathbf{r}_1 - \mathbf{r}_3|^3 = [2^2 + 1^2 + 4^2]^{3/2} = 21^{3/2}$$

Thus,

$$\mathbf{F}_1 = \frac{1}{4\pi\varepsilon_0}\left[\frac{10^{-5} \times 2 \times 10^{-4}}{6^{3/2}}(-\mathbf{a}_x - \mathbf{a}_y - 2\mathbf{a}_z) + \frac{10^{-5} \times 3 \times 10^{-4}}{21^{3/2}}(-2\mathbf{a}_x + \mathbf{a}_y - 4\mathbf{a}_z)\right]$$

$$\mathbf{F}_1 = \frac{-10^{-9}}{4\pi\varepsilon_0}[0.198\mathbf{a}_x + 0.105\mathbf{a}_y + 0.397\mathbf{a}_z]$$

and

$$\mathbf{F}_1 = [-1.79\mathbf{a}_x - 0.95\mathbf{a}_y - 3.57\mathbf{a}_z] \quad \text{(N)}$$

1.3 CHARGE CONFIGURATIONS

The most general charge configuration is the volume charge density distribution, $\rho_v(x, y, z)$, or simply ρ_v, in coulombs per cubic meter (C/m^3). All other configurations are special cases of volume charge density. The differential charge in a differential volume is $dQ = \rho_v\, dv$, so the total charge is

$$Q = \iiint_{\text{volume}} \rho_v(x, y, z)\, dx\, dy\, dz \quad \text{(C)} \tag{1.11}$$

where the limits on the triple integral must be such as to include *all* the charge. Some of the useful special cases of charge distributions are discussed below.

The point charge of Section 1.2 is a finite charge occupying zero volume, indicating an infinite volume charge density. Thus, the concept of a point charge is primarily a mathematical convenience. Its singular nature will become more apparent very shortly.

It is convenient to describe the point charge in terms of the Dirac delta function, or impulse function, which is not a function in the strict mathematical sense, but a *distribution.* We will continue to call it a function. It can be considered to be the limit of a wide variety of functions or other distributions if they have certain properties. Rather than becoming involved in a lengthy discussion of these properties, let us look at a couple of simple examples.

Consider the rectangular pulse $p_1(x)$ as shown in Figure 1.7(a). Its height or amplitude is $1/a$ while its width is a so that its area is unity regardless of the numerical value of a. Then, the unit impulse function $\delta(x)$ is defined as

$$\delta(x) = \lim_{a \to 0} p_1(x)$$

We can visualize the height approaching infinity while the width is simultaneously approaching zero. The increasing height, decreasing width,

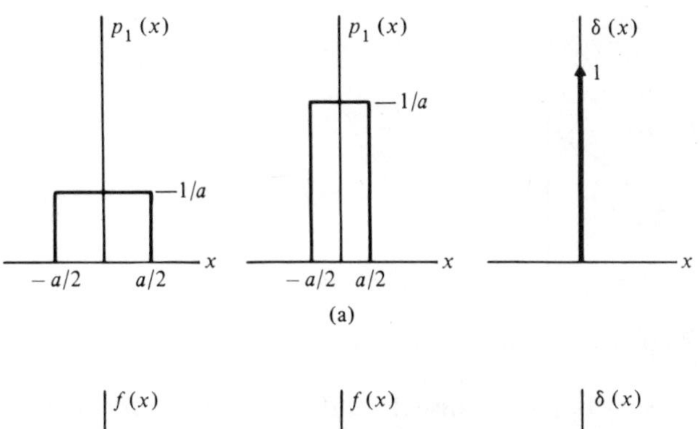

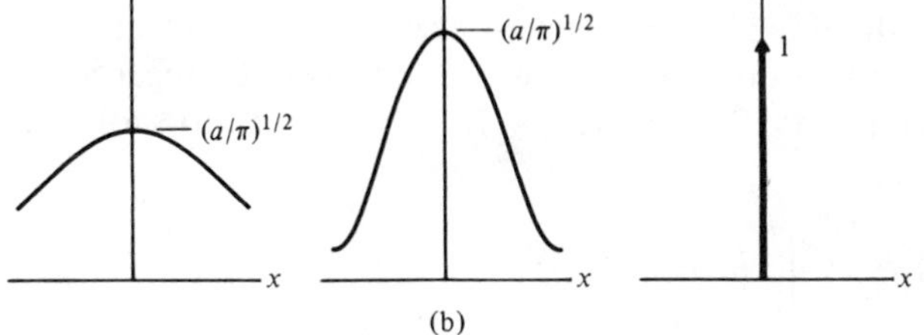

Figure 1.7. Examples of functions which may represent unit impulses in limiting form: (a) rectangular pulse, (b) continuous function, $f(x) = [a/\pi]^{1/2}e^{-ax^2}$.

and unity area are all characteristics of functions which may represent unit impulses in the limit. Another example is shown in Figure 1.7(b).

$$f(x) = \left[\frac{a}{\pi}\right]^{1/2} e^{-ax^2}$$

$$\lim_{a \to \infty} f(0) \to \infty, \qquad x = 0 \qquad \text{(increasing height)}$$

$$\lim_{a \to \infty} f(x) = 0, \qquad |x| > 0 \quad \text{(decreasing width)}$$

$$\int_{-\infty}^{\infty} f(x)\, dx = 1 \quad \text{(unity area)}$$

$$\lim_{a \to \infty} f(x) = \delta(x)$$

Notice that we consider $\delta(x)$ to be an even function; that is, $\delta(-x) = \delta(x)$. Some of the more useful properties of impulse functions are

$$\int_a^c f(x)\, \delta(x - b)\, dx = f(b), \qquad a < b < c \tag{1.12}$$

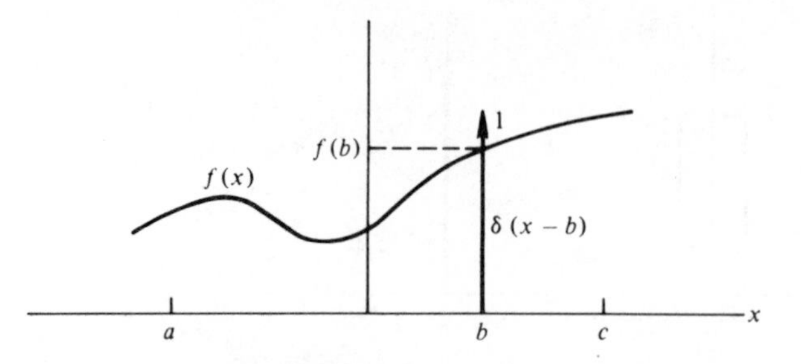

Figure 1.8. The sampling property of the unit impulse.

This is the *sampling* property of the impulse. It leads to a result that can only occur at $x = b$ since the impulse itself is zero everywhere except at $x = b$ where it is infinite but with unity area. This property may be regarded as defining the impulse function. See Figure 1.8. It is worthwhile to show that equation 1.12 is correct.

$$\int_a^c f(x)\,\delta(x-b)\,dx = \int_a^c f(b)\,\delta(x-b)\,dx$$

$$= f(b)\int_a^c \delta(x-b)\,dx$$

$$= f(b), \qquad a < b < c$$

$$\int_a^c f(x)\,\delta'(x-b)\,dx = -f'(b), \qquad a < b < c \tag{1.13}$$

$$\delta'(x-b) = \frac{d}{dx}[\delta(x-b)] \tag{1.14}$$

$$\int_a^c f(x)\,\delta^n(x-b)\,dx = (-1)^n f^n(b), \qquad a < b < c \tag{1.15}$$

$$\delta^n(x-b) = \frac{d^n}{dx^n}[\delta(x-b)], \qquad f^n(b) = \frac{d^n}{dx^n} f(x)\big|_{x=b} \tag{1.16}$$

$$\delta[f(x)] = \sum_{n=-\infty}^{\infty} \frac{1}{|df/dx|}\,\delta(x-b_n), \qquad f(b_n) = 0 \tag{1.17}$$

$$u(x) = \begin{cases} 1, & x > 0 \\ 0, & x < 0 \end{cases} \tag{1.18}$$

Figure 1.9. The unit step function.

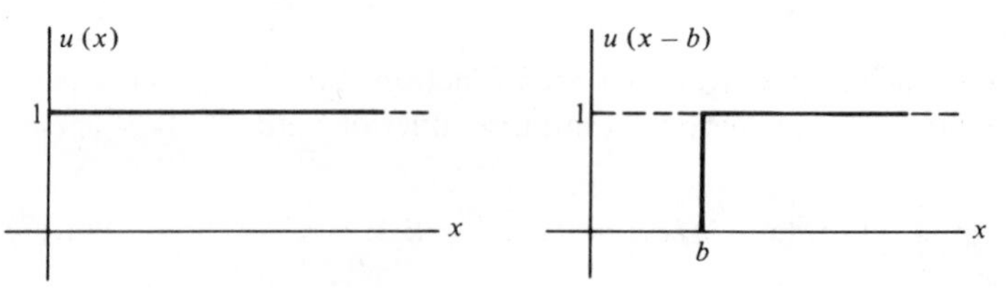

The last function is the unit step function which is related to the unit impulse function (see Figure 1.9) according to

$$\frac{d}{dx}[u(x)] = \delta(x) \tag{1.19a}$$

or

$$u(x) = \int_{-\infty}^{x} \delta(x')\, dx' \tag{1.19b}$$

$$u[f(x)] = \begin{cases} 1, & f(x) > 0 \\ 0, & f(x) < 0 \end{cases} \tag{1.20}$$

Notice that the dimension of an impulse function will be that of the reciprocal of its argument.

It is convenient at this point to note for future use that the cylindrical radial variable (ρ) and the spherical radial variable (r) *are never negative*, and so it is correct to state that $f(\rho) = f(\rho)u(\rho)$, or $g(r) = g(r)u(r)$. The latter forms are *necessary* if we are required to *differentiate* certain functions. That is, when we write $d/d\rho[f(\rho)]$, we mean

$$\frac{d}{d\rho}[f(\rho)u(\rho)] = f(0)\delta(\rho) + u(\rho)\frac{d}{d\rho}f(\rho)$$

since it is not true in general that $f(0) = 0$.

Let us now try to find a formulation that describes a point charge Q as a volume charge density ρ_v in coulombs per cubic meter. First of all, the point charge is located (only) at (x', y', z') with infinite density and zero volume but total charge Q. This certainly sounds like the description of an impulse, but a three-dimensional impulse. We are led to consider

$$\rho_v(x, y, z) = Q\, \delta(x - x')\, \delta(y - y')\, \delta(z - z') \quad (\mathrm{C/m^3}) \tag{1.21}$$

Notice that the dimensions are correct and the charge is located precisely at x', y', z'. Furthermore, using the sampling property (three times)

$$\iiint_{\text{volume}} \rho_v(x, y, z)\, dx\, dy\, dz = \iiint_{\text{volume}} Q\, \delta(x - x')\, \delta(y - y')\, \delta(z - z')\, dx\, dy\, dz$$
$$= Q \quad (\mathrm{C})$$

so long as the limits on the volume integration are large enough so that the point charge is enclosed, or, stated simply, so that (x', y', z') is within the volume. Equation 1.21 is thus the correct representation of a point charge as a general volume charge density in Cartesian coordinates because it is consistent with equation 1.11.

For cylindrical coordinates we have

$$\rho_v(\rho, \phi, z) = [Q\, \delta(\rho - \rho')\, \delta(\phi - \phi')\, \delta(z - z')]/\rho' \quad (\mathrm{C/m^3}) \tag{1.22}$$

and for spherical coordinates we have

$$\rho_v(r, \theta, \phi) = \frac{Q\,\delta(r - r')\,\delta(\theta - \theta')\,\delta(\phi - \phi')}{(r')^2 \sin\theta'} \quad (\text{C/m}^3) \tag{1.23}$$

These densities are correct because when they are multiplied by dv (using the appropriate coordinates) and the integration carried out (using the sampling property of the impulse function), the result is Q in both cases.

The line charge density, ρ_l (C/m) is a filament of charge having a length (which may mathematically be infinite) but no thickness. It too is a mathematical model since its volume charge density is infinite. In many cases, a finite diameter conducting wire bearing a charge distribution, which is either known or can be approximated at every point, can be represented by a filamentary line charge density. The charged wire must look like a filament to an observer who is located several diameters away from the wire. The differential charge in a differential length is $dQ = \rho_l\, dl$, thus the total charge for a line charge density is

$$Q = \int_{\text{length}} \rho_l\, dl \quad (\text{C}) \tag{1.24}$$

As in the case of the point charge, let us find a suitable volume charge representation for the line charge density. Suppose we have a z-dependent line charge density $\rho_l(z)$ running parallel to the z axis and passing through the point $(\rho', \phi', 0)$ in the $z = 0$ plane. Then,

$$\rho_v(\rho, \phi, z) = [\rho_l(z)\,\delta(\rho - \rho')\,\delta(\phi - \phi')]/\rho' \quad (\text{C/m}^3) \tag{1.25}$$

will agree with equations 1.11 and 1.24.

The surface charge density, ρ_s (C/m^2) is a layer of charge having an area (which may mathematically be infinite) but no thickness so that its volume density is also infinite. A thin copper sheet with a known charge distribution could be approximated by a surface charge density for an observer who agreed to remain several thicknesses away from the copper sheet. The differential charge for a differential area is $dQ = \rho_s\, ds$, thus the total charge for this case is

$$Q = \iint_{\text{surface}} \rho_s\, ds \quad (\text{C}) \tag{1.26}$$

Suppose that on the spherical surface $r = a$ there exists the surface charge density $\rho_s(\theta, \phi)$, then

$$\rho_v(r, \theta, \phi) = \rho_s(\theta, \phi)\,\delta(r - a) \quad (\text{C/m}^3) \tag{1.27}$$

will agree with equations 1.11 and 1.26.

Suppose that $\rho_s(\phi, z)$ exists on the infinite cylindrical surface $\rho = a$. Then, a correct expression for the volume charge density is

$$\rho_v(\rho, \phi, z) = \rho_s(\phi, z)\,\delta(\rho - a) \quad (\text{C/m}^3) \tag{1.28}$$

The differential charge is thus represented as dQ or $\rho_l\, dl$ or $\rho_s\, ds$ or $\rho_v\, dv$, and the last form must be employed when representing volume charge densities with impulse functions. Some examples using the results just derived will be helpful.

EXAMPLE 2

Consider a line charge density lying on the z axis between $-h$ and $+h$ with $\rho_l(z) = \rho_{l0} z^2/h^2$ as shown in Figure 1.10. The charge is greatest

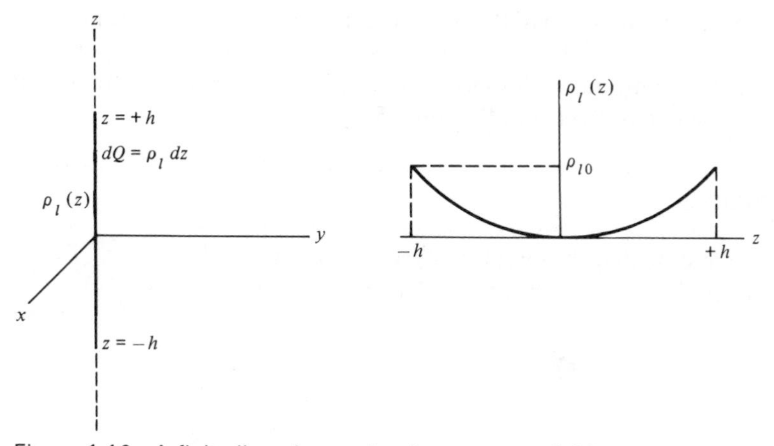

Figure 1.10. A finite line charge density $\rho_l = \rho_{l0} z^2/h^2$ located symmetrically on the z axis.

at the ends ($z = \pm h$) and smallest at the middle ($z = 0$). With no external restraining forces present, Coulomb's law predicts a distribution of this general nature.[2] (Why?). Application of equation 1.24 gives

$$Q = \int_{\text{length}} \rho_l\, dz = \int_{-h}^{h} \rho_{l0} \frac{z^2}{h^2}\, dz = \frac{2}{3} \rho_{l0} h \quad \text{(C)}$$

In terms of volume charge density

$$\rho_v(x, y, z) = \rho_{l0} \frac{z^2}{h^2} [u(z + h) - u(z - h)]\, \delta(x)\, \delta(y) \quad (\text{C/m}^3)$$

where the difference of two step functions has been employed to ensure that the charge exists for $-h \le z \le h$. Application of equations 1.11 and 1.12

[2] Section 5.5 deals more directly with this problem.

gives

$$Q = \int_{-\infty}^{\infty} \int_{-\infty}^{\infty} \int_{-\infty}^{\infty} \rho_v(x, y, z)\, dx\, dy\, dz = \int_{-\infty}^{\infty} \int_{-\infty}^{\infty} \int_{-h}^{h} \rho_{l0} \frac{z^2}{h^2}\, \delta(x)\, \delta(y)\, dx\, dy\, dz$$

or,

$$Q = \int_{-h}^{h} \rho_{l0} \frac{z^2}{h^2}\, dz$$

which agrees with the result above. Notice that using equation 1.24 gave us the answer much faster than using the impulse representation, the reason being, of course, that two of the three integrations were performed before we started with equation 1.24. On this basis alone we cannot justify the use of impulse functions, but an important concept can be expedited by their use. This will be demonstrated in Section 1.5.

EXAMPLE 3

Assume now that the circular cylindrical surface, $-h \le z \le h$, $\rho = a$, is covered with a surface charge density given by

$$\rho_s = \begin{cases} \dfrac{z^2}{h^2} \rho_{s0}, & -h \le z \le h, \quad \rho = a \\ 0, & |z| > h, \quad \rho \ne a \end{cases}$$

Also let the surfaces $z = \pm h$, $0 \le \rho \le a$ be covered with the surface charge density

$$\rho_s = \begin{cases} \rho_{s0}, & z = \pm h, \quad 0 \le \rho \le a \\ 0, & z = \pm h, \quad \rho > a \end{cases}$$

The charged surface (shown in Figure 1.11) is thus the surface of a closed circular cylindrical can of radius a and height $2h$. Using equation 1.26 with cylindrical coordinates,

$$Q = \int_0^{2\pi} \int_{-h}^{h} \left(\frac{z^2}{h^2} \rho_{s0} \right) a\, dz\, d\phi + 2 \int_0^{2\pi} \int_0^{a} (\rho_{s0}) \rho\, d\rho\, d\phi$$

$$Q = \frac{2\pi a \rho_{s0}}{h^2} \int_{-h}^{h} z^2\, dz + 4\pi\rho_{s0} \int_0^a \rho\, d\rho$$

and

$$Q = \tfrac{4}{3}\pi a \rho_{s0} h + 2\pi a^2 \rho_{s0} \quad \text{(C)}$$

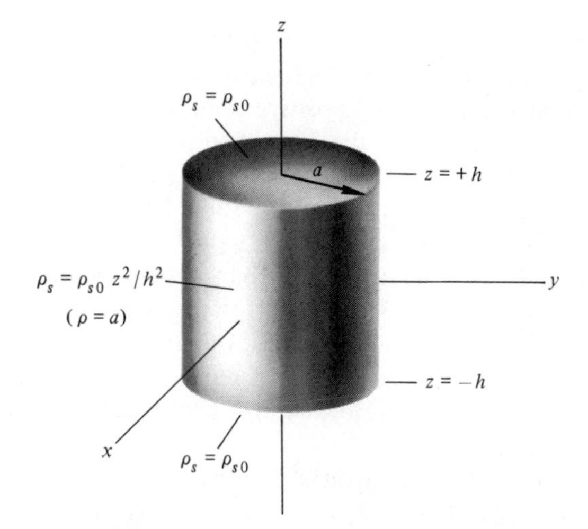

Figure 1.11. A cylindrical surface bearing the indicated surface charge densities.

EXAMPLE 4

Finally, suppose that the interior of the cylindrical can of Example 3 is filled with a volume charge density given by (no surface charge density)

$$\rho_v = \frac{\rho}{a}\frac{z^2}{h^2}\rho_{v0}$$

and there are no charges outside the can. Equation 1.11 gives

$$Q = \int_0^{2\pi}\int_0^a\int_{-h}^{h}\left(\frac{z^2}{h^2}\frac{\rho}{a}\rho_{v0}\right)\rho\, dz\, d\rho\, d\phi$$

$$Q = \frac{2\pi}{h^2 a}\rho_{v0}\int_0^a\int_{-h}^{h}\rho^2 z^2\, dz\, d\rho = \frac{2\pi\rho^3}{3h^2 a}\rho_{v0}\bigg|_0^a \int_{-h}^{h} z^2\, dz$$

and

$$Q = \frac{2\pi a^2}{3h^2}\rho_{v0}\frac{z^3}{3}\bigg|_{-h}^{h} = \frac{4\pi a^2 h}{9}\rho_{v0} \quad \text{(C)}$$

1.4 ELECTRIC FIELD INTENSITY

Coulomb's law tells us that there is a force between two charged particles. If one of the charges is fixed, while the other is moved from one point to another, the force on it may possibly change not only in magnitude, but also in direction. The charge that is being moved is under the influence of a *force field.* In describing this force field, it is much easier to simply let the charge that is being moved be a unit (+1 C) positive point charge. It may then be considered to be a test charge, and the force field it samples is the force on a unit positive point charge. This same test charge can be used to probe the force field of any charge configuration. The force field produced by an arbitrary charge distribution and sampled by a unit positive point charge is called the *electric field intensity*, **E**, of the charge distribution. In other words, the electric field of a charge distribution is the force it produces per unit positive point charge. The test charge must not disturb the charge distribution whose force field it is probing, and in this regard this situation is analogous to a voltage probe producing no loading effect when making measurements in an electric circuit.

In order to find the electric field of an arbitrarily located point charge, Q_1, we must find the force on the test charge, $Q_2 = 1$ (also arbitrarily located). Thus

$$\mathbf{F}_2 = \frac{1}{4\pi\varepsilon_0} \frac{Q_1(1)}{|\mathbf{r}_2 - \mathbf{r}_1|^2} \mathbf{a}_{R_{12}}$$

or

$$\frac{\mathbf{F}_2}{1} \equiv \mathbf{E}_2 = \frac{1}{4\pi\varepsilon_0} \frac{Q_1}{|\mathbf{r}_2 - \mathbf{r}_1|^2} \mathbf{a}_{R_{12}} \quad \text{(N/C)} \tag{1.29}$$

The subscripts in equation 1.29 are meaningful only to the case of two point charges. The definition of the electric field intensity, given in the previous paragraph, is a general one. With this fact in mind, it is much better to use a general notation. The following notation, which is used by many authors, will be used consistently throughout this textbook. This will enable the beginning student to avoid many of the difficulties that arise in problems involving integration with more than one variable. *Let unprimed coordinates denote the point where the field is being evaluated, and let primed coordinates denote the source* (*charge*) *coordinates.* As pictured in Figure 1.12, the distance between source point and field point is $R = |\mathbf{r} - \mathbf{r}'|$, and a unit vector from the source point to the field point is

$$\mathbf{a}_R = \frac{\mathbf{R}}{R} = \frac{\mathbf{r} - \mathbf{r}'}{|\mathbf{r} - \mathbf{r}'|}$$

Thus,

$$\mathbf{E}(x, y, z) = \frac{Q}{4\pi\varepsilon_0} \frac{\mathbf{a}_R}{|\mathbf{r} - \mathbf{r}'|^2} = \frac{Q}{4\pi\varepsilon_0} \frac{\mathbf{r} - \mathbf{r}'}{|\mathbf{r} - \mathbf{r}'|^3} \tag{1.30}$$

or simply

$$\boxed{\mathbf{E}(x, y, z) = \frac{Q}{4\pi\varepsilon_0 R^2} \mathbf{a}_R} \quad \text{(V/m)} \tag{1.31}$$

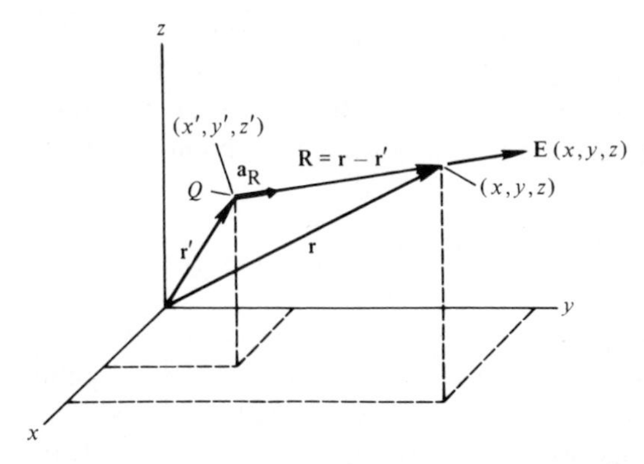

Figure 1.12. The location of a (source) point charge Q and a field point where **E** is to be determined.

The units of equation 1.31 will be newtons per coulomb (force per unit charge). We should be able to recall from circuit theory that a volt is a joule per coulomb and a joule (work or energy) is a newton-meter. Thus, a newton per coulomb is a volt per meter. The latter form is preferred, but the former is an obvious reminder of the definition of electric field intensity.

The term $|\mathbf{r} - \mathbf{r}'| = R$ occurs frequently in field theory, and it is convenient to list this term for rectangular, cylindrical, and spherical coordinates.

$$|\mathbf{r} - \mathbf{r}'| = R = [(x - x')^2 + (y - y')^2 + (z - z')^2]^{1/2} \quad \text{(rectangular)} \tag{1.32a}$$

$$[\rho^2 + (\rho')^2 - 2\rho\rho' \cos(\phi - \phi') + (z - z')^2]^{1/2} \quad \text{(cylindrical)} \tag{1.32b}$$

and

$$[r^2 + (r')^2 - 2rr' \cos\theta \cos\theta' - 2rr' \sin\theta \sin\theta' \cos(\phi - \phi')]^{1/2} \quad \text{(spherical)} \tag{1.32c}$$

Equation 1.32a is easily visualized geometrically in Figure 1.5, while equation 1.32b and equation 1.32c can be derived from equation 1.32a using the coordinate transformations in Figure 1.1.

It may seem that the mathematical forms being used here become unnecessarily complicated, and to some extent this may be true. There are, however, problems that we will encounter for which this notation is of tremendous value. It is much better to possess general forms from which special cases can be obtained, than to try to obtain more general forms from special cases.

1.5 SUPERPOSITION

The principle of superposition is of primary importance in any linear system. This is true whether we are dealing with a one-dimensional problem (perhaps an electric circuit where time is the one independent variable) or a multidimensional problem (perhaps an electrostatic field problem with x, y, and z as the independent variables). As was previously mentioned, Coulomb forces are additive, so that the electric field produced by N point charges is (by superposition)

$$E_N(x, y, z) = \frac{1}{4\pi\varepsilon_0} \sum_{n=1}^{N} \frac{Q_n(x'_n, y'_n, z'_n)}{|\mathbf{r} - \mathbf{r}'_n|^3} (\mathbf{r} - \mathbf{r}'_n) \quad \text{(V/m)} \tag{1.33}$$

where $\mathbf{r}'_n$ is the radial vector to the nth point charge. If the region containing the charge is described as being macroscopically "dense" with N point charges, *and if this region is finite in extent*, we may ascribe to every point in the region a volume charge density as in equation 1.11. We replace Q_n by ΔQ_n and then multiply and divide by the small volume Δv_n. That is,

$$\mathbf{E}_N(x, y, z) = \frac{1}{4\pi\varepsilon_0} \sum_{n=1}^{N} \frac{\Delta Q_n(x'_n, y'_n, z'_n)}{\Delta v_n} \frac{\Delta v_n(\mathbf{r} - \mathbf{r}'_n)}{|\mathbf{r} - \mathbf{r}'_n|^3} \tag{1.34}$$

We now let the volume, Δv_n, approach zero as the number of charge elements, N, approaches infinity. In the limit, $\Delta Q_n/\Delta v_n$ becomes the volume charge density $\rho_v = dQ/dv$, as before, while the sum becomes an integral in the usual way. This integral is (in general) over three primed variables, or, in other words, it is a *volume* integral. That is,

$$\mathbf{E}(x, y, z) = \lim_{\substack{\Delta v_n \to 0 \\ N \to \infty}} \frac{1}{4\pi\varepsilon_0} \sum_{n=1}^{N} \frac{\Delta Q_n}{\Delta v_n} \frac{(\mathbf{r} - \mathbf{r}'_n)}{|\mathbf{r} - \mathbf{r}'_n|^3} \Delta v_n \tag{1.35}$$

or

$$\mathbf{E}(x, y, z) = \frac{1}{4\pi\varepsilon_0} \iiint_{\text{vol}'} \rho_v(x', y', z') \frac{(\mathbf{r} - \mathbf{r}')}{|\mathbf{r} - \mathbf{r}'|^3} dv' \tag{1.36}$$

or simply

$$\boxed{\mathbf{E}(\mathbf{r}) = \frac{1}{4\pi\varepsilon_0} \iiint_{\text{vol}'} \rho_v(\mathbf{r}') \frac{\mathbf{a}_R}{R^2} dv'} \tag{1.37}$$

where $dv' = dx'\,dy'\,dz' = \rho'\,d\rho'\,d\phi'\,dz' = (r')^2 \sin\theta'\,dr'\,d\theta'\,d\phi'$ for rectangular, cylindrical, and spherical coordinates, respectively.[3]

Figure 1.13 shows a volume containing a charge density, $\rho_v(x', y', z')$ and a point (x, y, z) where the electric field is to be evaluated. In order to find the electric field produced by ρ_v, we must integrate throughout the volume (vol′) containing this charge density. Note particularly that the integration is performed on the *primed coordinates only*! In general this integration is over three space variables. Triple integrals of this type are usually

[3] See Figure 1.4.

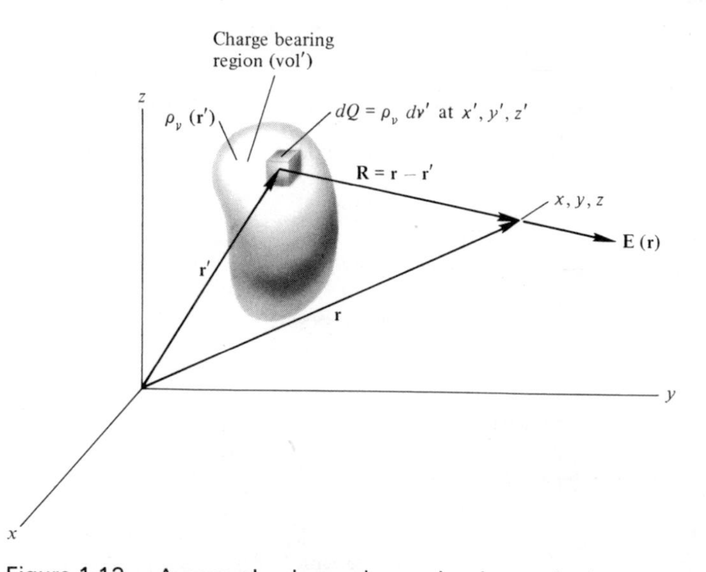

Figure 1.13. A general volume charge density producing an electric field. The vector **r**′ varies during the integration, but **r** is fixed.

very difficult. Quite often, in solving problems of a symmetrical nature, the triple integral will simplify to a double integral or even a single integral.

The process of integration is actually the result of taking the limit of a sum as equations 1.35 and 1.36 show. In performing calculations of the electric field with the aid of a digital computer, equation 1.35 would actually be used. That is, the computer, because of its discrete nature, integrates by calculating a large sum. With this in mind, equation 1.36 indicates that in order to find the electric field at some point in space, we sum the contributions to the field at the point in space from each of the many small charges which together constitute the charge distribution. This is a very fundamental concept.

We now intend to digress and spend some time developing a concept that is probably the most important in the analysis of linear systems of all types. Equation 1.37 can be obtained immediately if we simply recognize that it is a superposition integral or convolution type integral (Borel's theorem). According to these concepts, much used in linear systems where time is usually the only independent variable, the response of a linear system to *any* excitation can be found if the system response to unit impulse excitation, often called the *Green's function*, is known. Formally, the response is the *convolution* of the excitation with the impulse response. Consider first a simple one-dimensional example. Let $e(x)$ be the excitation and let $h(x, x')$ be the response at x due to a unit impulse applied at x'. The units of x are unimportant here. In electrostatic problems of the type considered in this chapter $h(x, x') = h(x - x')$. That is, for a *fixed system*, only the "interval" $(x - x')$ from the source (excitation) to the point of observation (response)

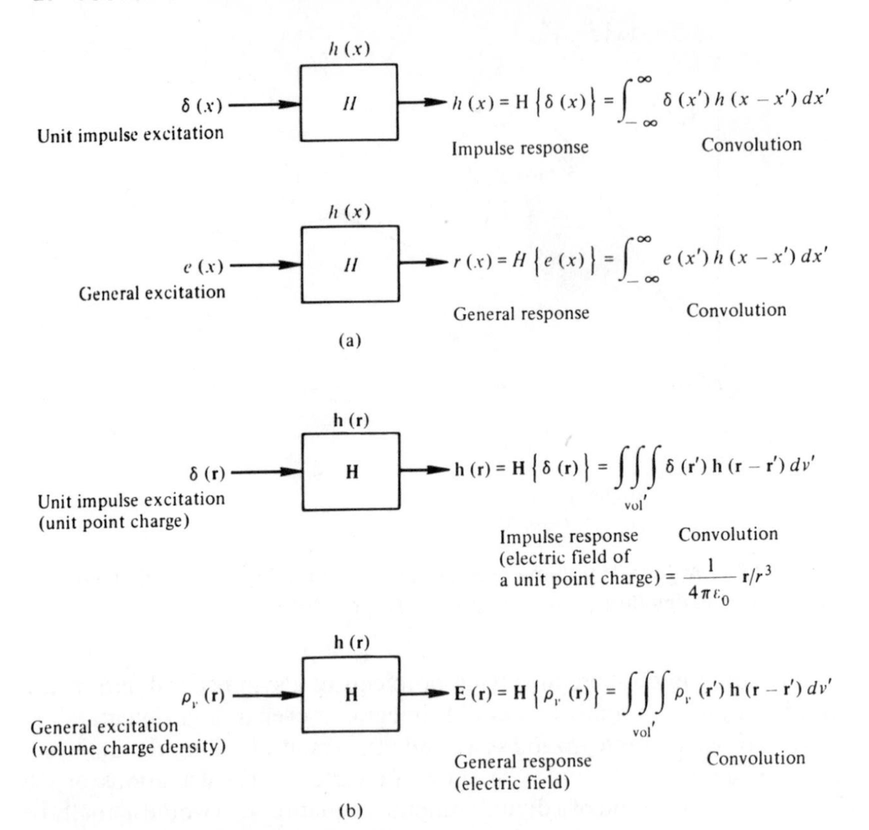

Figure 1.14. (a) The impulse response and general response for a one-dimensional linear system. (b) The impulse response and general response for the three-dimensional linear system, electrostatic field intensity.

matters. The system approach relates the response, $r(x)$, the excitation, and the system according to Figure 1.14(a). Thus, we have

$$r(x) = H\{e(x)\} \tag{1.38}$$

In words, the system operating on the excitation produces the response. Equation 1.12, however, gives $e(x)$ as (a superposition of impulses)

$$e(x) = \int_{-\infty}^{\infty} e(x')\, \delta(x - x')\, dx'$$

where the limits have been changed to ensure that the impulse is enclosed in order to perform its sampling operation. Therefore,

$$r(x) = H\left\{\int_{-\infty}^{\infty} e(x')\, \delta(x - x')\, dx'\right\} = \int_{-\infty}^{\infty} e(x')H\{\delta(x - x')\}\, dx'$$

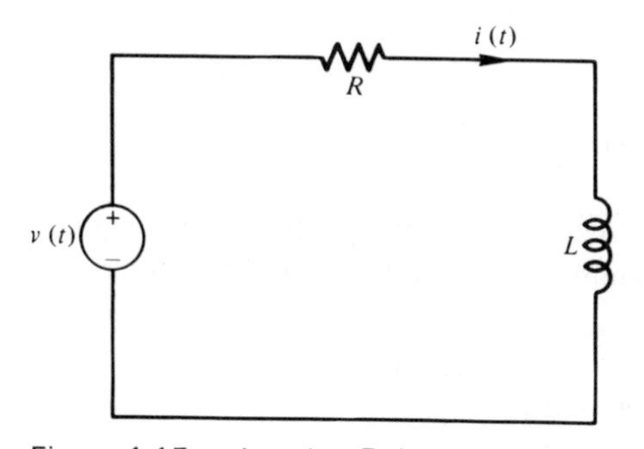

Figure 1.15. A series R-L circuit driven by an ideal voltage source.

since H operates on x, not x'. But $H\{\delta(x - x')\} = h(x, x') = h(x - x')$, the unit impulse response, so

$$r(x) = \int_{-\infty}^{\infty} e(x')h(x - x')\,dx' \tag{1.39}$$

The integral on the right side of equation 1.39 is the convolution of $e(x)$ with $h(x)$. *The general response of a fixed linear system is the convolution of the source function with the unit impulse response function, and thus the unit impulse response completely characterizes the system*! The reader should firmly grasp this idea before proceeding.

A familiar electric circuit example should be helpful. Consider the R-L series circuit driven by an ideal voltage source shown in Figure 1.15. The system differential equation is

$$v(t) = Ri(t) + L\frac{d}{dt}\,i(t)$$

where $v(t)$ is the source, $i(t)$ (current) is the desired response, and time t is the independent variable. The impulse response, or the response due to a unit impulse excitation $[v(t) = \delta(t)]$, is most easily obtained by simply finding the phasor solution of circuit theory. That is, for sinusoidal excitation in the steady state

$$V(\omega) = RI(\omega) + j\omega LI(\omega)$$

or

$$\frac{I(\omega)}{V(\omega)} \equiv H(\omega) = \frac{1}{L}\,\frac{1}{R/L + j\omega} \qquad \text{(transfer function)}$$

Then $h(t)$ is simply the inverse Fourier transform of $H(\omega)$. That is,

$$h(t) = \mathscr{F}^{-1}\{H(\omega)\} = \mathscr{F}^{-1}\left\{\frac{1}{L}\,\frac{1}{R/L + j\omega}\right\}$$

or, using Table E.1, Appendix E,

$$h(t) = u(t)\frac{1}{L}e^{-Rt/L}$$

It is easy to show that this is the correct solution by simply substituting $h(t)$ for $i(t)$ in the original differential equation with $v(t) = \delta(t)$. Thus, replacing t by $t - t'$,

$$h(t - t') = u(t - t')\frac{e^{-R(t-t')/L}}{L}$$

According to equation 1.39,

$$i(t) = \int_{-\infty}^{\infty} v(t')u(t - t')\frac{e^{-R(t-t')/L}}{L}\,dt'$$

$$= \frac{e^{Rt/L}}{L}\int_{-\infty}^{t} v(t')e^{Rt'/L}\,dt'$$

Now, given any reasonable $v(t)$, the last equation is a formal solution[4] for $i(t)$. Suppose the voltage source is an ideal battery in series with a switch which closes at $t = 0$; that is, $v(t) = V_0 u(t)$, where V_0 is the battery voltage. What is the current? We have

$$i(t) = \frac{e^{-Rt/L}}{L}\int_{-\infty}^{t} V_0 u(t')e^{Rt'/L}\,dt' = \frac{V_0}{L}e^{-Rt/L}\int_{0}^{t} e^{Rt/L}\,dt'$$

$$i(t) = \frac{V_0}{R}(1 - e^{-Rt/L})u(t)$$

a well-known answer.

Now consider equation 1.36 or 1.37 and Figure 1.14(b). The response is $E(x, y, z)$ (the field), the excitation is $\rho_v(x', y', z')$ (the source) and

$$\mathbf{h}(\mathbf{r} - \mathbf{r}') = \frac{1}{4\pi\varepsilon_0}\frac{\mathbf{r} - \mathbf{r}'}{|\mathbf{r} - \mathbf{r}'|^3} = \frac{1}{4\pi\varepsilon_0}\frac{\mathbf{a}_R}{R^2} \tag{1.40}$$

is the response at $\mathbf{r}$ due to a *unit* impulse (or a unit point charge) at $\mathbf{r}'$, or simply the unit impulse response. Notice that $\mathbf{h}(\mathbf{r}) = (1/4\pi\varepsilon_0)(\mathbf{a}_r/r^2)$ is the response at $\mathbf{r}$ due to a unit point charge at the origin, and for any *fixed* linear system $\mathbf{h}(\mathbf{r} - \mathbf{r}')$ is obtained from $\mathbf{h}(\mathbf{r})$ by replacing $\mathbf{r}$ with $\mathbf{r} - \mathbf{r}'$. Any doubt that this is the impulse response should be resolved by referring to equations 1.30 or 1.31 with $Q = 1$. Thus, equation 1.37 is just a superposition integral or convolution type integral of three dimensions. In future developments we will be able to take advantage of this knowledge and anticipate superposition type integrals.

[4] This solution is the *forced* response only, but any *free* response (complementary function) may be added to it to accommodate any initial conditions that might be present.

One special case can be disposed of almost immediately. Suppose the field point is far removed from the origin, while the charge density is finite in extent and relatively close to the origin. These conditions are satisfied if $r \gg r'$. Formally, if

$$r \gg r'$$

then

$$R = |\mathbf{r} - \mathbf{r}'| \approx r$$

$$\mathbf{R} = \mathbf{r} - \mathbf{r}' \approx \mathbf{r}$$

and

$$\mathbf{E}(\mathbf{r}) = \frac{1}{4\pi\varepsilon_0} \iiint_{\text{vol}'} \frac{\rho_v(\mathbf{r}')}{r^3}\, \mathbf{r}\, dv'$$

or

$$\mathbf{E}(\mathbf{r}) = \frac{\mathbf{r}}{4\pi\varepsilon_0 r^3} \iiint_{\text{vol}'} \rho_v(\mathbf{r}')\, dv'$$

It is helpful to verify the approximations being made by re-examining Figure 1.13. Now, using equation 1.11,

$$\mathbf{E}(\mathbf{r}) = \frac{Q\mathbf{r}}{4\pi\varepsilon_0 r^3}, \qquad r \gg r' \tag{1.41}$$

The interpretation of equation 1.41 is easy. From a large distance any finite charge distribution looks like a point charge at the origin with total charge Q C! See equation 1.31 for verification of this statement. Also, see Problem 17.

If the source is a surface charge density, there are several avenues of approach we may take in arriving at an equation for the electric field, but since we already have the impulse response for **E** in equation 1.40, the simplest thing we can do is use the superposition integral which immediately gives

$$E(r) = \frac{1}{4\pi\varepsilon_0} \iint_{s'} \frac{\rho_s(r')}{R^2}\, \mathbf{a}_R\, ds' \tag{1.42}$$

This is precisely what we predicted we would be able to do, and any other method will lead to the same result. Notice now that ρ_s apparently depends on only two of the three space variables. The third variable is hopefully constant on the given surface bearing the charge, or is eliminated by means of the equation of the surface. If the surface bearing the charge cannot be described analytically, then numerical integration must be performed.

In the case of a line charge density we can use the same principle once more to obtain

$$\mathbf{E}(\mathbf{r}) = \frac{1}{4\pi\varepsilon_0} \int_{l'} \frac{\rho_l(\mathbf{r}')}{R^2} \mathbf{a}_R \, dl' \tag{1.43}$$

The line charge density ρ_l apparently depends on only one of the space variables. The other two are hopefully constant, or at least capable of being eliminated by means of the equations for the path containing the line charge density.

We have seen in this section that the principle of superposition is very powerful indeed, for we *know* the electric field intensity for a unit positive charge from Coulomb's law (the unit impulse response), and we can *immediately* write the superposition integral for the electric field intensity for *any* charge distribution. If this charge distribution is known (even approximately), we can always use the digital computer to evaluate the integral.

1.6 FIELD OF AN INFINITE LINE CHARGE DENSITY

Now that we have armed ourselves with some rather complicated looking equations, let us first examine some simple looking special cases. The first to be treated is the nonphysical, but nevertheless useful, infinite line charge of uniform density ρ_l C/m shown in Figure 1.16(a). If this uniform line charge is placed on the z axis, we have azimuthal symmetry and z independence, so that $\mathbf{E}$ is independent of ϕ and z. This can be shown by means of a simple analogy. If we represent the uniform line charge by a very thin (but visible) uniform light source on the entire z axis, then an observer sees the same thing regardless of ϕ or z! We may then locate the field point on the x axis without loss of generality. Consider a differential charge $dQ = \rho_l \, dz'$ located at $z' = +h$ and another differential charge $dQ = \rho_l \, dz'$ located at $z' = -h$. Applying Coulomb's law to these charges shows that at a point on the x axis the z components of field cancel while the x components add. Since the entire charge can be divided into such pairs, we ultimately obtain $E_z = 0$ and $E_x = E_x(x)$. See Figure 1.16(b). Using equation 1.43 with Cartesian coordinates we have

$$\mathbf{E}(x, 0, 0) = \frac{1}{4\pi\varepsilon_0} \int_{-\infty}^{\infty} \frac{\rho_l(z')}{R^2} \mathbf{a}_R \, dz'$$

From Figure 1.12(a) we see that

$$\mathbf{R} = \mathbf{r} - \mathbf{r}' = \mathbf{a}_x x - \mathbf{a}_z z'$$

Thus,

$$\mathbf{E}(x, 0, 0) = \frac{\rho_l}{4\pi\varepsilon_0} \int_{-\infty}^{\infty} \frac{\mathbf{a}_x x - \mathbf{a}_z z'}{[x^2 + (z')^2]^{3/2}} \, dz'$$

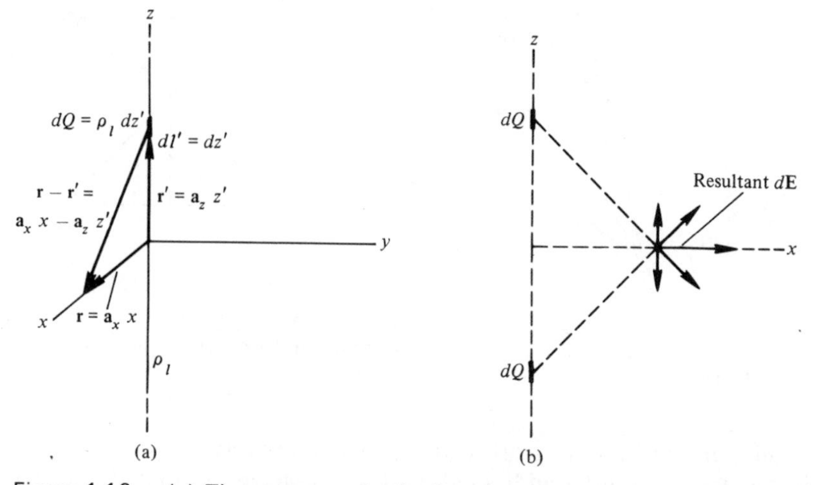

Figure 1.16. (a) The geometry for finding the electric field of a uniform line charge density. (b) Cancellation of dE_z by symmetry.

or

$$E_x(x, 0, 0) = \frac{\rho_l x}{4\pi\varepsilon_0} \int_{-\infty}^{\infty} \frac{dz'}{[x^2 + (z')^2]^{3/2}}$$

and

$$E_z(x, 0, 0) = \frac{-\rho_l}{4\pi\varepsilon_0} \int_{-\infty}^{\infty} \frac{z'\, dz'}{[x^2 + (z')^2]^{3/2}}$$

We have already deduced from symmetry arguments that $E_z = 0$, so the last integral is zero. On the other hand, the mathematics will show the same thing. The integrand of the integral for $E_z(x, 0, 0)$ is an *odd function* of z', meaning that [integrand (z')] = − [integrand $(-z')$]. Since the integral represents an area, and here we will have equal amounts of positive and negative area, the integral is zero! We are left with

$$E_x(x, 0, 0) = \frac{\rho_l x}{4\pi\varepsilon_0} \frac{z'}{x^2[x^2 + (z')^2]^{1/2}}\Bigg|_{-\infty}^{\infty}$$

$$= \lim_{z' \to \infty} \frac{\rho_l}{2\pi\varepsilon_0 x} \frac{1}{[1 + (x/z')^2]^{1/2}}$$

or

$$E_x(x, 0, 0) = \frac{\rho_l}{2\pi\varepsilon_0 x}$$

after performing the integration. In this case, if we move the field point off the x axis, the radial distance is ρ and the radial component is E_ρ, so the general result is

$$E_\rho(\rho, \phi, z) = \frac{\rho_l}{2\pi\varepsilon_0 \rho} \quad \text{(V/m)} \tag{1.44}$$

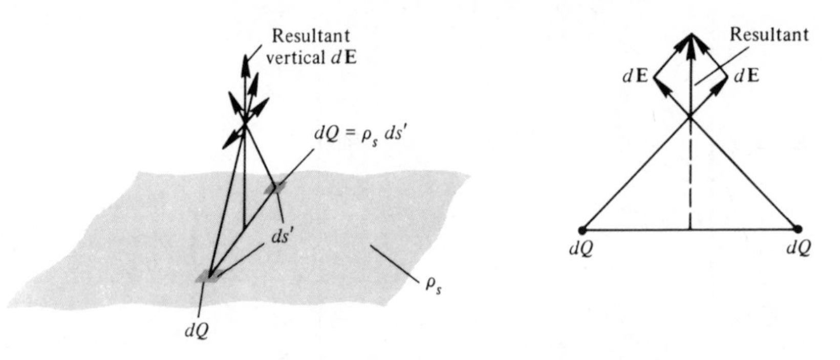

Figure 1.17. Symmetry properties of the infinite, uniform, plane surface charge density.

We could, of course, have worked this problem entirely in cylindrical coordinates with very little additional work. It is suggested that the reader carry out this calculation. Contours of constant electric field amplitude will be cylinders coaxial with the z axis such that if the radial distance, ρ, is doubled, the field is halved. Before passing on to another example let us take note of the fact that a *finite* line charge density is usually not uniformly distributed because of the Coulomb force of repulsion. Exactly how the charge would distribute itself is a topic for future concern. It was mentioned earlier that this was a useful distribution. Besides the benefits of the academic exercise, practical results for the electric field around a finite length, finite diameter, charged wire can be obtained from equation 1.44 if the field point is so close that the wire looks infinitely long, but, at the same time, far enough away so that the wire looks like a filament. See Problem 17.

1.7 FIELD OF A SHEET OF SURFACE CHARGE DENSITY

Consider an infinite sheet of uniform surface charge density, ρ_s, located in the $z = 0$ plane and a field point located above it ($z > 0$). This is another charge distribution whose field can be determined exactly with little difficulty. Symmetry conditions are such that we immediately conclude that the electric field must be the same in amplitude in any plane parallel to the surface charge. For every differential contribution to the electric field from a differential element of surface charge, we can find a symmetrically located element of charge of the same amount. These two elements of charge produce equal components of electric field perpendicular to the surface, but produce opposite amounts of electric field parallel to the surface. This symmetry is shown in Figure 1.17. Using the light source analogy of the previous section, we can imagine the entire $z = 0$ plane covered with a uniform light source. An observer located above this source will certainly see the same thing as far as his x or y coordinate is concerned. He will, in fact, also see the same thing as far as his z coordinate is concerned! We can now predict that the electric

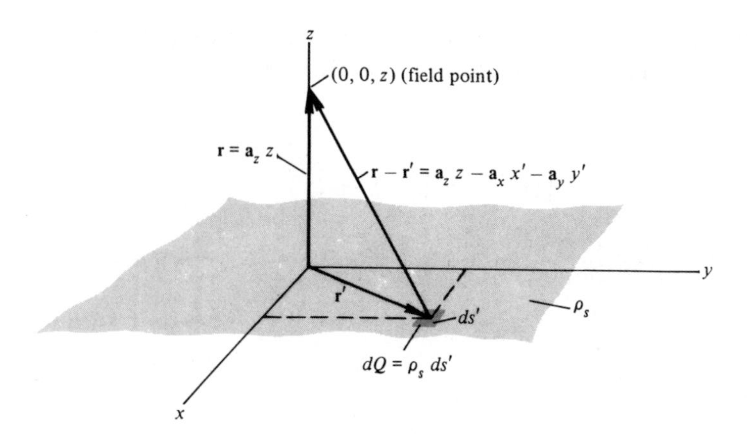

Figure 1.18. Geometry for finding the field of an infinite uniform surface charge density lying in the $z = 0$ plane.

field intensity has only a *uniform* z component. More light is shed on this subject when the integration is carried out.

The geometry of the problem is shown in Figure 1.18. Without loss of generality the field point is located on the z axis $[\mathbf{E}(0, 0, z) = \mathbf{E}(x, y, z)]$. From equation 1.42

$$\mathbf{E}(x, y, z) = \frac{1}{4\pi\varepsilon_0} \int_{-\infty}^{\infty} \int_{-\infty}^{\infty} \frac{\rho_s(\mathbf{a}_z z - \mathbf{a}_x x' - \mathbf{a}_y y')}{[(x')^2 + (y')^2 + z^2]^{3/2}} \, dx' \, dy'$$

or

$$E_x(x, y, z) = \frac{-\rho_s}{4\pi\varepsilon_0} \int_{-\infty}^{\infty} \int_{-\infty}^{\infty} \frac{x' \, dx' \, dy'}{[(x')^2 + (y')^2 + z^2]^{3/2}}$$

$$E_y(x, y, z) = \frac{-\rho_s}{4\pi\varepsilon_0} \int_{-\infty}^{\infty} \int_{-\infty}^{\infty} \frac{y' \, dx' \, dy'}{[(x')^2 + (y')^2 + z^2]^{3/2}}$$

and

$$E_z(x, y, z) = \frac{\rho_s z}{4\pi\varepsilon_0} \int_{-\infty}^{\infty} \int_{-\infty}^{\infty} \frac{dx' \, dy'}{[(x')^2 + (y')^2 + z^2]^{3/2}} \tag{1.45}$$

The integrals for E_x and E_y (the components parallel to the surface charge) are zero because the respective integrands are odd functions. Recall the similar situation that existed for the infinite line charge density of the previous section. The net area represented by the integrals for E_x and E_y is zero. This is the mathematical result of the symmetry conditions and was predicted. We are left with the double integral for E_z. If we look at this integral closely, we see that it is in the same form as that performed for the line charge in the preceding section. Then, we have

$$E_z(x, y, z) = \frac{\rho_s z}{4\pi\varepsilon_0} \int_{-\infty}^{\infty} \frac{y'}{[(x')^2 + z^2][(x')^2 + (y')^2 + z^2]^{1/2}} \Bigg|_{-\infty}^{\infty} dx'$$

$$= \lim_{y' \to \infty} \frac{\rho_s z}{2\pi\varepsilon_0} \int_{-\infty}^{\infty} \frac{dx'}{[(x')^2 + z^2]\{1 + [(x')^2 + z^2]/y'^2\}^{1/2}}$$

$$= \frac{\rho_s z}{2\pi\varepsilon_0} \int_{-\infty}^{\infty} \frac{dx'}{(x')^2 + z^2}$$

$$= \frac{\rho_s z}{2\pi\varepsilon_0} \frac{1}{z} \tan^{-1} \frac{x'}{z} \Bigg|_{-\infty}^{\infty}$$

$$= \frac{\rho_s}{2\pi\varepsilon_0} \left(\frac{\pi}{2} + \frac{\pi}{2}\right)$$

or

$$E_z(x, y, z) = \frac{\rho_s}{2\varepsilon_0} \quad \text{(V/m)} \tag{1.46}$$

This result is as predicted. What is the electric field for $z < 0$? This question is easily answered by simply noting that for $z = -\xi$, the field must be the same as that given by equation 1.46, except that it must be in the negative z direction (or directed away from the charge). Thus

$$E_z(x, y, -\xi) = -\frac{\rho_s}{2\varepsilon_0} \tag{1.47}$$

We now have

$$E_z(x, y, z) = \begin{cases} \rho_s/2\varepsilon_0, & z > 0 \\ -\rho_s/2\varepsilon_0, & z < 0 \end{cases} \tag{1.48}$$

This form is a little cumbersome, and we may make it more compact by utilizing the signum function.

$$\text{signum}(z) = \text{sgn}(z) \equiv 2u(z) - 1 \tag{1.49}$$

We may now write the concise result

$$E_z(x, y, z) = \frac{\rho_s}{2\varepsilon_0}[2u(z) - 1] = \frac{\rho_s}{2\varepsilon_0} \text{sgn}(z) \tag{1.50}$$

for the infinite sheet of uniform surface charge density in the $z = 0$ plane.

Useful results for a finite plane surface bearing a surface charge density can be obtained from equation 1.50 if the field point is located somewhere near the middle of the plane and not far removed from it. In this case the finite plane looks like an infinite plane. See Problem 17.

Before leaving this problem, it is worthwhile to expose (explicitly) a pitfall to the unwary investigator. In order to do this, let us rework the preceding problem in cylindrical coordinates (ignoring the simplifications due to the symmetry conditions). This might seem natural enough since cylindrical coordinates are apparently just as appropriate as are rectangular. We now have

$$\mathbf{r} = \mathbf{a}_z z, \qquad x' = \rho' \cos \phi'$$

$$\mathbf{r}' = \mathbf{a}_{\rho'} \rho', \qquad y' = \rho' \sin \phi'$$

$$ds' = \rho' \, d\rho' \, d\phi' \qquad (z' \text{ constant})$$

and

$$\mathbf{E}(\rho, \phi, z) = \frac{1}{4\pi\varepsilon_0} \int_0^\infty \int_0^{2\pi} \frac{\rho_s(\mathbf{a}_z z - \mathbf{a}_{\rho'} \rho')\rho' \, d\rho' \, d\phi'}{[(\rho')^2 \cos^2 \phi' + (\rho')^2 \sin^2 \phi' + z^2]^{3/2}}$$

or

$$\mathbf{E}(\rho, \phi, z) = \frac{\rho_s}{4\pi\varepsilon_0} \int_0^\infty \int_0^{2\pi} \frac{(\mathbf{a}_z z - \mathbf{a}_{\rho'} \rho')\rho' \, d\rho' \, d\phi'}{[(\rho')^2 + z^2]^{3/2}} \tag{1.51}$$

The subtle difficulty in equation 1.51 is that $\mathbf{a}_{\rho'}$ is *not* a constant unit vector! Its magnitude is certainly constant (one) but its direction depends on ϕ', the azimuth angle (a variable). The obvious cure is to represent $\mathbf{a}_{\rho'}$ in terms of constant unit vectors; namely, the rectangular unit vectors $\mathbf{a}_x$ and $\mathbf{a}_y$. This representation was mentioned in Section 1.1. It is

$$\mathbf{a}_{\rho'} = \mathbf{a}_x \cos \phi' + \mathbf{a}_y \sin \phi' \tag{1.52}$$

Then, equation 1.51 becomes

$$\mathbf{E}(\rho, \phi, z) = \frac{\rho_s}{4\pi\varepsilon_0} \int_0^\infty \int_0^{2\pi} \frac{(\mathbf{a}_z z - \mathbf{a}_x \cos \phi' - \mathbf{a}_y \sin \phi')\rho' \, d\rho' \, d\phi'}{[(\rho')^2 + z^2]^{3/2}}$$

or

$$E_x(\rho, \phi, z) = \frac{-\rho_s}{4\pi\varepsilon_0} \int_0^\infty \int_0^{2\pi} \frac{\rho' \cos \phi' \, d\rho' \, d\phi'}{[\rho')^2 + z^2]^{3/2}},$$

$$E_y(\rho, \phi, z) = \frac{-\rho_s}{4\pi\varepsilon_0} \int_0^\infty \int_0^{2\pi} \frac{\rho' \sin \phi' \, d\rho' \, d\phi'}{[(\rho')^2 + z^2]^{3/2}}$$

and

$$E_z(\rho, \phi, z) = \frac{\rho_s z}{4\pi\varepsilon_0} \int_0^\infty \int_0^{2\pi} \frac{\rho' \, d\rho' \, d\phi'}{[(\rho')^2 + z^2]^{3/2}}$$

Now, since the integral of $\cos \phi'$ or $\sin \phi'$ over a range of 2π rad is zero, E_x and E_y are (again) both zero. The third integral above gives

$$E_z(\rho, \phi, z) = \frac{\rho_s z}{2\varepsilon_0} \int_0^\infty \frac{\rho' \, d\rho'}{[(\rho')^2 + z^2]^{3/2}}$$

or

$$E_z(\rho, \phi, z) = \frac{\rho_s}{2\varepsilon_0}, \qquad z > 0,$$

as before! Failing to recognize that $\mathbf{a}_{\rho'}$ is not a constant vector has caused embarrassing moments for many of us!

1.8 A FINAL EXAMPLE

In concluding this chapter we will consider a final example in which a more general, but certainly not completely general, charge distribution exists. Let the charge density be $\rho_v = f(r)$ which still has a high degree of symmetry. Using equation 1.37 we have

$$\mathbf{E} = \frac{1}{4\pi\varepsilon_0} \iiint\limits_{\text{all space}} \frac{f(r')}{R^2} \mathbf{a}_R \, dv'$$

The symmetry in the charge density leads us to conclude that both E_θ and E_ϕ are zero, and that E_r depends only on r. This should be verified before proceeding. In this case the field point may be located on the z axis without loss of generality. It is left as an exercise[5] to show that the symmetry conditions lead to

$$E_r(r) = \frac{1}{2\varepsilon_0} \int_0^\pi \int_0^\infty \frac{f(r')(r - r' \cos \theta')(r')^2 \sin \theta' \, dr' \, d\theta'}{[r^2 + (r')^2 - 2rr' \cos \theta']^{1/2}}$$

We are now left with a rather formidable double integration to perform, even if $f(r)$ is simple [unless, of course, $f(r)$ is the charge density for a point charge at the origin]. For some functions the integration can be analytically performed, but for present purposes it is sufficient to note that the direct approach has apparently not lead to a simple solution. On the other hand, because of the high degree of symmetry present in this problem, a method to be introduced in the next chapter can easily treat it.

1.9 CONCLUDING REMARKS

In this chapter we have taken Coulomb's law and with it defined the electric field for the static case. The equations we obtained were formulated with vector analysis in a consistent and general way. By finding the electric

[5] See Problem 20.

field intensity of an arbitrarily located unit point charge we were able to find superposition integrals for the electric field of any charge distribution. The examples investigated included the point charge and the infinite line and surface charge densities, which are very special cases. We must remember that real practical problems will normally be much more difficult to solve than these.

If the particular charge distribution in a problem is known, or is given, then the formal solution to the problem of finding the resulting electric field is complete, being given by equation 1.37. Suppose that a conducting body is charged, or has some excess charge dumped on it. Coulomb's law assures us that these charges will arrange themselves into some final static distribution (because of the forces between the charges) where all of the forces are balanced. We must first somehow find the charge distribution before equation 1.37 can be used to find the field. Problems of this type normally lead to *integral equations* which are difficult to solve. The point being made here is that knowing the charge distribution is a tremendous advantage, but only in a few cases is there enough symmetry to ensure that we actually do know the charge distribution.

One question, which should have occurred to those familiar with general systems concepts, remains. We have found integral solutions for the electrostatic field intensity in terms of the charge density. These solutions must be the solution to a differential equation. We will determine this differential equation in the next chapter.

REFERENCES

Edminister, Joseph A. *Electromagnetics*, Schaum Outline Series. New York: McGraw-Hill, 1979. Many solved problems of the type encountered in the first nine chapters of this text are included.

Hayt, W. H., Jr. and Kemmerly, J. E. *Engineering Circuit Analysis*, 3rd ed. New York: McGraw-Hill, 1978. The concepts of linear system response, convolution, impulse response, and Fourier transforms are presented in Chapter 19 of this sophomore level textbook.

Johnson, D. E., Hilburn, J. L., and Johnson, J. R. *Basic Electric Circuit Analysis.* Englewood Cliffs, N.J.: Prentice-Hall, 1978. Chapters 17 and 18 of this textbook discuss Fourier and Laplace methods and the use of the impulse function.

Lighthill, M. J. *Fourier Analysis and Generalized Functions.* London: Cambridge University Press, 1960. The relation between impulse reponse and linear system response is discussed.

McQuistan, Richmond B. *Scalar and Vector Fields, a Physical Interpretation.* New York: Wiley, 1965. All of the topics of vector analysis (plus more) we shall need are included in this inexpensive paperback.

Spiegel, M. R. *Mathematical Handbook*, Schaum Outline Series. New York: McGraw-Hill, 1968. Many of the integrals we must evaluate are listed in this inexpensive paperback.

Spiegel, M. R. *Vector Analysis*, Schaum Outline Series. New York: McGraw-Hill, 1959. Many solved problems are included in this inexpensive paperback.

PROBLEMS

1. Point charges $Q_1 = Q_2 = 10^{-9}$ C are located at (1, 0, 0) and (0, 0, 1), respectively. Find the force on each charge.
2. Point charge $Q_1 = 10^{-9}$ C is located at (0, 0, 0), while point charge Q_2 is located at (0, 0, 1). If $E_z = 0$ at (2, 2, 2), find Q_2.
3. Assume that a line charge density is uniform and located symmetrically between $-h$ and $+h$ on the z axis. Find the cylindrical components of **E**. In an actual situation, can the line charge density be uniformly distributed?
4. An infinite line charge density $\rho_l = 10^{-9}$ C/m exists on the z axis, while a uniform surface charge density $\rho_s = 10^{-9}$ C/m^2 exists on the $x = 1$ plane. Find the locus for $\mathbf{E} = 0$.
5. A uniform line charge density exists on the entire z axis except for $-L \leq z \leq +L$. Find **E**.
6. A charge Q is dumped on a conducting spherical shell of radius a (negligible thickness). How will this charge reside? Find an integral for **E** for $r > a$.
7. A surface charge density exists for $a < \rho < b$, $z = 0$. If it is assumed to be uniformly distributed, find **E** at a point on the z axis. Comment on the assumption of a uniform density.
8. Obtain the field of an infinite surface density using the results of Problem 7.
9. Find the field from a uniform surface charge density lying in the plane $x - y + 2z = 3$. Find the field at (0, 0, 1) if $\rho_s = 10^{-9}$ C/m^2.
10. A cube 1 mm on a side has total charge of 10^{-9} C on each face, and its geometric center is at the origin. Find **E**(1, 0, 0).
11. A point charge is being accelerated according to $\mathbf{a} = 100\mathbf{a}_x + 10\mathbf{a}_z$. Find **E**.
12. A point charge Q with mass m is suspended in equilibrium above an infinite horizontal surface charge density, $\rho_s = 10^{-9}$ C/m^2. Find Q.
13. A pair of deflection plates produces a uniform electric field $\mathbf{E} = -\mathbf{a}_x E_x$. The upper plate is located by $x = d/2$. $0 \leq z \leq l$, while the lower plate is located by $x = -d/2, 0 \leq z \leq l$. A screen is erected at $z = L$. An electron enters at the origin with an initial velocity $\mathbf{u} = \mathbf{a}_z u_{z0}$. Find the x position of the electron when $z = l$ and $z = L$ (*electrostatic deflection system*). See Figure 1.19.
14. Verify that equation 1.23 is correct.
15. Using equations 1.21 and 1.37 explicitly, find the electric field at (2, 2, 2) from a point charge at (0, 0, 1). Use rectangular coordinates.
16. Find the electric field in the $z = 0$ plane if a point charge $+Q$ is located at (0, 0, h) while $-Q$ is located at (0, 0, $-h$).
17. A surface charge density 10^{-9} C/m^2 (assumed uniform) exists for $z = 0$, $-50 \leq x \leq 50$, $-0.05 \leq y \leq 0.05$. Using reasonable approximations find **E**(0, 0, z) when (a) $z = 10^{-3}$, (b) $z = 1$, (c) $z = 10^{+4}$

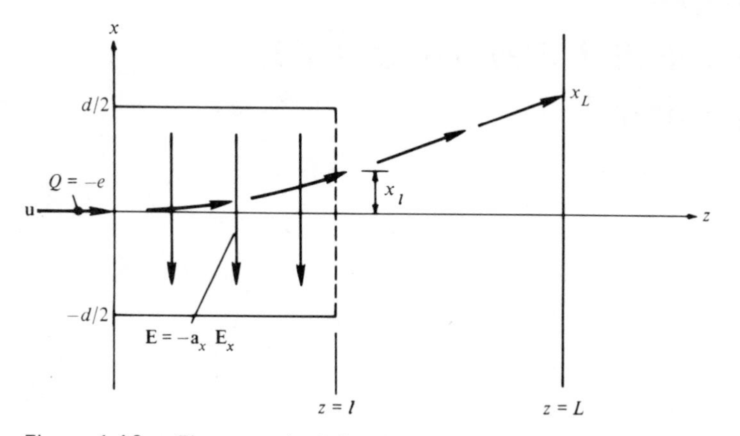

Figure 1.19. Electrostatic deflection system.

18. Evaluate

(*a*) $\int_{-1}^{+1} \delta(t)\, dt$ (b) $\int_{-1}^{+1} \delta(t^2)\, dt$ (c) $\int_0^2 \sin t \delta(\cos t)\, dt$

(d) $\int_{-\infty}^{\infty} f(x - x_0)\, \delta(x)\, dx$ (e) $\int_{-1}^{+1} u(x^2)\, dx$

19. Derive equation 1.13. Hint: integrate by parts.
20. Derive the integral formula for $E_r(r)$ in Section 1.8.
21. Derive equations 1.32a, 1.32b, and 1.32c.

Chapter 2
Gauss's Law and Flux Density

In this chapter we will introduce electric flux density and its simple relationship to the electric field in a vacuum. This enables us to consider electric flux and its relation to charge. Gauss's law expresses the latter relationship in an extremely simple manner, enabling us to solve many problems of a symmetrical nature. Gauss's law leads directly to Maxwell's first equation of electrostatics which may be expressed in integral or differential form. The differential form of Gauss's law involves the vector operation *divergence*, which is introduced and defined. We then have the relation between source and field in terms of a partial differential equation which was predicted in the conclusion of Chapter 1.

2.1 ELECTRIC FLUX AND FLUX DENSITY

Let us return to the infinite surface charge density ρ_s C/m^2. In Section 1.7 it was shown that if this charge density is located in the $z = 0$ plane, then the electric field it produces is

$$\mathbf{E} = \frac{\rho_s}{2\varepsilon_0}[2u(z) - 1]\mathbf{a}_z \tag{2.1}$$

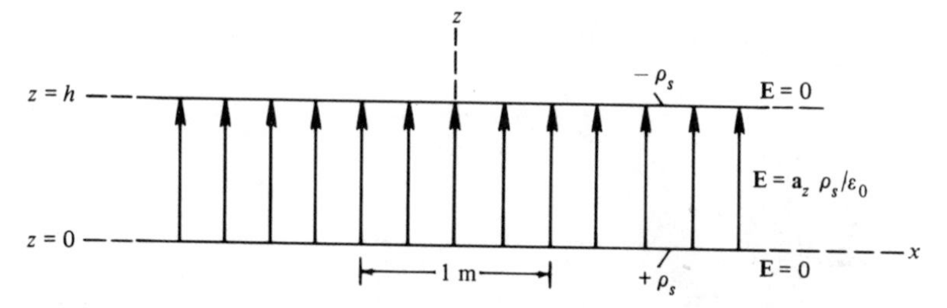

Figure 2.1. Uniform flux or field lines between infinite, parallel, and opposite surface charge densities.

If a second surface charge density, $-\rho_s$ C/m^2, is located at $z = h$, the field it (alone) produces is

$$\mathbf{E} = \frac{\rho_s}{2\varepsilon_0}[1 - 2u(z - h)]\mathbf{a}_z \tag{2.2}$$

The total field, produced by both charge distributions *acting together*, is by superposition,

$$\mathbf{E} = \frac{\rho_s}{\varepsilon_0}[u(z) - u(z - h)]\mathbf{a}_z \tag{2.3}$$

This configuration is shown in Figure 2.1.

Considering only a column 1 m wide, 1 m deep, and h m high in Figure 2.1, a charge of $Q = \rho_s(1)\,(1) = \rho_s$ C is located at the surface $z = 0$. Likewise, a charge of $Q = -\rho_s$ C is located at the surface $z = h$. For this column we may write

$$\varepsilon_0 E_z = \text{charge}, \qquad z = 0 \tag{2.4}$$

The interpretation is as follows. It is easy to visualize imaginary flow lines or *flux* lines emanating from the surface $z = 0$ (the positive charge) and terminating on the surface $z = h$ (the negative charge) as shown in Figure 2.2.

Figure 2.2. Relation between flux and charge for a section 1 m wide by 1 m deep from Figure 2.1.

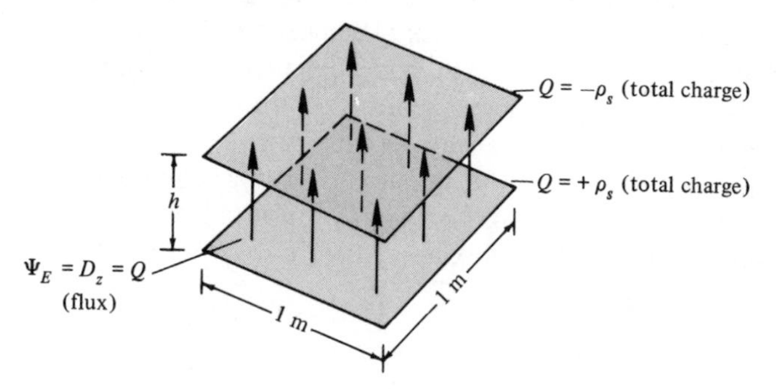

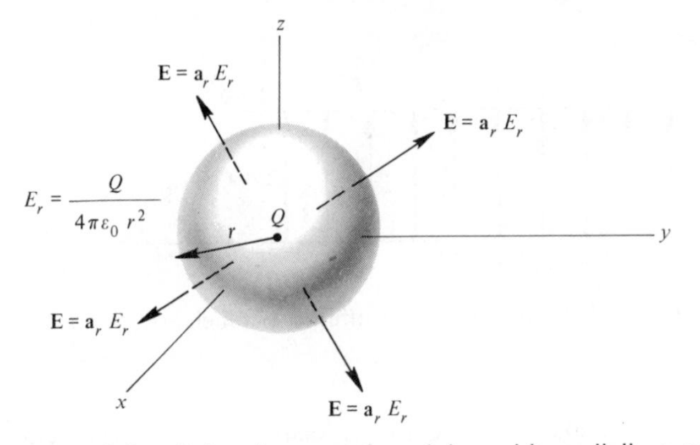

Figure 2.3. Point charge at the origin and its radially symmetric electric field.

Let us give this flux the symbol Ψ_E (electric flux) and tentatively equate it to the charge within 1 m^2 from which it emanates. In this way the flux will be measured in coulombs. From equation 2.4 we have

$$\Psi_E = Q = \varepsilon_0 E_z \quad \text{(C)}, \qquad z = 0 \tag{2.5}$$

The density of flux in the present example will be the flux divided by the area through which it flows, and will have the units of coulombs per square meter. We will give the flux density the symbol **D** since it will be a vector quantity in general. For the present case the flux density is uniform for $0 < z < h$ over the column since the charge is uniform. Therefore,

$$D_z = \frac{\Psi_E}{1} = \frac{Q}{1} = \varepsilon_0 E_z \quad (\text{C/m}^2) \tag{2.6}$$

Before attempting to draw any general conclusions from this last result, let us attempt a similar procedure for another geometry.

Consider an isolated point charge Q in a vacuum and located at the origin as shown in Figure 2.3. From equation 1.31, with $r' = 0$, we have

$$\mathbf{E}(r) = (Q/4\pi\varepsilon_0 r^2)\mathbf{a}_r \quad \text{(V/m)} \tag{2.7}$$

That is, the electric field is radial in direction and dependent only on the spherical radius r. Following the same procedure as in the previous example, we should expect lines of flux to emanate from the positive point charge Q and extend radially outward. Since there is no preferred direction (symmetry), the flux density will be the same for every direction at a fixed radius, r. The flux density, being a vector, will always point in the radial direction. Equating the flux passing through an imaginary sphere of radius r to the charge from which it emanates (as we did in the previous example), we have $\Psi_E = Q$. At the radius r the density of flux is

$$\frac{\Psi_E}{\text{surface area}} = \frac{\Psi_E}{4\pi r^2} = \frac{Q}{4\pi r^2}$$

since the surface area of a sphere is $4\pi r^2$. Since this flux density is always radially directed, we have

$$D_r = \frac{Q}{4\pi r^2} \quad (\text{C/m}^2)$$

or

$$\mathbf{D} = \frac{Q\mathbf{a}_r}{4\pi r^2} \tag{2.8}$$

Comparing equations 2.7 and 2.8, we again have the simple result

$$\mathbf{D} = \varepsilon_0 \mathbf{E} \tag{2.9}$$

Is it possible for **D** and **E** to be as simply related in a vacuum for all cases as equation 2.9 indicates? The answer is yes because we can take equation 2.8, let $Q = 1$ and we then have the unit impulse response for **D**, namely

$$\mathbf{h}(\mathbf{r}) = \frac{\mathbf{a}_r}{4\pi r^2} = \frac{\mathbf{r}}{4\pi r^3}$$

and replacing **r** by $\mathbf{r} - \mathbf{r}'$

$$\mathbf{h}(\mathbf{r} - \mathbf{r}') = \frac{\mathbf{r} - \mathbf{r}'}{4\pi|\mathbf{r} - \mathbf{r}'|^3} = \frac{\mathbf{a}_R}{4\pi R^2}$$

The former is the electric flux density at **r** from a unit point charge at the origin, while the latter is the electric flux density at **r** from a unit point charge at **r**′. Using the superposition integral gives the result

$$\boxed{\mathbf{D}(\mathbf{r}) = \frac{1}{4\pi} \iiint_{\text{vol}'} \frac{\rho_v(\mathbf{r}')}{R^2} \mathbf{a}_R \, dv'} \tag{2.10}$$

Comparing equations 1.37 and 2.10 shows that it is *always* true that $\mathbf{D} = \varepsilon_0 \mathbf{E}$ for a vacuum. Comparing impulse responses for **D** and **E** shows the same thing!

2.2 GAUSS'S LAW

The examples discussed in the previous section suggest strongly that there is a unique relationship between electric flux and electric charge. Does an equation as simple as $\Psi_E = Q$ always hold, regardless of how the charge is distributed?

A series of experiments performed by Faraday around 1837 gives us a very simple answer. Without going into the experiments themselves, except to mention that some were performed in a vacuum, or essentially a vacuum, we will simply quote the results, which are commonly called Gauss's law.

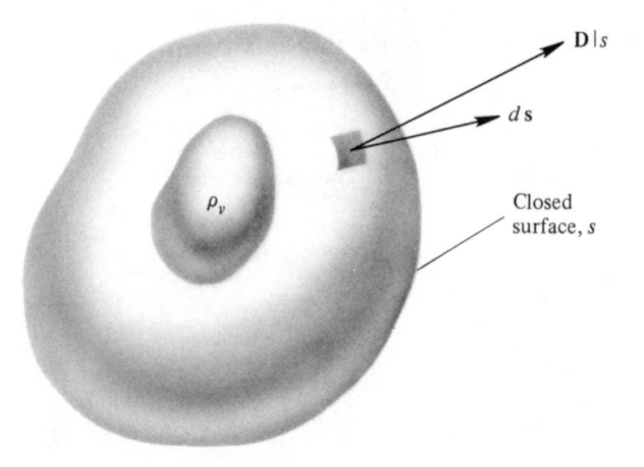

Figure 2.4. Flux density out of a closed surface surrounding a volume charge density.

The electric flux passing through any closed surface is equal to the charge enclosed by that surface.

In abbreviated form,

$$\text{flux out} = \text{net charge enclosed}$$

or

$$\boxed{\Psi_E = Q} \tag{2.11}$$

Gauss's law supports our suspicions.

As it is stated, Gauss's law is simplicity itself. Before rejoicing too much, however, let us formulate Gauss's Law mathematically. The left side of equation 2.11 is the flux out of a region, and can be written as the integral of the *normal* component of the flux density over *whatever closed surface* surrounding the region one wishes to consider. The differential flux at a point on the closed surface is $\mathbf{D}|_s \cdot d\mathbf{s}$, so

$$\boxed{\Psi_E = \oint\!\!\oint_s \mathbf{D}|_s \cdot d\mathbf{s}} \tag{2.12}$$

This arrangement is shown in Figure 2.4. Remember that the direction of $d\mathbf{s}$ is normal (outward) to the surface at the point and that the circle on the integral sign means we are concerned with a closed surface. From equation 1.11, we have

$$Q = \iiint_{\text{vol}} \rho_v dv \tag{2.13}$$

Then Gauss's law simply says

$$\oint\!\!\oint_s \mathbf{D}|_s \cdot d\mathbf{s} = \iiint_{\text{vol}} \rho_v \, dv \tag{2.14}$$

where the closed surface on the left side of equation 2.14 defines the volume on the right side of the equation, irrespective of whether or not all of the charge in a system is within the volume. In other words, the left side of equation 2.14 calculates the net flux out of the closed surface, while the right side calculates the net charge within the closed surface. A simple example is now in order.

EXAMPLE 1

Consider two concentric conducting spheres with radii a and b as shown in Figure 2.5. On the outer surface of the inner sphere we have placed

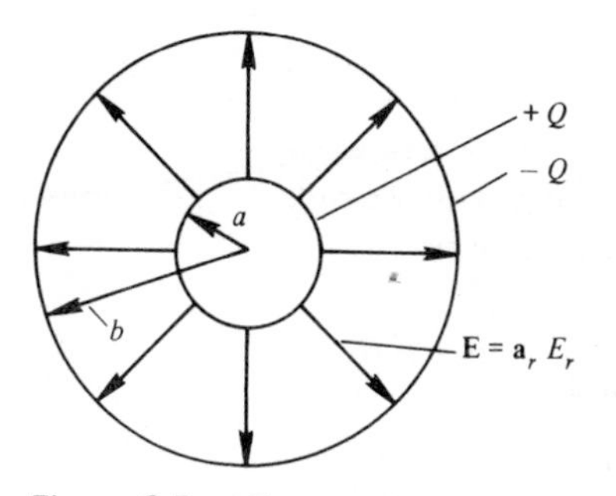

Figure 2.5. Concentric conducting spheres (capacitor).

a uniform[1] surface charge density $\rho_{sa} = Q/4\pi a^2$, while on the inner surface of the outer sphere we have (also uniform) $\rho_{sb} = -Q/4\pi b^2$. That is, we have $+Q$ C on the inner sphere and $-Q$ C on the outer sphere. This arrangement is very similar to Faraday's original experiment. The electric field and flux density obviously have radial symmetry. We desire a plot of E_r versus r. It is simpler in this case to consider the different regions separately.

1. $0 \le r < a$.

If we construct an imaginary spherical surface, or a "Gaussian surface," of radius $r(0 \le r < a)$, then there is *no* charge enclosed. Thus,

$$\Psi_E = \oint\!\!\oint \mathbf{D}|_s \cdot d\mathbf{s} = \oint\!\!\oint D_r \, ds = D_r \oint\!\!\oint ds = D_r(4\pi r^2) = Q_{\text{enc}} = 0$$

[1] It will ultimately be uniform no matter how we left it!

so

$$D_r = \frac{Q_{\text{enc}}}{4\pi r^2} = 0, \qquad r < a$$

Notice that **D** and $d\mathbf{s}$ are codirected, and D_r is *normal* to and *constant*[2] on the chosen Gaussian surface.

2. $a < r < b$.

We now construct a spherical Gaussian surface of radius $r(a < r < b)$. Then,

$$\Psi_E = \oiint \mathbf{D}\big|_s \cdot d\mathbf{s} = D_r(4\pi r^2) = Q_{\text{enc}} = +Q$$

so

$$D_r = \frac{Q}{4\pi r^2}$$

or

$$E_r = \frac{Q}{4\pi\varepsilon_0 r^2}, \qquad a < r < b$$

We note that this field is the same as that which would be produced by a point charge at the origin! We also note that

$$\lim_{r \to a} D_r = \frac{Q}{4\pi a^2} = \rho_{sa}$$

when r approaches a from $r > a$, and

$$\lim_{r \to b} D_r = \frac{Q}{4\pi b^2} = -\rho_{sb}$$

when r approaches b from $r < b$.

We finally construct a Gaussian surface of radius r, where $r > b$, then the *total* charge enclosed is $+Q - Q = 0$. Therefore,

$$\Psi_E = \oiint \mathbf{D}\big|_s \cdot d\mathbf{s} = D_r(4\pi r^2) = Q_{\text{enc}} = 0$$

or

$$D_r = E_r = 0, \qquad r > b$$

Thus, for *any* r we have

$$E_r = \frac{Q}{4\pi\varepsilon_0 r^2}[u(r-a) - u(r-b)]$$

[2] See Section 1.8.

In this case, the system has zero external field, or the outside is "shielded" from fields inside. This system is the spherical version of the parallel planar surface charge densities seen earlier. Both are shielded capacitors, and more will be said about them when capacitance is introduced in Chapter 4. A plot of E_r versus r is shown in Figure 2.6. The results,

$$D_r\big|_{r=a} = \rho_{sa} \quad \text{and} \quad D_r\big|_{r=b} = -\rho_{sb}$$

are particular *boundary conditions* about which we will have more to say at an appropriate time. As a final remark about this example, we should take care to notice that Gauss's law enabled us to arrive at the

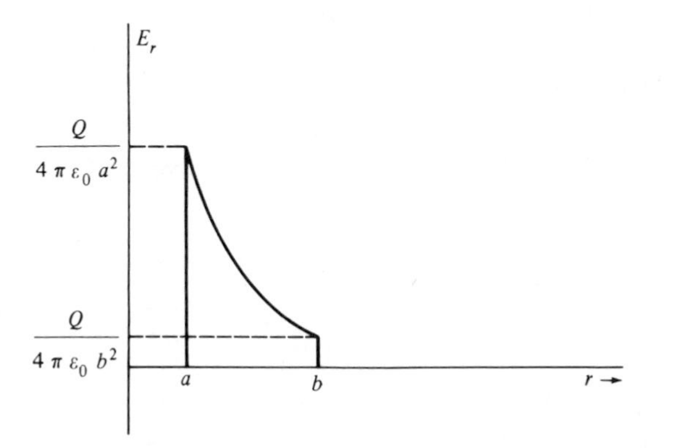

Figure 2.6. E_r versus r for the concentric capacitor of Figure 2.5.

correct flux density very quickly, but *only* because the flux was *always normal* to and *constant* on the chosen Gaussian surface. This fortunate arrangement allowed the integration to be only that for finding the surface area of the Gaussian surface.

EXAMPLE 2

As an additional example, suppose we have the same inner conducting sphere charged to $+Q$ surrounded by a uniform surface charge density ρ_{sb} at a radius $r = b$. This outer surface is nonconducting; that is, it is just a layer of charge $Q = \rho_{sb}(4\pi b^2)$. We desire to plot E_r versus r. (E_r and D_r again have spherical symmetry.)

1. $0 \leq r < a$. This is obviously the same as (1) in Example 1.
2. $a < r < b$. This is obviously the same as case (2) in Example 1 for $r < b$.
3. $r > b$.

We now construct a spherical Gaussian surface of radius r, where $r > b$. Then

$$D_r = \frac{\Psi_E}{4\pi r^2} = \frac{Q \text{ total}}{4\pi r^2} = \frac{Q + \rho_{sb}(4\pi b^2)}{4\pi r^2}$$

or

$$D_r = \frac{Q}{4\pi r^2} + \rho_{sb}\frac{b^2}{r^2} = \frac{\rho_{sa}a^2 + \rho_{sb}b^2}{r^2}, \qquad r > b$$

We also notice that

$$\lim_{r \to b} D_r = \frac{Q}{4\pi b^2} + \rho_{sb}$$

when r approaches b from $r > b$, whereas

$$\lim_{r \to b} D_r = \frac{Q}{4\pi b^2}$$

when r approaches b from $r < b$ (from Example 1). We are now able to plot E_r versus r as in Figure 2.7.

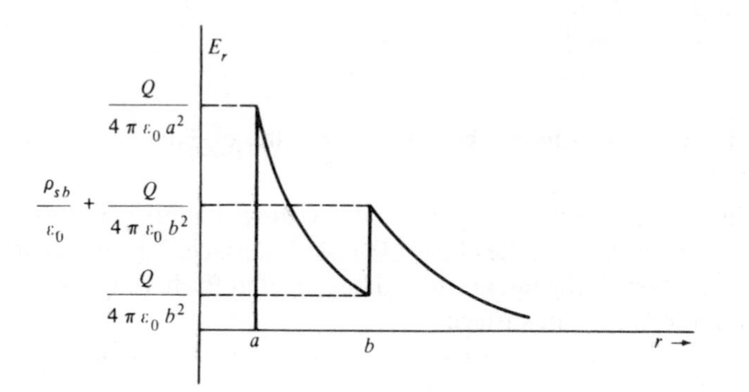

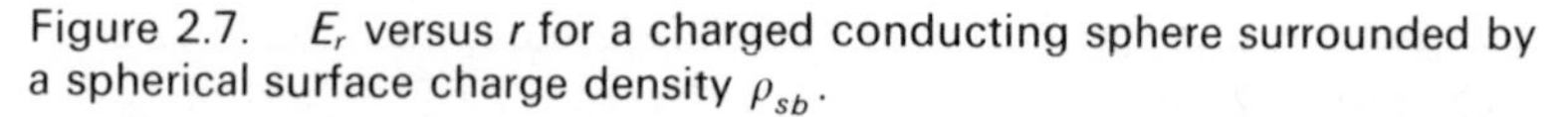

Figure 2.7. E_r versus r for a charged conducting sphere surrounded by a spherical surface charge density ρ_{sb}.

The difference between this example and Example 1 is apparent. We now have a nonzero field for $r > b$, and, furthermore, the flux density is discontinuous at the nonconducting surface charge layer by an amount equal to that surface charge density, ρ_{sb}. In other words

$$D_r\big|_{r=b+} - D_r\big|_{r=b-} = \rho_{sb}$$

Gauss's law enables us to find the electric field distribution about a charge distribution if enough symmetry is present. If we are able to pick a Gaussian surface where $\mathbf{D}$ is tangent to the surface, then $\mathbf{D}|_s \cdot d\mathbf{s}$ in equation 2.14 is zero. If $\mathbf{D}$ is normal to the surface, then $\mathbf{D}|_s \cdot d\mathbf{s}$ is

$D_n\,ds$, and D_n may be factored from inside the integral if, in addition, $|\mathbf{D}|$ is constant on the surface.

EXAMPLE 3

Recall that it required an integration to find the electric field distribution around the *infinite* uniform line charge density in Chapter 1. Symmetry arguments showed that $\mathbf{E} = \mathbf{a}_\rho E_\rho(\rho)$. That is, the only component of electric field was radial and dependent only on radial distance. For a Gaussian surface, construct a cylindrical can of radius ρ and height h. Let this surface be coaxial with the z axis and rest on the $z = 0$ plane as shown in Figure 2.8. We now know that $\mathbf{D}|_s \cdot d\mathbf{s} = 0$

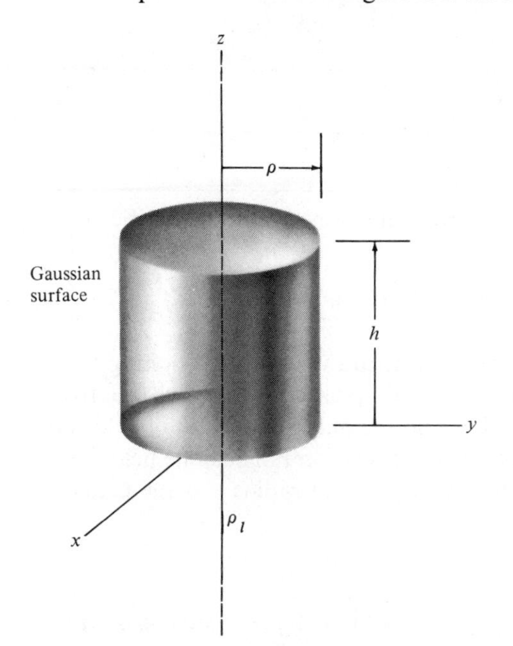

Figure 2.8. Gaussian surface for an infinite uniform line charge density.

for the top and bottom of the can, whereas $\mathbf{D}|_s \cdot d\mathbf{s} = D_\rho\,ds$ for the side. Gauss's law gives

$$\Psi_E = \int_0^{2\pi}\int_0^{h} D_\rho \rho \, d\phi \, dz = Q = \rho_l h$$

or

$$\rho_l h = D_\rho \int_0^{2\pi}\int_0^{h} \rho \, d\phi \, dz = 2\pi\rho h D_\rho$$

so

$$D_\rho = \frac{\rho_l}{2\pi\rho}$$

or

$$\mathbf{D} = \frac{\rho_l \mathbf{a}_\rho}{2\pi\rho}$$

which gives

$$\mathbf{E} = \frac{\rho_l \mathbf{a}_\rho}{2\pi\varepsilon_0 \rho} \tag{2.15}$$

This is the same as equation 1.44.

EXAMPLE 4

The last example in Chapter 1 treated a spherically symmetric charge distribution,

$$\rho_v = f(r) \quad (\text{C/m}^3) \tag{2.16}$$

The straightforward approach left us with a very difficult looking double integration to perform, but the use of Gauss's law gives us (relatively) no trouble. We reasoned earlier that the only component of field is the radial component, and it can only depend on r. Thus, the field is $\mathbf{E} = \mathbf{a}_r E_r(r)$ and we therefore choose a sphere of radius r as the Gaussian surface. Gauss's law gives

$$\Psi_E = \int_0^{2\pi}\int_0^{\pi} D_r r^2 \sin\theta \, d\theta \, d\phi = Q = \int_0^{2\pi}\int_0^{\pi}\int_0^{r} f(r) r^2 \sin\theta \, dr \, d\theta \, d\phi$$

or

$$D_r 4\pi r^2 = 4\pi \int_0^r f(r) r^2 \, dr$$

Thus,

$$E_r = \frac{1}{\varepsilon_0 r^2} \int_0^r f(r) r^2 \, dr$$

The required integration is now much simpler than before, and in most cases can easily be performed.

For example, if $\rho_v = e^{-br}$, we can integrate by parts, or, for those of us who are lazy, we can use integral tables, giving

$$E_r = -\frac{e^{-br}}{\varepsilon_0 br^2}\left[r^2 + \frac{2r}{b} + \frac{2}{b^2}\right] + \frac{2}{\varepsilon_0 r^2 b^3} \tag{2.17}$$

The use of Gauss's law has made this problem much simpler than when it was approached directly.

EXAMPLE 5

Another demonstration of the use of Gauss's law is provided by the following example. An electric field is given by

$$\mathbf{E} = \frac{1.5}{\varepsilon_0} x^2 y^2 \mathbf{a}_x + \frac{1}{\varepsilon_0} x^3 y \mathbf{a}_y \quad \text{(V/m)}$$

How much charge lies within a cube 2 m on a side if its geometric center is at the origin and its sides are parallel to the coordinate axes? Since

$$\mathbf{D} = 1.5x^2y^2\mathbf{a}_x + x^3y\mathbf{a}_y \quad (\text{C/m}^2)$$

we have

$$\Psi_E = \oint\!\!\oint \mathbf{D}|_s \cdot d\mathbf{s} = \oint\!\!\oint (1.5x^2y^2\mathbf{a}_x + x^3y\mathbf{a}_y)_s \cdot d\mathbf{s} = Q$$

Figure 2.9. Cubical surface centered at the origin and enclosing charge for finding Q, given **E** (or **D**).

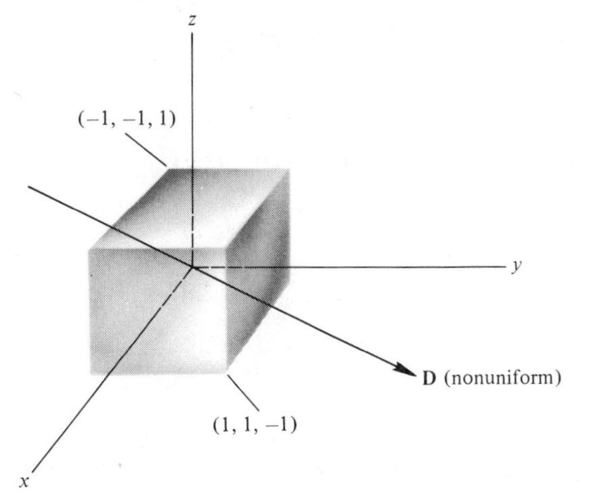

from Gauss's law. We want to calculate the flux out of the six faces of the cube shown in Figure 2.9. Then,

$$Q = \int_{-1}^{1}\int_{-1}^{1} (1.5x^2y^2\mathbf{a}_x + x^3y\mathbf{a}_y)|_{y=-1} \cdot (-\mathbf{a}_y\, dx\, dz)$$

$$+ \int_{-1}^{1}\int_{-1}^{1} (1.5x^2y^2\mathbf{a}_x + x^3y\mathbf{a}_y)|_{y=1} \cdot (\mathbf{a}_y\, dx\, dz)$$

$$+ \int_{-1}^{1}\int_{-1}^{1} (1.5x^2y^2\mathbf{a}_x + x^3y\mathbf{a}_y)|_{x=-1} \cdot (-\mathbf{a}_x\, dz\, dy)$$

$$+ \int_{-1}^{1}\int_{-1}^{1} (1.5x^2y^2\mathbf{a}_x + x^3y\mathbf{a}_y)|_{x=1} \cdot (\mathbf{a}_x\, dz\, dy)$$

$$+ \int_{-1}^{1}\int_{-1}^{1} (1.5x^2y^2\mathbf{a}_x + x^3y\mathbf{a}_y)|_{z=-1} \cdot (-\mathbf{a}_z\, dx\, dy)$$

$$+ \int_{-1}^{1}\int_{-1}^{1} (1.5x^2y^2\mathbf{a}_x + x^3y\mathbf{a}_y)|_{z=1} \cdot (\mathbf{a}_z\, dx\, dy)$$

Because of the dot product the last two terms are zero. (There is no flux out or in the top and bottom of the cube.) Then,

$$Q = \int_{-1}^{1}\int_{-1}^{1} + x^3\, dx\, dz + \int_{-1}^{1}\int_{-1}^{1} x^3\, dx\, dz$$

$$+ \int_{-1}^{1}\int_{-1}^{1} -1.5y^2\, dz\, dy + \int_{-1}^{1}\int_{-1}^{1} 1.5y^2\, dz\, dy$$

or

$$Q = 2\int_{-1}^{1} x^3\, dx + 2\int_{-1}^{1} x^3\, dx = 4\left.\frac{x^4}{4}\right|_{-1}^{+1} = 0$$

This result indicates that there is no *net* charge inside the cube. Emphasis is placed on the word net because there may be some charge inside the cube even though the net amount is zero. Suppose we calculate the charge within a cube 1 m on a side lying in the first octant. Then,

$$Q = \int_0^1 \int_0^1 (1.5x^2y^2\mathbf{a}_x + x^3y\mathbf{a}_y)|_{y=0} \cdot -\mathbf{a}_y \, dx \, dz$$

$$+ \int_0^1 \int_0^1 (1.5x^2y^2)\mathbf{a}_x + x^3y\mathbf{a}_y)|_{y=1} \cdot \mathbf{a}_y \, dx \, dz$$

$$+ \int_0^1 \int_0^1 (1.5x^2y^2\mathbf{a}_x + x^3y\mathbf{a}_y)|_{x=0} \cdot -\mathbf{a}_x \, dz \, dy$$

$$+ \int_0^1 \int_0^1 (1.5x^2y^2\mathbf{a}_x + x^3y\mathbf{a}_y)|_{x=1} \cdot \mathbf{a}_x \, dz \, dy$$

or

$$Q = \int_0^1 \int_0^1 x^3 \, dx \, dz + \int_0^1 \int_0^1 1.5y^2 \, dz \, dy$$

or

$$Q = \int_0^1 x^3 \, dx + \int_0^1 1.5y^2 \, dy$$

or

$$Q = \frac{x^4}{4}\bigg|_0^1 + \frac{y^3}{2}\bigg|_0^1 = \tfrac{3}{4}\text{C}$$

Thus, there is *some* charge inside the 8 m^3 cube even though the net amount is zero. The relation between flux and charge *at a point* is examined in the next section.

2.3 DIVERGENCE AND MAXWELL'S FIRST EQUATION

Gauss's law may be written

$$\oint\!\!\oint_s \mathbf{D}|_s \cdot d\mathbf{s} = Q = \iiint_{\text{vol}} \rho_v \, dv \tag{2.18}$$

which is simply a mathematical statement of the fact that flux out is charge enclosed. A very useful relationship in vector analysis is the divergence theorem[3] or Gauss's integral theorem which states that

$$\oiint \mathbf{A}\big|_s \cdot d\mathbf{s} \equiv \iiint_{\text{vol}} \nabla \cdot \mathbf{A} \, dv \tag{2.19}$$

where **A** is *any* vector and the closed surface on the left side of equation 2.19 has an interior whose volume is that on the right side of equation 2.19. The term $\nabla \cdot \mathbf{A}$ represents the *divergence of* **A** and is defined by

$$\nabla \cdot \mathbf{A} \equiv \lim_{\Delta v \to 0} \frac{1}{\Delta v} \oiint_{\Delta s} \mathbf{A}\big|_s \cdot d\mathbf{s} \tag{2.20}$$

∇ is a *vector operator* called the vector "del" operator and defined[4] as a differential operator by

$$\nabla \equiv \mathbf{a}_x \frac{\partial}{\partial x} + \mathbf{a}_y \frac{\partial}{\partial y} + \mathbf{a}_z \frac{\partial}{\partial z}$$

so that

$$\nabla \cdot \mathbf{A} = \left(\mathbf{a}_x \frac{\partial}{\partial x} + \mathbf{a}_y \frac{\partial}{\partial y} + \mathbf{a}_z \frac{\partial}{\partial z}\right) \cdot (\mathbf{a}_x A_x + \mathbf{a}_y A_y + \mathbf{a}_z A_z)$$

or

$$\nabla \cdot \mathbf{A} = \frac{\partial A_x}{\partial x} + \frac{\partial A_y}{\partial y} + \frac{\partial A_z}{\partial z}$$

for rectangular coordinates. In words, equation 2.20 states that the divergence of **A** (considering **A** to be a vector flux density, whatever it may actually be) at a point is the limit of the flux out of a small closed surface per unit volume as the volume inside the surface approaches zero. An incompressible fluid must have a velocity field, **u**, which has *no divergence* ($\nabla \cdot \mathbf{u} \equiv 0$). Fluid is neither created nor destroyed at any point and, thus, has no *sources* or sinks. The magnetic field, which we will encounter later, also has no sources or sinks because no magnetic charges have been found in nature. It too has no divergence. Such fields are said to be *solenoidal* or *sourceless*. The electric field, on the other hand, does have *sources* and sinks (the electric charges) and must have nonzero divergence at some point or points in space. This will be demonstrated in the next paragraph.

[3] The divergence theorem is derived in Appendix G.

[4] The del operator can also be defined as an integral operator

$$\nabla \circ - \equiv \lim_{\Delta v \to 0} \frac{1}{\Delta v} \oiint_{\Delta s} d\mathbf{s} \circ -$$

where $\circ$ denotes a dot, cross or an ordinary multiplication as, for example, the dot in equation 2.20.

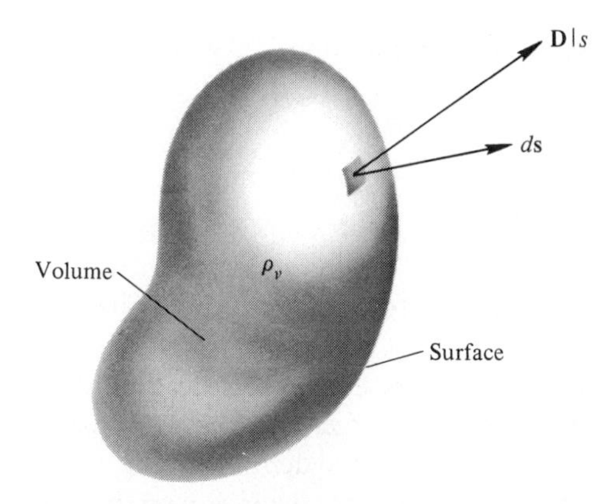

Figure 2.10. Surface and volume containing charge density, ρ_v.

In equation 2.18 we will let the surface on the left shrink until it just encloses the volume containing the charge density ρ_v. Note that equation 2.18 is still valid (Gauss's law still holds). This is pictured in Figure 2.10. It is still true in equation 2.18 that the surface on the left has an interior whose volume is that on the right. If we now apply equation 2.19 to equation 2.18, we obtain

$$\iint_s \mathbf{D}|_s \cdot d\mathbf{s} = \iiint_{\text{vol}} \nabla \cdot \mathbf{D} \, dv = \iiint_{\text{vol}} \rho_v \, dv \qquad (2.21)$$

In equation 2.21, the volumes are identical; so

$$\iiint_{\text{vol}} (\nabla \cdot \mathbf{D} - \rho_v) \, dv = 0 \qquad (2.22)$$

Now, equation 2.22 holds regardless of the limits of the integrals, that is, it holds for *any* finite volume. In this case, the integral can only be zero if the integrand itself is zero. Thus,

$$\boxed{\nabla \cdot \mathbf{D} = \rho_v} \qquad (2.23)$$

Equation 2.23 is known as Maxwell's first equation for electrostatics, and it states that the divergence of electric flux density *at a point* is equal to the volume charge density at the point. It is the point, or differential, form of Gauss's law.

Recall that in the conclusion of Chapter 1 it was pointed out that since we had found a superposition integral solution for $\mathbf{E}$ in terms of ρ_v, there must be a differential equation for ρ_v in terms of $\mathbf{E}$. That is, the integral solution of Chapter 1, equation 1.36, is a solution to a partial differential equation. We have now found that equation. It is equation 2.23 with $\mathbf{D} = \varepsilon_0 \mathbf{E}$, or

$$\nabla \cdot \mathbf{E} = \frac{\rho_v}{\varepsilon_0} \tag{2.24}$$

A direct proof that equation 1.36 is a solution to equation 2.24 is given in Appendix G.

Care must be exercised in using equation 2.24 when dealing with idealized (singular) sources such as the point charge, line charge density, and surface charge density. The infinite surface charge density on the $z = 0$ plane, for example, can be expressed as a volume charge density, $\rho_v = \rho_s \delta(z)$, and its field is given by equation 2.1.

$$\mathbf{E} = \frac{\rho_s}{2\varepsilon_0}[2u(z) - 1]\mathbf{a}_z \tag{2.25}$$

Since $\nabla \cdot \mathbf{E} = \partial E_z/\partial z = dE_z/dz$ in this case, and since $(d/dz)u(z) = \delta(z)$, we have

$$\nabla \cdot \mathbf{E} = \frac{d}{dz}\left\{\frac{\rho_s}{2\varepsilon_0}[2u(z) - 1]\right\}$$

or

$$\nabla \cdot \mathbf{E} = \frac{\rho_s \delta(z)}{\varepsilon_0} = \frac{\rho_v}{\varepsilon_0} \tag{2.26}$$

and Maxwell's equation holds.

We will now attempt to attach more physical meaning to equation 2.23 [other than that given below equation 2.20] by means of a frequently used semirigorous derivation. Consider a small boxlike surface with sides Δx, Δy, and Δz. Let the flux density $\mathbf{D}$ at the geometric center be given by

$$\mathbf{D} = D_1 \mathbf{a}_x + D_2 \mathbf{a}_y + D_3 \mathbf{a}_z \tag{2.27}$$

where D_1, D_2, and D_3 are the x, y, and z components of $\mathbf{D}$. This arrangement is shown in Figure 2.11. Since the box is small, D is nearly constant over the six faces making up the box. We now apply Gauss's law,

$$\oiint_s \mathbf{D}|_s \cdot d\mathbf{s} = Q \tag{2.28}$$

realizing that we have the sum of six double integrals; one for each face. Consider the integral over the surface closest to the reader:

$$\iint_{\text{front}} \approx \mathbf{D}|_{\text{front}} \cdot \mathbf{a}_x \, \Delta y \, \Delta z = D_x|_{\text{front}} \, \Delta y \, \Delta z \tag{2.29}$$

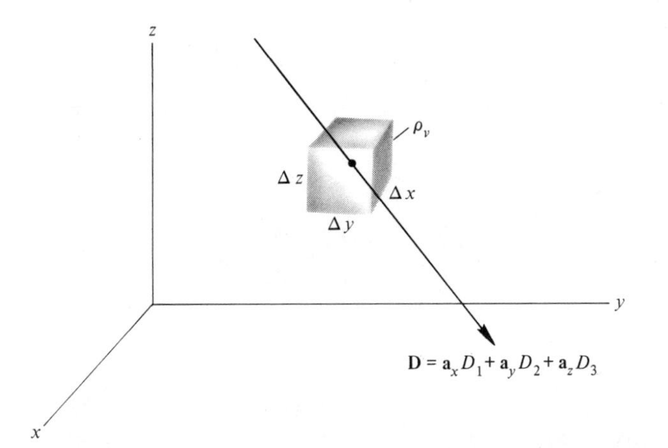

Figure 2.11. Incremental Gaussian surface for deriving Maxwell's equation.

Equation 2.29 is approximate because $\mathbf{D}|_{\text{front}}$ is *not*, in general, constant over this face. In terms of the given flux density at the center of the box, we may write

$$D_x|_{\text{front}} \approx D_1 + \frac{\Delta x}{2} \frac{\partial D_x}{\partial x} \tag{2.30}$$

In words, equation 2.30 states that D_x on the front face is approximately equal to D_x at the center of the box (i.e., D_1) plus the rate of change of D_x with x times the distance ($\Delta x/2$) over which this change occurs. It is worth mentioning that equation 2.30 may be obtained more rigorously as the first two terms in the Taylor's series expansion for D_x about the center of the box. We now have

$$\iint_{\text{front}} \approx \left(D_1 + \frac{\Delta x}{2} \frac{\partial D_x}{\partial x}\right) \Delta y \, \Delta z \tag{2.31}$$

Proceeding in like manner for the back face, we have

$$\iint_{\text{back}} \approx \mathbf{D}|_{\text{back}} \cdot -\mathbf{a}_x \Delta y \, \Delta z = -D_x|_{\text{back}} \Delta y \, \Delta z$$

where

$$D_x|_{\text{back}} \approx D_1 - \frac{\Delta x}{2} \frac{\partial D_x}{\partial x}$$

so that

$$\iint_{\text{back}} \approx \left(-D_1 + \frac{\Delta x}{2} \frac{\partial D_x}{\partial x}\right) \Delta y \, \Delta z \tag{2.32}$$

Thus

$$\iint_{\text{front}} + \iint_{\text{back}} \approx \frac{\partial D_x}{\partial x} \Delta x \, \Delta y \, \Delta z \tag{2.33}$$

For all six faces

$$\oint\!\!\oint_s \mathbf{D}\big|_s \cdot d\mathbf{s} = Q \approx \left(\frac{\partial D_x}{\partial x} + \frac{\partial D_y}{\partial y} + \frac{\partial D_z}{\partial z} \right) \Delta v \tag{2.34}$$

or

$$\frac{\partial D_x}{\partial x} + \frac{\partial D_y}{\partial y} + \frac{\partial D_z}{\partial z} \approx \frac{Q}{\Delta v} = \frac{\oint\!\!\oint_s \mathbf{D} \cdot d\mathbf{s}}{\Delta v} \tag{2.35}$$

If we now take the limit as Δv approaches zero, we obtain exactly[5]

$$\frac{\partial D_x}{\partial x} + \frac{\partial D_y}{\partial y} + \frac{\partial D_z}{\partial z} = \lim_{\Delta v \to 0} \frac{Q}{\Delta v} = \lim_{\Delta v \to 0} \frac{\oint\!\!\oint_s \mathbf{D} \cdot d\mathbf{s}}{\Delta v} \tag{2.36}$$

From our previous definitions of ρ_v and $\nabla \cdot \mathbf{D}$, we immediately obtain

$$\frac{\partial D_x}{\partial x} + \frac{\partial D_y}{\partial y} + \frac{\partial D_z}{\partial z} = \rho_v = \nabla \cdot \mathbf{D} \tag{2.37}$$

Thus the Cartesian form for $\nabla \cdot \mathbf{D}$ must be (as originally asserted)

$$\nabla \cdot \mathbf{D} = \frac{\partial D_x}{\partial x} + \frac{\partial D_y}{\partial y} + \frac{\partial D_z}{\partial z} \tag{2.38}$$

The cylindrical and spherical coordinate forms for $\nabla \cdot \mathbf{D}$ are now listed for completeness.[6] They are also found in Appendix A.

$$\nabla \cdot \mathbf{D} = \frac{1}{\rho} \frac{\partial}{\partial \rho} (\rho D_\rho) + \frac{1}{\rho} \frac{\partial D_\phi}{\partial \phi} + \frac{\partial D_z}{\partial z} \tag{2.39}$$

and

$$\nabla \cdot \mathbf{D} = \frac{1}{r^2} \frac{\partial}{\partial r} (r^2 D_r) + \frac{1}{r \sin \theta} \frac{\partial}{\partial \theta} (\sin \theta \, D_\theta) + \frac{1}{r \sin \theta} \frac{\partial D_\phi}{\partial \phi} \tag{2.40}$$

As can be seen from the preceding equations, the divergence of a vector quantity is a scalar involving the partial derivative of a particular component with respect to the variable associated with that component. A simple

[5] We have not proved that our approximations are exact in the limit.

[6] See problem 22.

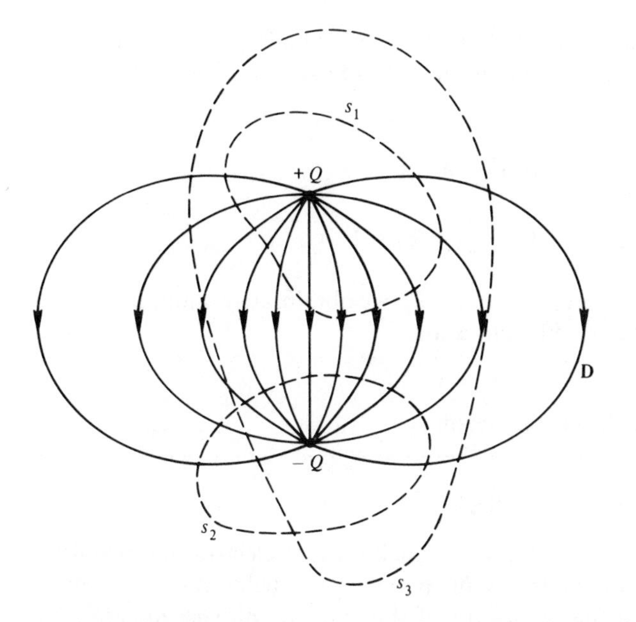

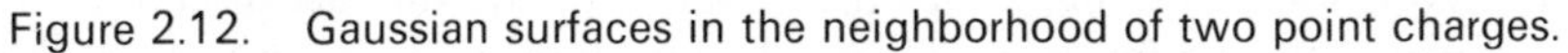

Figure 2.12. Gaussian surfaces in the neighborhood of two point charges.

example will lend more physical significance to this concept. Consider positive and negative point charges, $+Q$ and $-Q$, respectively, as shown in Figure 2.12. A closed surface around $+Q$, such as s_1, shows that (with Gauss's law) there is a net flux out and a net positive charge inside. It also shows that the flux lines appear to *diverge* from within, indicating that the divergence of **D** is nonzero at some point within s_1 (a source). Flux lines appear to converge inside s_2, indicating nonzero divergence at some point within s_2 (a sink). A nonzero divergence in a region is intimately associated with the presence of charge in that region, and this is explicitly stated in Maxwell's equation,

$$\nabla \cdot \mathbf{D} = \rho_v$$

or

$$\nabla \cdot \mathbf{E} = \frac{\rho_v}{\varepsilon_0} \tag{2.41}$$

since $\mathbf{D} = \varepsilon_0 \mathbf{E}$ and ε_0 is a scalar constant.

EXAMPLE 6

As a final example, let us consider again the problem concerning the charge distribution, $\rho_v = e^{-br}$, whose field we found previously. The field for this distribution is given by equation 2.17. Assume for present purposes that this field is known, and we want to find the source (charge) producing it. According to equation 2.41, we can find ρ_v as

$$\rho_v = \varepsilon_0(\nabla \cdot \mathbf{E})$$

So, using equation 2.40, with $\mathbf{E} = \mathbf{a}_r E_r(r)$,

$$\rho_v = \frac{\varepsilon_0}{r^2} \frac{\partial}{\partial r} (r^2 E_r)$$

When equation 2.17 is substituted for E_r, we obtain, after multiplication by r^2 and the necessary differentiation,

$$\rho_v = e^{-br}$$

which is obviously the correct result.

2.4 CONCLUDING REMARKS

Gauss's law enables us to solve very quickly those electrostatic problems where a high degree of symmetry in the charge distribution is present. It should be emphasized that problems of this type are all special cases, but nevertheless, very useful cases. They are important because, first of all, the solution to a realistic problem can often be found as an approximation to the solution for the special case. Secondly, many of the techniques we have studied and learned to use are general and can be applied to more realistic problems.

Gauss's law can be written as an equation involving integrals and is inherently concerned with *regions*. On the other hand, Maxwell's equation, as we derived it, is a partial differential equation, and is really nothing more than the point form of Gauss's law. The important results of this chapter are summarized below.

$$\Psi_E = \oiint_s \mathbf{D}|_s \cdot d\mathbf{s} = \iiint_{\text{vol}} \nabla \cdot \mathbf{D}\, dv = Q$$

$$= \iiint_{\text{vol}} \rho_v\, dv = \iint_s \rho_s\, ds = \int_l \rho_l\, dl \tag{2.42}$$

$$\mathbf{D} = \varepsilon_0 \mathbf{E} \qquad \text{(free space)} \tag{2.43}$$

$$\nabla \cdot \mathbf{D} = \rho_v \tag{2.44}$$

REFERENCES

Boast, W. B. *Vector Fields*. New York: Harper & Row, 1964. Sketches of the fields for many of the commonly used examples are given.

Hayt, W. H., Jr. *Engineering Electromagnetics*, 3rd ed. New York: McGraw-Hill, 1974. Many of our topics are covered at the same level in this textbook.

Krause, J. D. and Carver, K. R. *Electromagnetics*, 2nd ed. New York: McGraw-Hill, 1973. A slightly more advanced and comprehensive textbook. Students should have little difficulty in understanding the material.

McQuistan (see References for chapter 1). Divergence and the divergence theorem are discussed.

Plonsey, R. and Collin, R. E. *Principles and Applications of Electromagnetic Fields.* New York: McGraw-Hill, 1961. A classical approach to electromagnetics on a slightly higher level.

PROBLEMS

1. Find the electric field at (1, 1, 1) produced by an infinite line charge density, $\rho_l = 10^{-9}$ C/m, on the z axis with an infinite surface charge density on the $z = 0$ plane. The surface charge density is $\rho_s = 10^{-9}$ C/m^2.
2. Find the charge distributions which produce the following electric flux densities (below). Refer to the paragraph below equation 1.20.

   (a) $\mathbf{D} = \dfrac{1}{r^2} u(r)\mathbf{a}_r$

   (b) $\mathbf{D} = \dfrac{1}{\rho} u(\rho)\mathbf{a}_\rho$

3. What volume charge density produces the linearly increasing electric field $\mathbf{E} = \mathbf{E}_0 z a_z$?
4. An electric field

$$\mathbf{E} = E_0(\sin x\mathbf{a}_x + \cos x\mathbf{a}_y)e^{-y}$$

   exists in free space. Find the volume charge density everywhere.
5. Show that Maxwell's equation is satisfied for:

   (a) A point charge at the origin.
   (b) An infinite line charge density on the z axis.
   (c) An infinite surface charge density on the $z = 0$ plane.

6. A volume charge distribution is given by

$$\rho_v = \rho_{v0} e^{-\alpha\rho} \quad (\text{C/m}^3)$$

   Find **E**.
7. The $z = 0$ plane contains a surface charge density $\rho_s = y^2$. Determine the electric flux, Ψ_E, leaving the following closed surfaces.

   (a) A cube centered at the origin with 1-m sides parallel to the coordinate axes.
   (b) A 1-m-radius sphere centered at the origin.
   (c) A cylinder whose axis is on the x axis from $x = 0$ to $x = 1$ and whose radius is 1.

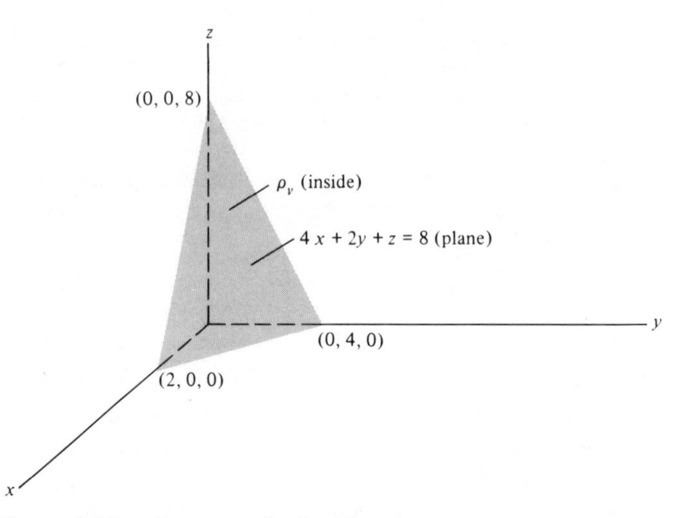

Figure 2.13. Geometry for Problem 14.

8. A volume charge density is given by

$$\rho_v = \rho_{v0}\left(1 - \frac{r^2}{a^2}\right)u(a - r)$$

Find **E** everywhere.

9. Show that $\nabla \cdot \mathbf{E} = \rho_v/\varepsilon_0$ for Problem 8.
10. A uniform surface charge density ρ_{s0} covers the $r = a$ sphere. A point charge Q is located at the origin and then displaced by a small amount Δx. How much work is required to do this?
11. If $\nabla \cdot \mathbf{D} = 0$ everywhere, what does this mean? If $\nabla \cdot \mathbf{D} = 0$ within a finite volume, what does this mean?
12. Using equation 2.38, 2.39, or 2.40, find $\nabla \cdot \mathbf{C}$ if

    (a) $\mathbf{C} = x^2y\mathbf{a}_x + yz\mathbf{a}_y + 10\mathbf{a}_z$

    (b) $\mathbf{C} = \frac{1}{\rho^2}\mathbf{a}_\rho + \frac{\sin\phi}{\rho}\mathbf{a}_\phi + \frac{1}{\rho}\mathbf{a}_z$

    (c) $\mathbf{C} = \frac{1}{r^2}\mathbf{a}_r + r\cos\phi\mathbf{a}_\theta + \cos\theta\mathbf{a}_\phi$

    (d) $\mathbf{C} = x^2\mathbf{a}_r + \sin\theta\mathbf{a}_\theta + \mathbf{a}_\phi$

    (e) $\mathbf{C} = x\cos\phi\mathbf{a}_\rho + y\sin\phi\mathbf{a}_\phi + \mathbf{a}_z$

    (f) $\mathbf{C} = x\mathbf{a}_r = 10\mathbf{a}_z$

13. Using the definition of the ∇ operator, given below equation 2.20:

    (a) Find $\nabla\alpha$, if $\alpha = x^2y$.

    (b) Find $\nabla \times \mathbf{C}$, if $\mathbf{C} = x^2y\mathbf{a}_x + yz\mathbf{a}_y + 10\mathbf{a}_z$.

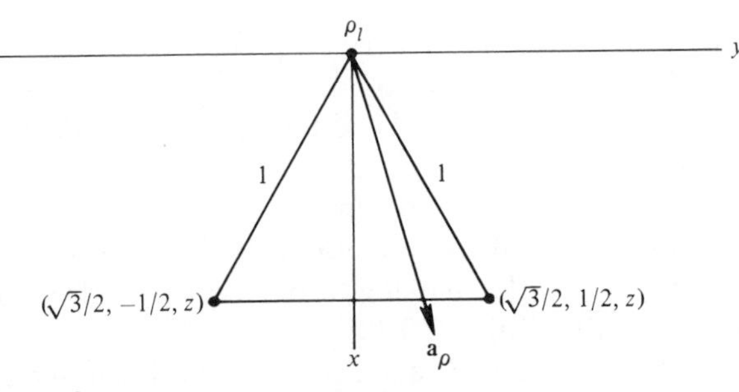

Figure 2.14. Geometry for Problem 15.

14. A nonuniform charge density $\rho_v = x^2y\ 10^{-9}\ \text{C/m}^3$ exists within the region bounded by the planes $x = 0$, $y = 0$, $z = 0$ and $4x + 2y + z = 8$. Find the flux out of the sphere $r = 8$. What is the electric field at (2, 2, 1000)? See Figure 2.13.
15. Find the flux out of the plane $x = \sqrt{3}/2$ bounded by $-0.5 \leq y \leq +0.5$, $0 \leq z \leq 1$ due to a uniform infinite line charge density, ρ_l, on the z axis.

    (a) Find the flux by making the indicated calculation explicitly.
    (b) Can you think of a simpler way of arriving at the answer? See Figure 2.14.
16. Rotate the surface in Problem 15 by 45° and recalculate the flux out. See Figure 2.15.
17. A point charge Q is located at the origin. How much flux passes through the spherical surface:

    (a) $0 \leq \theta \leq \pi/4, 0 \leq \phi \leq 2\pi$
    (b) $\pi/4 \leq \theta \leq 3\pi/4, -\pi/2 \leq \phi \leq \pi/2$

Figure 2.15. Geometry for Problem 16.

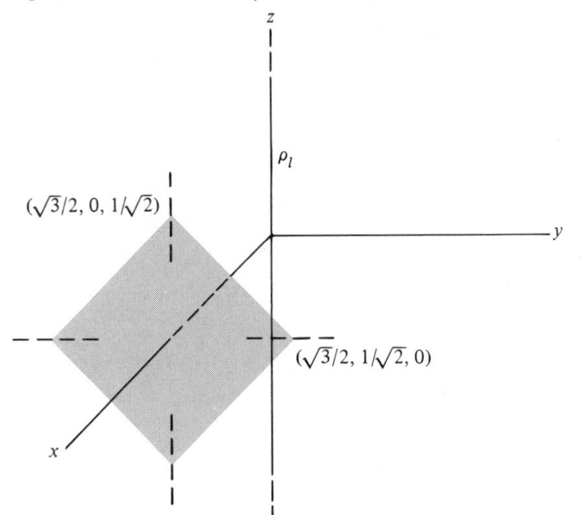

18. Starting with equation 2.20 derive equation 2.39.
19. A coaxial capacitor consists of long concentric conducting circular cylinders of radii a, b ($a < b$). If the system is "shielded" and ρ_{sa} is the surface charge density at $\rho = a$, what is ρ_{sb}?
20. Find **E** for $a < \rho < b$ in Problem 19.
21. Assume that a flux density vector field (not electric) is given by $\mathbf{C} = \mathbf{a}_\phi/\rho$. Find the flux passing through the triangle whose corners are located at (0.5, 0, 0), (1, 0, 0.5), and (1, 0, −0.5).
22. Starting with equation 2.38 and using the coordinate and component transformations in Appendix A:

    (a) Obtain equation 2.39.
    (b) Obtain equation 2.40.

Chapter 3
Scalar Electric Potential

In Chapters 1 and 2 we developed methods for determining electrostatic fields from (known) fixed charge distributions. These methods followed directly from Coulomb's law and Gauss's law, respectively. The integral for **E** which was developed from Coulomb's law is general, but may become very cumbersome and difficult to evaluate in certain cases. Gauss's law is simple to use, but is only useful for determining fields in those cases where a high degree of symmetry in the charge distribution occurs. These cases are usually already well understood.

The *potential difference*, familiar from circuit theory, will be obtained from the electrostatic field by means of a *line integral*. The line integral of the electrostatic field taken around a closed path, called the *circulation* of **E**, will be found to be zero. This property has other important consequences. The techniques developed in this chapter are certainly not unique to electrostatics, and occur in other areas of engineering and physics. The circulation of fluid velocity, for example, is important in fluid mechanics and aerodynamics.

In this chapter we will develop yet another method for determining the electrostatic field. This method involves the calculation of an auxiliary *scalar* potential function (a scalar *field*) from which we will be able to find the electrostatic field by (relatively) simple differentiation. The equation relating the vector field to its scalar potential has the same mathematical form as that relating heat flux to temperature in a (heat) conducting body. This form occurs frequently in the mathematics of engineering.

It is convenient in this chapter to introduce and define the vector operations *gradient* and *curl*. With these vector operations we are able to derive new differential equations involving the field quantities.

3.1 ENERGY

Coulomb's law tells us that a force exists between two charged bodies. If these bodies have been somehow placed in their respective positions, it will require an external force, and hence external work, to move one of them. This external work, or energy, may be positive or negative, depending on whether or not the external force is attempting to oppose that given by Coulomb's law. For example, if a positive point charge Q_1, is located at the origin and a positive point charge Q_2 is located on the positive z axis, then the natural tendency would be for Q_1 to move in the minus z direction, and for Q_2 to move in the positive z direction (mutual repulsion—Coulomb's law). If we now consider Q_1 to be fixed and attempt to move Q_2, we find that our external source (whatever it may be) must supply a positive amount of energy to move Q_2 toward Q_1. Conversely, if Q_2 moves away from Q_1, the external source supplies a negative amount (or *receives* a positive amount) of energy. We now need to formulate this process in a systematic and general way.

The electric field was defined as the force on a *unit* positive charge, so the force on *any* point charge Q in an arbitrary electric field is

$$\boxed{\mathbf{F} = Q\mathbf{E}} \quad \text{(N)} \tag{3.1}$$

It is this force which must be overcome if an external force is to move the charge Q. More precisely, if we desire to move the charge an incremental distance, $\Delta\mathbf{l} = \mathbf{a}_l\,\Delta l$, then the component of force which must be overcome is $\mathbf{F}\cdot\mathbf{a}_l = Q\mathbf{E}\cdot\mathbf{a}_l$, so that the external scalar force required is

$$F_{\text{ext}} = -Q\mathbf{E}\cdot\mathbf{a}_l \tag{3.2}$$

Since work is force times distance, the incremental *work done by the external source* is

$$\Delta W = Q\mathbf{E}\cdot\mathbf{a}_l\,\Delta l = -Q\mathbf{E}\cdot\Delta\mathbf{l} \quad \text{(J)} \tag{3.3}$$

The unit of force is the newton, and the unit of work is the joule. Suppose that the external force moves the point charge from an initial point P_i to a final point P_f along a prescribed path C as shown in Figure 3.1. The total work

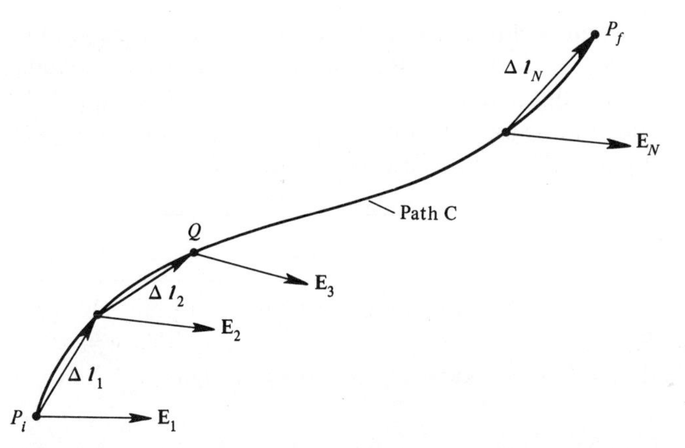

Figure 3.1. Calculating the approximate work required to move a point charge Q from P_i to P_f along path C by means of incremental straight line segments.

required can be approximated as a finite sum of incremental amounts of work, each being given by equation 3.3. Thus,

$$W \approx -Q \sum_{n=1}^{N} \mathbf{E}_n \cdot \Delta \mathbf{l}_n$$

where $\mathbf{E}_n$ is the electric field intensity at the nth point on the path. If we let N approach infinity in such a way that $\Delta \mathbf{l}_n$ approaches zero for any n, the limit is an exact integral expression for W, called a *line integral* in vector analysis,

$$W = -Q \int_{P_i}^{P_f} \mathbf{E} \cdot d\mathbf{l} \quad \text{(J)} \tag{3.4}$$

For an arbitrary vector $\mathbf{A}$, we would expect that

$$\int_{P_i}^{P_f} \mathbf{A} \cdot d\mathbf{l}$$

depends on the path chosen in arriving at P_f from P_i. It would take a very special vector field $\mathbf{A}$ in order for the preceding integral to be independent of the path. As we will see, the electrostatic electric field $\mathbf{E}$ *is* such a special field.

In working with line integrals similar to equation 3.4, it is important to recognize at the outset that $d\mathbf{l}$ is *always*[1]

$$d\mathbf{l} = \mathbf{a}_x\, dx + \mathbf{a}_y\, dy + \mathbf{a}_z\, dz \quad \text{(Cartesian coordinates)} \tag{3.5}$$

$$d\mathbf{l} = \mathbf{a}_\rho\, d\rho + \mathbf{a}_\phi \rho\, d\phi + \mathbf{a}_z\, dz \quad \text{(cylindrical coordinates)} \tag{3.6}$$

and

$$d\mathbf{l} = \mathbf{a}_r\, dr + \mathbf{a}_\theta r\, d\theta + \mathbf{a}_\phi r \sin\theta\, d\phi \quad \text{(spherical coordinates)} \tag{3.7}$$

[1] See Section 1.1 or Appendix A.

Once a coordinate system is chosen for a particular problem, then $d\mathbf{l}$ must be equation 3.5, 3.6, 3.7, or a suitable form from some other orthogonal coordinate system. The point here is that we *do not have* a choice about the *sign of* $d\mathbf{l}$. The direction of integration along the path in equation 3.4 is taken care of completely by the limits on the integral, P_i and P_f.

EXAMPLE 1

Suppose we desire to find the work required to move a point charge of 5 C from (0, 0, 1) to (1, 1, 1) along the arc of the parabola $y = x^2$, $z = 1$ in the field $\mathbf{E} = 2y\mathbf{a}_x + 2x\mathbf{a}_y + \mathbf{a}_z$ as shown in Figure 3.2. Since

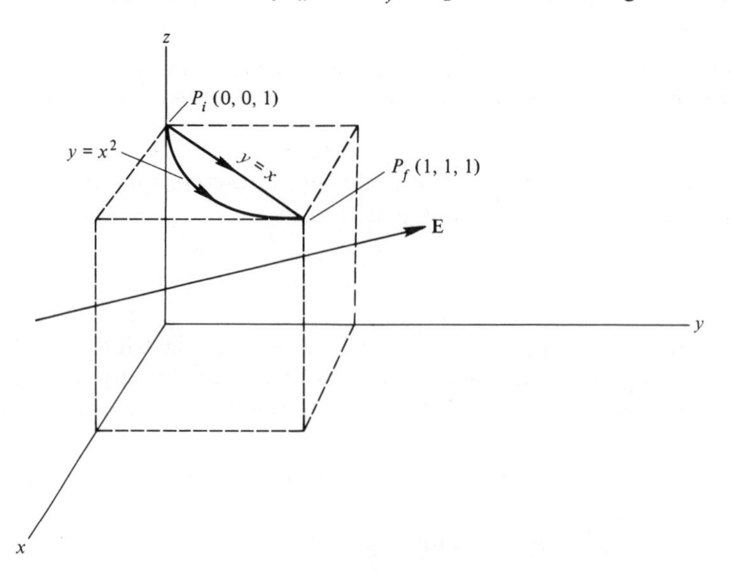

Figure 3.2. Geometry for calculating work (Example 1).

Cartesian coordinates are obviously appropriate, we have

$$\begin{aligned}\mathbf{E}\cdot d\mathbf{l} &= (2y\mathbf{a}_x + 2x\mathbf{a}_y + \mathbf{a}_z)\cdot(\mathbf{a}_x\,dx + \mathbf{a}_y\,dy + \mathbf{a}_z\,dz)\\ &= 2y\,dx + 2x\,dy + dz\end{aligned}$$

and

$$W = -5\left[\int_0^1 2y\,dx + \int_0^1 2x\,dy + \int_1^1 dz\right]$$

The last integral above is zero ($dz = 0$); so

$$W = -5\left[\int_0^1 2y\,dx + \int_0^1 2x\,dy\right]$$

We now have the choice of eliminating either y or x by means of the equation, $y = x^2$, for the path. Here, we will use $y = x^2$ and $dy = 2x\,dx$, so

$$W = -5\int_0^1 2x^2\,dx - 5\int_0^1 4x^2\,dx = -10\left.\frac{x^3}{3}\right|_0^1 - 20\left.\frac{x^3}{3}\right|_0^1$$

$$W = -10 \quad (\text{J})$$

The minus sign indicates that the external source *receives* energy.

If the path is the straight line, $y = x$, $z = 1$, instead of the parabolic path, then we use $y = x$ and $dy = dx$. Then

$$W = -5\int_0^1 2x\,dx - 5\int_0^1 2x\,dx = -10\left.\frac{x^2}{2}\right|_0^1 - \left.\frac{10x^2}{2}\right|_0^1 = -10 \quad (\text{J})$$

which is the same answer as before. It appears that the line integral of **E** is indeed independent of the path. This example offers evidence, but no proof, that the line integral is independent of the path. The proof will be given at an appropriate time.

3.2 POTENTIAL DIFFERENCE AND POTENTIAL

It is a simple matter now to define the *potential difference between* P_f and P_i as the work required of an external source in moving a *unit* positive charge from P_i to P_f in the field **E**. From equation 3.4, with $Q = 1$,

$$\boxed{\Phi_{fi} \equiv \Phi_f - \Phi_i = -\int_{P_i}^{P_f} \mathbf{E}\cdot d\mathbf{l}} \quad (\text{J/C or V}) \tag{3.8}$$

As indicated by equation 3.8, the symbol for potential is Φ, and it is the work per unit charge (J/C). As a matter of fact, in the *electrostatic* case (which concerns us here), the *potential difference* is the same as the *voltage difference* of circuit theory. In the dynamic case, to be considered later, the relation between voltage and potential will be reexamined. Notice that the potential difference $\Phi_f - \Phi_i = \Phi_{fi}$ can be interpreted as the absolute *potential* at P_f (i.e., Φ_f) minus the *absolute potential* at P_i (i.e., Φ_i). We must now explain what is meant by absolute potential.

The absolute potential at a point, $\Phi(x, y, z)$, is simply the potential difference between that point and some *reference* point where the potential is zero, commonly called the "ground" or "earth." These terms may be confusing. Some examples will clarify the terminology being employed here.

EXAMPLE 2

Consider a point charge Q located at the origin in free space. As we have shown, the electric field produced by this source is

$$\mathbf{E} = \frac{Q}{4\pi\varepsilon_0 r^2}\,\mathbf{a}_r$$

Let us now find the potential difference between two points, a and b, $(b > a)$ lying along a spherical radial line (θ, Φ constant) from the origin as shown in Figure 3.3. From equation 3.8,

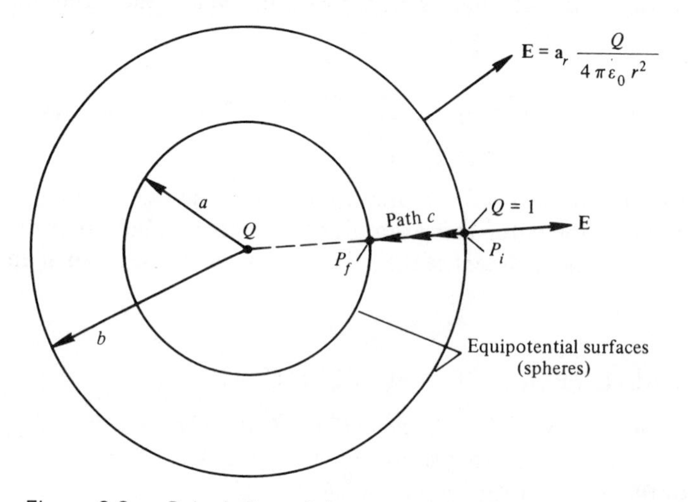

Figure 3.3. Calculation of the potential difference $\Phi_f - \Phi_i = \Phi(a) - \Phi(b)$ between points a and b along a radial path c for a point charge at the origin.

$$\Phi_f - \Phi_i = \Phi(a) - \Phi(b) = -\int_b^a \mathbf{E} \cdot d\mathbf{l}$$

or

$$\Phi(a) - \Phi(b) = -\int_b^a \frac{Q}{4\pi\varepsilon_0 r^2}\,\mathbf{a}_r \cdot \mathbf{a}_r\,dr = -\frac{Q}{4\pi\varepsilon_0}\int_b^a \frac{dr}{r^2}$$

Then,

$$\Phi(a) - \Phi(b) = +\left.\frac{Q}{4\pi\varepsilon_0 r}\right|_b^a$$

or

$$\Phi(a) - \Phi(b) = \frac{Q}{4\pi\varepsilon_0}\left[\frac{1}{a} - \frac{1}{b}\right] \quad \text{(V)} \tag{3.9}$$

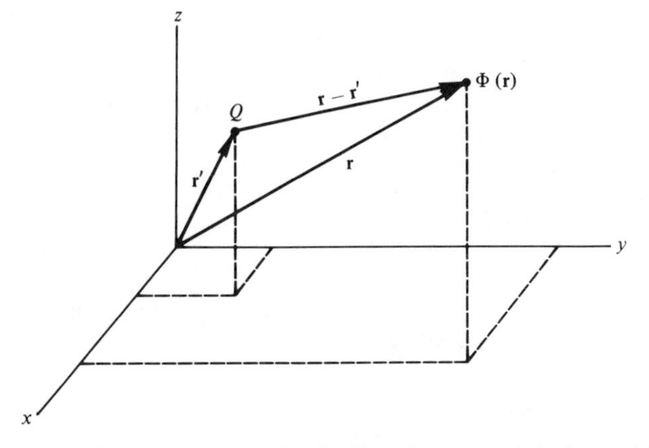

Figure 3.4. Geometry for finding the potential of an arbitrarily located point charge.

Since $b > a$, the potential difference is positive indicating that it takes positive work to move a unit positive charge toward the origin where Q is located. This is, of course, rather obvious. Inspection of equation 3.9 reveals that if $b \to \infty$, the term containing b disappears. This implies that the potential, or absolute potential, at infinity for a point charge at the origin is zero. These conditions meet those specified for finding the absolute potential at *any* finite point a. Thus,

$$\Phi(a) = \frac{Q}{4\pi\varepsilon_0 a}$$

is the absolute potential at a. Since the point a is perfectly general in this problem, we may write

$$\Phi(\mathbf{r}) = \frac{Q}{4\pi\varepsilon_0 r} \quad \text{(V)} \tag{3.10}$$

as the potential at any radial distance r for a point charge at the origin *with zero potential referred to infinity*. Notice from equation 3.10 that the potential on a sphere centered at the origin is constant, and such a surface is called an *equipotential* surface.

If the point charge Q is not located at the origin, but is located at $\mathbf{r}'$, then

$$\Phi(\mathbf{r}) = \frac{Q}{4\pi\varepsilon_0 R} \quad \text{(V)} \tag{3.11}$$

This is shown in Figure 3.4, where

$$R = |\mathbf{r} - \mathbf{r}'| = [(x - x')^2 + (y - y')^2 + (z - z')^2]^{1/2}$$

as before. What are the equipotential surfaces in this case?

EXAMPLE 3

Let us now perform the same calculations for the uniform infinite line charge density ρ_l C/m. The field is

$$\mathbf{E} = \frac{\rho_l}{2\pi\varepsilon_0 \rho} \mathbf{a}_\rho$$

The potential difference between two points a and b ($b > a$) lying along a cylindrical radial line (ϕ, z constant) is

$$\Phi_f - \Phi_i = \Phi(a) - \Phi(b) = -\int_b^a \mathbf{E} \cdot d\mathbf{l}$$

or

$$\Phi(a) - \Phi(b) = -\int_b^a \frac{\rho_l}{2\pi\varepsilon_0 \rho} \mathbf{a}_\rho \cdot \mathbf{a}_\rho \, d\rho = -\frac{\rho_l}{2\pi\varepsilon_0} \int_b^a \frac{d\rho}{\rho}$$

Then,

$$\Phi(a) - \Phi(b) = -\frac{\rho_l}{2\pi\varepsilon_0} \ln \rho \Big|_b^a \tag{3.12}$$

(*Note*: ln represents a natural logarithm.)

or

$$\Phi(a) - \Phi(b) = \frac{\rho_l}{2\pi\varepsilon_0} \ln \frac{b}{a} \quad \text{(V)} \tag{3.13}$$

Since $b > a$, $\Phi(a) - \Phi(b)$ is positive, as it should be. If we let b approach infinity, the potential difference is infinite. Thus, a cylinder of infinite radius is *not* suitable as a reference for absolute potential! The proper way to overcome this difficulty *in a general way* is to define the absolute potential $\Phi(r)$ according to

$$\boxed{\Phi(r) = -\int_l \mathbf{E} \cdot d\mathbf{l} + C} \tag{3.14}$$

where the constant C is chosen according to whatever reference we desire. Then, for the infinite line charge,

$$\Phi(\rho) = -\frac{\rho_l}{2\pi\varepsilon_0} \ln \rho + C$$

Let us choose C so that $\Phi(\rho) = 0$ when $\rho = 1$. Then it follows immediately that $C = 0$ and

$$\Phi(\rho) = -\frac{\rho_l}{2\pi\varepsilon_0} \ln \rho \tag{3.15}$$

for any ρ. Equipotential surfaces are cylinders coaxial with the line charge. Also, using equation 3.15,

$$\Phi(a) - \Phi(b) = -\frac{\rho_l}{2\pi\varepsilon_0}(\ln a - \ln b) = \frac{\rho_l}{2\pi\varepsilon_0} \ln \frac{b}{a}$$

as in equation 3.13. *Thus, the absolute potential is arbitrary to within an additive constant* C. For the point charge at the origin, which we considered previously,

$$\Phi(r) = \frac{Q}{4\pi\varepsilon_0 r} + C$$

and

$$\Phi(a) - \Phi(b) = \frac{Q}{4\pi\varepsilon_0 a} + C - \frac{Q}{4\pi\varepsilon_0 b} - C = \frac{Q}{4\pi\varepsilon_0}\left(\frac{1}{a} - \frac{1}{b}\right)$$

as in equation 3.9.

3.3 THE POTENTIAL INTEGRAL

The potential at $\mathbf{r}$ due to a unit point charge at $\mathbf{r}'$ is the unit impulse response for scalar electrostatic potential and from equation 3.11 it is given by ($Q = 1$)

$$h(\mathbf{r} - \mathbf{r}') = \frac{1}{4\pi\varepsilon_0|\mathbf{r} - \mathbf{r}'|} = \frac{1}{4\pi\varepsilon_0 R} \tag{3.16}$$

The superposition integral for the potential due to any charge distribution can be written down once equation 3.16 has been obtained. We have used this concept several times in Chapter 1 and Chapter 2 (e.g., in finding integral expressions for **E** and **D**).

$$\boxed{\Phi(\mathbf{r}) = \frac{1}{4\pi\varepsilon_0} \iiint_{\text{vol}'} \frac{\rho_v(\mathbf{r}')}{R}\, dv'} \quad \text{(V)} \tag{3.17}$$

It also follows that for surface charge densities the superposition integral will be

$$\Phi(\mathbf{r}) = \frac{1}{4\pi\varepsilon_0} \iint_{s'} \frac{\rho_s(\mathbf{r}')}{R}\, ds' \quad \text{(V)} \tag{3.18}$$

and for line charge densities

$$\Phi(\mathbf{r}) = \frac{1}{4\pi\varepsilon_0} \int_{l'} \frac{\rho_l(\mathbf{r}')}{R} \, dl' \quad \text{(V)} \tag{3.19}$$

We now have the concise general results.

$$\Phi(\mathbf{r}) = \frac{1}{4\pi\varepsilon_0} \iiint_{\text{vol}'} \frac{\rho_v(\mathbf{r}')}{R} \, dv' \tag{3.17}$$

and

$$\mathbf{E}(\mathbf{r}) = \frac{1}{4\pi\varepsilon_0} \iiint_{\text{vol}'} \frac{\rho_v(\mathbf{r}')}{R^2} \mathbf{a}_R \, dv' \tag{1.37}$$

Obviously, there should be some simple relation between $\mathbf{E}$ and Φ besides that given by equation 3.14. In fact, when equations 3.17 and 1.37 are compared, we suspect that $\mathbf{E}$ can be obtained from Φ by some kind of vector differentiation. In order to find this relation, let us introduce the gradient operation of vector analysis. Using the vector del operator,

$$\nabla = \mathbf{a}_x \frac{\partial}{\partial x} + \mathbf{a}_y \frac{\partial}{\partial y} + \mathbf{a}_z \frac{\partial}{\partial z} \tag{3.20}$$

$$\text{grad } \alpha(x, y, z) = \nabla\alpha(x, y, z) = \mathbf{a}_x \frac{\partial \alpha}{\partial x} + \mathbf{a}_y \frac{\partial \alpha}{\partial y} + \mathbf{a}_z \frac{\partial \alpha}{\partial z}$$

or

$$\text{grad}' \beta(x', y', z') = \nabla' \beta(x', y', z') = \mathbf{a}_x \frac{\partial \beta}{\partial x'} + \mathbf{a}_y \frac{\partial \beta}{\partial y'} + \mathbf{a}_z \frac{\partial \beta}{\partial z'}$$

for Cartesian coordinates. Notice that ∇ is the vector del operator indicating operations with respect to x, y, and z (unprimed), whereas ∇' indicates the same operation with respect to x', y', and z'. We will give physical interpretations of the gradient later. For present purposes it is only necessary to recognize that[2]

$$\nabla \frac{1}{R} = -\nabla' \frac{1}{R} = -\frac{\mathbf{R}}{R^3} = -\frac{\mathbf{a}_R}{R^2} \tag{3.21}$$

where, lest we forget,

$$R = |\mathbf{r} - \mathbf{r}'| = [(x - x')^2 + (y - y')^2 + (z - z')^2]^{1/2} \tag{3.22}$$

It is straightforward to verify equation 3.21 in rectangular coordinates. Substituting equation 3.21 into equation 1.37 gives

$$\mathbf{E}(\mathbf{r}) = \frac{1}{4\pi\varepsilon_0} \iiint_{\text{vol}'} \rho_v(r') \left[-\nabla \frac{1}{R} \right] dv' \tag{3.23}$$

[2] See Problem 11.

but since ∇ operates on x, y, and z, not on x', y', z',

$$\mathbf{E}(\mathbf{r}) = -\nabla\left[\frac{1}{4\pi\varepsilon_0}\iiint\limits_{\text{vol}'}\frac{\rho_v(\mathbf{r}')}{R}\,dv'\right] \tag{3.24}$$

We recognize from equation 3.17 that the bracketed term of equation 3.24 is $\Phi(\mathbf{r})$, so

$$\boxed{\mathbf{E}(\mathbf{r}) = -\nabla\Phi(\mathbf{r})} \tag{3.25}$$

or

$$\mathbf{E}(\mathbf{r}) = -\nabla[\Phi(\mathbf{r}) + C] \tag{3.26}$$

since $\nabla C \equiv 0$ if C is a constant. Thus, a constant C added to the potential to change the reference, for example, has no effect at all on $\mathbf{E}$. It should also be noted that the potential *must be* a *continuous* function, for, otherwise, the gradient operation will create impulses of electric field, and this is contrary to experience!

The implications of equation 3.25 are obvious. If we can find Φ with equation 3.17 (which is much simpler than equation 1.37) or, hopefully, by some simpler method, then we can find $\mathbf{E}$ by means of the gradient operation of equation 3.25. The gradient operation is usually very simple, so any difficulties will most likely occur in finding Φ. This is the first encounter we have had with solving a problem by means of an auxiliary potential function.

We are now in a position to prove that (as we suspected) the line integral of $\mathbf{E}$ is independent of the chosen path. Equation 3.8, with $\mathbf{E} = -\nabla\Phi$, gives

$$\Phi_f - \Phi_i = -\int_{P_i}^{P_f}\mathbf{E}\cdot d\mathbf{l} = +\int_{P_i}^{P_f}\nabla\Phi\cdot d\mathbf{l}$$

Now, using Cartesian coordinates,

$$\nabla\Phi\cdot d\mathbf{l} = \left(\mathbf{a}_x\frac{\partial\Phi}{\partial x} + \mathbf{a}_y\frac{\partial\Phi}{\partial y} + \mathbf{a}_z\frac{\partial\Phi}{\partial z}\right)\cdot(\mathbf{a}_x\,dx + \mathbf{a}_y\,dy + \mathbf{a}_z\,dz)$$

or

$$\nabla\Phi\cdot d\mathbf{l} = \frac{\partial\Phi}{\partial x}dx + \frac{\partial\Phi}{\partial y}dy + \frac{\partial\Phi}{\partial z}dz \tag{3.27}$$

It is easy to recognize the right side of equation 3.27 as the differential, $d\Phi$; that is,

$$d\Phi = \frac{\partial\Phi}{\partial x}dx + \frac{\partial\Phi}{\partial y}dy + \frac{\partial\Phi}{\partial z}dz \tag{3.28}$$

This enables us to write

$$\Phi_f - \Phi_i = \int_{P_i}^{P_f}d\Phi = \Phi\Big|_{P_i}^{P_f} = \Phi_f - \Phi_i \tag{3.29}$$

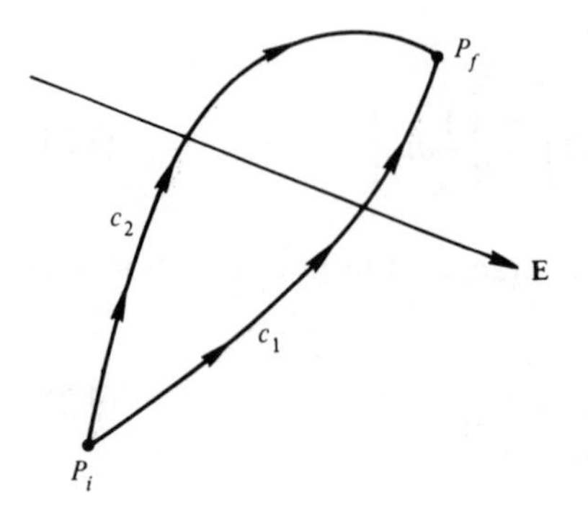

Figure 3.5. Two arbitrary paths for evaluating $\int_{P_i}^{P_f} \mathbf{E} \cdot d\mathbf{l}$.

and so the line integral of **E** is indeed independent of the path and depends only on the limits or end points! We now have proved that

$$\underset{c_1}{\int_{P_i}^{P_f}} \mathbf{E} \cdot d\mathbf{l} = \underset{c_2}{\int_{P_i}^{P_f}} \mathbf{E} \cdot d\mathbf{l}$$

where c_1 and c_2 are different paths, as in Figure 3.5. Then,

$$\underset{c_1}{\int_{P_i}^{P_f}} \mathbf{E} \cdot d\mathbf{l} - \underset{c_2}{\int_{P_i}^{P_f}} \mathbf{E} \cdot d\mathbf{l} = 0 = \underset{c_1}{\int_{P_i}^{P_f}} \mathbf{E} \cdot d\mathbf{l} + \underset{c_2}{\int_{P_f}^{P_i}} \mathbf{E} \cdot d\mathbf{l}$$

The last two integrals (together) represent the line integral of **E** taken around a *closed* path. Indicating this by a circle through the integral symbol, we have

$$\boxed{\oint \mathbf{E} \cdot d\mathbf{l} = 0} \tag{3.30}$$

Thus, the line integral of **E** around a *closed* path, called the *circulation of* **E**, is identically zero. This result occurs because $\mathbf{E} = -\nabla\Phi$.

Stoke's theorem[3] of vector analysis states that the line integral of *any* vector around a *closed* path, c, is equal to the integral over *any* of the possible *open* surfaces (all of which are bounded by the closed path c) of the normal (to s) component of the *curl* of the vector. Consider the spherical surface $r = a$ centered at the origin. The circle $r = a$, $\theta = \pi/2$ could represent a closed path c, and in this case the upper hemisphere $r = a$, $0 \leq \theta \leq \pi/2$ would be one open surface bounded by c. The lower hemisphere $r = a$, $\pi/2 \leq \theta \leq \pi$ and the plane circular surface $0 \leq r \leq a$, $\theta = \pi/2$ are two other

[3] Stoke's theorem is derived in Appendix G.

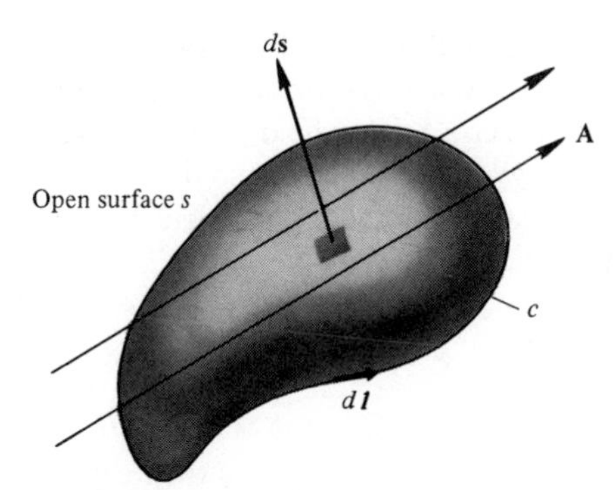

Figure 3.6. Line integral contour, c, and open surface, s, for Stoke's theorem.

open surfaces bounded by c. In terms of an equation for an arbitrary vector, **A**, Stoke's theorem is

$$\oint_c \mathbf{A} \cdot d\mathbf{l} = \iint_s \nabla \times \mathbf{A} \cdot d\mathbf{s} \tag{3.31}$$

where $\nabla \times \mathbf{A}$ is the curl of **A** or the rot **A** (rotation of **A**). An arbitrary field **A**, closed path c, and open surface s are shown in Figure 3.6. Problem 14 deals with the use of equation 3.31. Notice particularly the positive sense of $d\mathbf{l}$ and $d\mathbf{s}$ as shown in Figure 3.6. A right-hand rule can be devised to account for the proper directions. If the thumb of the right hand is placed along the path c so that it extends in the direction of $d\mathbf{l}$, and if the fingers of the right hand then pierce the surface s, then the fingers of the right hand point in the direction of $d\mathbf{s}$. The differential vector area $d\mathbf{s}$ is normal to the surface at every point.

A physical interpretation of the curl of **A** will be given later. Applying Stoke's theorem to equation 3.30 gives

$$0 \equiv \oint_c \mathbf{E} \cdot d\mathbf{l} = \iint_s (\nabla \times \mathbf{E}) \cdot d\mathbf{s} \tag{3.32}$$

Since equation 3.32 is valid for *any* of the possible open surfaces bounded by *any* c, it follows that

$$\boxed{\nabla \times \mathbf{E} \equiv 0} \tag{3.33}$$

A field, such as **E** in electrostatics, which has no curl, is said to be a lamellar or irrotational field. The earth's gravitational field is irrotational. Equation 3.33 provides us with a simple test for determining whether or not a given field is a possible electrostatic field. We merely need to find its curl. If the curl of the given field is identically zero, then it is a possible electrostatic field, and furthermore, as a consequence, its line integral is independent of the path. Since the line integral around a closed path is zero, the work done in moving a unit test charge in an electrostatic field from some point along some path and back to the same starting point is zero. Such a field (**E** here) is said to be a *conservative* field since work or energy is conserved in moving around a closed path. As mentioned above, the earth's gravitational field has the same properties. We note here that the equations

$$\mathbf{E} = -\nabla\Phi \tag{3.25}$$

and

$$\nabla \times \mathbf{E} = 0 \tag{3.33}$$

are consistent, for substituting equation 3.25 into equation 3.33 gives

$$\nabla \times (-\nabla\Phi) = 0$$

or

$$\nabla \times (\nabla\Phi) = 0 \tag{3.34}$$

An identity from vector analysis tells us that the curl of the gradient (see Appendix A) of *any* scalar is zero, so equation 3.34 is obviously correct.

It would be useful at this point to return to Example 1 where the work required to move a 5-C point charge from 0, 0, 1 to 1, 1, 1 was asked for. Since this is a conservative field, we can pick any path we want. Let us then pick the straight line segments 0, 0, 1 to 1, 0, 1 and 1, 0, 1 to 1, 1, 1 giving

$$W = -5\left\{\int_0^1 0\,dx + \int_0^1 2\,dy\right\} = -10 \quad \text{(J)}$$

as before. How did we know that **E** was conservative?

We now know *both* the curl of **E** and the divergence of **E**. Specification of both $\nabla \times \mathbf{E}$ and $\nabla \cdot \mathbf{E}$ uniquely determines **E**. It is helpful if we summarize the differential equations and auxiliary equations we have developed so far. These are

$$\nabla \times \mathbf{E} = 0 \tag{3.35a}$$

$$\nabla \cdot \mathbf{E} = \frac{\rho_v}{\varepsilon_0} \tag{3.35b}$$

$$\mathbf{E} = -\nabla\Phi \tag{3.35c}$$

$$\mathbf{D} = \varepsilon_0 \mathbf{E} \tag{3.35d}$$

and may be included in what we loosely call Maxwell's equations. The derivation of this set of equations has followed mathematical lines, and has certainly left us with the desired results, but with very little feeling as to the physical picture. Before considering examples of the use of these equations, let us interpret the gradient and curl as we promised.

3.4 GRADIENT AND CURL

The gradient of a scalar field (for example Φ) is a vector that lies in the direction for which the scalar field is changing most rapidly. The magnitude of the gradient *is* the greatest rate of change of the scalar field. A scalar field, in particular a two-dimensional scalar electrostatic potential field, Φ, is shown in Figure 3.7 in terms of contours of constant Φ, called equipotentials. The direction of the greatest rate of change of Φ will be *normal* to the equipotentials since the rate of change of Φ is zero *along* or tangent to the equipotentials. Put another way, a positive point charge will be accelerated in a direction perpendicular to the equipotential, and of the two possible directions (both perpendicular to the equipotential), the correct one is that of decreasing potential ($\mathbf{E} = -\nabla\Phi$). There will be no force acting to accelerate a point charge along (tangent to) an equipotential. Thus,

$$\textbf{grad}\ \Phi = \mathbf{a}_n \frac{d\Phi}{dn} \tag{3.36}$$

Consider the potential $\Phi(x, y, z)$ at a point $P(x, y, z)$ and $\Phi(x + \Delta x, y + \Delta y, z + \Delta z)$ at a nearby point $Q(x + \Delta x, y + \Delta y, z + \Delta z)$. If Δl is the distance between P and Q, then

$$\begin{aligned} \Delta\Phi &= \Phi(x + \Delta x, y + \Delta y, z + \Delta z) - \Phi(x, y, z) \\ &\approx \frac{\partial\Phi}{\partial x}\Delta x + \frac{\partial\Phi}{\partial y}\Delta y + \frac{\partial\Phi}{\partial z}\Delta z \end{aligned}$$

Higher order infinitesimals that will disappear in the limiting process to follow have been dropped. Then

$$\frac{\Delta\Phi}{\Delta l} \approx \frac{\partial\Phi}{\partial x}\frac{\Delta x}{\Delta l} + \frac{\partial\Phi}{\partial y}\frac{\Delta y}{\Delta l} + \frac{\partial\Phi}{\partial z}\frac{\Delta z}{\Delta l}$$

and

$$\frac{d\Phi}{dl} = \lim_{\Delta l \to 0}\frac{\Delta\Phi}{\Delta l} = \frac{\partial\Phi}{\partial x}\frac{dx}{dl} + \frac{\partial\Phi}{\partial y}\frac{dy}{dl} + \frac{\partial\Phi}{\partial z}\frac{dz}{dl}$$

This quantity is the rate of change of Φ with respect to distance at point P in a direction toward Q and is thus called the *directional derivative*. The directional derivative may also be written

$$\frac{d\Phi}{dl} = \left(\mathbf{a}_x\frac{\partial\Phi}{\partial x} + \mathbf{a}_y\frac{\partial\Phi}{\partial y} + \mathbf{a}_z\frac{\partial\Phi}{\partial z}\right)\cdot\left(\mathbf{a}_x\frac{dx}{dl} + \mathbf{a}_y\frac{dy}{dl} + \mathbf{a}_z\frac{dz}{dl}\right)$$

or

$$\frac{d\Phi}{dl} = \left[\mathbf{a}_x\frac{\partial\Phi}{\partial x} + \mathbf{a}_y\frac{\partial\Phi}{\partial y} + \mathbf{a}_z\frac{\partial\Phi}{\partial z}\right]\cdot\frac{d\mathbf{l}}{dl}$$

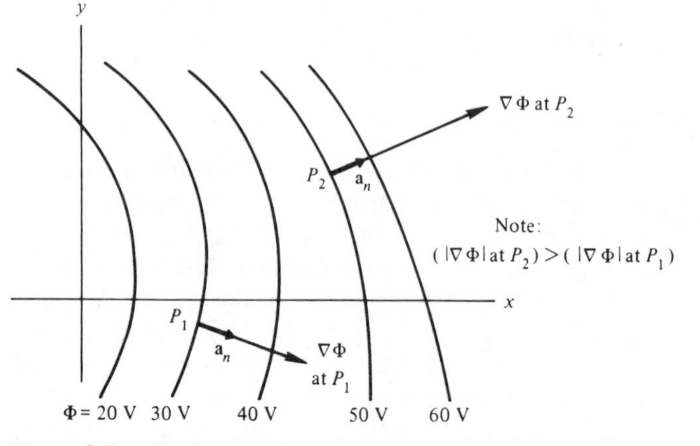

Figure 3.7. Equipotential contours and the gradient for a two-dimensional potential field.

using equation 3.5. Notice that $d\mathbf{l}/dl = \mathbf{a}_l$ is a *unit* vector. From the discussion in the preceding paragraph it is apparent that the directional derivative is maximum when $\mathbf{a}_l = \mathbf{a}_n$, that is, when it is taken in a direction *normal* to the equipotential surface at P (as in Figure 3.7). Thus,

$$\left[\frac{d\Phi}{dl}\right]_{\max} = \left[\mathbf{a}_x \frac{\partial \Phi}{\partial x} + \mathbf{a}_y \frac{\partial \Phi}{\partial y} + \mathbf{a}_z \frac{\partial \Phi}{\partial z}\right] \cdot \mathbf{a}_n$$

It is now obvious that the maximum directional derivative takes place in the direction of, and has the magnitude of the bracketed term in the last equation. From the definition in the preceding paragraph, the bracketed term is the gradient of the scalar field Φ, so

$$\mathbf{grad}\ \Phi = \nabla\Phi = \mathbf{a}_x \frac{\partial \Phi}{\partial x} + \mathbf{a}_y \frac{\partial \Phi}{\partial y} + \mathbf{a}_z \frac{\partial \Phi}{\partial z} \quad \text{(Cartesian)} \tag{3.37}$$

This result has already been used in equations 3.23 and 3.24. For cylindrical and spherical coordinates,[4] we have

$$\nabla\Phi = \mathbf{a}_\rho \frac{\partial \Phi}{\partial \rho} + \mathbf{a}_\phi \frac{1}{\rho}\frac{\partial \Phi}{\partial \phi} + \mathbf{a}_z \frac{\partial \Phi}{\partial z} \quad \text{(cylindrical)} \tag{3.38}$$

and

$$\nabla\Phi = \mathbf{a}_r \frac{\partial \Phi}{\partial r} + \mathbf{a}_\theta \frac{1}{r}\frac{\partial \Phi}{\partial \theta} + \mathbf{a}_\phi \frac{1}{r \sin\theta}\frac{\partial \Phi}{\partial \phi} \quad \text{(spherical)} \tag{3.39}$$

[4] See Problem 25.

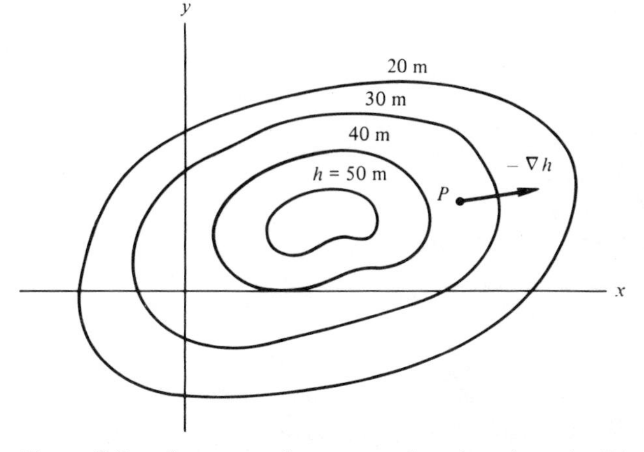

Figure 3.8. Contours of constant elevation, h, and $-\nabla h$.

It is important to recognize that, in general, Φ varies from point to point. A glance at a map having elevation contours (equipotentials) shows this very well. Figure 3.8 shows a hill with elevation contours around it. We now ask ourselves a question. "If we place a small ball at P, in what direction will it roll and what is the magnitude of the force acting to accelerate it in that direction?" This is equivalent to asking what is $-\text{grad}\ h(x, y)$? It should be clear now that the electrostatic field $\mathbf{E}$ is normal to equipotentials in the direction of decreasing Φ, since $\mathbf{E} = -\nabla\Phi$.

Temperature is a term that is familiar to us. The temperature in a region may vary from point to point, and can, therefore, be described as a three-dimensional scalar field. Surfaces of constant temperature (the "equipotentials") are called *isothermal* surfaces. The heat flux vector $\mathbf{s}$ (W/m^2) at a point in a conducting body is given by Fourier's heat conduction law: $\mathbf{s} = -K\nabla T$, where T is the temperature and K the thermal conductivity. Thus, heat flows in a direction opposite to the positive temperature gradient, or, put more simply, heat flows from a region of higher temperature to a region of lower temperature. It also follows that the heat flux field $\mathbf{s}$ is *irrotational* or, $\nabla \times \mathbf{s} = 0$ (Why?).

It was mentioned earlier that the earth's gravitational field is irrotational, so it too must be derivable from a gravitational scalar potential field. Problem 10 at the end of this chapter deals with this item.

We can now conclude that equipotential surfaces for the infinite line charge density are cylinders coaxial with the line charge with decreasing amplitude as ρ increases. The exact functional dependence will be re-examined shortly. In the same way (as we have already seen), concentric spheres are the equipotential surfaces for a point charge. The electrostatic field will always be perpendicular to the equipotential surfaces. What are the equipotential surfaces for a pair of oppositely charged parallel planes?

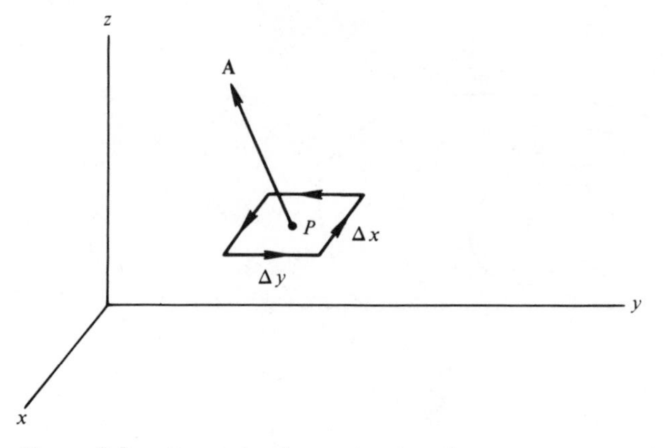

Figure 3.9. Geometry for evaluating the z component of the curl of **A**.

As in the case of the divergence, the gradient of a scalar field at a point P may be defined in integral form by[5]

$$\nabla\Phi \equiv \lim_{\Delta v \to 0} \frac{1}{\Delta v} \oiint_s \Phi \, d\mathbf{s} \tag{3.40}$$

where Δv is a small volume surrounding P, and s is the surface area of Δv.

The component in the $\mathbf{a}_n$ direction (general) of the curl of **A** at a point P may be defined as the limit of the circulation of **A** per unit area as the area $(\mathbf{a}_n \, \Delta s)$ approaches zero. Thus,

$$(\text{curl } \mathbf{A})_n = \lim_{\Delta s \to 0} \frac{1}{\Delta s} \oint_c \mathbf{A} \cdot d\mathbf{l} \tag{3.41}$$

where $\Delta\mathbf{s} = \mathbf{a}_n \, \Delta s$, and the closed path c defines the area Δs. The unit vector $\mathbf{a}_n$ is normal to the surface Δs, and in a direction determined by the right-hand rule as explained below equation 3.31 and in Figure 3.6.

For example, following a procedure similar to that used in arriving at a formula for divergence, we may describe a rectangle parallel to the x-y plane as shown in Figure 3.9. At the center of the rectangle (whose sides are Δx, Δy), $\mathbf{A} = \mathbf{a}_x A_x + \mathbf{a}_y A_y + \mathbf{a}_z A_z$. It is easy to show that since

$$A_y\big|_{x+\Delta x/2} = A_y + \frac{\partial A_y}{\partial x}\frac{\Delta x}{2} + \cdots, \qquad A_x\big|_{y+\Delta y/2} = A_x + \frac{\partial A_x}{\partial y}\frac{\Delta y}{2} + \cdots$$

and

$$A_y\big|_{x-\Delta x/2} = A_y - \frac{\partial A_y}{\partial x}\frac{\Delta x}{2} + \cdots, \qquad A_x\big|_{y-\Delta y/2} = A_x - \frac{\partial A_x}{\partial y}\frac{\Delta y}{2} + \cdots$$

[5] Refer to footnote 4, Section 2.3.

by their Taylor series expansion,

$$\oint \mathbf{A} \cdot d\mathbf{l} = \left(\frac{\partial A_y}{\partial x} - \frac{\partial A_x}{\partial y}\right) \Delta x \, \Delta y$$

Higher order terms that would disappear in the limiting process have simply been dropped. In the limit then,

$$(\text{curl } \mathbf{A})_z = \lim_{\Delta x \, \Delta y \to 0} \left(\frac{\partial A_y}{\partial x} - \frac{\partial A_x}{\partial y}\right) \frac{\Delta x \, \Delta y}{\Delta x \, \Delta y}$$

or

$$(\text{curl } \mathbf{A})_z = \frac{\partial A_y}{\partial x} - \frac{\partial A_x}{\partial y}$$

Repeating this procedure for rectangles parallel to the x-z plane and y-z plane gives

$$(\text{curl } \mathbf{A})_y = \frac{\partial A_x}{\partial z} - \frac{\partial A_z}{\partial x}$$

and

$$(\text{curl } \mathbf{A})_x = \frac{\partial A_z}{\partial y} - \frac{\partial A_y}{\partial z}$$

Then, the curl of $\mathbf{A}$, symbolized $\nabla \times \mathbf{A}$, may be written for Cartesian coordinates as

$$\nabla \times \mathbf{A} = \mathbf{a}_x\left(\frac{\partial A_z}{\partial y} - \frac{\partial A_y}{\partial z}\right) + \mathbf{a}_y\left(\frac{\partial A_x}{\partial z} - \frac{\partial A_z}{\partial x}\right) + \mathbf{a}_z\left(\frac{\partial A_y}{\partial x} - \frac{\partial A_x}{\partial y}\right) \quad (3.42)$$

or, in a form more easily remembered,

$$\nabla \times \mathbf{A} = \begin{vmatrix} \mathbf{a}_x & \mathbf{a}_y & \mathbf{a}_z \\ \dfrac{\partial}{\partial x} & \dfrac{\partial}{\partial y} & \dfrac{\partial}{\partial z} \\ A_x & A_y & A_z \end{vmatrix} \quad (3.43)$$

For cylindrical and spherical coordinates[6]

$$\nabla \times \mathbf{A} = \mathbf{a}_\rho\left(\frac{1}{\rho}\frac{\partial A_z}{\partial \phi} - \frac{\partial A_\phi}{\partial z}\right) + \mathbf{a}_\phi\left(\frac{\partial A_\rho}{\partial z} - \frac{\partial A_z}{\partial \rho}\right)$$
$$+ \mathbf{a}_z\left(\frac{1}{\rho}\frac{\partial(\rho A_\phi)}{\partial \rho} - \frac{1}{\rho}\frac{\partial A_\rho}{\partial \phi}\right) \quad \text{(cylindrical)} \quad (3.44)$$

[6] See Problem 26.

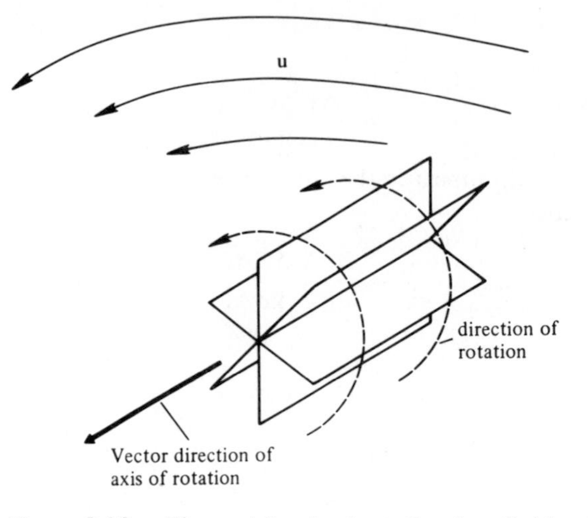

Figure 3.10. The paddlewheel rotating in a fluid velocity field and the right-hand rule.

and

$$\nabla \times \mathbf{A} = \frac{\mathbf{a}_r}{r \sin \theta}\left[\frac{\partial(A_\phi \sin \theta)}{\partial \theta} - \frac{\partial A_\theta}{\partial \phi}\right] + \frac{\mathbf{a}_\theta}{r}\left[\frac{1}{\sin \theta}\frac{\partial A_r}{\partial \phi} - \frac{\partial(rA_\phi)}{\partial r}\right] + \frac{\mathbf{a}_\phi}{r}\left[\frac{\partial(rA_\theta)}{\partial r} - \frac{\partial A_r}{\partial \theta}\right] \quad \text{(spherical)} \tag{3.45}$$

In order to attach some physical significance to the curl of a vector, we will employ the small "paddlewheel" as suggested by Skilling.[7] Let the vector field be a fluid velocity field, regardless of what it is physically. Place the small paddlewheel in this velocity field and move it about. For every point in the field, the paddlewheel axis should be oriented in all possible directions. The maximum angular velocity of the paddlewheel at a point is proportional to the curl, while the axis of the paddlewheel points in the *direction* of the curl according to the right-hand rule. That is, if the fingers of the right hand point in the direction of the rotation of the paddlewheel blades, then the thumb of the right hand points in the direction of the axis of rotation or curl. This is demonstrated in Figure 3.10. If the paddlewheel does not rotate, the vector field is *irrotational, or has zero curl*! A simple example will help clarify this paddlewheel concept.

[7] See References at end of chapter.

EXAMPLE 4

Suppose $\mathbf{A} = Kz\mathbf{a}_x$, where K is a constant. This field is sketched in Figure 3.11 for $z > 0$. It might be analogous to water flow in a river. It

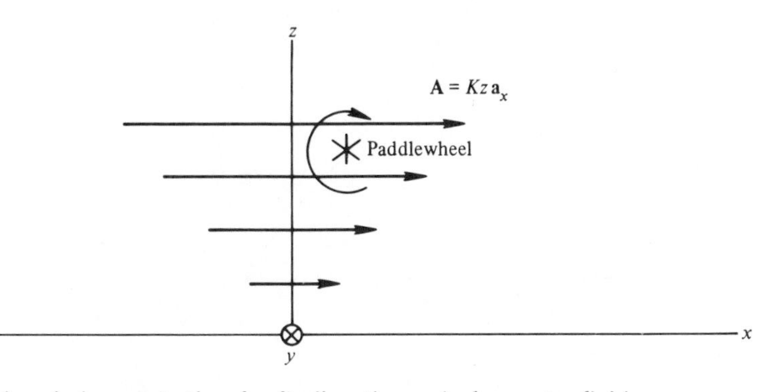

Figure 3.11. A demonstration for finding the curl of a vector field using a small paddlewheel.

is obvious that the paddlewheel rotates in a clockwise direction (regardless of z) and the angular velocity of the paddlewheel is proportional to K (the slope) regardless of z. Furthermore, the maximum angular velocity occurs when the paddlewheel axis is parallel to the y axis and is zero when the paddlewheel axis is parallel to the x or z axis. According to the right-hand rule the direction of the curl is that of $+\mathbf{a}_y$.

Analytically, using equation 3.42, we have

$$\nabla \times \mathbf{A} = +\mathbf{a}_y \frac{\partial A_x}{\partial z} = \mathbf{a}_y K$$

which certainly agrees with our observations with the paddlewheel! Problem 12 at the end of this chapter deals further with the paddlewheel, curl, and irrotational (or perhaps rotational) fields.

3.5 SOME SIMPLE EXAMPLES

EXAMPLE 5

As a first example let us calculate the potential field of an infinite line charge density, ρ_l, by using equation 3.19. Then

$$\Phi(x, y, z) = \frac{1}{4\pi\varepsilon_0} \int_{-\infty}^{\infty} \frac{\rho_l(z')\, dz'}{R}$$

for Figure 3.12. Notice that Φ is independent of ϕ and z. Choose $z = 0$.

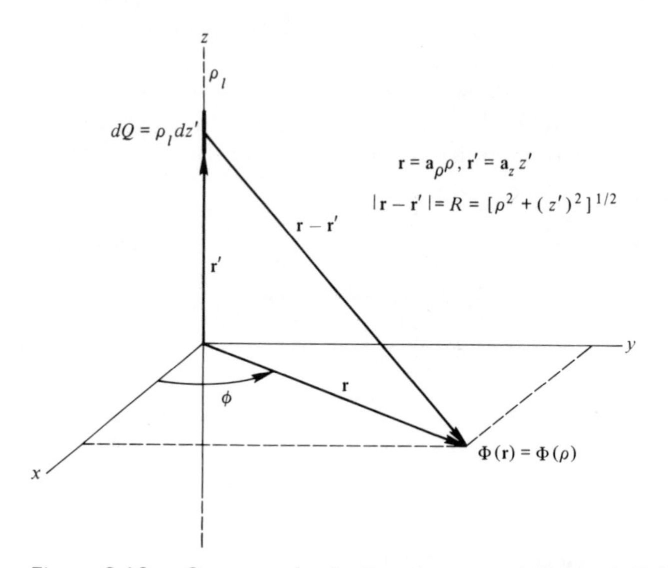

Figure 3.12. Geometry for finding the potential of an infinite line charge on the z axis.

Cylindrical coordinates are most convenient, so

$$R = [\rho^2 + (z')^2]^{1/2}$$

and

$$\Phi(\rho) = \frac{\rho_l}{4\pi\varepsilon_0} \int_{\infty}^{\infty} \frac{dz'}{[\rho^2 + (z')^2]^{1/2}}$$

or

$$\Phi(\rho) = \frac{\rho_l}{2\pi\varepsilon_0} \int_{0}^{\infty} \frac{dz'}{[\rho^2 + (z')^2]^{1/2}} = \frac{\rho_l}{2\pi\varepsilon_0} \ln [z' + \sqrt{\rho^2 + (z')^2} \Big|_0^{\infty}$$

This may be written

$$\Phi(\rho) = -\frac{\rho_l}{2\pi\varepsilon_0} \ln \rho + \lim_{u \to \infty} \frac{\rho_l}{2\pi\varepsilon_0} \ln [u + (u^2 + \rho^2)^{1/2}]$$

The first term on the right is the correct potential according to equation 3.15 (zero potential at $\rho = 1$), while the limit on the right does not exist! This is rather puzzling, until we remember that the potential is always arbitrary to within an additive constant. No added constant was used in equation 3.19 so it will calculate the potential in such a way as to force it to vanish at infinity. Put another way, the use of equation 3.19 requires a source of *finite* extent so the potential can vanish at infinity, and the infinite line charge density does not meet these conditions! Of course we could add a constant to our present solution, but it would have to take on the value $-\infty$ to remove the last term in the preceding equation. This is not to our liking, so we must be content with

$$\Phi(\rho) = -\frac{\rho_l}{2\pi\varepsilon_0} \ln \rho + C$$

for which the potential is zero at $\rho = 1$ when $C = 0$ but approaches minus infinity as ρ increases and approaches plus infinity as ρ decreases. An investigation of the infinite surface charge density ρ_s in the x-z plane reveals the same behavior as far as the potential is concerned.

EXAMPLE 6

As a second example, suppose we desire the potential on the z axis for a filamentary circular loop of uniform line charge density ρ_l, if it lies in the x-y plane and is centered on the z axis. This is one of the rare cases where a charge density of *finite* extent will be uniformly distributed. (Can you think of another case where this happens?) The geometry is shown in Figure 3.13. The loop radius is a. From equation 3.19, using

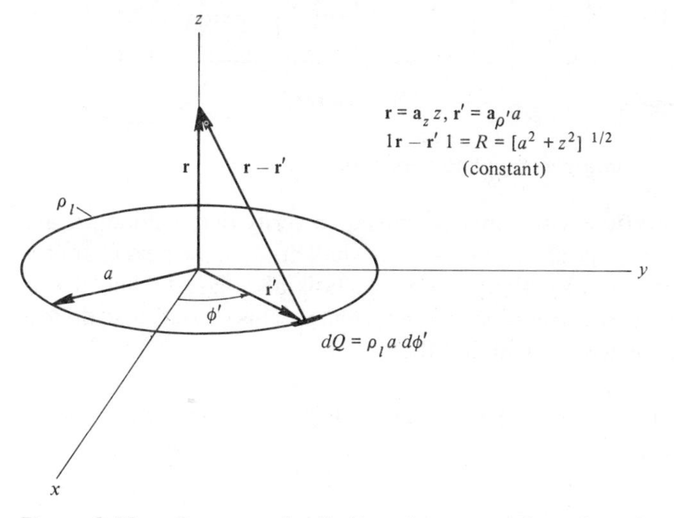

Figure 3.13. Geometry for finding the potential on the axis of a filamentary circular charged loop.

cylindrical coordinates,

$$\Phi(0, 0, z) = \frac{\rho_l}{4\pi\varepsilon_0} \int_0^{2\pi} \frac{a\, d\phi'}{(a^2 + z^2)^{1/2}} = \frac{\rho_l a}{4\pi\varepsilon_0 (a^2 + z^2)^{1/2}} \int_0^{2\pi} d\phi'$$

So,

$$\Phi(0, 0, z) = \frac{\rho_l a}{2\varepsilon_0 (a^2 + z^2)^{1/2}} \quad \text{(V)}$$

Thus, $\Phi(0, 0, 0) = \rho_l/2\varepsilon_0$, a very simple form indeed. On the other hand, for large z, $\Phi(0, 0, z) \approx \rho_l a/2\varepsilon_0 z = \rho_l 2\pi a/4\pi\varepsilon_0 z = Q/4\pi\varepsilon_0 z$. This form shows that from a large distance the loop looks like a point charge because the potential on the z axis from a point charge at the origin has an identical form (see equation 3.10). This result is not surprising, and can be shown to hold in general by taking the superposition integral form, equation 3.17, and letting $R \approx r$.

EXAMPLE 7

For another example, consider the large parallel plate capacitor as shown in Figure 3.14. The parallel plates are conductors, and as such are

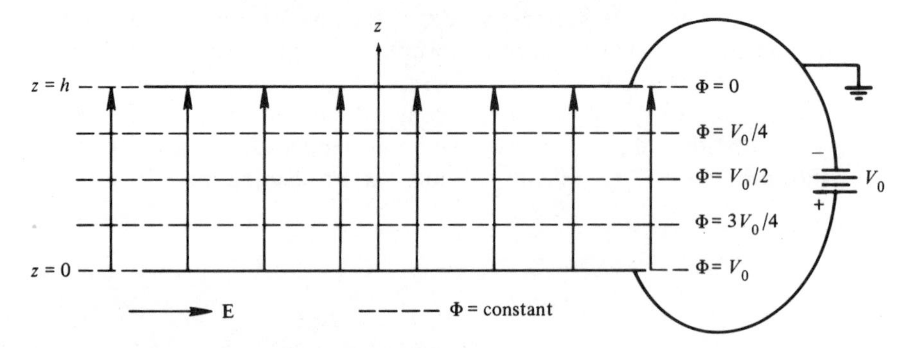

Figure 3.14. Large parallel plate capacitor.

equipotentials (Φ = constant). (It is easy to verify that conductors are equipotentials in the static case, and we shall do so in the next chapter.) The potential between the parallel plates is also easy to find if this problem is treated as a *boundary value problem*. This we shall also do in a later chapter, where we will find that

$$\Phi(z) = \frac{V_0}{h}(h - z)[u(z) - u(z - h)] + V_0 u(-z) \quad \text{(V)}$$

Then

$$\mathbf{E} = -\nabla\Phi = -\mathbf{a}_z \frac{\partial \Phi}{\partial z} = \mathbf{a}_z \frac{V_0}{h}[u(z) - u(z - h)]$$

or simply,

$$E_z = \frac{V_0}{h}[u(z) - u(z - h)]$$

Comparing this result with that of equation 2.3,

$$E_z = \frac{\rho_s}{\varepsilon_0}[u(z) - u(z - h)]$$

indicates there must be a surface charge density of

$$\rho_s = \frac{V_0}{h} \varepsilon_0 \quad (\text{C/m}^2)$$

on the bottom plate and

$$\rho_s = -\frac{V_0}{h} \varepsilon_0$$

on the top plate. This will be discussed further in the next chapter.

3.6 ENERGY AND ENERGY DENSITY

In this section we will find the potential energy present in a system of point charges. To do this, we merely need to calculate the work required to position these charges into whatever final arrangement they attain. We already know that the potential energy of a point charge, Q, is the product of the charge and the potential, for this was the way in which we originally defined the potential. In other words, if a point charge Q_n is transported from infinity to a position $\mathbf{r}_n$ in the presence of other point charges (with their associated fields and potentials), the work done or the potential energy is

$$W_n = Q_n \Phi(\mathbf{r}_n) \quad (\text{J}) \tag{3.46}$$

We now calculate the energy required to assemble these point charges in their final configuration. With *no* charges present, the first charge, Q_1, may be moved from infinity to its final position with no work required. The work required to position Q_2 is (refer to Figure 3.15)

$$W_2 = Q_2 \Phi_{21} = Q_2 \frac{Q_1}{4\pi\varepsilon_0 |\mathbf{r}_2 - \mathbf{r}_1|} = Q_2 \frac{Q_1}{4\pi\varepsilon_0 R_{21}}$$

according to equation 3.46. The work required to position Q_3 is

$$W_3 = Q_3 \Phi_{31} + Q_3 \Phi_{32} = Q_3 \frac{Q_1}{4\pi\varepsilon_0 R_{31}} + Q_3 \frac{Q_2}{4\pi\varepsilon_0 R_{32}}$$

If there are N charges in the assembly, the total work to position them is

$$W_E = \frac{1}{4\pi\varepsilon_0} \sum_{m=2}^{N} \sum_{n=1}^{N-1} \frac{Q_m Q_n}{R_{mn}}, \qquad \begin{cases} m \neq n \\ m > n \end{cases} \tag{3.47}$$

We do not allow $m = n$, so that infinite "self-energy" terms are excluded ($R_{nn} = 0$). Equation 3.47 represents a triangular array (matrix), the first row containing one term,

$$\frac{Q_2 Q_1}{4\pi\varepsilon_0 R_{21}}$$

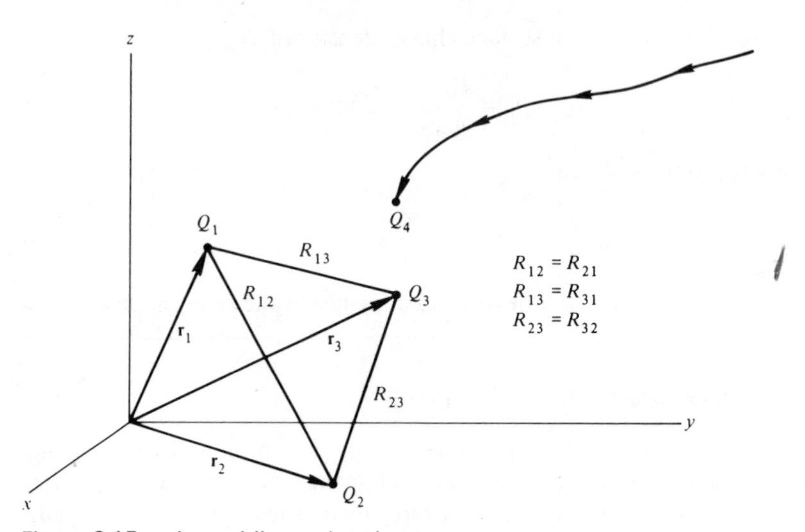

Figure 3.15. Assembling point charges.

the second term containing two terms, the third three, and so on. The triangular array may be almost completed if the same number of terms consisting of

$$\frac{Q_1 Q_2}{4\pi\varepsilon_0 R_{12}} + \frac{Q_1 Q_3}{4\pi\varepsilon_0 R_{13}} + \frac{Q_1 Q_4}{4\pi\varepsilon_0 R_{14}} + \cdots$$

$$+ \frac{Q_2 Q_3}{4\pi\varepsilon_0 R_{23}} + \frac{Q_2 Q_4}{4\pi\varepsilon_0 R_{24}} + \cdots$$

$$+ \frac{Q_3 Q_4}{4\pi\varepsilon_0 R_{34}} + \cdots$$

is added. Now $R_{mn} = R_{nm}$, since $|\mathbf{r}_m - \mathbf{r}_n| = |\mathbf{r}_n - \mathbf{r}_m|$, and so every term added is already present. The new array, then just totals $2W_E$ and is a complete $N \times N$ array, *except for the missing main diagonal* (no "self-energy" terms). Then,

$$2W_E = \frac{1}{4\pi\varepsilon_0} \sum_{m=1}^{N} \sum_{n=1}^{N} \frac{Q_m Q_n}{R_{mn}}, \qquad m \neq n \tag{3.48}$$

or

$$W_E = \frac{1}{8\pi\varepsilon_0} \sum_{m=1}^{N} \sum_{n=1}^{N} \frac{Q_m Q_n}{|\mathbf{r}_m - \mathbf{r}_n|}, \qquad m \neq n \quad \text{(J)} \tag{3.49}$$

If we next multiply and divide by $\Delta v_m \, \Delta v_n$, and then take the limit as the number of charges (in a finite region) increases without limit, we obtain

$$W_E = \frac{1}{2} \iiint_{\text{vol}} \iiint_{\text{vol}'} \frac{\rho_v(x, y, z)\rho_v(x', y', z')}{4\pi\varepsilon_0 |\mathbf{r} - \mathbf{r}'|} \, dx \, dy \, dz \, dx' \, dy' \, dz' \tag{3.50}$$

The limiting procedure used here is exactly the same as that used to obtain the integral for **E**.

Either of the integrals in equation 3.50 may be recognized (equation 3.17) as the potential, so equation 3.50 may be written

$$\boxed{W_E = \frac{1}{2} \iiint_{\text{vol}} \rho_v(x, y, z)\Phi(x, y, z)\, dx\, dy\, dz \qquad \text{(J)}} \tag{3.51}$$

We calculate Φ from ρ_v, using equation 3.17, and use equation 3.51 to calculate W_E. This is a rather lengthy procedure. Furthermore, equation 3.51 may be difficult to use in some cases. For example, consider an isolated point charge located at $\mathbf{r}_1$. Since ρ_v must be represented as a three-dimensional impulse, and Φ is infinite at $\mathbf{r}_1$, the energy calculated by equation 3.51 will be infinite! This occurs because the missing main diagonal "self-energy" terms were inserted in passing from equation 3.49 to equation 3.50 by means of the limiting process. Example 8 examines this difficulty more closely. Equation 3.51 should cause no difficulty with continuous charge distributions if the integration is carried out over *all* the charge.

The total energy, W_E, may be viewed somewhat differently if we eliminate ρ_v in equation 3.51. This may be done with the aid of Maxwell's equation,

$$\rho_v = \nabla \cdot \mathbf{D} \tag{3.52}$$

(already derived). Substituting equation 3.52 into equation 3.51 gives

$$W_E = \frac{1}{2} \iiint_{\text{vol}} \Phi(\nabla \cdot \mathbf{D})\, dv \tag{3.53}$$

We now need to use an identity of vector analysis[8]

$$\alpha(\nabla \cdot \mathbf{A}) \equiv \nabla \cdot (\alpha \mathbf{A}) - \mathbf{A} \cdot (\nabla \alpha) \tag{3.54}$$

which when applied to equation 3.53 gives

$$W_E = \frac{1}{2} \iiint_{\text{vol}} \nabla \cdot (\Phi \mathbf{D})\, dv - \frac{1}{2} \iiint_{\text{vol}} \mathbf{D} \cdot (\nabla \Phi)\, dv \tag{3.55}$$

We now perform two steps simultaneously. We use the divergence theorem on the first integral of equation 3.55 and substitute $\nabla\Phi = -\mathbf{E}$ in the second integral. Then,

$$W_E = \frac{1}{2} \oiint_{s} (\Phi \mathbf{D}) \cdot d\mathbf{s} + \frac{1}{2} \iiint_{\text{vol}} (\mathbf{D} \cdot \mathbf{E})\, dv \tag{3.56}$$

[8] See Appendix A and Problem 24.

The only restriction on the surface over which the first integration of equation 3.56 is carried out is that *all* the charges are inside (see equations 3.50 or 3.51). We assumed in arriving at equation 3.50 from equation 3.49 that all the charges were in a finite region. Then, we are justified in letting the surface in equation 3.56 become an infinite sphere. For any reasonable charge distribution[9] the integrand of the surface integral is zero since Φ decreases asymptotically as $1/r$, while $|\mathbf{D}|$ decreases as $1/r^2$ and $|d\mathbf{s}|$ increases as r^2. Thus, the surface integral vanishes for any real problem, and

$$\boxed{W_E = \frac{1}{2} \iiint_{\text{vol}} (\mathbf{D} \cdot \mathbf{E})\, dv} \quad \text{(J)} \tag{3.57}$$

It is only natural to call the integrand of equation 3.57 an energy density. That is,

$$w_E = \frac{\mathbf{D} \cdot \mathbf{E}}{2} \quad (\text{J/m}^3) \tag{3.58}$$

or

$$w_E = \tfrac{1}{2}\varepsilon_0 |\mathbf{E}|^2 \quad (\text{J/m}^3) \tag{3.59}$$

This interpretation leads to the reasoning behind the statement that "the energy is stored in the electric field."

EXAMPLE 8

An interesting example will now be given. Consider two point charges arbitrarily located as shown in Figure 3.16. Equation 3.49 gives

$$W_E = \frac{1}{2}\left(\frac{1}{4\pi\varepsilon_0}\right)\left[\frac{Q_1 Q_2}{|\mathbf{r}_1 - \mathbf{r}_2|} + \frac{Q_2 Q_1}{|\mathbf{r}_2 - \mathbf{r}_1|}\right]$$

or, since $|\mathbf{r}_1 - \mathbf{r}_2| = |\mathbf{r}_2 - \mathbf{r}_1|$,

$$W_E = \frac{Q_1 Q_2}{4\pi\varepsilon_0 |\mathbf{r}_1 - \mathbf{r}_2|} \quad \text{(J)} \tag{3.60}$$

We would next like to calculate W_E from equation 3.57, so we must first calculate w_E from equation 3.59. For point P,

$$\mathbf{E} = \frac{1}{4\pi\varepsilon_0}\left[\frac{Q_1 \mathbf{a}_1}{|\mathbf{r} - \mathbf{r}_1|^2} + \frac{Q_2 \mathbf{a}_2}{|\mathbf{r} - \mathbf{r}_2|^2}\right]$$

[9] The infinite line charge density is an exception.

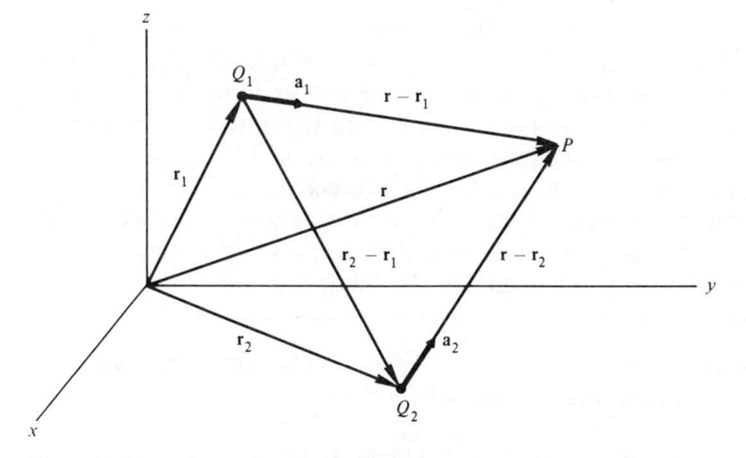

Figure 3.16. Geometry for finding the energy of two point charges.

as shown in Figure 3.16. Then,

$$\mathbf{E}\cdot\mathbf{E} = \left(\frac{1}{4\pi\varepsilon_0}\right)^2\left[\frac{Q_1^2}{|\mathbf{r}-\mathbf{r}_1|^4} + \frac{Q_2^2}{|\mathbf{r}-\mathbf{r}_2|^4} + \frac{2Q_1Q_2\mathbf{a}_1\cdot\mathbf{a}_2}{|\mathbf{r}-\mathbf{r}_1|^2|\mathbf{r}-\mathbf{r}_2|^2}\right]$$

and

$$w_E = \left(\frac{1}{4\pi\varepsilon_0}\right)^2\frac{\varepsilon_0}{2}\left[\frac{Q_1^2}{|\mathbf{r}-\mathbf{r}_1|^4} + \frac{Q_2^2}{|\mathbf{r}-\mathbf{r}_2|^4} + \frac{2Q_1Q_2\mathbf{a}_1\cdot\mathbf{a}_2}{|\mathbf{r}-\mathbf{r}_1|^2|\mathbf{r}-\mathbf{r}_2|^2}\right]$$

Then, equation 3.57 gives ($\mathbf{D} = \varepsilon_0\mathbf{E}$)

$$W_E = \frac{Q_1^2}{32\pi^2\varepsilon_0}\iiint_{\text{vol}}\frac{dv}{|\mathbf{r}-\mathbf{r}_1|^4} + \frac{Q_2^2}{32\pi^2\varepsilon_0}\iiint_{\text{vol}}\frac{dv}{|\mathbf{r}-\mathbf{r}_2|^4}$$

$$+ \frac{Q_1Q_2}{16\pi^2\varepsilon_0}\iiint_{\text{vol}}\frac{\mathbf{a}_1\cdot\mathbf{a}_2}{|\mathbf{r}-\mathbf{r}_1|^2|\mathbf{r}-\mathbf{r}_2|^2}\,dv \tag{3.61}$$

The third integral can be evaluated explicitly, so that

$$W_E = \frac{Q_1^2}{32\pi^2\varepsilon_0}\iiint_{\text{vol}}\frac{dv}{|\mathbf{r}-\mathbf{r}_1|^4} + \frac{Q_2^2}{32\pi^2\varepsilon_0}\iiint_{\text{vol}}\frac{dv}{|\mathbf{r}-\mathbf{r}_2|^4} + \frac{Q_1Q_2}{4\pi\varepsilon_0|\mathbf{r}_1-\mathbf{r}_2|} \tag{3.62}$$

The third term of equation 3.62 is the interaction energy in agreement with equation 3.60, but what are the first two terms of equation 3.62?

If we are required to calculate the energy of an *isolated* point charge located at $\mathbf{r}_1$, using equation 3.57, the equation will be identical with the first term of equation 3.62. Evidently, the first two terms of equation 3.62 are the "self-energy" terms. Using equation 3.51 instead of equation 3.57 gives the same result as that given by equation 3.62. See Problem 21. Ordinarily, we are content to calculate the energy stored in a charge-free region using equation 3.57. That is, we use equation 3.57 and exclude the "self-energy" terms to agree with the original development. This is a rather subtle situation of adding and subtracting infinity!

In order to emphasize this point let us calculate the energy of an *isolated* point charge at the origin. From equation 3.57,

$$W_E = \frac{\varepsilon_0}{2} \iiint_{\text{vol}} \mathbf{E} \cdot \mathbf{E} \, dv$$

Now,

$$\mathbf{E} = \mathbf{a}_r \frac{Q}{4\pi\varepsilon_0 r^2}$$

so

$$\mathbf{E} \cdot \mathbf{E} = \frac{Q^2}{16\pi^2 \varepsilon_0^2 r^4}$$

and

$$W_E = \frac{Q^2}{32\pi^2 \varepsilon_0} \iiint_{\text{vol}} \frac{dv}{r^4}$$

We choose spherical coordinates and integrate on r from a to b $(b > a)$. Then

$$W_E = \frac{Q^2}{32\pi^2 \varepsilon_0} \int_a^b \int_0^\pi \int_0^{2\pi} \frac{r^2 \sin\theta \, dr \, d\theta \, d\phi}{r^4}$$

or

$$W_E = \frac{Q^2}{8\pi\varepsilon_0} \int_a^b \frac{dr}{r^2} = \frac{Q^2}{8\pi\varepsilon_0}\left(\frac{1}{a} - \frac{1}{b}\right)$$

As b goes to infinity (a fixed), W_E goes to $+Q^2/8\pi\varepsilon_0 a$. However, with b fixed, letting a approach zero gives an infinite energy (the self-energy). This will occur using equation 3.57 when the charge is included in the volume where the integration is carried out. In calculating the energy in the neighborhood of point charges we must take care to avoid including the charges themselves if we desire finite answers.

3.7 CONCLUDING REMARKS

In this chapter we have developed the concept of energy, or more specifically, electrostatic energy. We also defined electrostatic potential in terms of this energy and were easily able to find a superposition integral for the potential once the potential for an arbitrarily located unit positive charge was found. The electrostatic field is irrotational and any such field has the following properties.

1. Its circulation is identically zero.
2. It is the (−) gradient of a scalar potential field.
3. Its curl is identically zero.
4. It is conservative.

A knowledge of the energy in the region around charged bodies is very important in the concept of capacitance. In this chapter we developed the basic tools necessary for finding the energy. In the following chapters we will use these tools.

REFERENCES

Durney, C. H. and Johnson, C. C. *Introduction to Modern Electromagnetics.* New York: McGraw-Hill, 1969. A more detailed coverage of the same material as in this text book.

McQuistan (see References for Chapter 1). Gradient and curl are well explained and illustrated.

Popović, B. D. *Introductory Engineering Electromagnetics.* Reading, Mass.: Addison-Wesley, 1971. The level is about the same as that of this textbook, but the detail and coverage is greater.

Skilling, H. H. *Fundamentals of Electric Waves.* New York: Wiley, 1948. Skilling's discussion of curl and the paddlewheel idea begins on page 23.

PROBLEMS

1. Find the work required to transport an electron from (1, 1, 1) to (7, 2, −1) in the field $\mathbf{E} = y\mathbf{a}_x + x\mathbf{a}_y$.
2. A point charge Q is located at the origin. The potential at (1, 0, 0) is 20 V, while the potential at (0, 2, 0) is 10 V. Find Q.
3. If $\mathbf{E} = \rho/(\rho^2 + a^2)\mathbf{a}_\rho$, find the potential at any point, if
   (a) $\Phi = 0$ at $\rho = 1$
   (b) $\Phi = 10$ at $\rho = a$
4. A conducting sphere of radius a has $\Phi = 0$ and has a total charge of 10^{-9} C on its surface. What is the potential at $r = 2a$?
5. Describe the equipotential surfaces for (a) the infinite line charge density on the z axis, (b) the point charge, (c) the infinite surface charge density on the $z = 0$ plane, (d) an r dependent spherical volume charge density, (e) a uniform spherical volume charge

density, and (f) an infinite cylindrical volume charge density with ρ dependence.

6. Show that the $z = 0$ plane is an equipotential surface for Fig. 4.5.
7. A line charge density, *assumed* to be given by $\rho_{l0} z^2[u(z + h) - u(z - h)]$, exists on the z axis.

   (a) Set up the integral for $\Phi(\rho, \phi, z)$
   (b) Find $\Phi(\rho, \phi, z)$ for $\rho, z \gg h$.

8. Determine which of the following fields are possible electrostatic fields.

   (a) $\mathbf{F} = y\mathbf{a}_x + x\mathbf{a}_y$
   (b) $\mathbf{F} = y\mathbf{a}_x - x\mathbf{a}_y$
   (c) $\mathbf{F} = 2r/(r^2 + a^2)^2\mathbf{a}_r$
   (d) $\mathbf{F} = (z/2\rho)\mathbf{a}_\rho$

9. (a) Calculate the energy stored per unit length for a coaxial cable, given the electric field

$$E_\rho = -\frac{V_0}{\rho \ln(a/b)}[u(\rho - a) - u(\rho - b)]$$

   where (see Figure 4.12)

$$V_0 = \text{potential difference, } \Phi_{ab}$$
$$a = \text{radius of inner conductor}$$
$$b = \text{radius of outer conductor}$$

   (b) Equate the energy to $\frac{1}{2} CV_0^2$ and solve for C, the capacitance.

10. Newton's law of gravity is

$$\mathbf{F}_1 = G\frac{m_1 m_2(\mathbf{r}_2 - \mathbf{r}_1)}{|\mathbf{r}_1 - \mathbf{r}_2|^3}$$

   where

$$G = 6.664 \times 10^{-11} \qquad (\text{m}^3/\text{kg} \cdot \text{s}^2)$$

   If the masses of the earth and moon are 5.98×10^{24} kg and 7.35×10^{22} kg and their centers are separated by 3.848×10^8 m on the average, find:

   (a) The force of attraction between the earth and moon. Assume the origin of a coordinate system at the center of the earth (whose radius is 6.371×10^6 m) and find:
   (b) The force on a point mass m at the earth's surface.
   (c) The acceleration due to gravity, $\mathbf{g}$, at the earth's surface.
   (d) The gravitational potential field, Φ_g, (zero reference at $r \to \infty$).
   (e) The energy required to move a point mass from the earth's surface to infinity.
   (f) The escape velocity if the supplied energy is kinetic energy.

11. Verify equation 3.21 using Cartesian coordinates.
12. Use the small paddlewheel to explore the following fields, and determine if the curl is zero. Verify the results analytically. If the curl is not zero, try to interpret the results physically.

(a) $\mathbf{F} = \dfrac{20\mathbf{a}_r}{r}$

(b) $\mathbf{F} = \dfrac{\mathbf{a}_x x + \mathbf{a}_y y}{(x^2 + y^2)^{1/2}}$

(c) $\mathbf{F} = 10\mathbf{a}_c$ ($\mathbf{a}_c$ is a constant unit vector)

(d) $\mathbf{F} = \dfrac{\mathbf{a}_\phi}{\rho}$

(e) $\mathbf{F} = [u(z) - u(z - d)]\mathbf{a}_y$

13. Verify that equation 3.34 holds for any Φ. Use Cartesian coordinates.
14. Verify that Stoke's theorem holds for (a) $\mathbf{F} = \mathbf{a}_\rho/\rho$ and (b) $\mathbf{F} = \mathbf{a}_\phi/\rho = \mathbf{a}_\phi[u(\rho)/\rho]$. The path c is a circular path of radius a in the $z = 0$ plane.
15. The electric field for a uniform charge density over the entire $z = 0$ plane is $\mathbf{E} = (\rho_s/2\varepsilon_0)[2u(z) - 1]\mathbf{a}_z$. Find $\Phi(z)$ such that $\Phi(0) = 0$.
16. The temperature in a region is given by $T = 10x^2yz + 5$. Specify the direction of the heat flux vector by means of a unit vector.
17. Calculate the energy stored in the parallel plate capacitor of Example 7 for a plate area of 1 m^2. Assume that the field does not fringe at the plate edges. Use equation 3.57.
18. Repeat Problem 17 using equation 3.51 and

$$\rho_v = \frac{V_0 \varepsilon_0}{h}[\delta(z) - \delta(z - h)]$$

$$\Phi = V_0 u(-z) + \frac{V_0}{h}(h - z)[u(z) - u(z - h)]$$

19. Calculate the electric field produced by an infinite line charge density using the results of Example 5.
20. Calculate the energy stored in the region $0 \le z \le 1$, $a \le \rho \le b$, $0 \le \phi \le 2\pi$ for the infinite line charge density of Problem 19 using equation 3.57
21. Use equation 3.51 to derive results similar to those given by equation 3.62. Integrate over all space and use impulsive representations for the two point charges.
22. Calculate the energy stored in the region $a < r < b$ for Example 1, Chapter 2 using equation 3.57.
23. Given the continuous field $\mathbf{E} = \mathbf{a}_r E_0 e^{-r}$, (a) find Φ, (b) find ρ_v, (c) calculate W_E using equation 3.51 and (d) calculate W_E using equation 3.57.

24. (a) Show that equation 3.54 holds for Cartesian coordinates.
    (b) Repeat for cylindrical coordinates.
    (c) Repeat for spherical coordinates.

25. Starting with equation 3.37 and using the coordinate and component transformations in Appendix A;
    (a) Obtain equation 3.38.
    (b) Obtain equation 3.39.

26. Starting with equation 3.42 and using the coordinate and component transformations in Appendix A:
    (a) Obtain equation 3.44.
    (b) Obtain equation 3.45.

Chapter 4
Materials and Material Discontinuities with Electrostatic Fields

In previous chapters we have purposely avoided problems involving conductors (except for some simple cases), and the only *dielectric* encountered was free space for which $\varepsilon_0 \approx 1/36\pi \times 10^{-9}$ F/m. In this chapter we will be able to consider conductors and more general dielectrics because of the experience we have gained in the first three chapters. We will find that conductors cause no difficulties in the electrostatic case. The electrical properties of the classic electrostatic dipole will be determined because the behavior of a dipole under the influence of an external electric (force) field will aid our understanding of the process of the polarization of a dielectric when acted upon by an external field. A thorough treatment of materials requires the use of quantum theory, but a qualitative treatment, sufficient for our purposes, will be given in this chapter. This qualitative treatment gives the usual quantitative results for simple dielectrics (for which the permittivity is a scalar constant). The consideration of more complex dielectrics may require the use of *tensor* analysis and this will not be pursued here.

Material discontinuities (a special case of inhomogeneous materials) introduce boundary surfaces. Conditions at these boundaries will be studied by means of Gauss's law and the conservative property of the electrostatic field. In this way a simple set of *boundary conditions* for conductors and dielectrics will be derived for future use. Finally, we are able to define the "circuit" parameter capacitance and consider examples of finding its value.

4.1 CONDUCTOR PROPERTIES

We may define a conductor as a material for which an applied electric field will cause charge motion. This rather obvious statement has immediate consequences for us. First, the electrostatic electric field in a conductor is zero. This follows since charges in motion do not represent an electrostatic situation. Secondly, the conductor must be an equipotential in the static case. If $\mathbf{E} = 0$ in the conductor, then Φ must be constant because $\mathbf{E} = -\nabla\Phi$ (the gradient of a constant is zero). Thus, Φ is constant at every point *in* and *on* a conductor in the static case.

The definition we have given for conductors (above) completely excludes the consideration of charge motion or electric current. We wish to do this for present purposes only. There are nonstatic situations we will examine later (uniform motion of charge in a resistor accompanied by a uniform electric field, for example) which can be treated by static methods even though there is charge motion.

We now wish to examine the behavior of electrostatic fields in the neighborhood of conducting bodies.

EXAMPLE 1

As a simple example, consider a spherical conducting solid ball centered at the origin with a total excess charge Q released (somehow) inside. The Coulomb forces of mutual repulsion ensure that this charge ultimately resides statically on the surface in the form of a *uniform* surface charge density,

$$\rho_s = \frac{Q}{4\pi a^2} \tag{4.1}$$

for a sphere with a radius a. Gauss's law (with a sphere of radius $r > a$ as the Gaussian surface) tells us immediately that for free space surrounding the ball

$$\mathbf{E} = \mathbf{a}_r \frac{Q}{4\pi\varepsilon_0 r^2}, \qquad r > a \tag{4.2}$$

or

$$E_r = \frac{Q}{4\pi\varepsilon_0 r^2} = -(\nabla\Phi)_r = -\frac{\partial\Phi}{\partial r}$$

Then,

$$\Phi = \frac{Q}{4\pi\varepsilon_0 r}, \qquad r > a \tag{4.3}$$

for a zero potential reference at infinity. At the conductor surface, $r = a$, Φ is the same everywhere and is

$$\Phi = \frac{Q}{4\pi\varepsilon_0 a}$$

Then, the potential must be this same constant everywhere inside the sphere. A plot of Φ and E_r versus r is shown in Figure 4.1.

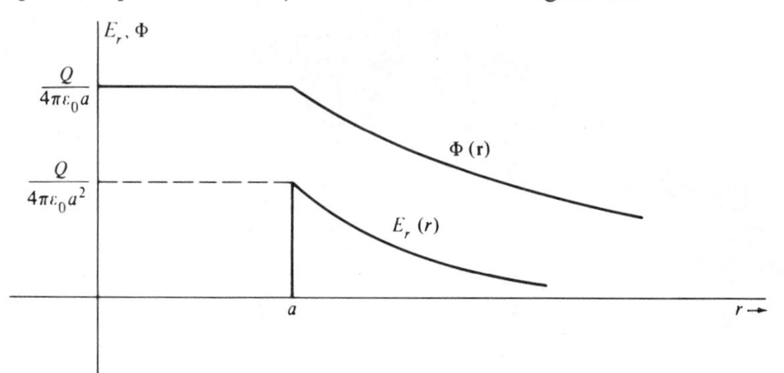

Figure 4.1. $E_r(r)$ and $\Phi(r)$ versus r for a charged, conducting sphere of radius $a > 1$.

It is interesting to note that

$$E_r|_{r=a} = \frac{Q}{4\pi\varepsilon_0 a^2} = \frac{\rho_s}{\varepsilon_0}$$

or

$$D_r|_{r=a} = \rho_s \quad (\text{C/m}^2) \tag{4.4}$$

Equation 4.4 represents a boundary condition for which we will soon be able to make a general statement. Finally, notice that for any $r > a$ it is impossible to tell the difference between this uniform spherical surface charge density and a simple point charge at the origin insofar as **E** is concerned.

Let us next formalize the conductor characteristics demonstrated by the preceding example for conductor-free space interfaces. Consider such an interface as shown in Figure 4.2. It is very convenient to decompose the field (**E** or **D**) into components normal and tangent to the interface at some point P. Now consider a small rectangle L m by Δw m surrounding the point P as shown in Figure 4.3. If we attempt to find the circulation $\oint \mathbf{E} \cdot d\mathbf{l}$ around the rectangle, we can neglect the contributions from the *normal* component of field, E_n, at the sides of length Δw since we will ultimately let Δw approach zero. That is, in this particular investigation, we are primarily interested in the behavior of E_t at P. There is also no contribution to $\mathbf{E} \cdot d\mathbf{l}$

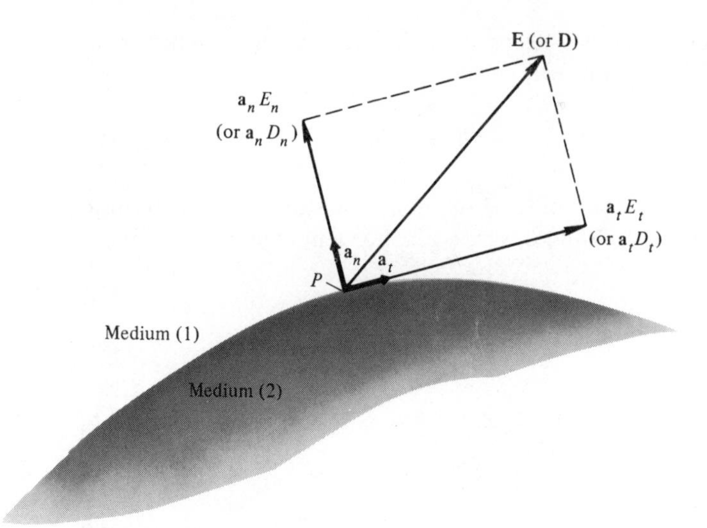

Figure 4.2. Normal and tangential components of **E** (or **D**) at an interface between two media.

along the side of length L *in the conductor*, for we have already found that *electrostatic* **E** is zero in *any* conductor. The remaining contribution to the circulation is that from $\mathbf{E} \cdot d\mathbf{l}$ along the length L in the free space region. Since we know that

$$\oint \mathbf{E} \cdot d\mathbf{l} = 0 \tag{4.5}$$

the only conclusion we can reach is that **E**, or more precisely E_t, is zero along the interface. Since $\mathbf{D} = \varepsilon_0 \mathbf{E}$ for free space, it immediately follows that $D_t = 0$ at the interface. For tangential components then the desired *boundary conditions* for a free space-conductor boundary are simply

$$\boxed{D_t = E_t = 0} \tag{4.6}$$

Figure 4.3. Small rectangular path for finding $\oint \mathbf{E} \cdot d\mathbf{l}$ at a free space-conductor interface.

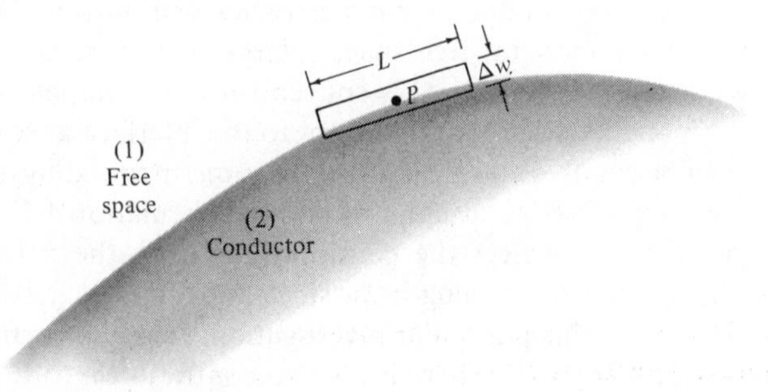

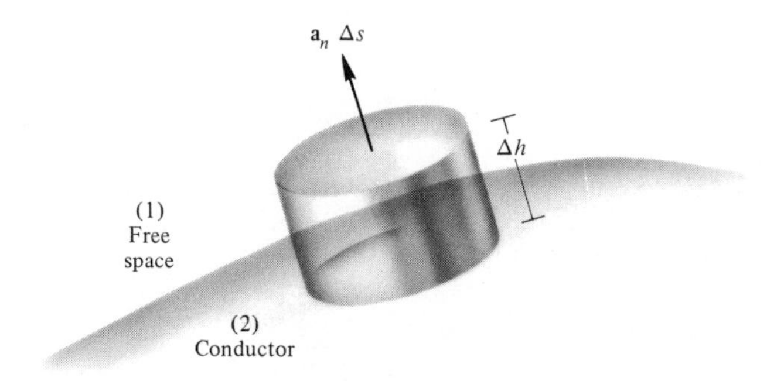

Figure 4.4. Circular cylindrical Gaussian surface for determining the relations between normal components, free space-conductor interface.

In order to examine the behavior of the normal components, consider the small cylindrical "pill box" surrounding the point P as shown in Figure 4.4. We will apply Gauss's law to its surface. In this case, we are interested in normal components so that contributions to the flux out $(\mathbf{D} \cdot d\mathbf{s})$ of the side from tangential components may be neglected for vanishingly small Δh. The flux out of the bottom cap in the conductor is zero since $\mathbf{D} = \mathbf{E} = 0$ in the conductor. Since we know that (Gauss's law)

$$\oint\!\!\oint_s \mathbf{D} \cdot d\mathbf{s} = Q \tag{4.7}$$

and the only contribution is the flux out of the top cap in free space, we have

$$D_n \, \Delta s = Q = \rho_s \, \Delta s$$

for Δh approaching zero. It immediately follows that

$$\boxed{D_n = \varepsilon_0 E_n = \rho_s} \qquad (\mathrm{C/m^2})$$

are the desired *boundary conditions* for a free space-conductor boundary. That is, the normal component of electric flux density (out) at the surface is equal to the surface charge density there. The discontinuity is easy to see in Figure 4.1.

EXAMPLE 2

Referring to Example 1 of this section, what is ρ_s, if the potential of the conducting sphere is known to be -10 V (zero reference at infinity)? From equation 4.3,

$$-10 = \frac{Q}{4\pi\varepsilon_0 a}$$

or

$$Q = -40\pi\varepsilon_0 a$$

So, from equation 4.1,

$$\rho_s = \frac{-40\pi\varepsilon_0 a}{4\pi a^2} = \frac{-10\varepsilon_0}{a}$$

It also follows that ($D_n = \varepsilon_0 E_n = \rho_s$)

$$E_r = \frac{\rho_s}{\varepsilon_0}, \qquad (r = a)$$

or

$$E_r = \frac{-10}{a}, \qquad (r = a)$$

4.2 THE ELECTROSTATIC DIPOLE

EXAMPLE 3

The electrostatic dipole consists of a pair of closely spaced positive and negative point charges as shown in Figure 4.5. A knowledge of

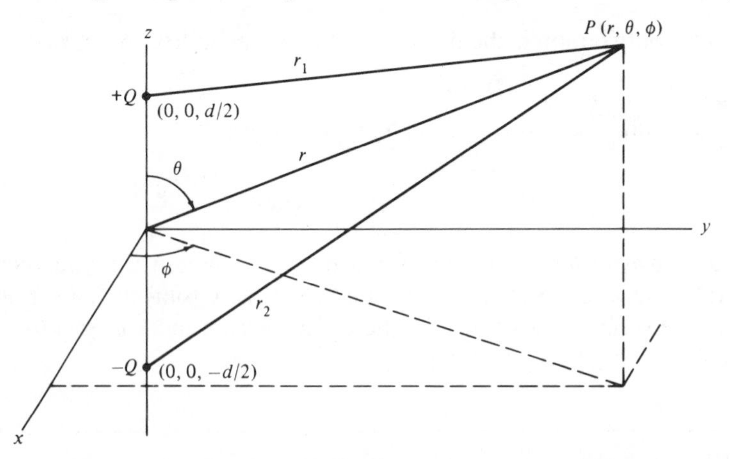

Figure 4.5. The electrostatic dipole on the z axis at the origin.

dipole fields is important in understanding the behavior of dielectrics under the influence of an applied electric field. By superposition, the potential at some point, P is

$$\Phi = \frac{Q}{4\pi\varepsilon_0}\left(\frac{1}{r_1} - \frac{1}{r_2}\right) \tag{4.9}$$

or

$$\Phi = \frac{Q}{4\pi\varepsilon_0 r_1 r_2}(r_2 - r_1) \quad \text{(V)} \tag{4.10}$$

where r_1 and r_2 are given in Figure 4.5. Equation 4.9 is an exact form, but a rather awkward one. If the point P is sufficiently far from the origin (large r), we know that any finite system of charges will look like a point charge. This concept leaves us with zero net charge for large r, and is thus useless for present purposes. It is obvious that r, r_1, and r_2 are all essentially parallel for fairly large r. In this case, as see in Figure 4.6,

$$r_2 - r_1 \approx d \cos\theta$$

and

$$r_1 r_2 \approx r^2$$

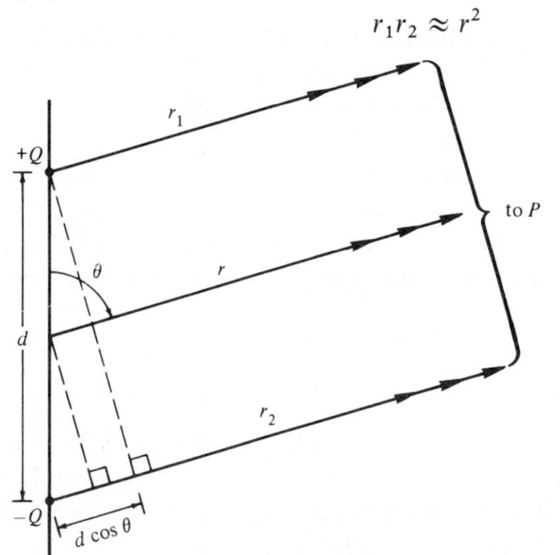

Figure 4.6. Simplified geometry for the dipole when $d \ll r$.

So, a good approximation to equation 4.10 is

$$\Phi = \frac{Qd\cos\theta}{4\pi\varepsilon_0 r^2} \tag{4.11}$$

for intermediate distances, $r \gg d$. Using equation 4.11 and $\mathbf{E} = -\nabla\Phi$ gives

$$\mathbf{E} = \frac{Qd}{4\pi\varepsilon_0 r^3}(\mathbf{a}_r 2\cos\theta + \mathbf{a}_\theta \sin\theta) \tag{4.12}$$

At this point, we introduce a concept which will prove to be useful to us. The dipole moment, **p**, is defined as a vector, directed from the minus charge toward the positive charge, having a magnitude Qd. If we let d carry the vector sign, then

$$\mathbf{p} = Q\mathbf{d} \quad (\text{C}\cdot\text{m}) \tag{4.13}$$

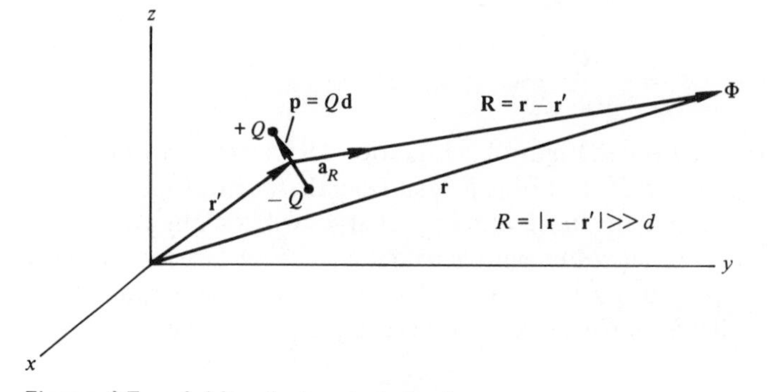

Figure 4.7. Arbitrarily located dipole.

Now consider a dipole that is arbitrarily oriented and arbitrarily located. A little thought reveals that in this case the dipole potential is

$$\Phi = \frac{1}{4\pi\varepsilon_0} \frac{\mathbf{p} \cdot \mathbf{a}_R}{R^2} \tag{4.14}$$

where, as usual, $\mathbf{a}_R$ is directed from the dipole to the field point and R is the distance from the dipole to the field point. The geometry is shown in Figure 4.7.

4.3 DIELECTRIC PROPERTIES

In this section we will present a simplified picture of dielectric behavior. We would like, if possible, to be able to use most of the results we have obtained up to this point without making major modifications to them. The goal of this section is a formulation that allows us to do this. We are not in a position here to examine the behavior of every atom in a dielectric which is subjected to the forces of an applied electric field. That is, we are not interested in the *microscopic* behavior within the dielectric. We are interested in the *macroscopic* behavior, and, in this sense, intend to utilize equation 4.14 for the dipole potential. The charge separation d is large microscopically, but small macroscopically. This concept places a limit on the validity of our field theory. Most modern electrical engineering curricula include a course or courses on the electrical properties of materials where a more detailed study may be pursued. We will only give a brief treatment here.

Materials can be broadly classified into four types.

1. If the characteristics of the material do not depend on *position* (x, y, z), the material is said to be *homogeneous*. The atmosphere above us is obviously inhomogeneous because of its varying density, humidity, temperature, or ionization.
2. If the characteristics of the material are independent of the *direction* of the vector fields, the material is *isotropic*.

3. If the parameters of the material do not depend on the *magnitude* of the field quantities, then the material is linear. A *ferroelectric* material is nonlinear because its electrical properties depend on the strength of the applied field.
4. In general the parameters of a material may be frequency-dependent. For example, the permittivity (to be introduced shortly) of polystyrene depends slightly on frequency. This dependence is discussed in more detail in Chapter 13.

It is easy to see why forces must depend on the presence of matter. Consider the simplified picture of some *isolated* charges in the midst of the neutral atoms of some material. The negative electron clouds and the positive nuclei of the atoms experience opposite forces due to the fields of the isolated charges. We can consider pairs of positive and negative charges in the material as acting like the dipoles of the previous section. The action of the field of the isolated charges is to tend to align the dipole moments in the direction of the field. The material is then said to be polarized or distorted from its normal equilibrium state where the dipole alignment is random. The two charges making up the individual dipoles are now "bound" charges. All dielectrics have the ability to store energy when polarized, and in this sense the situation is analogous to a spring storing energy when distorted by an external force.

The electrostatic dipole is described in terms of its moment by equation 4.13, $\mathbf{p} = Q\mathbf{d}$. If we now recognize that the electric dipole is a source of strength or moment Qd, then a unit would have a moment $\mathbf{a}_d$, a unit vector directed along $\mathbf{d}$. Then the (macroscopic) unit impulse response for the potential of the dipole, from equation 4.14, is

$$h(\mathbf{r} - \mathbf{r}') = \frac{1}{4\pi\varepsilon_0} \frac{\mathbf{a}_d \cdot \mathbf{a}_R}{R^2} \tag{4.15}$$

Based on the discussion in the first paragraph of this section, the dipole is essentially a point source and we should be able to use equation 4.15 as its impulse response. We have faced this situation before. For example, we found the potential for a volume charge density from the potential of a unit point charge by way of the superposition integral. The same thing should be possible here if we employ the proper density function. Since the impulse source is dipole moment, the density function must be dipole moment per unit volume, and it must have dimensions C/m^2. We logically call this function the *polarization vector* $\mathbf{P}$. The superposition integral for the potential of a dielectric in terms of $\mathbf{P}$ is immediately obtained as

$$\Phi(\mathbf{r}) = \frac{1}{4\pi\varepsilon_0} \iiint_{\text{vol}'} \frac{\mathbf{P}(\mathbf{r}') \cdot \mathbf{a}_R}{R^2} \, dv' \tag{4.16}$$

In equation 4.16 we are treating $\mathbf{P}$ like any other field quantity, but it should be visualized as an *average* value taken over *many* molecules, and it has the effect of giving the dipoles a net alignment or polarization from their otherwise random arrangement. Notice that we continue to use ε_0, the permittivity of free space, since we are including the dielectric effects in the molecular dipoles.

In equation 3.21 we saw that

$$\frac{\mathbf{a}_R}{R^2} = \nabla' \frac{1}{R} = -\nabla \frac{1}{R}$$

Substituting this result in equation 4.16 we have

$$\Phi(\mathbf{r}) = \frac{1}{4\pi\varepsilon_0} \iiint_{\text{vol}'} \mathbf{P}(r') \cdot \nabla' \frac{1}{R}\, dv' \tag{4.17}$$

A vector identity, equation 3.54,

$$\mathbf{A} \cdot \nabla\alpha = \nabla \cdot (\alpha\mathbf{A}) - \alpha\nabla \cdot \mathbf{A}$$

is needed once more. Since it applies to primed coordinates or unprimed coordinates, its use in equation 4.17 with primed coordinates gives

$$\Phi(\mathbf{r}) = \frac{1}{4\pi\varepsilon_0} \iiint_{\text{vol}'} -\frac{\nabla' \cdot \mathbf{P}(\mathbf{r}')}{R}\, dv' + \frac{1}{4\pi\varepsilon_0} \iiint_{\text{vol}'} \nabla' \cdot \frac{\mathbf{P}(\mathbf{r}')}{R}\, dv' \tag{4.18}$$

Recall the divergence theorem,

$$\iiint_{\text{vol}'} \nabla' \cdot \mathbf{A}\, dv' = \oiint_{s'} \mathbf{A} \cdot d\mathbf{s}' = \oiint_{s'} \mathbf{A} \cdot \mathbf{a}_n\, ds'$$

where $\mathbf{a}_n$ is a unit vector normal to and out of s'. We apply it to the last term of equation 4.18. This gives

$$\Phi(\mathbf{r}) = \frac{1}{4\pi\varepsilon_0} \iiint_{\text{vol}'} \frac{-\nabla' \cdot \mathbf{P}(\mathbf{r}')}{R}\, dv' + \frac{1}{4\pi\varepsilon_0} \oiint_{s'} \frac{\mathbf{P}(\mathbf{r}')}{R} \cdot \mathbf{a}_n\, ds' \tag{4.19}$$

Now compare equation 4.19 to equations 3.17 and 3.18. This comparison is very helpful. It reveals that $-\nabla \cdot \mathbf{P}$ has the effect of an isolated volume charge density, while $\mathbf{P} \cdot \mathbf{a}_n$ has the effect of a (closed) surface charge density. On this basis, we are justified in writing

$$-\nabla \cdot \mathbf{P} = \rho_v^b \tag{4.20}$$

and

$$\mathbf{P} \cdot \mathbf{a}_n = \rho_s^b \tag{4.21}$$

These are the *bound* charges referred to previously.

We hope to obtain a new simple relation between **D** and **E**, like $\mathbf{D} = \varepsilon_0 \mathbf{E}$ (free space), but more general. Inside the dielectric material we have free space permittivity ε_0 and *both* bound charges and isolated charges. The point form of Gauss's law may be written in terms of **E** to show this explicitly.

$$\nabla \cdot \varepsilon_0 \mathbf{E} = \text{total volume charge density}$$

or

$$\nabla \cdot \varepsilon_0 \mathbf{E} = \rho_v + \rho_v^b \tag{4.22}$$

This equation is still correct even though a dielectric material is present. All of the charge has been accounted for, and it may therefore be treated as though it existed in free space. Equation 4.20 may be substituted into equation 4.22, giving

$$\nabla \cdot \varepsilon_0 \mathbf{E} = \rho_v - \nabla \cdot \mathbf{P}$$

or

$$\nabla \cdot (\varepsilon_0 \mathbf{E} + \mathbf{P}) = \rho_v \tag{4.23}$$

The right side of equation 4.23 is isolated charge density only, so we now have a way of eliminating bound charge density, leaving its effect in the new definition of the relation between **D** and **E**. We can still accept

$$\nabla \cdot \mathbf{D} = \rho_v \tag{4.24}$$

as being the correct form for the point version of Gauss's law. In this case, the new relation between **D** and **E** is obtained by comparing equation 4.23 and equation 4.24.

$$\boxed{\mathbf{D} = \mathbf{P} + \varepsilon_0 \mathbf{E}} \tag{4.25}$$

Essentially then, a new term, the polarization **P**, or dipole moment per unit volume is added when a dielectric or polarized material is present.

In order to be able to utilize equation 4.25 with the results of previous chapters, we must know the relation between the electric field, **E**, and the polarization, **P**, which results. As has been previously indicated, this relation will depend on the material. In anisotropic materials **E** and **P** are not parallel and are therefore not simply related. In *ferroelectric* materials, the relation between **P** and **E** is nonlinear and may also depend on the past history of the sample, indicating a hysteresis effect. In most engineering applications, the material is isotropic and linear (or, in the interest of simplicity, is assumed to be so) in which case **P** and **E** are parallel and simply related. This relationship is

$$\boxed{\mathbf{P} = \chi_E \varepsilon_0 \mathbf{E}} \tag{4.26}$$

where χ_E is the electric susceptibility of the material. A simpler relation results if we define

$$\chi_E \equiv \varepsilon_R - 1 \tag{4.27}$$

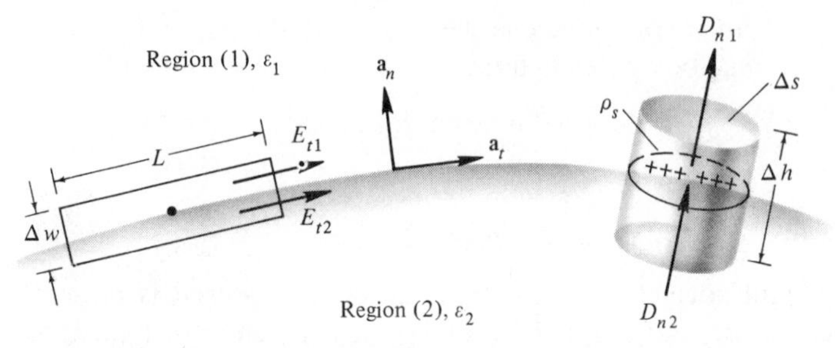

Figure 4.8. Geometry for determining boundary conditions, dielectric-dielectric interface.

where ε_R (dimensionless) is the *relative permittivity or dielectric constant* of the material. A table of values of ε_R for common materials is given in Appendix B.

Equation 4.26, for linear and isotropic materials may now be written

$$\mathbf{P} = (\varepsilon_R - 1)\varepsilon_0 \mathbf{E} \quad (\text{C/m}^2) \tag{4.28}$$

and from equation 4.25,

$$\boxed{\mathbf{D} = \varepsilon \mathbf{E}} \quad (\text{C/m}^2) \tag{4.29}$$

where

$$\boxed{\varepsilon = \varepsilon_0 \varepsilon_R} \quad (\text{F/m}) \tag{4.30}$$

The scalar constant ε is the permittivity of the material. Thus, for materials which are linear, isotropic, and homogeneous, the only change required in order to use previous equations is that of replacing ε_0 by $\varepsilon = \varepsilon_0 \varepsilon_R$. This is the desired result.

Polyethylene is a popular dielectric that finds much use in the construction of transmission lines. Its dielectric constant is approximately $\varepsilon_R \approx 2.25$, so that the permittivity we must use when working with this material is $\varepsilon = \varepsilon_0 \varepsilon_R = (2.25/36\pi) \times 10^{-9}$ F/m. Water has the much higher dielectric constant of about 80. Dielectrics also have loss, and will absorb energy when under the influence of time varying fields. The microwave oven is based on this fact. The behavior of dielectrics in the presence of sinusoidal time varying fields is discussed again in Chapter 13.

It should be mentioned, although we will pursue it no further here, that in an anisotropic material each component of **D** depends on *every* component of **E**. In using the relation $\mathbf{D} = \varepsilon \mathbf{E}$, ε is no longer a scalar constant, but is, in general, a nine-component quantity, called a *tensor*.

The boundary conditions for dielectrics are as easy to obtain as those for conductors. Consider Figure 4.8 where region (1) of permittivity ε_1 and region (2) of permittivity ε_2 are separated by an interface. This situation may be viewed as a special case of inhomogeneity. If we calculate the circulation, $\oint \mathbf{E} \cdot d\mathbf{l}$, around the small rectangle in Figure 4.8, we can neglect the contributions along the sides, Δw, from the normal component of the field because we will eventually let Δw approach zero. Then, for vanishingly small Δw, and for $\oint \mathbf{E} \cdot \mathbf{dl} = 0$, we have $E_{t1} L - E_{t2} L = 0$, or

$$E_{t1} = E_{t2} \tag{4.31a}$$

or more concisely,

$$\boxed{\mathbf{a}_n \times (\mathbf{E}_1 - \mathbf{E}_2) = 0} \tag{4.31b}$$

where $\mathbf{a}_n$ is a *unit normal vector* as shown in Figure 4.8. Thus, the *tangential* components of electric field are *continuous* across the interface. Now, since

$$\mathbf{D}_1 = \varepsilon_1 \mathbf{E}_1 \quad \text{and} \quad \mathbf{D}_2 = \varepsilon_2 \mathbf{E}_2$$

it follows immediately that

$$D_{t1} = \varepsilon_1 E_{t1}, \qquad D_{t2} = \varepsilon_2 E_{t2}$$

and

$$\frac{D_{t1}}{\varepsilon_1} = \frac{D_{t2}}{\varepsilon_2} \tag{4.32}$$

Thus, the *tangential* components of electric flux density are *discontinuous* across the interface.

The contributions to the flux out ($\mathbf{D} \cdot d\mathbf{s}$) of the side of the "pill box" in Figure 4.8 may be neglected if we let Δh approach zero. We are then left with the flux out the top and bottom due to the normal components of flux density. Gauss's law gives

$$D_{n1}\,\Delta S - D_{n2}\,\Delta S = \Delta Q = \rho_s\,\Delta S$$

or

$$D_{n1} - D_{n2} = \rho_s \tag{4.33a}$$

or more concisely,

$$\boxed{\mathbf{a}_n \cdot (\mathbf{D}_1 - \mathbf{D}_2) = \rho_s} \tag{4.33b}$$

The surface charge density ρ_s is some external charge somehow placed on the interface and is not *bound* charge, for this has already been accounted for when we use ε_1 and ε_2. In the *usual* case, this ρ_s is not present, and

$$D_{n1} = D_{n2}, \qquad (\rho_s = 0) \tag{4.34}$$

so the *normal* components of flux density are usually *continuous*. It follows immediately that

$$\frac{E_{n1}}{\varepsilon_2} = \frac{E_{n2}}{\varepsilon_1}, \qquad (\rho_s = 0) \tag{4.35}$$

and so the *normal* components of electric field intensity are usually *discontinuous*.

Notice that equation 4.31a or 4.31b will agree with equation 4.6 since $E_{t2} = D_{t2} = 0$ (Figure 4.3) in the conductor, and equation 4.33a or 4.33b will agree with equation 4.8 since $D_{n2} = E_{n2} = 0$ (Figure 4.4) in the conductor.

EXAMPLE 4

An example is in order. Suppose the $z = 0$ plane is the locus of a plane interface for which $z > 0$ has $\varepsilon_1 = 5\varepsilon_0$ and $z < 0$ has $\varepsilon_2 = 3\varepsilon_0$. If it is known that $\mathbf{E}_2 = 10\mathbf{a}_x + 20\mathbf{a}_z$, find $\mathbf{D}_2, \mathbf{D}_1$, and $\mathbf{E}_1$. We immediately obtain

$$\mathbf{D}_2 = \varepsilon_2 \mathbf{E}_2 = \varepsilon_0(30\mathbf{a}_x + 60\mathbf{a}_z)$$

The tangential **E** and normal **D** are continuous, so

$$E_{x1} = E_{t1} = E_{t2} = 10 \quad \text{and} \quad D_{z1} = D_{n1} = D_{n2} = +60\varepsilon_0$$

from equations 4.31 and 4.34. Also, from equations 4.32 and 4.35,

$$D_{x1} = \frac{\varepsilon_1}{\varepsilon_2} D_{x2} = \frac{5}{3}(30\varepsilon_0) = 50\varepsilon_0$$

and

$$E_{z1} = \frac{\varepsilon_2}{\varepsilon_1} E_{z2} = \frac{3}{5}(+20) = +12$$

Thus,

$$\mathbf{D}_1 = \varepsilon_0(50\mathbf{a}_x + 60\mathbf{a}_z)$$

and

$$\mathbf{E}_1 = 10\mathbf{a}_x + 12\mathbf{a}_z$$

This result is shown in Figure 4.9.

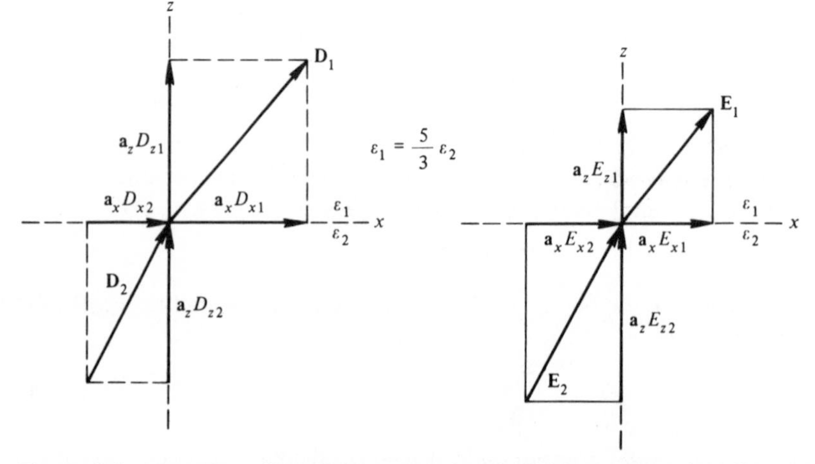

Figure 4.9. Resultant flux density and electric field at a plane interface.

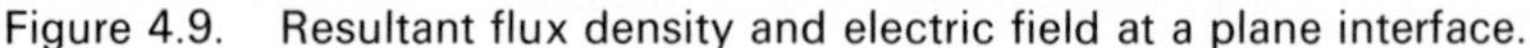

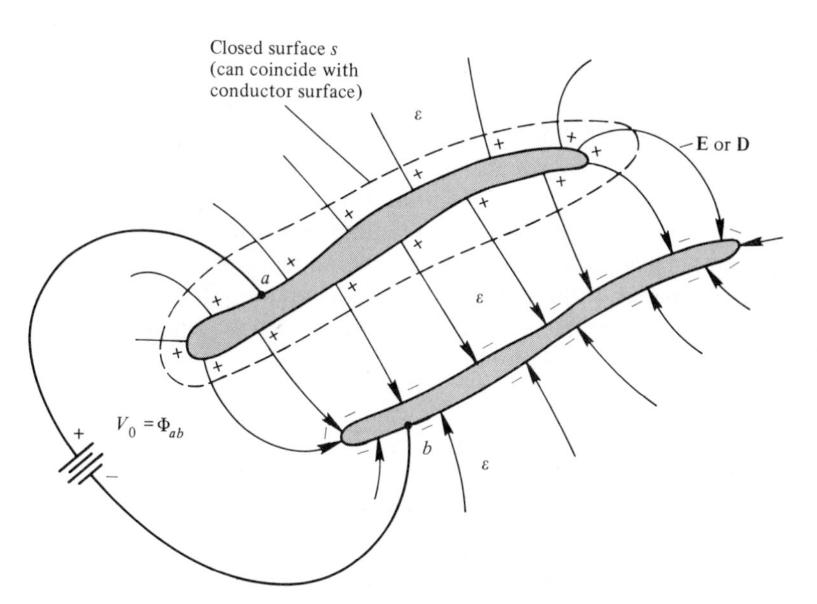

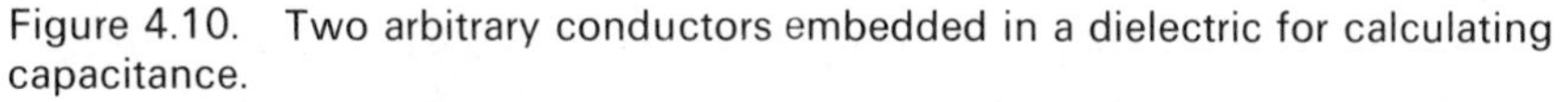
Figure 4.10. Two arbitrary conductors embedded in a dielectric for calculating capacitance.

4.4 CAPACITANCE

Consider two conductors of arbitrary shape surrounded by a dielectric as shown in Figure 4.10. A battery has been connected so that the conductors carry equal magnitude but opposite sign charges. These charges will appear as surface charges, and the electric field will be normal outward from the positively charged conductor and normal inward toward the negatively charged conductor (Why?). The capacitance of the system is defined as the ratio of the (total) charge on the positively charged conductor to the potential difference $\Phi_{ab} = \Phi_a - \Phi_b$. Point a is taken to be on the positively charged conductor so that Φ_{ab} is positive, making the capacitance, C, a positive quantity. Thus,

$$\boxed{C \equiv \frac{Q_a}{\Phi_{ab}}} \quad \text{(F)} \tag{4.36}$$

Gauss's law enables us to write equation 4.36 as

$$C = \frac{\oint\!\!\oint_s \varepsilon \mathbf{E} \cdot d\mathbf{s}}{\Phi_{ab}} = \frac{-\oint\!\!\oint_s \varepsilon \nabla \Phi \cdot d\mathbf{s}}{\Phi_{ab}} \tag{4.37}$$

Equation 4.37 is useful in calculating the capacitance when numerical techniques are employed to calculate the potential. It is also useful when resistance is considered in Chapter 6.

EXAMPLE 5

Returning to Example 7 in Section 3.5, where the problem of infinite, parallel conducting planes separated by a dielectric (air) h m thick was considered, we found that the potential difference was $V_0 = E_z h$ and $E_z = \rho_s/\varepsilon_0$. Thus, $V_0 = \rho_s h/\varepsilon_0$. (See Figure 3.14.) If we consider now a *finite* parallel plane system where the linear dimensions of the planes are large compared to the separation d, we have the system shown in Figure 4.11. This is *not* the same geometry with which we began, but the

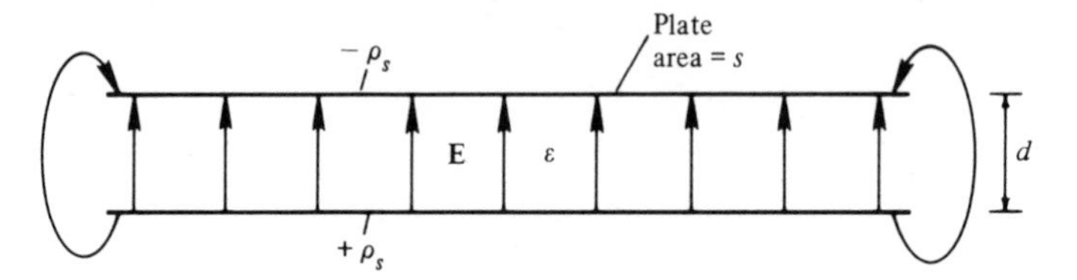

Figure 4.11. Finite, parallel conducting planes.

changes are only slight. The fringing (How can we show that the field "fringes"?) of the field at the edges may be safely neglected as playing a minor role in the overall picture under the given conditions. The permittivity of the dielectric within the plates is $\varepsilon = \varepsilon_R \varepsilon_0$, not ε_0, but we found in the preceding section that for a simple dielectric our previously developed equations are all valid if we merely replace ε_0 by ε. Thus, the capacitance is given by the well-known formula ($h = d$)

$$C = \frac{Q}{\Phi_{ab}} = \frac{\rho_s s}{\rho_s d/\varepsilon} = \frac{\varepsilon s}{d} \quad \text{(F)} \tag{4.38}$$

The energy stored in the capacitor is

$$W_E = \frac{1}{2} \iiint_{\text{vol}} \varepsilon E^2 \, dv = \frac{\varepsilon}{2} \int_0^s \int_0^d \frac{\rho_s^2}{\varepsilon^2} \, dz \, ds$$

or

$$W_E = \frac{1}{2} \frac{\rho_s^2}{\varepsilon} sd$$

This may be arranged into more familiar (circuit) forms to give

$$W_E = \frac{1}{2} C V_0^2 = \frac{1}{2} Q V_0 = \frac{1}{2} \frac{Q^2}{C}$$

where $V_0 = \Phi_{ab}$ is the potential difference.

EXAMPLE 6

Because of the inherent symmetry (which we previously utilized) we can very easily obtain the capacitance of a spherical capacitor, consisting of concentric conducting spheres separated by a dielectric. Recall Faraday's experiment with Gauss's law. Using equation 3.9, with ε_0 replaced by ε,

$$C = \frac{Q}{(Q/4\pi\varepsilon)(1/a - 1/b)}$$

or

$$C = \frac{4\pi\varepsilon}{1/a - 1/b} \quad \text{(F)} \tag{4.39}$$

EXAMPLE 7

The coaxial cable is a very popular shielded transmission line consisting of coaxial conducting cylinders. The inner conductor has a radius of a m while the outer conductor has an inner radius of b m as shown in Figure 4.12. The cable has been charged by connecting the positive

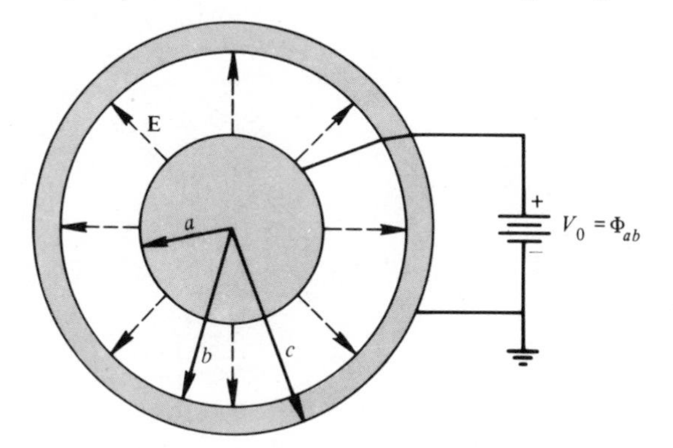

Figure 4.12. Coaxial, conducting cylinders.

terminal of a battery with potential V_0 to the inner conductor and connecting the negative (grounded) terminal to the outer conductor. The total charge on the inner conductor will be $Q = 2\pi a L \rho_{sa}$ for a length L, while the total charge on the outer conductor will be $-Q = +2\pi b L \rho_{sb}$. The charge densities ρ_{sa} and ρ_{sb} will be uniform if the cable is very long and end effects are neglected. Symmetry arguments are then exactly the same as those used for the uniform filamentary line charge density on the z axis (investigated in Chapter 1). They show that $\mathbf{E} = \mathbf{a}_\rho E_\rho(\rho)$. Using a coaxial cylinder of radius ρ as a Gaussian surface shows that $E_\rho = 0$ for $0 \leq \rho < a$ and for $\rho > b$. Thus, the coaxial cable is a shielded system. For $a < \rho < b$, Gauss's law gives

$$\Psi_E = D_\rho \int_0^L \int_0^{2\pi} \rho \, d\phi \, dz = Q_{enc} = 2\pi a L \rho_{sa}$$

$$D_\rho 2\pi\rho L = 2\pi a L \rho_{sa}$$

and

$$D_\rho = \frac{a}{\rho} \rho_{sa}$$

Therefore,

$$E_\rho = \frac{a\rho_{sa}}{\rho\varepsilon}$$

if the dielectric has a permittivity ε.

The potential is given by

$$\Phi = -\int E_\rho \, d\rho + C_1 = \frac{a\rho_{sa}}{\varepsilon} \int -\frac{d\rho}{\rho} + C_1$$

and

$$\Phi = -\frac{a\rho_{sa}}{\varepsilon} \ln \rho + C_1$$

Since $\Phi = 0$ when $\rho = b$, we have

$$\Phi = -\frac{a\rho_{sa}}{\varepsilon} \ln \frac{\rho}{b} = \frac{a\rho_{sa}}{\varepsilon} \ln \frac{b}{\rho}$$

and since $\Phi = V_0 = \Phi_{ab}$ when $\rho = a$, we have

$$\Phi_{ab} = \frac{a\rho_{sa}}{\varepsilon} \ln \frac{b}{a}$$

The capacitance is

$$C = \frac{Q}{\Phi_{ab}} = \frac{2\pi a L \rho_{sa}}{(a\rho_{sa}/\varepsilon) \ln (b/a)}$$

or

$$C = \frac{2\pi\varepsilon L}{\ln (b/a)} \quad \text{(F)} \tag{4.40}$$

For a coax with $a = 0.2 \times 10^{-2}$ m, $b = 10^{-2}$ m, and a polyethylene dielectric ($\varepsilon_R = 2.25$), equation 4.40 gives a value of 77.7×10^{-12} F/m, or 77.7 pF/m (picafarads per meter length). This is a typical value for many commercially available cables. The results for the field quantities $E_\rho(\rho)$ and $\Phi(\rho)$ inside the coax will be considered again in the next chapter. They are

$$E_\rho(\rho) = \frac{a\rho_{sa}}{\varepsilon}\left(\frac{1}{\rho}\right), \qquad a \leq \rho \leq b \tag{4.41}$$

and

$$\Phi(\rho) = -\frac{a\rho_{sa}}{\varepsilon}\ln\frac{\rho}{b}, \qquad a \leq \rho \leq b \tag{4.42}$$

EXAMPLE 8

Another practical geometry is the two-wire line. Let us approach this problem by first finding the potential of two (filamentary) line charges as shown in Figure 4.13. We have shown in Chapter 3 that the potential

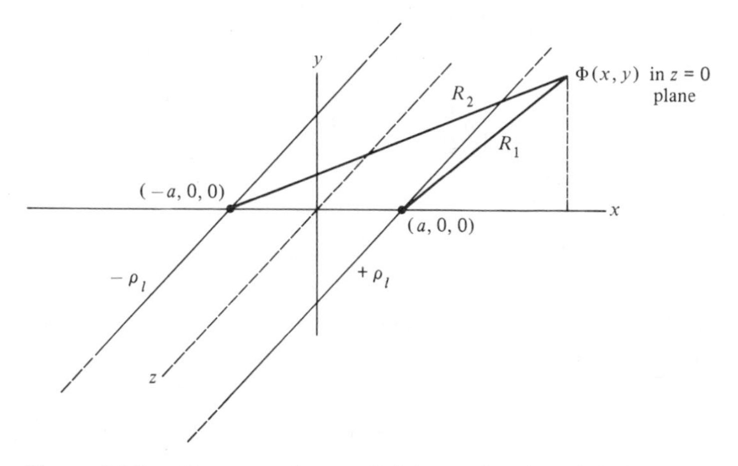

Figure 4.13. Geometry for parallel (opposite sign) infinite line charges.

of an infinite line charge on the z axis is $\Phi(\rho) = -(\rho_l/2\pi\varepsilon)\ln\rho + C$, so picking a zero reference at $\rho = b$ gives $\Phi = -(\rho_l/2\pi\varepsilon)\ln(\rho/b)$. In terms of Figure 4.13, the potential of *both* line charges is by superposition

$$\Phi = -\frac{\rho_l}{2\pi\varepsilon}\left[\ln\frac{R_1}{b_1} - \ln\frac{R_2}{b_2}\right] = -\frac{\rho_l}{2\pi\varepsilon}\ln\frac{R_1 b_2}{R_2 b_1}$$

where b_1 and b_2 are constants. In order to obtain a symmetrical balanced system, we would like the potential to be zero in the $x = 0$ plane.

Inspection of the preceding equation reveals that this is accomplished if $b_1 = b_2$. In this case,

$$\Phi = -\frac{\rho_l}{2\pi\varepsilon}\ln\frac{R_1}{R_2} = -\frac{\rho_l}{2\pi\varepsilon}\ln\left[\frac{(x-a)^2+y^2}{(x+a)^2+y^2}\right]^{1/2}$$

or

$$\Phi = \frac{\rho_l}{4\pi\varepsilon}\ln\frac{(x+a)^2+y^2}{(x-a)^2+y^2} \tag{4.43}$$

We next need to know what the equipotential surfaces are. These are not too difficult to determine if we use an intermediate step. Inspection of equation 4.43 reveals that the potential is constant, $\Phi = V_0$, if the argument of the logarithm is a constant, called C_1. Thus,

$$C_1 = \frac{(x+a)^2+y^2}{(x-a)^2+y^2} \tag{4.44}$$

The preceding equation can be rearranged to

$$\left[x - a\frac{C_1+1}{C_1-1}\right]^2 + y^2 = \left[\frac{2a\sqrt{C_1}}{C_1-1}\right]^2$$

Thus, equipotentials are circular cylinders whose axes are located at

$$\left[a\frac{C_1+1}{C_1-1}, 0\right]$$

and whose radii are

$$\frac{2a\sqrt{C_1}}{C_1-1}, \qquad \text{for } x > 0$$

Let x_0, 0 be the center of the equipotential and r_0 be the radius of the equipotential cylinder. Thus,

$$x_0 = a\frac{C_1+1}{C_1-1} \tag{4.45}$$

and

$$r_0 = \frac{2a\sqrt{C_1}}{C_1-1}$$

If we solve for the intermediate parameter C_1, or more precisely, $\sqrt{C_1}$, in the last two equations, we obtain

$$\sqrt{C_1} = \frac{x_0 + \sqrt{x_0^2 - r_0^2}}{r_0}$$

Solving for a in the same equations gives

$$a = \sqrt{x_0^2 - r_0^2} \tag{4.46}$$

Thus from equations 4.43, 4.44, and 4.45 we have

$$V_0 = \frac{\rho_l}{4\pi\varepsilon} \ln C_1 = \frac{\rho_l}{2\pi\varepsilon} \ln \sqrt{C_1}$$

or

$$V_0 = \frac{\rho_l}{2\pi\varepsilon} \ln \frac{x_0 + \sqrt{x_0^2 - r_0^2}}{r_0} \tag{4.47}$$

The situation for $x < 0$ can be determined from symmetry. Figure 4.14 shows the equipotentials for a typical situation.

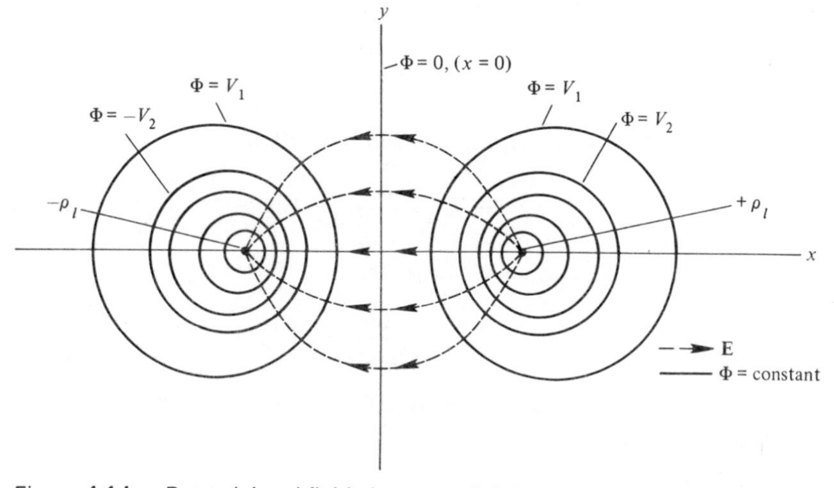

Figure 4.14. Potential and field about parallel (opposite sign) infinite line charges.

We are now able to find the capacitance for a two-wire line of conductors having a center-to-center spacing of $D = 2x_0$ and diameter $d = 2r_0$ as shown in Figure 4.15. The potential difference is $V_0 - (-V_0) = 2V_0$, or

$$2V_0 = \frac{\rho_l}{\pi\varepsilon} \ln \frac{x_0 + \sqrt{x_0^2 - r_0^2}}{r_0} \tag{4.48}$$

Figure 4.15. Charged two-wire line.

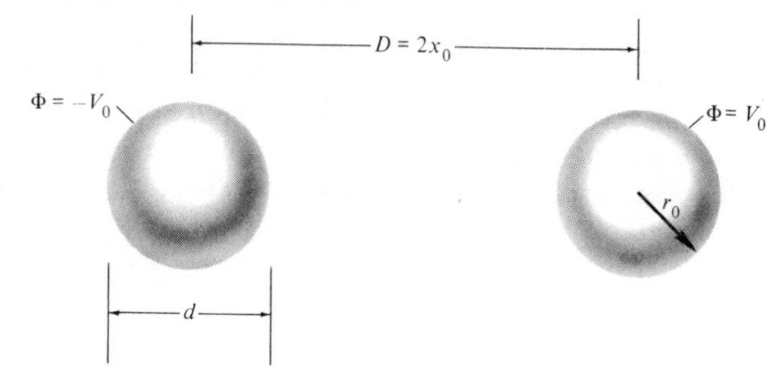

The charge on the conductor for which $\Phi = V_0$ is $\rho_l L$ (for a length L in the z direction) as can either be determined by Gauss's law, or the boundary condition on the normal component of **D**. Therefore,

$$C = \frac{\rho_l L}{(\rho_l/\pi\varepsilon) \ln [(x_0 + \sqrt{x_0^2 - r_0^2})/r_0]}$$

or

$$C = \frac{\pi\varepsilon L}{\ln [(x_0 + \sqrt{x_0^2 - r_0^2})/r_0]} \tag{4.49}$$

This can be written more compactly.

$$C = \frac{\pi\varepsilon L}{\cosh^{-1} (x_0/r_0)} \tag{4.50}$$

In terms of D and d equation 4.50 can be written

$$C = \frac{\pi\varepsilon L}{\cosh^{-1} (D/d)} \quad \text{(F)} \tag{4.51}$$

If $D \gg d$, then equation 4.49 shows that

$$C \approx \frac{\pi\varepsilon L}{\ln (2D/d)} \quad \text{(F)} \tag{4.52}$$

The capacitance of single wire of radius r_0 and length L located a distance h above a ground plane is

$$C = \frac{2\pi\varepsilon L}{\cosh^{-1} (h/r_0)} \approx \frac{2\pi\varepsilon L}{\ln (2h/r_0)} \quad \text{(F)} \tag{4.53}$$

Why?

Several things about the two-wire line are worth mentioning. First, the mathematics would have been more elegant (and less understandable) had we been working in *bicylindrical coordinates*, for in this system the equipotentials are orthogonal surfaces where *one* of the variables is constant. This system, in fact, allows us to calculate the capacitance of the two-wire line with equal or unequal conductor radii, the coax and the coax with an offset inner conductor,[1] all with one equation! Secondly, the two-wire line is inherently a balanced line and is normally used with balanced loads such as a dipole antenna. It can be operated as a shielded system if it is encased in, and insulated from, a grounded conductor. The equations we just derived do not apply in this case, however. The formulas (for C) for the geometries mentioned in this paragraph can be found in handbooks.

[1] See Problem 24.

A two-wire line with $D = 10^{-2}$ m and $d = 0.2 \times 10^{-2}$ m in air has a capacitance of 12.12 pF/m by equation 4.51. This is considerably less than the comparable coax of the previous example, but we included no dielectric (other than air) in our calculation, and all two-wire lines must be mechanically supported by a dielectric of some type. This will increase the capacitance.

EXAMPLE 9

Let us next consider a geometry whose capacitance takes a form that is probably familiar to us from circuit theory. This geometry is what is commonly called two capacitors in parallel. Figure 4.16 shows a parallel

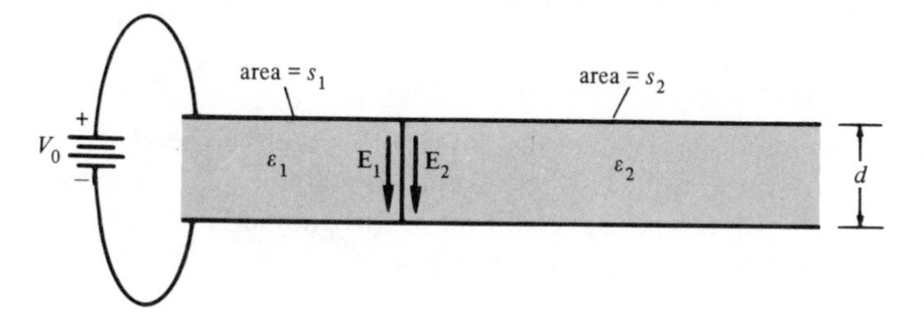

Figure 4.16. Two-dielectric (parallel) capacitor.

plate capacitor partitioned into two sections. If we apply known boundary conditions, the solution to this problem is not too difficult to obtain. At the interface between the two dielectrics, the tangential electric field is continuous so, $E_1 = E_2$. Thus,

$$\frac{D_1}{\varepsilon_1} = \frac{D_2}{\varepsilon_2} \quad \text{or} \quad \frac{\rho_{s1}}{\varepsilon_1} = \frac{\rho_{s2}}{\varepsilon_2}$$

In terms of the charges on the upper plate,

$$\frac{Q_1}{s_1\varepsilon_1} = \frac{Q_2}{s_2\varepsilon_2}, \qquad Q = Q_1 + Q_2 = E_1(s_1\varepsilon_1 + s_2\varepsilon_2)$$

Since $E_1 = E_2 = V_0/d$, where V_0 is the potential difference, we have

$$V_0 = E_1 d$$

Thus, the capacitance is

$$C = \frac{s_1\varepsilon_1 + s_2\varepsilon_2}{d} = \frac{s_1\varepsilon_1}{d} + \frac{s_2\varepsilon_2}{d} \quad \text{(F)} \tag{4.54}$$

A comparison of equation 4.54 with equation 4.38 reveals that we have the effect of $C = C_1 + C_2$.

EXAMPLE 10

We have (effectively) two capacitors in series in Figure 4.17. Whether

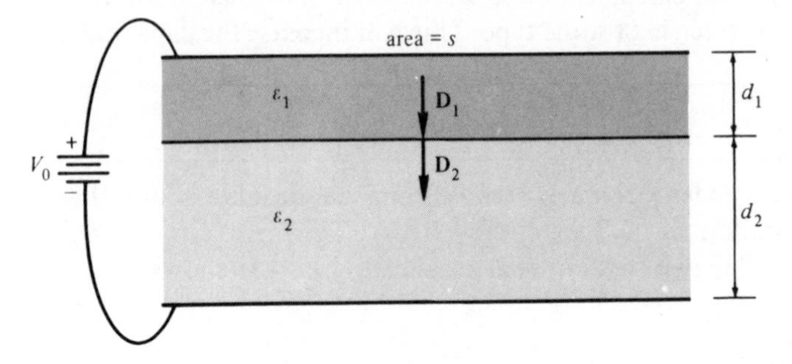

Figure 4.17. Two-dielectric (series) capacitor.

there is, or is not, a *thin* conducting plane at the interface between the dielectrics is unimportant insofar as the capacitance is concerned (Why?). The known boundary condition in this geometry is $D_1 = D_2$, or $\varepsilon_1 E_1 = \varepsilon_2 E_2$. Now, it is easy to show that

$$V_0 = V_1 + V_2 = E_1 d_1 + E_2 d_2$$

It is also true that the total charge on the upper conductor is $Q = s\rho_s = sD_1 = s\varepsilon_1 E_1$. Therefore,

$$C = \frac{s\varepsilon_1 E_1}{E_1 d_1 + E_2 d_2} = \frac{s\varepsilon_1 E_1}{\varepsilon_1(E_1 d_1/\varepsilon_1 + E_1 d_2/\varepsilon_2)}$$

or

$$C = \frac{1}{d_1/\varepsilon_1 s + d_2/\varepsilon_2 s} \quad \text{(F)} \tag{4.55}$$

which is effectively

$$C = \frac{1}{1/C_1 + 1/C_2} = \frac{C_1 C_2}{C_1 + C_2}$$

It should be emphasized that fringing effects have been neglected in the preceding problems. This was done in the case of the simple parallel plate capacitor of Example 5, and should always be considered as being reasonable when the thickness of the capacitor is small compared to its other dimensions.

4.5 CONCLUSIONS

In this chapter we have investigated the electrical properties of conductors and simple dielectrics under static conditions. The charges were always fixed. A somewhat oversimplified picture of dielectric behavior showed that for most cases we simply need to replace ε_0 with ε and use the equations already developed. For more complex media, other modifications are necessary. With this information, the boundary conditions at the interface between different media are easily obtained. These boundary conditions will be used in a more direct way in the next chapter. Capacitance was defined and determined for some practical geometries. The capacitance depends on the geometry and the permittivity of the *material* surrounding the conducting bodies. This parameter is useful in circuit theory as well as the transmission line theory which appears in later chapters of this text.

REFERENCES

Dekker, A. J. *Electrical Engineering Materials*. Englewood Cliffs, N.J.: Prentice-Hall, 1959. A short and readable book on the electrical properties of materials, including dielectrics.

Fink, D. G. and Carroll, J. M. *Standard Handbook for Electrical Engineers*, 10th ed. New York: McGraw-Hill, 1968.

Matsch, L. W. *Capacitors, Magnetic Circuits, and Transformers*. Prentice-Hall, 1964. Capacitors are discussed in Chapter 2.

Ramo, S., Whinnery, J. R. and Van Duzer, T. *Fields and Waves in Communications Electronics*. New York: Wiley, 1965. Primarily a graduate level textbook, a discussion of polarization paralleling that herein is included in Chapter 2.

PROBLEMS

1. Prove that the field inside a closed hollow conducting surface is zero when exposed to an external field. There are no charges inside the closed surface before the external field is applied. This is electrostatic shielding.
2. Prove that a closed hollow conducting surface does not shield its exterior from charges inside.
3. The potential on a closed conducting surface is $\Phi = V_0$. There are no charges inside. What is the potential inside the surface? Use the results of Problem 1.
4. The electric field at a point on a conductor surface is given by $\mathbf{E} = 0.3\mathbf{a}_x + 0.4\mathbf{a}_y$.

   (a) Find a unit vector normal to the surface at the point.
   (b) What is the magnitude of the surface charge density at the point?

5. Find the dielectric constant for a material for which the flux density is three times the polarization.
6. The region $z > 0$ is free space, while the region $z < 0$ has $\varepsilon_R = 9$. The uniform electric field for $z > 0$ is 10 V/m and in a direction $\theta = 30°$ and $\phi = 30°$. Find $\mathbf{D}$ and $\mathbf{E}$ everywhere in rectangular coordinates (see Figure 4.18).

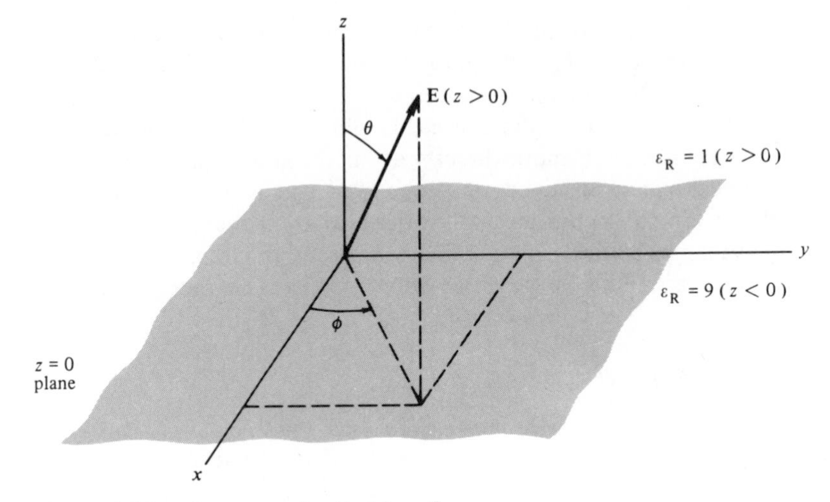

Figure 4.18. Geometry for Problem 6.

7. A dielectric sphere of radius a and dielectric constant ε_R is centered at the origin. An electric field, $\mathbf{E} = E_0 \mathbf{a}_z$, *without the dielectric sphere*, is applied. Show that a solution for the potential is

$$\Phi(r, \theta) = \begin{cases} -\dfrac{3rE_0 \cos\theta}{\varepsilon_R + 2}, & r \le a \\ -rE_0 \cos\theta + \dfrac{a^3 E_0}{r^2} \dfrac{\varepsilon_R - 1}{\varepsilon_R + 2} \cos\theta, & r \ge a \end{cases}$$

8. Find the electric field, $a \le \rho \le b$, for a coaxial cable by finding the potential first. What determines the maximum potential difference, $\Phi_a - \Phi_b$?
9. Find the capacitance per unit length of a coaxial cable if $a = 0.5$ cm, $b = 2$ cm, and $\varepsilon_R = 1$. Repeat if the dielectric is polyethylene.
10. The capacitance per unit length of a two-(round) wire line is 30 pF/m and the center-to-center conductor spacing is 2 cm. What is the wire radius? $\varepsilon_R = 1$.
11. A transmission line is made of parallel conducting strips 2 cm wide and spaced 0.5 cm apart. What is the capacitance per unit length, approximately? $\varepsilon_R = 4$.
12. A parallel plate capacitor has plates of area 10^{-2} m^2 spaced 10^{-2} m apart. The relative permittivity varies linearly from 1 to 11. Find the capacitance.
13. An air dielectric parallel plate capacitor is charged. Find the energy required to separate the plates by an additional amount. What is the force of attraction between the plates?
14. A parallel plate capacitor has square plates and a solid dielectric and is charged. If the dielectric is withdrawn a small amount, find the force on the dielectric.

15. A parallel plate capacitor is charged with a battery and the plate separation is increased by a factor of 3 with the battery still connected. Determine by what factor each of the following factors is altered: (a) Φ_d, (b) C, (c) E, (d) D, (e) Q, (f) ρ_s, and (g) W_E.
16. Repeat the previous problem if the battery is disconnected before the spacing is increased.
17.

$|\mathbf{E}_1| = 10$ v/m

30°

(1) $\varepsilon_1 = \varepsilon_0$

(2) $\varepsilon_2 = 4\varepsilon_0$

10 cm

(3) $\varepsilon_3 = \varepsilon_0$

Figure 4.19. A pair of parallel, infinite interfaces.

Find $\mathbf{E}_2$ and $\mathbf{E}_3$ for the geometry in Figure 4.19.

18. Find the capacitance of the two-dielectric coaxial cable shown in Figure 4.20.

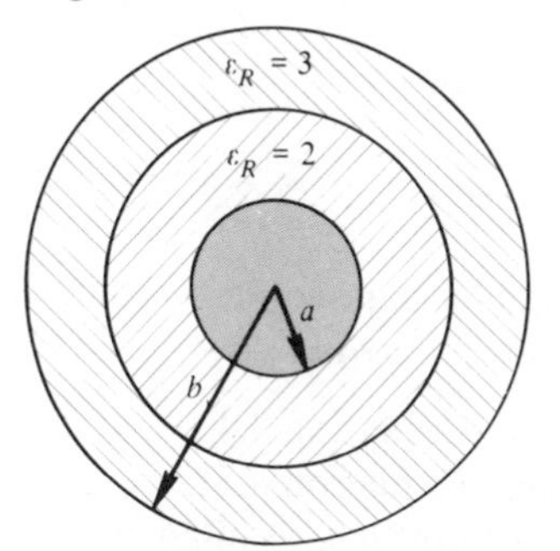

Figure 4.20. A two-dielectric coaxial cable. Inner conductor radius $a = 0.5$ cm. Outer conductor radius $b = 1.5$ cm. Inner dielectric radius = 1 cm.

19. A dipole $\mathbf{p} = Qd\mathbf{a}_z$ is located symmetrically on the z axis. An electric field, $\mathbf{E} = -\mathbf{a}_y E_0$, is applied. Find the energy required to rotate the dipole to a stable position. Hint: Show that $dW_E = |\mathbf{T}|\, d\theta$.
20. As a crude model of the earth and a layer of charged clouds above, consider a pair of large parallel conducting plates with the lower plate (earth) at zero potential and the upper plate (cloud) at some negative potential.

    (a) Sketch equipotentials and electric field lines.
    (b) Let a conducting cone with a small apex angle rest on the lower plane and extend upward so that the apex is about midway

between the two parallel planes. Sketch equipotentials and electric field lines.

(c) Explain how a lightning rod works.

21. Verify that $\mathbf{E} = -\nabla\Phi$ and $\nabla \cdot \mathbf{E} = \rho_v/\varepsilon$ for the coaxial cable of Example 7. Use unit step functions to help express $\mathbf{E}$ and Φ.
22. For the two-wire line of Figure 4.15 show that

(a) $$\Phi = \frac{V_0}{2\cosh^{-1} D/d} \ln \frac{(x+a)^2 + y^2}{(x-a)^2 + y^2}$$

(b) $$E_x = \frac{2V_0 a}{\cosh^{-1} D/d} \frac{x^2 - a^2 - y^2}{[(x+a)^2 + y^2][(x-a)^2 + y^2]}$$

(c) $$E_y = \frac{4V_0 a}{\cosh^{-1} D/d} \frac{xy}{[(x+a)^2 + y^2][(x-a)^2 + y^2]}$$

23. If $D = 3 \times 10^{-2}$ m, $d = 2 \times 10^{-2}$ m, and $2V_0 = 100$ V, find E_x at (0, 100) for Problem 22.
24. For two parallel circular hollow conductors of radii r_1 and r_2 ($r_1 < r_2$) and center-to-center spacing D the capacitance is

$$C = \frac{2\pi\varepsilon L}{\sinh^{-1} a/r_1 \pm \sinh^{-1} a/r_2} \quad \text{(F)}$$

where

$$a = \frac{1}{2D}[D^4 - 2D^2(r_2^2 + r_1^2) + (r_2^2 - r_1^2)^2]^{1/2}$$

and the $-$ sign applies if the smaller conductor is within the larger conductor, but the $+$ sign applies otherwise.

(a) Show that this result applies to equation 4.40.
(b) Show that this result applies to equation 4.51.
(c) Calculate the capacitance per unit length of an eccentric cable if $r_1 = 1$ cm, $r_2 = 2$ cm, and $D = 0.5$ cm (air). See Figure 4.21.

Figure 4.21. Geometry for Problem 24.

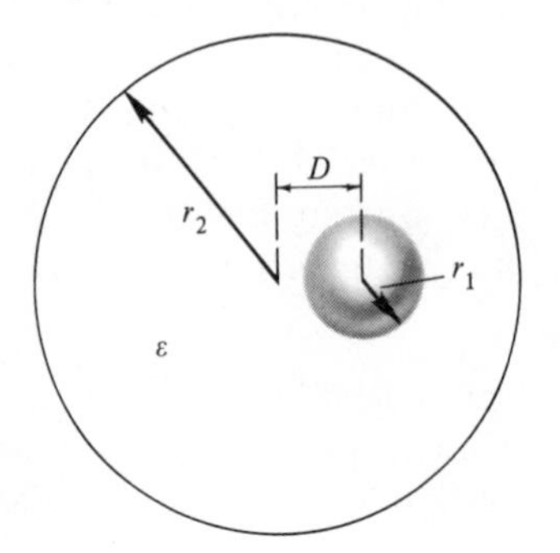

25. (a) Show that

$$E_\rho = \frac{\Phi_{ab}}{\ln b/a} \frac{1}{\rho}$$

for the coaxial cable.

(b) If $a = 1$ cm, $b = 2$ cm, and $\varepsilon = \varepsilon_0$, what maximum voltage can be used with the coaxial cable? Assume that air breaks down for $E > 3 \times 10^6$ V/m.

Chapter 5
Boundary Value Problems

We have solved electrostatic problems in the preceding chapters by many methods. It may come as a surprise, after spending four chapters with these problems to learn that there are other methods of solution still available to us. Of the remaining methods, we will only consider two because of their fundamental importance. The methods still remaining are also important, but are beyond the scope of the treatment.

The equations of the previous chapters lead almost immediately to Laplace's equation and Poisson's equation. These equations are second order partial differential equations, and are important in other areas of engineering and physics. As mentioned earlier, the steady-state diffusion equation for the temperature in a heat conducting body is Laplace's equation. Gravitational potential can be formulated in the same manner and is another example. We already have methods to solve these equations, but in this chapter we wish to treat them as a special class of problem called a *boundary value problem.* In a boundary value problem the potential (or perhaps its normal derivative) is specified over a given boundary and the potential is to be found within the boundary.

5.1 POISSON'S AND LAPLACE'S EQUATIONS

We have shown in the preceding chapter that

$$\mathbf{D} = \varepsilon \mathbf{E} \tag{5.1}$$

We also have the differential equation (point) form of Gauss's law,

$$\nabla \cdot \mathbf{D} = \rho_v \tag{5.2}$$

and the equation for **E** in terms of the potential,

$$\mathbf{E} = -\nabla \Phi \tag{5.3}$$

Substituting equation 5.1 into equation 5.2 gives

$$\nabla \cdot \mathbf{E} = \frac{\rho_v}{\varepsilon} \tag{5.4}$$

if ε is a scalar constant[1] (simple dielectric). Substituting equation 5.3 into equation 5.4,

$$\nabla \cdot \nabla \Phi = -\frac{\rho_v}{\varepsilon} \tag{5.5}$$

Since the operator ∇ is given by

$$\nabla \equiv \mathbf{a}_x \frac{\partial}{\partial x} + \mathbf{a}_y \frac{\partial}{\partial y} + \mathbf{a}_z \frac{\partial}{\partial z}$$

it is easy to show that

$$\nabla \cdot \nabla \Phi \equiv \nabla^2 \Phi = -\frac{\rho_v}{\varepsilon} = \frac{\partial^2 \Phi}{\partial x^2} + \frac{\partial^2 \Phi}{\partial y^2} + \frac{\partial^2 \Phi}{\partial z^2} \tag{5.6}$$

in Cartesian coordinates. For completeness we have

$$\nabla^2 \Phi = \frac{1}{\rho} \frac{\partial}{\partial \rho}\left(\rho \frac{\partial \Phi}{\partial \rho}\right) + \frac{1}{\rho^2}\left(\frac{\partial^2 \Phi}{\partial \phi^2}\right) + \frac{\partial^2 \Phi}{\partial z^2} \tag{5.7}$$

in circular cylindrical coordinates, and

$$\nabla^2 \Phi = \frac{1}{r^2} \frac{\partial}{\partial r}\left(r^2 \frac{\partial \Phi}{\partial r}\right) + \frac{1}{r^2 \sin \theta} \frac{\partial}{\partial \theta}\left(\sin \theta \frac{\partial \Phi}{\partial \theta}\right) + \frac{1}{r^2 \sin^2 \theta} \frac{\partial^2 \Phi}{\partial \phi^2} \tag{5.8}$$

in spherical coordinates. ∇^2 is called the *Laplacian* operator.

Equation 5.5 or 5.6 is *Poisson's* equation of mathematical physics,

$$\boxed{\nabla^2 \Phi = -\frac{\rho_v}{\varepsilon}} \qquad \text{(Poisson's equation)} \tag{5.9}$$

[1] See Problem 6 and Section 5.4.

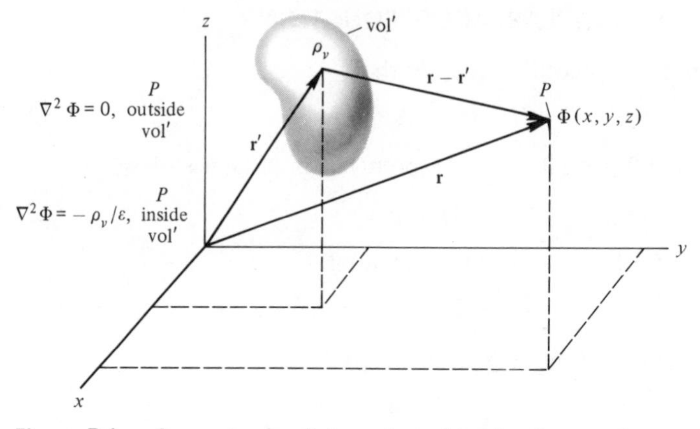

Figure 5.1. Geometry for Poisson's and Laplace's equation.

If this differential equation is homogeneous; that is, if the *region of interest* is free of *excess charge*, we have *Laplace's* equation of mathematical physics,

$$\boxed{\nabla^2\Phi = 0} \qquad \text{(Laplace's equation)} \tag{5.10}$$

We now inquire as to what problem these equations apply. Consider Figure 5.1 where a region containing a perfectly general volume charge density is shown. We already have, in fact, a solution to the problem of finding the potential at some general point P. This solution is the superposition integral, equation 3.17, with ε_0 replaced by ε (also called the Helmholtz integral, potential integral, or convolution integral),

$$\Phi(\mathbf{r}) = \frac{1}{4\pi\varepsilon} \iiint\limits_{\text{vol}'} \frac{\rho_v(\mathbf{r}')}{R}\, dv' \tag{5.11}$$

Evidently then, equation 5.11 is a solution to Laplace's equation, equation 5.10, if *the point P is not included in the charge bearing region* (vol′). If P is included in the region containing the charge, then equation 5.11 is a solution to Poisson's equation, equation 5.9. It is not necessary for a treatment at this point to *prove* the truth of the preceding statements, but this proof is included in Appendix G for the interested reader.

Armed with a solution to Laplace's and Poisson's equation, one might wonder why we should not stop here. The integrations in equation 5.11 may be difficult to perform in many cases, even if ρ_v is known. In many problems ρ_v is not even known, but other conditions on the boundary of some closed surface may be known. It is this latter problem, properly called a boundary value problem, in which we are primarily interested here.

Let us now consider solutions to Laplace's equation. As a first step, we should inquire as to what given information is *necessary* in order to actually obtain a solution. Secondly, we should inquire as to whether or not our solution is the *only* solution. Both of these questions are answered by a *uniqueness theorem* which we shall now prove.

Laplace's equation, equation 5.10, is $\nabla^2\Phi = 0$. Let us assume there are *two* solutions to this differential equation. Then,

$$\nabla^2\Phi_1 = 0$$

and (5.12)

$$\nabla^2\Phi_2 = 0$$

if the two solutions are Φ_1, and Φ_2, respectively. It is obviously true that

$$\nabla^2(\Phi_1 - \Phi_2) = 0 \tag{5.13}$$

since Φ_1 and Φ_2 represent solutions to Laplace's equation in a charge-free region. This region is *bounded* by some closed surface s. We next need to use the vector identity, equation 3.54,

$$\mathbf{A} \cdot \nabla\alpha \equiv \nabla \cdot (\alpha\mathbf{A}) - \alpha\nabla \cdot \mathbf{A}$$

which we integrate throughout the volume bounded by s.

$$\iiint_{\text{vol}} \mathbf{A} \cdot \nabla\alpha \, dv = \iiint_{\text{vol}} \nabla \cdot (\alpha\mathbf{A}) \, dv - \iiint_{\text{vol}} \alpha\nabla \cdot \mathbf{A} \, dv \tag{5.14}$$

The first term on the right can be changed to a surface integral by means of the divergence theorem.

$$\iiint_{\text{vol}} \mathbf{A} \cdot \nabla\alpha \, dv = \oint\!\!\oint_{s} \alpha\mathbf{A} \cdot d\mathbf{s} - \iiint_{\text{vol}} \alpha\nabla \cdot \mathbf{A} \, dv \tag{5.15}$$

We now specify that

$$\mathbf{A} \equiv \nabla(\Phi_1 - \Phi_2) \quad \text{and} \quad \alpha \equiv \Phi_1 - \Phi_2$$

Then,

$$\iiint_{\text{vol}} \nabla(\Phi_1 - \Phi_2) \cdot \nabla(\Phi_1 - \Phi_2) \, dv = \oint\!\!\oint_{s} (\Phi_1 - \Phi_2)\nabla(\Phi_1 - \Phi_2) \cdot d\mathbf{s}$$

$$- \iiint_{\text{ool}} (\Phi_1 - \Phi_2)\nabla^2(\Phi_1 - \Phi_2) \, dv \tag{5.16}$$

The last term in equation 5.16 is zero because of equation 5.13; so,

$$\iiint_{\text{vol}} |\nabla(\Phi_1 - \Phi_2)|^2 \, dv = \oint\!\!\oint_{s} (\Phi_1 - \Phi_2)\nabla(\Phi_1 - \Phi_2) \cdot d\mathbf{s}$$

$$= \oint\!\!\oint_{s} (\Phi_1 - \Phi_2)\frac{\partial(\Phi_1 - \Phi_2)}{\partial n} \, ds \tag{5.17}$$

since $\nabla\alpha \cdot d\mathbf{s} = (\partial\alpha/\partial n)\, ds$, where $\partial/\partial n$ is the normal (outward) derivative at s.[2] On the closed boundary surface s we have $\Phi_1 = \Phi_2$, for we certainly cannot have *different boundary conditions* in one problem. Hence, the right side of equation 5.17 is zero. Thus,

$$\iiint_{\text{vol}} |\nabla(\Phi_1 - \Phi_1)|^2 \, dv = 0 \tag{5.18}$$

The integrand of equation 5.18 is always positive, so the *only* way equation 5.18 can be satisfied is for

$$\nabla(\Phi_1 - \Phi_2) \equiv 0 \tag{5.19}$$

Equation 5.19 requires that

$$\Phi_1 - \Phi_2 = \text{constant} \tag{5.20}$$

and since $\Phi_1 = \Phi_2$ on s, this constant must be zero everywhere. Thus,

$$\Phi_1 = \Phi_2 \tag{5.21}$$

everywhere *inside* and *on* s.

We have answered the second part of our original inquiry. Once we have found a solution to Laplace's equation (in a region) which fits the boundary conditions, it is the *only* solution. This solution may be obtained by *any* method whatsoever.

What information is necessary in order to find a solution? The answer to this question is more subtle. We originally stated that the right side of equation 5.17 was zero because $\Phi_1 = \Phi_2$ on s. This is true, but this term may be zero for other reasons also. A closer inspection of equation 5.17 reveals that the term in question is zero (and the proof thus valid) if

$$\left.\begin{aligned}
&\Phi_1 = \Phi_2 \text{ on } s \\
&\text{or} \\
&\frac{\partial\Phi_1}{\partial n} = \frac{\partial\Phi_2}{\partial n} \text{ on } s \\
&\text{or} \\
&\Phi_1 = \Phi_2 \text{ on } s_1, \qquad \frac{\partial\Phi_1}{\partial n} = \frac{\partial\Phi_2}{\partial n} \text{ on } s_2, \qquad \text{where } s = s_1 + s_2
\end{aligned}\right\} \tag{5.22}$$

The reader can verify that the results of the preceding paragraphs apply to Poisson's equation as well as to Laplace's equation. Thus, if (1) we specify the potential on a closed surface (Dirichlet boundary conditions), or if (2) we specify $\partial\Phi/\partial n$ (normal derivative) on a closed surface (Neumann boundary conditions), or, finally, if we specify Φ on part of the closed surface and $\partial\Phi/\partial n$

[2] Refer to the comments on the directional derivative in Section 3.4.

on the remaining part of the closed surface (mixed boundary conditions), we can have a unique solution. Thus, these are the conditions necessary to find a unique solution.

5.2 ONE-DIMENSIONAL SOLUTIONS

As a first example, we will solve a problem we have already solved.

EXAMPLE 1

Consider the parallel plate capacitor shown in Figure 5.2. The battery

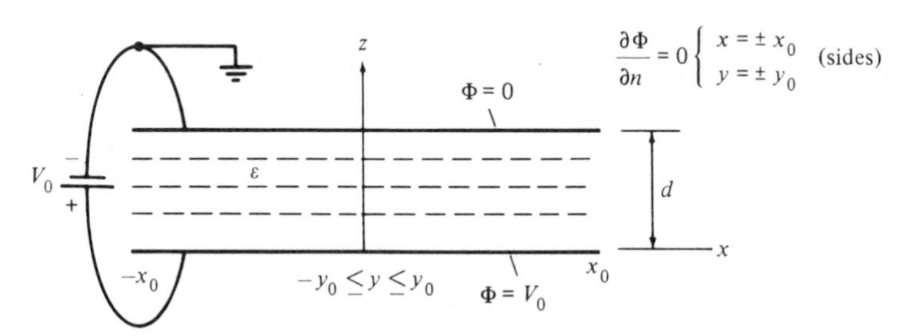

Figure 5.2. Geometry for finding the potential of a parallel plate capacitor, neglecting fringing.

and ground connections assure that the potential on both plates is known. We know from the uniqueness theorem, however, that more than one solution can be found which satisfies Laplace's equation between the plates and the boundary conditions on the plates! The reason for this lack of uniqueness, is, of course, that we have not yet specified boundary conditions over a *closed* surface. In solving this problem previously we *assumed* no fringing of the fields at the four sides. In order to avoid solving the very difficult problem of a real capacitor of finite dimensions, and to obtain the same solution as before, we will retain the original assumption. The sides are located at $= \pm x_0$ and $y = \pm y_0$. In order to have no fringing, the normal derivatives of the potential at the sides must be zero, therefore

$$\frac{\partial \Phi}{\partial x} = 0 \text{ at } x = \pm x_0 \quad \text{and} \quad \frac{\partial \Phi}{\partial y} = 0 \text{ at } y = \pm y_0 \tag{5.23}$$

The other boundary conditions are

$$\Phi = 0, \qquad z = d \tag{5.24}$$

and

$$\Phi = V_0, \qquad z = 0 \tag{5.25}$$

The boundary conditions are thus *mixed*.

Laplace's equation is

$$\nabla^2\Phi = \frac{\partial^2\Phi}{\partial x^2} + \frac{\partial^2\Phi}{\partial y^2} + \frac{\partial^2\Phi}{\partial z^2} = 0$$

In order to satisfy equations 5.23, we require $\partial\Phi/\partial x = \partial\Phi/\partial y = 0$ for *all* x and y. This reduces the problem to a one-dimensional version, and Laplace's equation becomes

$$\nabla^2\Phi = \frac{\partial^2\Phi}{\partial z^2} = \frac{d^2\Phi}{dz^2} = 0 \tag{5.26}$$

The solution is easily obtained by two successive integrations as

$$\Phi = Az + B$$

and from equation 5.25, $B = V_0$. Using equation 5.24,

$$0 = Ad + V_0$$

or

$$A = -\frac{V_0}{d}$$

Then,

$$\Phi = -V_0\frac{z}{d} + V_0 = V_0\left(1 - \frac{z}{d}\right), \qquad 0 \le z \le d$$

or

$$\Phi = V_0 u(-z) + V_0\left(1 - \frac{z}{d}\right)[u(z) - u(z-d)] \tag{5.27}$$

as can be seen in Figure 5.3. The electric field is

$$\mathbf{E} = -\nabla\Phi = -\mathbf{a}_z\frac{\partial\Phi}{\partial z} = \mathbf{a}_z\frac{V_0}{d}[u(z) - u(z-d)] \tag{5.28}$$

Figure 5.3. Potential and volume charge density for the parallel plate capacitor of Figure 5.2.

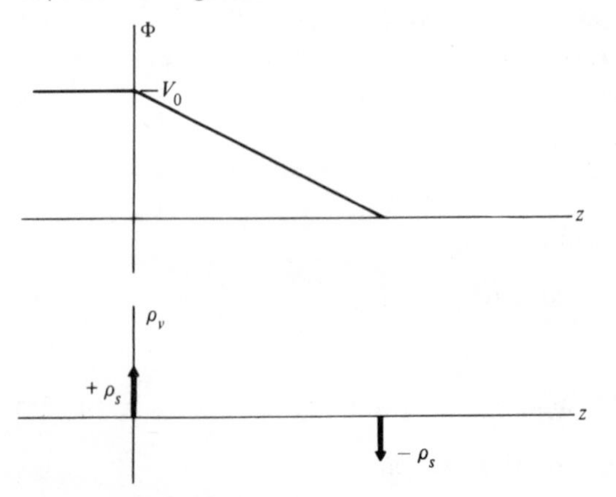

so

$$D_z = +\frac{\varepsilon V_0}{d} = \rho_s, \qquad (z = 0^+) \tag{5.29}$$

Therefore, as before,

$$C = \frac{\rho_s S}{V_0} = \frac{\varepsilon S}{d} \quad \text{(F)} \tag{5.30}$$

It is convenient at this point to demonstrate that Poisson's equation holds. The Laplacian operator gives

$$\nabla^2 \Phi = \frac{d^2\Phi}{dz^2} = \frac{d}{dz}\left\{\frac{d\Phi}{dz}\right\} = \frac{d}{dz}\left\{-\frac{V_0}{d}[u(z) - u(z-d)]\right\}$$

where equation 5.28 has been utilized for $d\Phi/dz$. Therefore,

$$\nabla^2 \Phi = -\left(\frac{V_0}{d}\right)[\delta(z) - \delta(z-d)]$$

Eliminating V_0 in the last equation by means of equation 5.29 gives

$$\nabla^2 \Phi = -\frac{1}{\varepsilon}[\rho_s\,\delta(z) - \rho_s\,\delta(z-d)] \tag{5.31}$$

The bracketed term in equation 5.31 is the correct volume charge density as can be seen in Figure 5.3.

EXAMPLE 2

As another example, consider the uniformly charged conducting sphere of radius a. Let the potential on the sphere be V_0. From the inherent symmetry, we have $\partial/\partial\theta = \partial/\partial\phi = 0$. From equation 5.8,

$$\nabla^2 \Phi = \frac{1}{r^2}\frac{\partial}{\partial r}\left(r^2 \frac{\partial \Phi}{\partial r}\right) = \frac{1}{r^2}\frac{d}{dr}\left(r^2 \frac{d\Phi}{dr}\right) = 0 \tag{5.32}$$

Thus,

$$r^2 \frac{d\Phi}{dr} = C_1 \quad \text{or} \quad \frac{d\Phi}{dr} = \frac{C_1}{r^2}$$

so

$$\Phi = -\frac{C_1}{r} + C_2 \tag{5.33}$$

We require that the potential vanish at infinity, so $C_2 = 0$. The potential on the sphere at $r = a$ is V_0, so,

$$V_0 = -\frac{C_1}{a} \quad \text{or} \quad C_1 = -aV_0$$

or

$$\Phi = \frac{V_0 a}{r}, \qquad r \geq a \tag{5.34}$$

or

$$\Phi = V_0 u(a - r) + V_0 u(r - a)a/r$$

and

$$\begin{aligned}\mathbf{E} = -\nabla\Phi &= -\mathbf{a}_r \frac{\partial \Phi}{\partial r} \\ &= -\mathbf{a}_r V_0\{-\delta(r - a) + a[-u(r - a)/r^2 + \delta(r - a)/r]\} \\ &= -\mathbf{a}_r V_0\{-\delta(r - a) - au(r - a)/r^2 + \delta(r - a)\}\end{aligned}$$

or

$$E_r = \frac{aV_0}{r^2} u(r - a) \tag{5.35}$$

Now,

$$\rho_s = D_r\big|_{r=a^+} = \frac{\varepsilon V_0}{a}, \qquad D_r\big|_{r=a^-} = 0$$

so that the total charge on the surface $r = a^+$ is

$$Q = 4\pi a^2 \rho_s = 4\pi a \varepsilon V_0 \tag{5.36}$$

and

$$C = 4\pi a \varepsilon \quad \text{(F)} \tag{5.37}$$

which agrees with equation 4.39 with $b \to \infty$. What were the boundary conditions?

Now, from equation 5.35 and the equation preceding it,

$$\frac{d\Phi}{dr} = -V_0 au(r - a)/r^2$$

$$r^2 \frac{d\Phi}{dr} = -V_0 au(r - a)$$

$$\frac{d}{dr}\left(r^2 \frac{d\Phi}{dr}\right) = -V_0 a\, \delta(r - a)$$

So Poisson's equation becomes

$$\nabla^2\Phi = -V_0\left(\frac{a}{r^2}\right) \delta(r - a) = -\left(\frac{V_0}{a}\right) \delta(r - a)$$

Eliminating V_0 by means of equation 5.36 gives

$$\nabla^2 \Phi = -\frac{1}{\varepsilon}[\rho_s\, \delta(r-a)] \tag{5.38}$$

and the bracketed term is ρ_v.

EXAMPLE 3

For another example, let us reconsider the coaxial cable of Figure 4.12. Here, we assume the simplest solution for $\partial/\partial\phi = \partial/\partial z = 0$. Laplace's equation is thus

$$\nabla^2 \Phi = \frac{1}{\rho}\frac{\partial}{\partial \rho}\left(\rho \frac{\partial \Phi}{\partial \rho}\right) = \frac{1}{\rho}\frac{d}{d\rho}\left(\rho \frac{d\Phi}{d\rho}\right) = 0 \tag{5.39}$$

Then,

$$\rho \frac{d\Phi}{d\rho} = C_1 \quad \text{or} \quad \frac{d\Phi}{d\rho} = \frac{C_1}{\rho}$$

so

$$\Phi = C_1 \ln \rho + C_2 \tag{5.40}$$

The boundary conditions are

$$\Phi = 0, \qquad \rho = b \tag{5.41}$$

and

$$\Phi = V_0, \qquad \rho = a \tag{5.42}$$

Then,

$$C_2 = -C_1 \ln b$$

so

$$\Phi = C_1 \ln \rho - C_1 \ln b = C_1 \ln \frac{\rho}{b}$$

Applying the second boundary condition,

$$V_0 = C_1 \ln\left(\frac{a}{b}\right) \quad \text{or} \quad C_1 = \frac{V_0}{\ln(a/b)}$$

so that,

$$\Phi = \frac{V_0}{\ln(a/b)} \ln\left(\frac{\rho}{b}\right), \qquad a \le \rho \le b$$

or

$$\Phi = \frac{V_0}{\ln (a/b)} \ln \left(\frac{\rho}{b}\right) [u(\rho - a) - u(\rho - b)] + V_0 u(a - \rho) \quad (5.43)$$

The electric field is

$$\mathbf{E} = -\nabla\Phi = -\mathbf{a}_\rho \frac{\partial \Phi}{\partial \rho}$$

or

$$E_\rho = -\frac{V_0}{\rho \ln (a/b)} [u(\rho - a) - u(\rho - b)] \quad (5.44)$$

and

$$\rho_{sa} = D_\rho \big|_{\rho = a^+} = \frac{-\varepsilon V_0}{a \ln (a/b)} = \frac{\varepsilon V_0}{a \ln (b/a)}$$

while

$$\rho_{sb} = -D_\rho \big|_{\rho = b^-} = \frac{\varepsilon V_0}{b \ln (a/b)} = \frac{-\varepsilon V_0}{b \ln (b/a)}$$

For a length L, the capacitance is

$$C = \frac{2\pi\varepsilon L}{\ln b/a} \quad \text{(F)} \quad (5.45)$$

which agrees with equation 4.40.

Then,

$$\nabla^2\Phi = \frac{1}{\rho}\frac{d}{d\rho}\left(\rho \frac{d\Phi}{d\rho}\right)$$

Using equation 5.44 for $d\Phi/d\rho$,

$$\begin{aligned}
\nabla^2\Phi &= \frac{1}{\rho}\frac{d}{d\rho}\left\{\frac{V_0}{\ln (a/b)} [u(\rho - a) - u(\rho - b)]\right\} \\
&= \frac{1}{\rho}\frac{V_0}{\ln (a/b)} [\delta(\rho - a) - \delta(\rho - b)] \\
&= \frac{V_0}{\ln (a/b)} \left[\frac{\delta(\rho - a)}{a} - \frac{\delta(\rho - b)}{b}\right]
\end{aligned}$$

Eliminating V_0/a and V_0/b in the last equation by means of the equations below equation 5.44 gives

$$\nabla^2\Phi = -\frac{1}{\varepsilon} [\rho_{sa}\, \delta(\rho - a) + \rho_{sb}\, \delta(\rho - b)] \quad (5.46)$$

Again, we can identify $\rho_v = \rho_{sa}\,\delta(\rho - a) + \rho_{sb}\,\delta(\rho - b)$ as the correct volume charge density. Remember that $\rho_{sa} = -\rho_{sb}(b/a)$ so that the total charge per unit length is the same in magnitude, but opposite in sign, on each cylinder (see Figure 5.4).

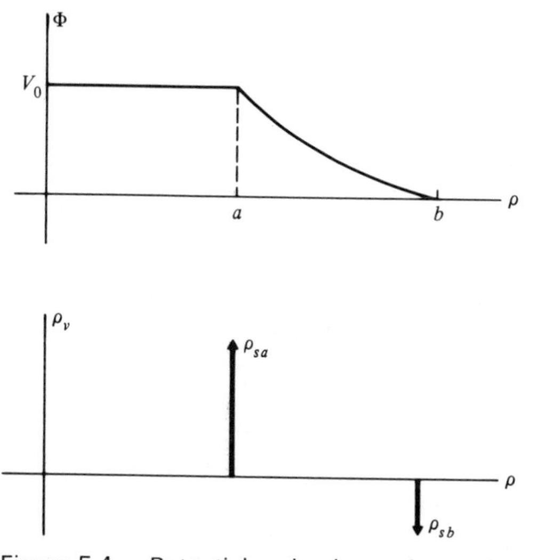

Figure 5.4. Potential and volume charge density for the coaxial cable of Figure 4.12.

The examples so far have been straightforward, principally because they were all one-dimensional. One might expect that the two-wire line would be easily solved if we followed the same procedure as in the preceding examples. Inspection of the geometry of the two-wire line reveals that none of the three common coordinate systems are appropriate. As pointed out in Chapter 4, the most useful coordinate system here is the bicylindrical one, and, indeed in this system, the problem can be solved by the same methods we have previously used. Since most of us are not familiar with bicylindrical coordinates,[3] we shall not follow that course here.

5.3 A TWO-DIMENSIONAL EXAMPLE OF A BOUNDARY VALUE PROBLEM USING ANALYTICAL METHODS

EXAMPLE 4

Let us now attempt to solve the *two*-dimensional potential problem shown in Figure 5.5. We desire the potential inside the rectangle or very

[3] See Moon and Spencer in the References at the end of the chapter.

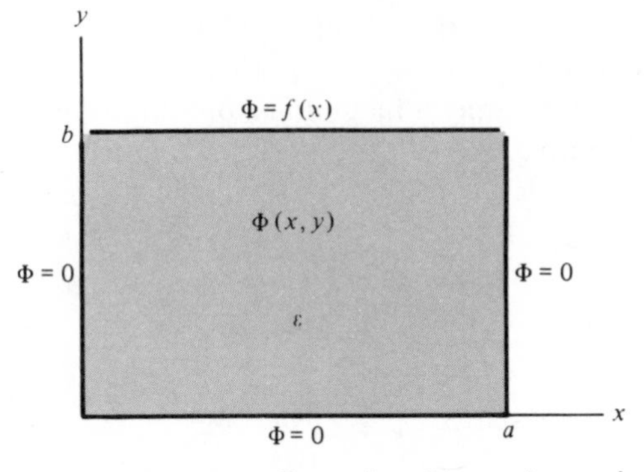

Figure 5.5. Two-dimensional boundary value problem (potential).

long trough shown if $\partial/\partial z = 0$. The boundary conditions are

$$\begin{aligned} \Phi &= 0, & x &= 0 \\ \Phi &= 0, & x &= a \\ \Phi &= 0, & y &= 0 \\ \Phi &= f(x), & y &= b \end{aligned} \tag{5.47}$$

Laplace's equation is

$$\nabla^2\Phi = \frac{\partial^2\Phi}{\partial x^2} + \frac{\partial^2\Phi}{\partial y^2} = 0$$

Notice that even though the problem is two-dimensional, it is still somewhat a special case in that there are no charges enclosed and ε is a constant.

The question now is: how do we solve this partial differential equation of *two* variables? Up to this point, our differential equations have contained only one variable, and we were able to obtain a solution by simple integration. This technique is no longer possible. Fortunately, a standard technique for solving Laplace's equation in 11 standard coordinate systems is available to us. Our three common coordinate systems (as well as the bicylindrical one mentioned above) are included in this list. This technique is called the method of *separation of variables.* For this method we assume a solution in the form of a product. Each term in the product is assumed to be a function of only *one* of the variables. Thus, in the present problem we assume

$$\Phi(x, y) = X(x)Y(y) \tag{5.48}$$

Substituting equation 5.48 into Laplace's equation,

$$0 = Y\frac{d^2X}{dx^2} + X\frac{d^2Y}{dy^2} \tag{5.49}$$

Division of equation 5.49 by $\Phi = XY$ gives

$$\frac{1}{X}\frac{d^2X}{dx^2} = -\frac{1}{Y}\frac{d^2Y}{dy^2} \tag{5.50}$$

The left side of equation 5.50 is a function of x only, while the right side is a function of y only. Under these conditions, each side must be a constant. We choose

$$\frac{1}{X}\frac{d^2X}{dx^2} = -\alpha^2 \tag{5.51}$$

and

$$\frac{1}{Y}\frac{d^2Y}{dy^2} = +\alpha^2 \tag{5.52}$$

The particular choice of a constant made here is only a matter of convenience, as will be seen.

We now have the pair of *ordinary* differential equations,

$$\frac{d^2X}{dx^2} + \alpha^2 X = 0 \tag{5.53}$$

and

$$\frac{d^2Y}{dy^2} - \alpha^2 Y = 0 \tag{5.54}$$

The solutions to these well-known equations are

$$X = A\cos\alpha x + B\sin\alpha x \tag{5.55}$$

and

$$Y = C\cosh\,\alpha y + D\sinh\alpha y \tag{5.56}$$

That is, the variation is sinusoidal in one direction and exponential in the other and these variations can be swapped by merely changing the signs in front of α^2 in equations 5.51 and 5.52. The original choice was made because we require functions of x which have at least two zeros to satisfy the boundary conditions. The exponential (or hyperbolic) form cannot do this, and using a little foresight, we picked the correct sign initially to avoid changing it later. Then,

$$\Phi(x, y) = (A\cos\alpha x + B\sin\alpha x)(C\cosh\alpha y + D\sinh\alpha y) \tag{5.57}$$

We now return to the boundary conditions of equations 5.47, where we notice that the first of these will be satisfied if $A \equiv 0$. (Why?) Then,

$$\Phi(x, y) = \sin\alpha x\,(BC\cosh\alpha y + BD\sinh\alpha y) \tag{5.58}$$

The second boundary condition will be satisfied if $\alpha = n\pi/a$, for sin $(n\pi x/a)$ is zero for $x = 0$ and $x = a$ (if $n = 1, 2, 3, \ldots$). Thus, we require

$$\alpha = \frac{n\pi}{a}, \qquad n = 1, 2, 3, \ldots \tag{5.59}$$

Notice that $n = 0$ gives a trivial ($\Phi \equiv 0$) solution. We now have

$$\Phi(x, y) = \sin\frac{n\pi x}{a}\left(E\cosh\frac{n\pi y}{a} + F\sinh\frac{n\pi y}{a}\right) \tag{5.60}$$

where $E \equiv BC$ and $F \equiv BD$. The α given by equation 5.59 are called *eigenvalues* or *characteristic values.*

The third boundary condition will be satisfied if $E \equiv 0$ (Why?), giving

$$\Phi(x, y) = F\sin\frac{n\pi x}{a}\sinh\frac{n\pi y}{a} \tag{5.61}$$

It is worthwhile at this point to go back and make sure that equation 5.61 satisfies Laplace's equation and the first three boundary conditions. It does. We are left with one boundary condition and one unknown constant (F). It is useful to take a long look at this boundary condition.

$$\Phi(x, b) = f(x) \tag{5.62}$$

We cannot simply substitute $y = b$ into equation 5.61 and then equate equation 5.61 to $f(x)$. In other words, *it is not true in general that*

$$f(x) = \left[F\sinh\frac{n\pi b}{a}\right]\sin\frac{n\pi x}{a}$$

because the left side may be any function of x, while the right side is a known function of x for each n. The difficulty here occurs because we have not used all the freedom available to us in equation 5.61. In other words, a more general solution than equation 5.61 is a superposition,

$$\Phi(x, y) = \sum_{n=1}^{\infty} F_n\sin\frac{n\pi x}{a}\sinh\frac{n\pi y}{a} \tag{5.63}$$

since a solution exists for each $n = 1, 2, 3, \ldots$. Equation 5.63 represents a superposition of solutions, each one of which is in the form of equation 5.61 but containing a different eigenvalue. The right side of equation 5.63 is an infinite series of characteristic functions or *eigenfunctions.*[4] If we *now* use the last boundary condition, we obtain

$$f(x) = \sum_{n=1}^{\infty}\left[F_n\sinh\frac{n\pi b}{a}\right]\sin\frac{n\pi x}{a} \tag{5.64}$$

[4] Each eigenfunction can have its own amplitude insofar as equation 5.63 is concerned, hence a subscript is used on F.

or

$$f(x) = \sum_{n=1}^{\infty} B_n \sin \frac{n\pi x}{a} \tag{5.65}$$

with

$$B_n = F_n \sinh \frac{n\pi b}{a} \tag{5.66}$$

or

$$F_n = \frac{B_n}{\sinh (n\pi b/a)} \tag{5.67}$$

Equation 5.65 should be familiar, because its right side has the form of a *Fourier trigonometric series*! In particular, it is a series of sine terms (odd functions), and can represent an *odd* periodic function with period $2a$. See Figure 5.6. The coefficients, B_n, for such a series are given by the

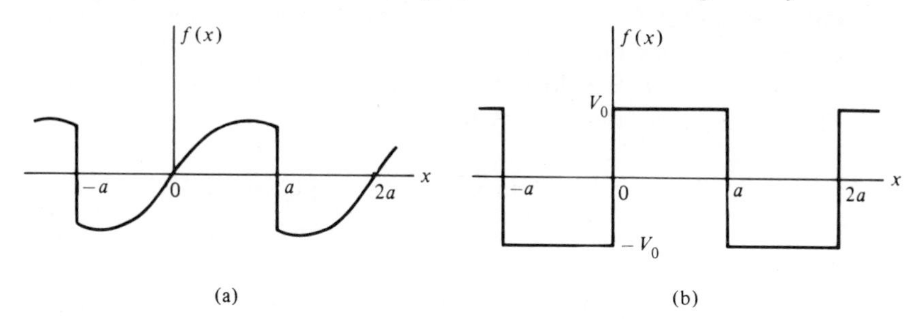

Figure 5.6. (a) An odd periodic function of x. (b) The particular odd $f(x)$ to represent V_0 in $0 < x < a$.

Euler formulas. Using Appendix D, equations D-2 and D-3, with $L = 2a$, we obtain $A_n \equiv 0$ [since $f(x)$ must be expanded as an *odd* function], and

$$B_n = \frac{1}{a} \int_{-a}^{a} f(x) \sin \left(\frac{n\pi x}{a} \right) dx \tag{5.68}$$

or, since $f(x) \sin (n\pi x/a)$ is an *even* function (Why?)

$$B_n = \frac{2}{a} \int_{0}^{a} f(x) \sin \left(\frac{n\pi x}{a} \right) dx \tag{5.69}$$

Our formal solution, equation 5.63, with equation 5.67 and equation 5.69 is

$$\Phi(x, y) = \sum_{n=1}^{\infty} \left[\frac{2}{a} \int_{0}^{a} f(x) \sin \left(\frac{n\pi x}{a} \right) dx \right] \frac{\sinh (n\pi y/a)}{\sinh (n\pi b/a)} \sin \left(\frac{n\pi x}{a} \right) \tag{5.70}$$

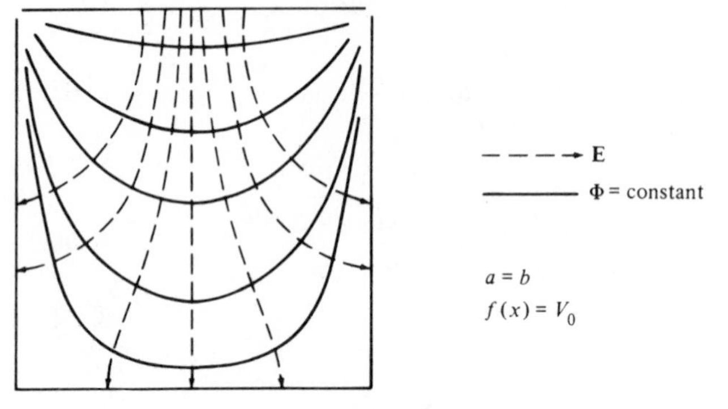

Figure 5.7. Equipotentials and electric field for a particular two-dimensional boundary value problem.

For a particular example, let $f(x) = V_0$ in $0 < x < a$ for Figure 5.5. That is, the two sides and bottom are grounded, while the top is insulated from the sides and raised to a potential V_0. The $f(x)$ we *must* use is shown in Figure 5.6(b). Notice that $f(x)$ will always be periodic when expanded in a Fourier trigonometric series, but we are only interested in its value (that is, V_0) in the interval $0 < x < a$. We do not care what values it assumes outside this interval. Using equation 5.69, we have

$$B_n = \frac{2V_0}{a}\int_0^a \sin\frac{n\pi x}{a}\,dx = \frac{2V_0}{n\pi}[1-(-1)^n]$$

so, from equation 5.70,

$$\Phi(x, y) = \frac{2V_0}{\pi}\sum_{n=1}^{\infty}[1-(-1)^n]\frac{\sinh(n\pi y/a)}{n\sinh(n\pi b/a)}\sin\frac{n\pi x}{a} \tag{5.71}$$

Suppose $a = b$ and we desire the potential at $x = y = a/2$. Then,

$$\Phi\left(\frac{a}{2}, \frac{a}{2}\right) = \frac{2V_0}{\pi}\sum_{n=1}^{\infty}[1-(-1)^n]\frac{\sinh(n\pi/2)}{n\sinh n\pi}\sin\frac{n\pi}{2} \tag{5.72}$$

or

$$\Phi\left(\frac{a}{2}, \frac{a}{2}\right) = \frac{2V_0}{\pi}(0.398 - 0.006 + \cdots)$$

or since the infinite series in equation 5.72 converges to $\pi/8$, we obtain the *exact* result

$$\Phi\left(\frac{a}{2}, \frac{a}{2}\right) = \frac{V_0}{4} \tag{5.73}$$

This particular series converges very rapidly. A sketch of the equipotentials and electric field lines is shown in Figure 5.7. Notice that the electric field lines and lines of constant potential are perpendicular (Why?) How would you calculate the charge on the boundaries? Is the total charge on the top plate finite?

The example worked out in this section was chosen to demonstrate the method of separation of variables in rectangular coordinates with which most of us are familiar. In other coordinate systems, we may meet other functions, such as the Bessel function and the Legendre function for circular cylindrical and spherical coordinates, respectively. If the problem is one-dimensional or two-dimensional, we may only encounter familiar functions such as the natural logarithm. Several of the latter type problems are given at the end of this chapter, but the former are reserved for more advanced texts and are not treated here.

5.4 A NUMERICAL METHOD FOR SOLVING BOUNDARY VALUE PROBLEMS

The boundary value problem in which charge is enclosed, or in which ε is not a constant, is very difficult to treat analytically except for a few special cases. Problem 7 deals with one of these special cases. The vacuum diode of Chapter 6 is treated analytically, but it is a special case. In this section we will consider a region that is still isotropic, but not necessarily homogeneous or linear.

Simultaneous solution of equations 5.1, 5.2, and 5.3 using equations A.10 for $\mathbf{\nabla} \cdot \varepsilon \mathbf{\nabla} \Phi$ gives

$$\varepsilon \nabla^2 \Phi + \mathbf{\nabla} \Phi \cdot \mathbf{\nabla} \varepsilon = -\rho_v \tag{5.74}$$

Equation 5.74 can be approximated by a finite difference form, but it is easier and more informative to consider Gauss's law, which is just the integral equivalent of equation 5.74.

$$\oiint \varepsilon \mathbf{\nabla} \Phi \cdot d\mathbf{s} = -Q \tag{5.75}$$

Let the region of interest be subdivided into a number of cubes d m on a side. These cubes or cells may be thought of as the basic building blocks of the material within the boundaries. The cells are distinguished by a uniform ρ_v and ε throughout the cell and Φ measured at the center of the cell. We are seeking a formula for the potential at the center of a cell in terms of the potential, permittivity, and charge of the six cells adjacent to the six faces of the first cell. Figure 5.8 shows two views of a cube (numbered 0) and its six neighbors (numbered 1 through 6). Considering only cube 0 and cube 1, as in Figure 5.9, the flux out of cube 0 must equal the flux into cube 1. For small d the requested flux is

$$\Psi_E = D_x s = \varepsilon E_x s = -\varepsilon (\mathbf{\nabla} \Phi)_x \, d^2$$

So we have (approximately)

$$\varepsilon_0 \left[\frac{\Phi_0 - \Phi_{01}}{d/2} \right] d^2 = \varepsilon_1 \left[\frac{\Phi_{01} - \Phi_1}{d/2} \right] d^2 \tag{5.76}$$

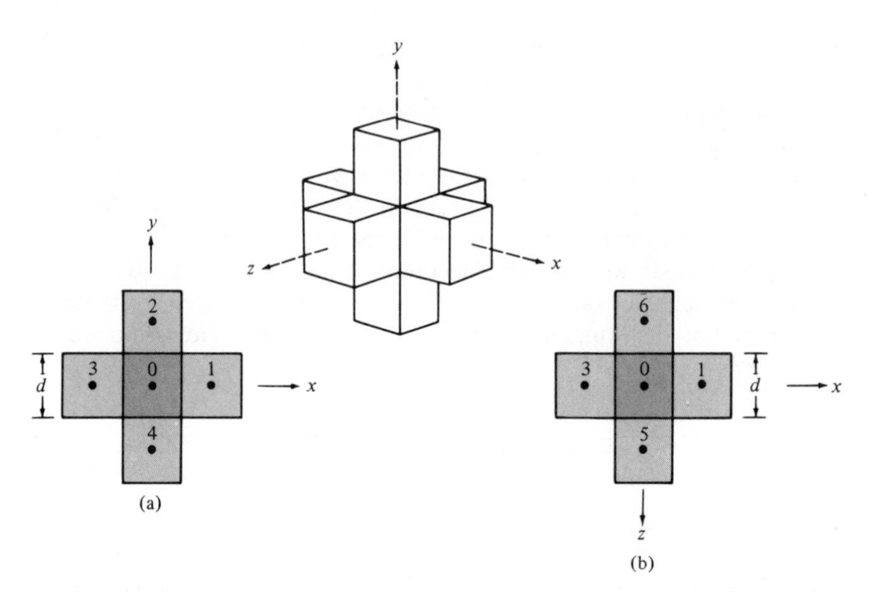

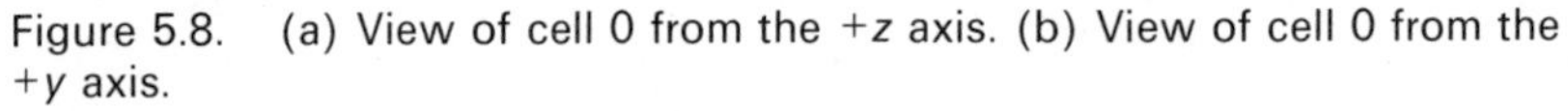

Figure 5.8. (a) View of cell 0 from the $+z$ axis. (b) View of cell 0 from the $+y$ axis.

where Φ_{01} is the potential at the interface (Figure 5.9) and the appropriate difference form has been used to approximate (minus) the gradient of Φ. Solving for the interface potential, we have

$$\Phi_{01} = \frac{\varepsilon_0 \Phi_0 + \varepsilon_1 \Phi_1}{\varepsilon_0 + \varepsilon_1} \tag{5.77}$$

Substituting equation 5.77 into the left side of equation 5.76 gives

$$2d \frac{\varepsilon_0 \varepsilon_1}{\varepsilon_0 + \varepsilon_1} (\Phi_0 - \Phi_1) = \frac{(2\,d\varepsilon_0)(2\,d\varepsilon_1)}{2\,d\varepsilon_0 + 2\,d\varepsilon_1} (\Phi_0 - \Phi_1) \tag{5.78}$$

for the flux out of cube 0 into cube 1. For all six faces, Gauss's law gives

$$\sum_{n=1}^{6} \frac{(2\,d\varepsilon_0)(2\,d\varepsilon_n)}{2\,d\varepsilon_0 + 2\,d\varepsilon_n} (\Phi_0 - \Phi_n) = \rho_0\, d^3 \tag{5.79}$$

where ρ_0 is the uniform charge density in cube 0. Solving for Φ_0 we obtain

$$\boxed{\Phi_0 = \frac{\rho_0\, d^3 + \displaystyle\sum_{n=1}^{6} \frac{(2\,d\varepsilon_0)(2\,d\varepsilon_n)}{2\,d\varepsilon_0 + 2\,d\varepsilon_n} \Phi_n}{\displaystyle\sum_{n=1}^{6} \frac{(2\,d\varepsilon_0)(2\,d\varepsilon_n)}{2\,d\varepsilon_0 + 2\,d\varepsilon_n}}} \tag{5.80}$$

Notice, first, that equation 5.79 leads almost immediately to a simple resistive circuit analog (via Kirchhoff's current law) as shown in Figure 5.10.

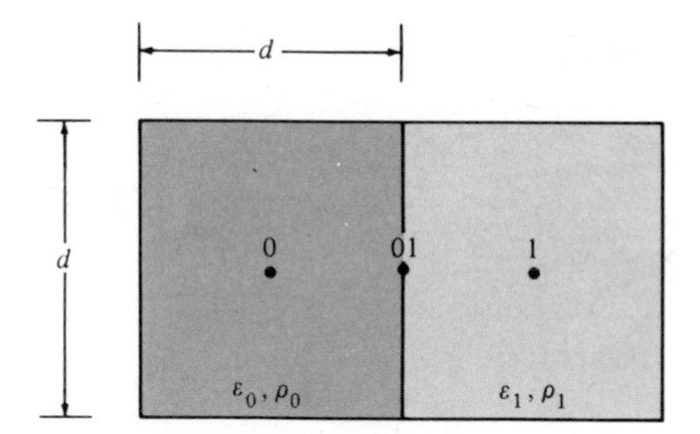

Figure 5.9. Cell 0 and 1.

Second, if the region is charge free and homogeneous (Laplacian), then equation 5.80 reduces to

$$\Phi_0 = \frac{1}{6} \sum_{n=1}^{6} \Phi_n, \qquad (\rho_v = 0, \quad \varepsilon_0 = \varepsilon_n = \varepsilon) \tag{5.81}$$

the well-known[5] result that the potential is the average of the potential at six equally spaced adjacent points.

Figure 5.10. Electric circuit analog for $\nabla \cdot (\varepsilon \nabla \Phi) = -\rho_v$ or $\oiint \varepsilon \nabla \Phi \cdot d\mathbf{s} = -Q$.

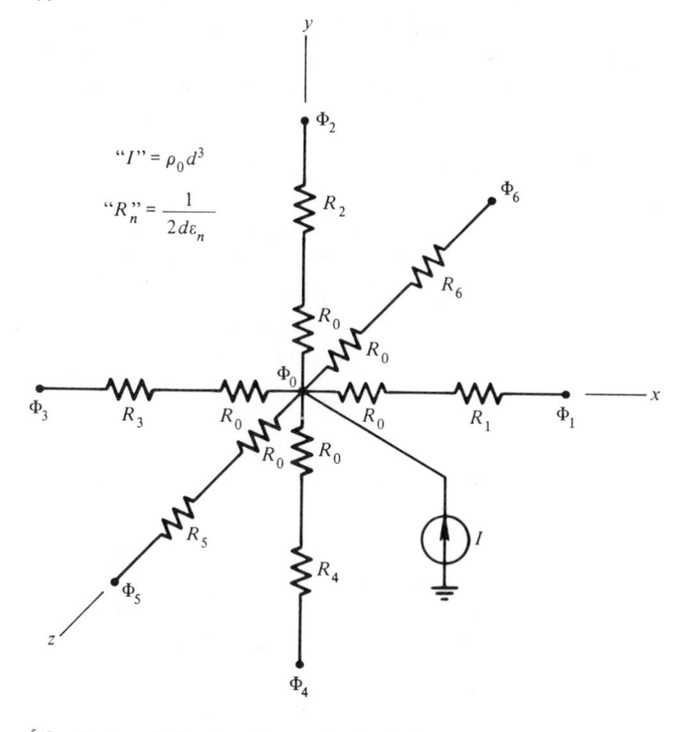

[5] See Paris and Hurd or Silvester in the References at the end of the chapter.

The two-dimensional form for equation 5.80 is

$$\Phi_0 = \frac{\rho_0 d^2 l + \sum_{n=1}^{4} \frac{(2l\varepsilon_0)(2l\varepsilon_n)}{2l\varepsilon_0 + 2l\varepsilon_n} \Phi_n}{\sum_{n=1}^{4} \frac{(2l\varepsilon_0)(2l\varepsilon_n)}{2l\varepsilon_0 + 2l\varepsilon_n}} \tag{5.82}$$

where l is the depth. The one-dimensional form is

$$\Phi_0 = \frac{\rho_0 dlW + \sum_{n=1}^{2} \frac{(2l'\varepsilon_0)(2l'\varepsilon_n)}{2l'\varepsilon_0 + 2l'\varepsilon_n} \Phi_n}{\sum_{n=1}^{2} \frac{(2l'\varepsilon_0)(2l'\varepsilon_n)}{2l'\varepsilon_0 + 2l'\varepsilon_n}} \tag{5.83}$$

where W is the width and $l' = lW/d$. Both of the preceding equations can be algebraically simplified for the purpose of calculation, but in doing so some of the physical insight is lost.

Some remarks concerning boundary conditions are in order. First, points of known or specified potential (Dirichlet conditions) must lie on field points (i.e., at the center of the cubes) because of the nature of the formulation. Second, derivatives of potential are calculated in terms of the difference of potential between two adjacent points, so points where the normal derivative is specified (Neumann conditions) should lie midway between field points. With only a finite number of cubes available for numerical calculation, the above conditions are somewhat restrictive. This is part of the price we pay for simplicity.

EXAMPLE 5

In order to demonstrate how this method is used to obtain an approximate solution, let us reconsider Example 4. In particular, consider Figure 5.7. A low-order solution is sufficient for present purposes, so only nine (actually, only six) unknown field points are chosen. The problem is two-dimensional, and, $\varepsilon_n = \varepsilon$, $\rho_v = 0$, so equation 5.82 becomes

$$\Phi_0 = \frac{1}{4} \sum_{n=1}^{4} \Phi_n \tag{5.84}$$

Figure 5.11 shows the boundary and field points, while Figure 5.12 shows the electric circuit analog. Because of the infinitesimal gaps at the upper corners, the potential there cannot be specified, but it is not necessary to do so. Symmetry conditions indicate that $\Phi_1 = \Phi_3$, $\Phi_4 = \Phi_6$ and $\Phi_7 = \Phi_9$, leaving six unknowns, Φ_1, Φ_2, Φ_4, Φ_5, Φ_7, and Φ_8. Equation 5.84 is applied to these six unknowns using the appropriate subscripts, giving six equations in six unknowns. These equations

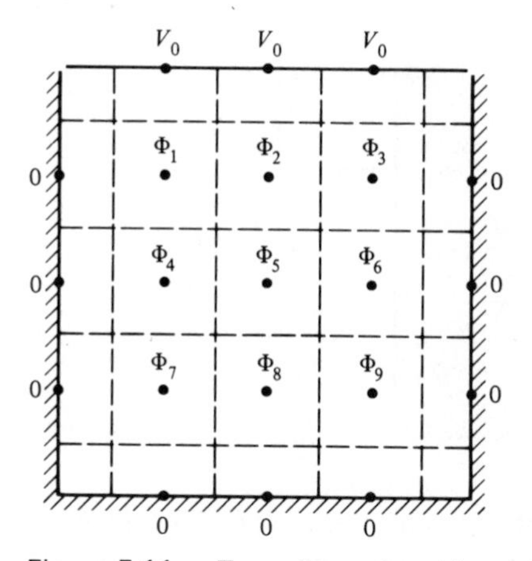

Figure 5.11. Two-dimensional boundary value problem arranged for a low-order numerical solution.

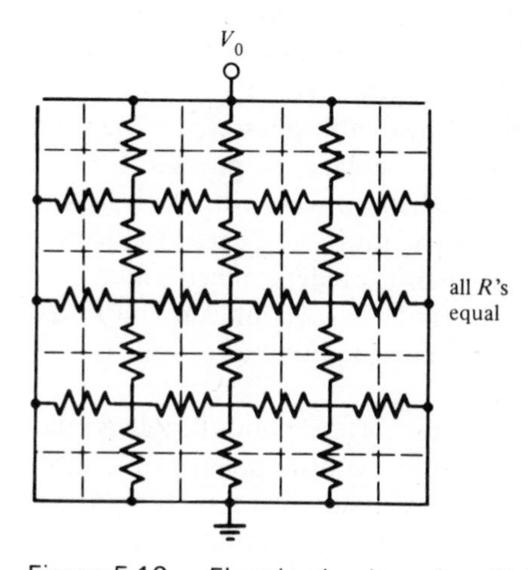

Figure 5.12. Electric circuit analog (finite element model) for Fig. 5.11.

and their solutions are shown below. These solutions can be compared

$$\begin{aligned}
4\Phi_1 &= 0 + V_0 + \Phi_2 + \Phi_4 \\
4\Phi_2 &= \Phi_1 + V_0 + \Phi_1 + \Phi_5 \\
4\Phi_4 &= 0 + \Phi_1 + \Phi_5 + \Phi_7 \\
4\Phi_5 &= \Phi_4 + \Phi_2 + \Phi_4 + \Phi_8 \\
4\Phi_7 &= 0 + \Phi_4 + \Phi_8 + 0 \\
4\Phi_8 &= \Phi_1 + \Phi_5 + \Phi_7 + 0
\end{aligned}
\quad \Longrightarrow \quad
\begin{aligned}
\Phi_1 &= 0.4286 V_0 \\
\Phi_2 &= 0.5268 V_0 \\
\Phi_4 &= 0.1875 V_0 \\
\Phi_5 &= 0.2500 V_0 \\
\Phi_7 &= 0.0714 V_0 \\
\Phi_8 &= 0.0982 V_0
\end{aligned}$$

to those given by the exact solution, equation 5.71. For example, $\Phi_5 = 0.25 V_0$ which agrees exactly with equation 5.73, but $\Phi_4 = 0.1875 V_0$ and equation 5.71 gives the exact answer $0.18198 V_0$ (3% error).

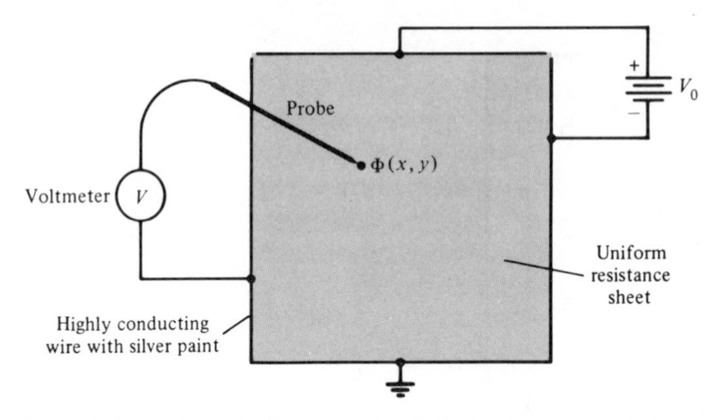

Figure 5.13. Electric circuit analog (infinite element model) for Figure 5.6.

Any method can be used to solve the simultaneous equations, and when the number of these equations is large, a digital computer will normally be used. Since each equation contains at most four unknowns (for the two-dimensional case) an iterative method is attractive. In this scheme the unknown potentials are first obtained as reasonable guesses, and then equation 5.84 is used to calculate new values to replace the guesses. If the new values are used as soon as they are calculated, the process will converge rapidly in most cases. This procedure is repeated until two successive values of potential for each point differ by less than a predetermined value or calculation error. When the calculation error is insignificant, more exact solutions can only be obtained by increasing the number of cells (i.e., field points).

It is easy to imagine that when the number of field points becomes infinite, the mesh of resistors in Figure 5.12 becomes a continuous resistive sheet as shown in Figure 5.13. Since we can easily calculate the capacitance per unit length (the exact capacitance per unit length is infinite in Figure 5.6) using the numerical techniques of this section, we should be able to calculate the resistance per unit length of its finite element model. More importantly, the reverse process allows us to build an exact resistive analog and measure the potential (voltage) at *any* point we wish! The resistance per unit length of the exact two-dimensional resistive analog and the capacitance per unit length of the corresponding exact two-dimensional potential problem are simply related. The truth of this statement will be demonstrated in Chapter 6.

EXAMPLE 6

Consider the square coaxial cable of Figure 5.14. Because of the symmetry there are only five unknown potentials, and they can be determined from five simultaneous equations, which are generated by

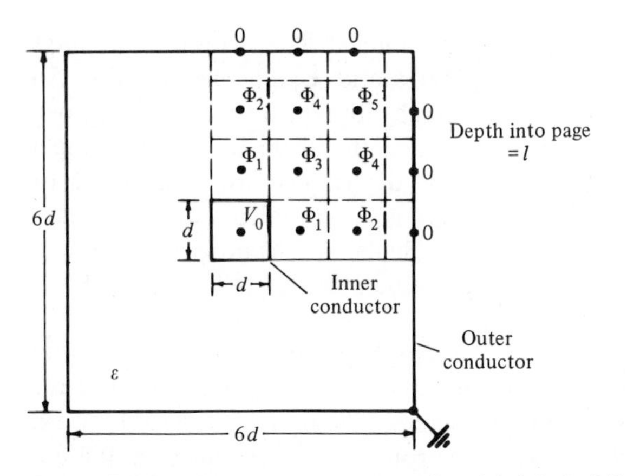

Figure 5.14. A square coaxial cable with labeled field points.

equation 5.84. The solutions are easily found by any method. Using substitution gives

$$\Phi_1 = \frac{140}{322} V_0, \qquad \Phi_2 = \frac{56}{322} V_0, \qquad \Phi_3 = \frac{91}{322} V_0, \qquad \Phi_4 = \frac{42}{322} V_0,$$

$$\Phi_5 = \frac{21}{322} V_0$$

If Q_i is the total charge on the inner conductor and ρ_{si} is the surface charge density on one face of the inner conductor, then (see equation 4.37)

$$C \equiv \frac{Q_i}{V_0} = \frac{\rho_{si}(4\,dl)}{V_0} = \frac{D_{ni}(4\,dl)}{V_0} = \frac{\varepsilon(V_0 - \Phi_1)/d}{V_0}(4\,dl)$$

or

$$C = 4l\varepsilon \frac{V_0 - \Phi_1}{V_0} = 2.26 l\varepsilon \tag{5.85}$$

Notice that a finite difference form was again used for the gradient, and because only one field point adjacent to each face of the inner conductor was used, the charge was easy to calculate. When more field points are used to improve the approximation, the calculation of C will become more lengthy, but will only involve sums of terms like that in equation 5.85. Increasing the number of field points should give a much more accurate answer. Notice also that instead of treating the potential at the center as a known quantity, we could treat it as unknown (i.e., Φ_0) and let the charge ($\rho_0\, d^2 l$) at the center conductor be the known quantity. The capacitance by this method, using equation 5.82 with $\varepsilon_n = \varepsilon_0 = \varepsilon$, is $\rho_0\, d^2 l/\Phi_0 = 2.26 l\varepsilon$, as in equation 5.85.

Suppose the region of interest is not linear. Then a table, or perhaps some analytic expression, relating $|\mathbf{D}|$, ε, and $|\mathbf{E}|$ must be available. An interative scheme can be devised whereby initial values of the unknown potential are guessed and then $\mathbf{E}$ is calculated from $\mathbf{E} = -\nabla\Phi$ (using a difference form). The permittivity is determined from the table or analytic expression. Next, equation 5.80 is used to calculate new values of potential from which $|\mathbf{E}|$ and ε are recalculated. The process is repeated until it converges. In calculating $|\mathbf{E}|$, remember that $|\mathbf{E}| = (E_x^2 + E_y^2 + E_z^2)^{1/2}$.

It was mentioned in the paragraph preceding Example 5 that because of the nature of the formulation, field points must lie at the center of the cubes, and cannot, therefore, lie on an interface between two different dielectrics. This may be troublesome in some cases. It is not difficult to overcome this problem in those cases where it is necessary to do so. If a field point must be placed on the interface between two dielectrics, then the potential at this point can be calculated by equation 5.77. On the other hand, a second method can be devised to treat field points on an interface.

Imagine eight cubes occupying the octants about an origin (point 0). Let the locus of the six adjacent field points remain the same as in the original formulation. Refer to Figure 5.10 for the location of the field points. Now, if each octant is occupied by a cube (d^3), and if the permittivity in each cube is uniform (but possibly different from cube to cube), we have a situation where field points lie on interfaces.

We can derive a difference equation for the potential at point 0 using Gauss's law as before, but this is not necessary. It is much easier to obtain the desired equation by visualizing the change in Figure 5.10. Instead of two "resistances" in series for each leg as in Figure 5.10, we will now have four "resistances" in parallel for each leg. These "resistances" arise from each of the fours cubes that share the line 0-n from field point 0 to the field point n, and they have the general form $d/[\varepsilon(d/2)^2] = 4/(d\varepsilon)$. Thus, the "*conductance*" in each leg is $(d\varepsilon^{(1)} + d\varepsilon^{(2)} + d\varepsilon^{(3)} + d\varepsilon^{(4)})/4 = d\langle\varepsilon_{0n}\rangle$, where $\langle\varepsilon_{0n}\rangle$ is the *average* permittivity for the four cubes that share the line 0-n. The resistance in each leg is $1/(d\langle\varepsilon_{0n}\rangle)$. Notice that the "current" of the current source is the flux out or charge enclosed, and, if the uniform charge density in a cube d m on a side and centered at point 0 (as in the first method) is ρ_0, the charge enclosed is $\rho_0 d^3$ as in the first method. Notice that the cube in the preceding sentence is *not* one of the cubes occupying the eight octants about point 0. Using Figure 5.15 and elementary circuit theory,

$$\boxed{\Phi_0 = \frac{\rho_0 d^3 + d\sum_{n=1}^{6} \Phi_n\langle\varepsilon_{0n}\rangle}{d\sum_{n=1}^{6}\langle\varepsilon_{0n}\rangle}} \tag{5.86}$$

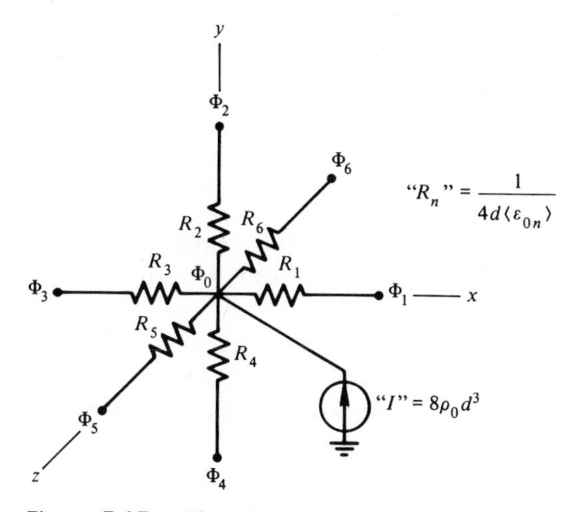

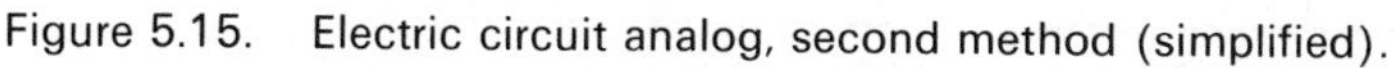

Figure 5.15. Electric circuit analog, second method (simplified).

For the two-dimensional case we have

$$\Phi_0 = \frac{d^2 l\, \rho_0 + l \sum_{n=1}^{4} \Phi_n \langle \varepsilon_{0n} \rangle}{l \sum_{n=1}^{4} \langle \varepsilon_{0n} \rangle} \tag{5.87}$$

where l is the depth, ρ_0 is the charge density for the parallelepiped ($d^2 l$) about point 0, and $\langle \varepsilon_{0n} \rangle$ is the average permittivity of the *two quadrants* sharing the line 0-n. The two-dimensional resistive analog is not difficult to sketch completely, and is shown in Figure 5.16. For

Figure 5.16. Two-dimensional electric circuit analog, second method.

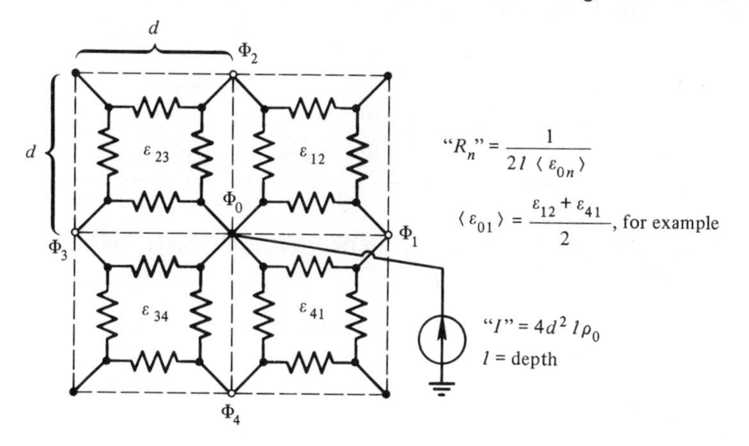

one-dimension we have

$$\Phi_0 = \frac{dlW\rho_0 + \dfrac{lW}{d}\sum_{n=1}^{2}\Phi_n\varepsilon_{0n}}{\dfrac{lW}{d}\sum_{n=1}^{2}\varepsilon_{0n}} \tag{5.88}$$

where W is the width, ρ_0 is the charge density about point 0, and ε_{0n} is the permittivity on the line 0-n. Notice that the last three equations can be simplified.

We can compare the two numerical methods in a simple problem for a charge-free region where they both apply. Consider the two-dielectric capacitor problem shown in Figure 5.17. The equations and

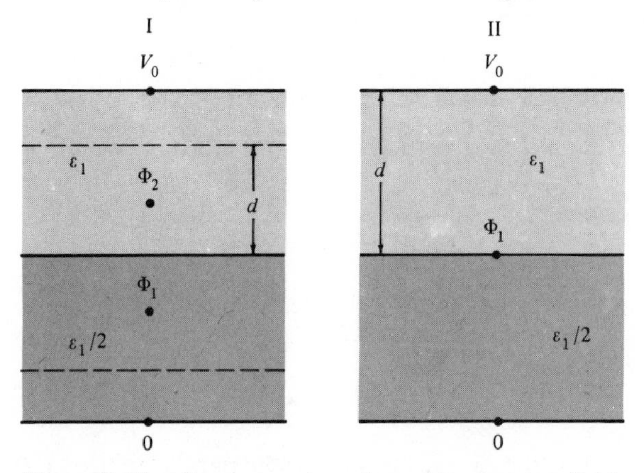

Figure 5.17. Two capacitors in series showing field point loci, both methods.

solutions are shown below. The reader can verify that the two methods agree, and furthermore, both give *exact* results for the potential at the chosen points (Why?). It is also worth mentioning, as a reminder, that the results for this particular problem solved by method II can also be obtained directly from equation 5.77.

I			*II*
$\Phi_1 = \frac{4}{7}\Phi_2$		$\Phi_1 = \frac{4}{9}V_0$	
$\Phi_2 = \frac{2}{5}\Phi_1 + \frac{3}{5}V_0$	$\Longrightarrow$	$\Phi_2 = \frac{7}{9}V_0$	$\Phi_1 = \frac{2}{3}V_0$

5.5 A NUMERICAL METHOD FOR SOLVING AN INTEGRAL EQUATION

We have mentioned several times in previous chapters that solutions to electrostatic field problems are simple (in a formal sense, at least) *if the charge* (source) *distribution is known everywhere.* If the charge distribution is not known, an integral equation may result. More specifically, the Helmholtz superposition integral,

$$\Phi(\mathbf{r}) = \frac{1}{4\pi\varepsilon} \iiint_{\text{vol}'} \frac{\rho_v(\mathbf{r}')}{R}\, dv'$$

is a general solution to Poisson's equation,

$$\nabla^2\Phi(\mathbf{r}) = -\frac{\rho_v(\mathbf{r})}{\varepsilon}$$

but if ρ_v is not known, it appears as an unknown under the integral symbol, giving an *integral equation* for ρ_v in terms of Φ if we can find a locus where Φ is known. It is worthwhile at this point to consider an example of an approximate numerical solution to Poisson's equation using an integral equation.

EXAMPLE 7

Consider a conducting solid cylinder of radius a and length $2h$ as shown in Figure 5.18. Let the cylinder be charged so that its surface is an

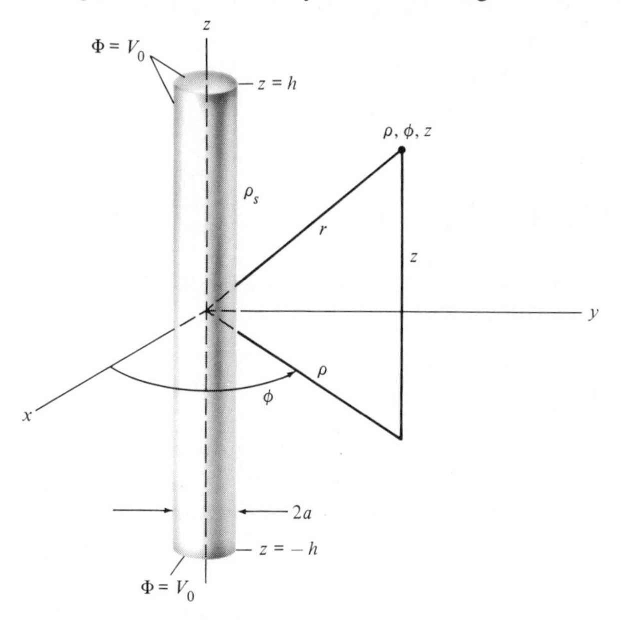

Figure 5.18. A charged conducting cylinder of length $2h$ and radius a with an unknown surface charge density, ρ_s.

equipotential, $\Phi = V_0$. This problem has been solved exactly by successive *conformal transformations*, but a treatment of this method is beyond the scope of the material in this book and we will not discuss it. In order to expedite matters we assume that a is small and $h \gg a$.

In this way we can safely neglect the effect of the end caps of the cylinder, and we are left with a cylinder that approximates a filamentary wire. If we let the field point, denoted as usual by **r**, lie on the lateral surface of the cylinder where $\Phi = V_0$, the Helmholtz integral becomes an approximate integral equation,

$$V_0 = \left(\frac{1}{4\pi\varepsilon_0}\right) \iint_{s'} \frac{\rho_s(\mathbf{r}')}{R}\, ds' \begin{cases} \rho = a, & -h \le z \le h, \quad 0 \le \phi \le 2\pi \\ \rho' = a, & -h \le z' \le h, \quad 0 \le \phi' \le 2\pi \end{cases}$$

if $\varepsilon = \varepsilon_0$. Notice that we intend to use cylindrical coordinates for obvious reasons. Symmetry conditions show that the surface charge density, ρ_s, depends on z', not on ϕ'. The last equation can then be written

$$V_0 = \frac{1}{4\pi\varepsilon_0} \int_{-h}^{h} \rho_s(z') \int_0^{2\pi} \frac{1}{R}\, a\, d\phi'\, dz'$$

The potential is independent of ϕ for any field point, so we choose $\phi = 0$ for convenience. Using equation 1.17b with $\rho' = a$, $\rho = a$, and $\phi = 0$ gives

$$R = |\mathbf{r} - \mathbf{r}'| = [2a^2 - 2a^2 \cos\phi' + (z - z')^2]^{1/2}$$

$$= \left[4a^2 \sin^2\left(\frac{\phi'}{2}\right) + (z - z')^2\right]^{1/2}$$

Then,

$$V_0 = \frac{a}{4\pi\varepsilon_0} \int_{-h}^{h} \rho_s(z') \int_0^{2\pi} \left[4a^2 \sin^2\left(\frac{\phi'}{2}\right) + (z - z')^2\right]^{-1/2} d\phi'\, dz'$$

Now, it can be shown (after much labor) that the inner integral may be replaced with the approximation

$$\int_0^{2\pi} \left[4a^2 \sin^2\left(\frac{\phi'}{2}\right) + (z - z')^2\right]^{-1/2} d\phi' \approx 2\pi[a^2 + (z - z')^2]^{-1/2}$$

so long as a is small. A little thought about this approximation, which avoids one integral, reveals that it is equivalent to calculating the potential at $\rho = a$ as if it arose from a filamentary line source at $\rho' = 0$. Then, we have a simpler form.

$$V_0 = \frac{a}{2\varepsilon_0} \int_{-h}^{h} \frac{\rho_s(z')\, dz'}{[a^2 + (z - z')^2]^{1/2}}, \qquad -h \le z \le h$$

It should be emphasized that the last equation is only an approximation, but it is a good one if a is small compared to h and we agree to be satisfied with a low-order solution.

We must now find a method for solving the integral equation we have derived. Fortunately, there is a method available, generally called the *method of moments*,[6] which can treat a very wide class of problems in field theory. This method has become very popular in recent years. It is well beyond the scope of the present treatment to go into the details of the general theory of the method, but the present example will serve to show how it is used.

The surface charge density, $\rho_s(z')$, is the unknown, so we assume it can be expanded into a sum of N linearly independent trial functions. That is,

$$\rho_s(z') = \sum_{n=1}^{N} a_n f_n(z')$$

If the order of summation and integration can be interchanged in the integral equation, then

$$\frac{2\varepsilon_0 V_0}{a} = \sum_{n=1}^{N} a_n \int_{-h}^{h} \frac{f_n(z')\,dz'}{[a^2 + (z - z')^2]^{1/2}}, \qquad -h \le z \le h$$

The end result of this scheme is to convert the last equation into a set of N simultaneous equations for the N unknown $f_n(z')$ so that the digital computer can do the real labor for us. With this in mind we should take a close look at the integral in the last equation which will in many cases be evaluated by the computer also. It is

$$I_n = \int_{-h}^{h} \frac{f_n(z')\,dz'}{[a^2 + (z - z')^2]^{1/2}}.$$

When $z = z'$, the denominator is very small (we have already assumed that a is small) and the integrand highly peaked. Some computer time may be saved if we can remedy this by rearranging the integrand. Fortunately, this can be done for the present problem by a rather neat scheme which is applicable when $f_n(z')$ is a continuous function.

$$I_n = \int_{-h}^{h} \frac{f_n(z') - f_n(z) + f_n(z)}{[a^2 + (z - z')^2]^{1/2}}\,dz'$$

$$I_n = f_n(z) \int_{-h}^{h} \frac{dz'}{[a^2 + (z - z')^2]^{1/2}} + \int_{-h}^{h} \frac{f_n(z') - f_n(z)}{[a^2 + (z - z')^2]^{1/2}}\,dz'$$

$$I_n = f_n(z) \ln \frac{z + h + [a^2 + (z + h)^2]^{1/2}}{z - h + [a^2 + (z - h)^2]^{1/2}}$$

$$+ \int_{-h}^{h} \frac{f_n(z') - f_n(z)}{[a^2 + (z - z')^2]^{1/2}}\,dz'$$

The integrand of the remaining integral is zero when $z = z'$ and is not highly peaked. It should be relatively easy to evaluate numerically.

[6] See Harrington in the References at the end of the chapter.

Simultaneous equations are generated by multiplying both sides of the integral equation by some function of z, $g_m(z)$, $m = 1, 2, \ldots, N$ and integrating from $-h$ to h on z. Many functions may be candidates, but the only one considered here because of its common sense appeal is $g_m(z) = \delta(z - z_m)$, where δ is the Dirac delta function. To make a longer story short, this choice forces the integral equation to hold at N points, z_m, in $-h \leq z \leq h$. The logic here is that the larger values of N will give more exact values for ρ_s. In our present problem we can pick our N "matching points," z_m in $0 \leq z \leq h$ since the left side of the integral equation is an even function of z.

The integral equation may now be written

$$\frac{2\varepsilon_0 V_0}{a} = \sum_{n=1}^{N} a_n \int_{-h}^{h} \frac{f_n(z')\,dz'}{[a^2 + (z_m - z')^2]^{1/2}}, \qquad m = 1, 2, \ldots, N$$

$$\frac{2\varepsilon_0 V_0}{a} = \sum_{n=1}^{N} a_n \left\{ f_n(z_m) \ln \frac{(z_m + h) + [a^2 + (z_m + h)^2]^{1/2}}{(z_m - h) + [a^2 + (z_m - h)^2]^{1/2}} + \int_{-h}^{h} \frac{f_n(z') - f_n(z_m)}{[a^2 + (z_m - z')^2]^{1/2}}\,dz' \right\}$$

or

$$\frac{2\varepsilon_0 V_0}{a} = \sum_{n=1}^{N} a_n I_{mn}, \qquad m = 1, 2, \ldots, N$$

This, then, is a set of N simultaneous equations in N unknowns, which may be written

$$\frac{2\varepsilon_0 V_0}{a} = a_1 I_{11} + a_2 I_{12} + a_3 I_{13} + \cdots + a_N I_{1N}$$

$$\frac{2\varepsilon_0 V_0}{a} = a_1 I_{21} + a_2 I_{22} + a_3 I_{23} + \cdots + a_N I_{2N}$$

$$\frac{2\varepsilon_0 V_0}{a} = a_1 I_{31} + a_2 I_{32} + a_3 I_{33} + \cdots + a_N I_{3N}$$

$$\vdots$$

$$\frac{2\varepsilon_0 V_0}{a} = a_1 I_{N1} + a_2 I_{N2} + a_3 I_{N3} + \cdots + a_N I_{NN}$$

In more elegant matrix form this can be written

$$[V_m] = [I_{mn}][a_n]$$

The formal solution for the N unknown a_n's, equivalent to any method for solving for them, is

$$[a_n] = [I_{mn}]^{-1}[V_m]$$

The only limits on N are the capability of the computer to handle the required matrix inversion and the choice of $f_n(z')$.

Hopefully, a low-order (small N) approximate solution can be obtained. Whether or not this is possible depends not only on how the matching points are chosen, but also on what trial functions, $f_n(z')$ are chosen. In other words, a good guess as to what $\rho_s(z')$ is like should help in choosing $f_n(z')$. A poor choice may give a divergent solution. In the present problem, we expect $\rho_s(z')$ to be largest near the ends and smallest in the middle due to the Coulomb forces of repulsion. The derivative of $\rho_s(z')$ at $z' = 0$ should be zero. Many possible candidates for $f_n(z')$ exist, but a simple choice that meets the stated conditions is

$$f_n(z') = \left(\frac{z'}{h}\right)^{2n}$$

which are even (polynomial) functions of z'. Results for this choice of $f_n(z')$ for $h = 1$ m, $a = 10^{-2}$ m, equally spaced matching points $z_m = (m - 1/2)h/N$, $2\varepsilon_0 V_0/a = 1$, $N = 3, 6$ are shown in Figure 5.19.

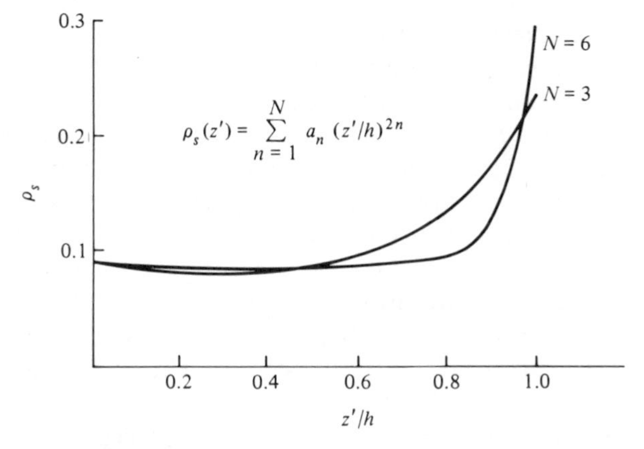

Figure 5.19. Approximate surface charge density, conducting cylinder, $h = 1$, $a = 10^{-2}$, $2\varepsilon_0 V_0/a = 1$, equally spaced matching points $z_m = (m - \frac{1}{2})h/N$. Polynomial approximation.

Results are pretty much as expected with the higher order solution being more accurate over most of the range. The exact solution (not shown) shows an infinite charge density at the end of the cylinder,[7] and we would have a difficult time reproducing this behavior with the present method in a low-order form.

At the risk of boring the reader, we will mention one more trial function which, in our present problem, removes the necessity for any numerical integration. Suppose $f_n(z')$ is represented by side-by-side rectangular pulses

[7] This always occurs at "knife" edges.

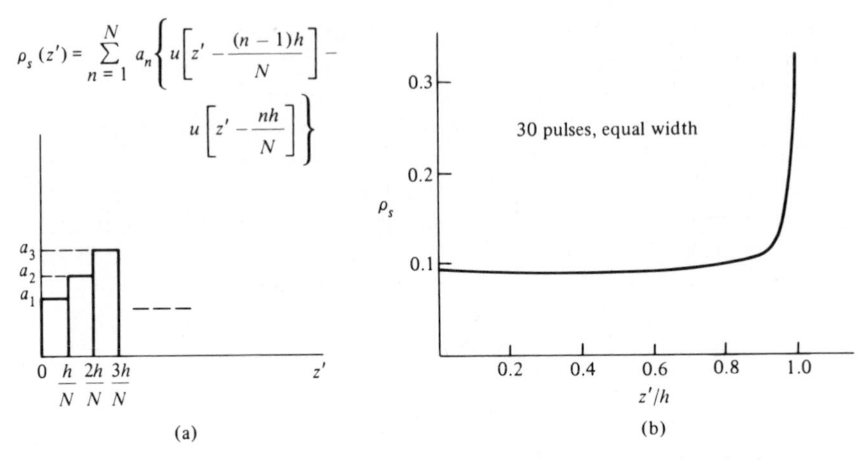

Figure 5.20. (a) $\rho_s(z')$, equal width pulses (stair-step approximation). (b) Approximate surface charge density, conducting cylinder, $h = 1$ m, $a = 10^{-2}$ m, $2\varepsilon_0 V/a = 1$. Stair-step approximation.

of height a_n. These pulses are in a real sense a form of prelimit weighted impulses and the integral equation (a sum) is the prelimit form of the superposition integral as it would appear if we had derived it naturally from basic concepts. Furthermore, since $f_n(z')$ is now constant over each pulse, the original form for I_n will contain an integrand of the form $[a^2 + (z - z_n')^2]^{-1/2}$ in every case, and this form integrates exactly to the logarithmic form we saw appear in the other development. This form of trial function gives a stair-step approximation to ρ_s. It can only be accurate in a relatively high-order (N) solution, and N is ultimately limited by the computer and the approximations we have made. The representation of $\rho_s(z')$ with equal width pulses is shown in Figure 5.20(a), while the result for this case with $z_m = (m - \frac{1}{2})h/N$ and $N = 30$ is shown in Figure 5.20(b).

We mention that in general there is no guarantee of convergence with the method presented in this section, but it is, nevertheless, very attractive because the computer does most of the real work. There are very many field type problems (not just electromagnetic) that can now be solved approximately, regardless of whether or not the boundaries fit a standard coordinate system. These problems required too many man-hours before the advent of the computer. Differential equations (treated as difference equations) and integrodifferential equations can also be treated directly by the method.

5.6 IMAGE THEORY

Sometimes a problem can be solved by a technique called the *method of images* or *image theory*, which amounts to finding an equivalent problem whose solution we already know. An example serves best to show how the method works.

EXAMPLE 8

A point charge located at $(0, 0, d/2)$ above an infinite grounded conducting plane (located at $z = 0$) is shown in Figure 5.21(a). What is

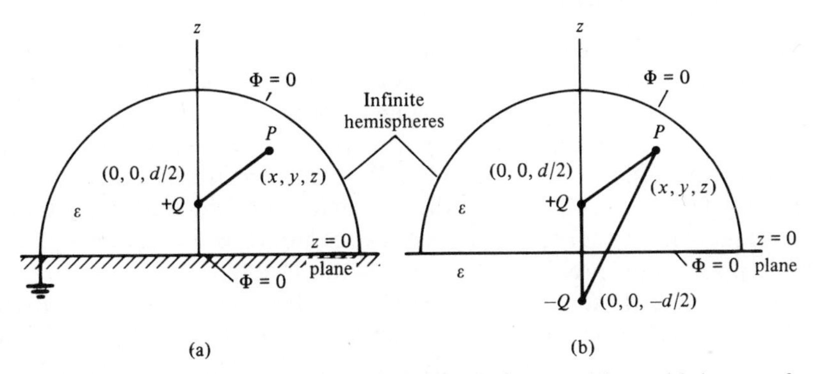

Figure 5.21. (a) Original problem (b) Equivalent problem with image of the original charge.

the potential at any point P $(z > 0)$? That is, what is $\Phi(x, y, z)$? Notice that we require that the potential vanish at infinity on the infinite hemisphere. Now consider the electrostatic dipole located in the same medium (ε) with no ground plane as shown in Figure 5.21(b). It is easy to show by means of superposition that the potential at P is given by

$$\Phi(x, y, z) = \frac{Q}{4\pi\varepsilon}\left\{\frac{1}{[x^2 + y^2 + (z - d/2)^2]^{1/2}} - \frac{1}{[x^2 + y^2 + (z + d/2)^2]^{1/2}}\right\} \tag{5.89}$$

For any point in the $z = 0$ plane equation 5.89 gives $\Phi(x, y, 0) \equiv 0$. Equation 5.89 also shows that the potential on the infinite hemisphere vanishes. Comparing Figure 5.21(a) and Figure 5.21(b), we see that *for* $z > 0$ the boundary conditions are the same, the media are the same, and the sources are the same. Therefore, according to the uniqueness theorem, the solutions must be the same *for* $z > 0$. That is, the original problem with only one charge and a ground plane can be replaced by another problem with the same charge and its (negative) *image* and no ground plane whose solution is given by equation 5.89.

What is the potential for $z < 0$ in Figure 5.21(a)? What is the surface charge density at $(x, y, 0+)$ in Figure 5.21(a)? See Problem 9. Other examples of the use of image theory will be found in the problems.

There are other techniques available for solving certain electrostatic problems. Among these techniques is the use of *conformal transformations.* It is beyond the scope of this treatment to consider this method.

5.7 CONCLUSIONS

In this chapter we have considered solutions to Poisson's and Laplace's equations, primarily applied to boundary value problems. In a few cases, simple integration is sufficient to obtain a solution. Three simple examples of this were given because, not only were they simple, but two of them are used as common transmission lines. We will meet them again in later chapters. There are only two one-dimensional geometries remaining insofar as rectangular, cylindrical, and spherical coordinates are concerned, and these appear in the problems at the end of this chapter.

A simple and rather general method for obtaining numerical solutions to Poisson's or Laplace's equation was developed. It can treat those problems where ε is a scalar but not necessarily a constant. In addition, charges may be included in the region of interest. Two simple examples were given where the resulting simultaneous equations were solved by hand. More precise resolution will require the use of a digital computer. The technique is by nature ideally suited to mutually perpendicular plane boundaries, but it can also treat curved surfaces in a stair-step manner. This again requires the use of many field points for accurate resolution, and, thus, the use of a digital computer.

In order to show once more how the computer can serve as an invaluable aid in solving certain electrostatic field problems, an example was given which used the method of moments to solve (approximately) an integral equation. The computer was used to calculate a set of integrals, and also to invert a matrix to solve for the unknowns in the problem. The techniques employed in the method are very flexible, and a wide variety of problems can be treated.

REFERENCES

Adams, A. T. *Electromagnetics for Engineers.* New York: Ronald Press, 1971. This book contains many examples of numerical methods for solving field problems.

Harrington, R. F. *Field Computation by Moment Methods.* New York: Macmillan, 1968. This book is devoted entirely to numerical methods for treating field problems.

Moon, P. and Spencer, D. E. *Field Theory for Engineers.* New York: Van Nostrand, 1961. A highly mathematical book. Excellent for boundary value problems in coordinate systems other than the three usual ones.

Paris, D. T. and Hurd, F. K. *Basic Electromagnetic Theory.* New York: McGraw-Hill, 1969. The two-dimensional trough problem is treated analytically and numerically in Chapter 3.

Salvadori, M. G., and Baron, M. L. *Numerical Methods in Engineering*, 2nd ed. Englewood Cliffs, N.J.: Prentice-Hall, 1961.
Silvester, P. *Modern Electromagnetic Fields*. Englewood Cliffs, N.J.: Prentice-Hall, 1968. Some good examples of numerical methods are included.

PROBLEMS

1. State explicitly the boundary conditions that were *assumed* in solving for the fields of a parallel plate capacitor.
2. Two potential functions, $\Phi_1(z)$ and $\Phi_2(z)$, satisfy Laplace's equation and the boundary conditions, $\Phi(z_1) = V_1$ and $\Phi(z_2) = V_2$. Yet $\Phi_1(z) \not\equiv \Phi_2(z)$. Explain.
3. Show that the functions, $\Phi_1 = 2x$ and $\Phi_2 = \ln[e^{2x}\sqrt{(1+x)/(1-x)}] - \tanh^{-1} x$, satisfy Laplace's equation and the boundary conditions $\Phi(0) = 0$, $\Phi(\frac{1}{2}) = 1$, $\partial\Phi/\partial y = 0$ and $\partial\Phi/\partial z = 0$. Explain.
4. Two semi-infinite conducting planes are inclined at an angle of ϕ_0 with respect to each other. At the apex, the two planes do not quite touch, so that a battery of potential V_0 can be connected between them. (See Figure 5.22.)

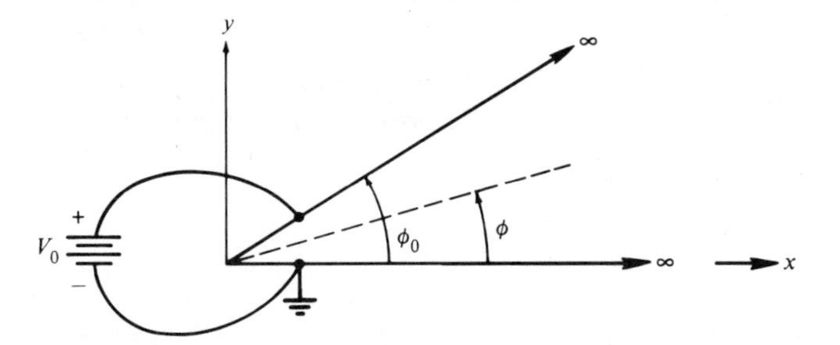

Figure 5.22. A new one-dimensional boundary value problem (inclined planes).

   (a) State a complete set of boundary conditions for finding Φ.
   (b) Find Φ anywhere between the planes.

5. If

$$\Phi = V_0 \frac{\ln(\tan\theta/2)}{\ln(\tan\theta_0/2)}$$

   where V_0 and θ_0 are constants, show that Laplace's equation is satisfied. What boundary value problem has been solved?

6. In a charge-free, but inhomogeneous, region, what differential equation must Φ satisfy?
7. Use a one-dimensional form of the answer to the previous problem to solve Problem 12, Chapter 4, completely.
8. If $\Phi = V_0 e^{-r/a}$, find ρ_v and Q. ($\varepsilon = \varepsilon_0$)
9. (a) What is the surface charge density at $(x, y, 0^+)$ in Figure 5.21(a)?
   (b) What is the total charge on the $z = 0^+$ plane in Figure 5.21(a)?

10. For the quadrupole consisting of point charges, $+Q$ at $(d/\sqrt{2}, d/\sqrt{2}, 0)$ $-Q$ at $(-d/\sqrt{2}, d/\sqrt{2}, 0)$, $+Q$ at $(-d/\sqrt{2}, -d/\sqrt{2}, 0)$, and $-Q$ at $(d/\sqrt{2}, -d/\sqrt{2}, 0)$:

    (a) Find the exact expression for the potential.
    (b) For the region $x \geq 0$, $y \geq 0$ what is the equivalent problem according to image theory?

11. Show that the problem of an infinite line charge density, ρ_l, located at $(\rho_1, 0, z)$ parallel to the z axis outside an infinite grounded conducting cylinder of radius a, is equivalent (for $\rho > a$) to the problem of the same original line charge density, no conducting cylinder, and an image line charge density $\rho_{li} = -\rho_l$ located at $(a^2/\rho_1, 0, z)$ parallel to the z axis (*image principle*).
12. Solve the problem of a point charge outside a grounded conducting sphere of radius a. Results are similar to those of Problem 11.
13. What is the potential of the system consisting of a point charge above an infinite conducting plane raised to a potential V_0?
14. How many image charges would be necessary to solve the problem of a point charge between infinite, parallel, and grounded conducting planes?

15. (a) What are the differential equations for Φ and the boundary conditions for **E** for the problem shown in Figure 5.23(a)? Use cylindrical coordinates.

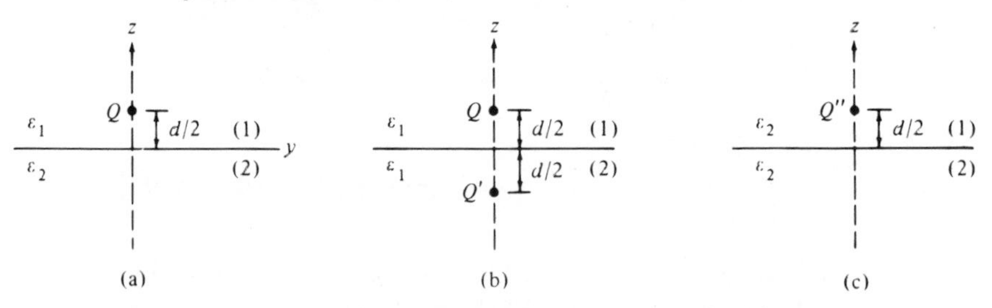

Figure 5.23. (a) Point charge and a plane interface between two dielectrics. (b) Equivalent problem for $z > 0$. (c) Equivalent problem for $z < 0$.

    (b) Show that this problem is equivalent to that shown in Figure 5.23(b) for $z > 0$ and to that shown in Figure 5.23(c) for $z < 0$ if

$$Q' = Q\frac{\varepsilon_1 - \varepsilon_2}{\varepsilon_1 + \varepsilon_2} \quad \text{and} \quad Q'' = Q\frac{2\varepsilon_2}{\varepsilon_1 + \varepsilon_2}$$

This is another example of the use of the *image principle*.

16. If $\rho_v = Ae^{-\rho/a}u(a - \rho)$, $\varepsilon = \varepsilon_0$:

    (a) Find **E** everywhere.
    (b) Show that Poisson's and Laplace's equations are satisfied.

17. If $\rho_v = Ae^{-r/a}u(a - r)$, $\varepsilon = \varepsilon_0$:
    (a) Find **E** everywhere.
    (b) Show that Poisson's and Laplace's equations are satisfied.
18. What is the surface charge density on the upper plate ($\Phi = V_0$) of Figure 5.5?
19. Find **E** and Φ for the boundary value problem shown in Figure 5.24. (Poisson's equation.) E_0 is a constant.

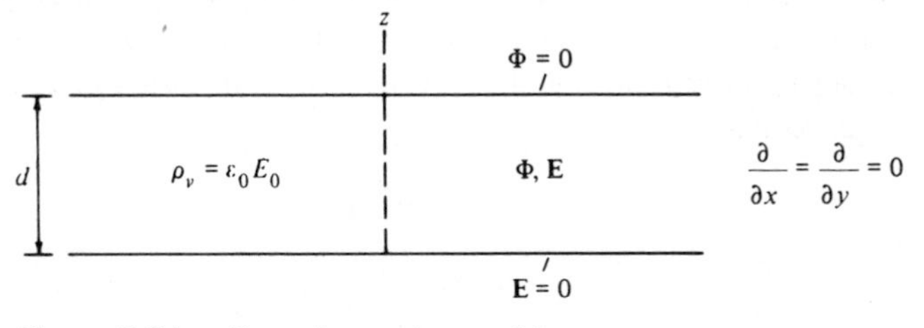

Figure 5.24. Boundary value problem.

20. Find a solution for the potential in Figure 5.5 if $f(x) = 10 \sin \pi x/a$.
21. Solve the two dimensional boundary value problem of Figure 5.5 if the boundary conditions are

$$\Phi = 0, \qquad x = 0, x = a$$

$$\Phi = V_0, \qquad y = 0, y = b$$

    Hint: Remember superposition.
22. (a) Find the volume charge density for Problem 4.
    (b) Find the surface charge density on the conducting plane at $\phi = \phi_0$ by using part (a).
    (c) Find the surface charge density on the conducting plane at $\phi = \phi_0$ by using $\rho_s^+ = -D_\phi|_{\phi=\phi_0}$.
23. Show that the capacitance for Figure 5.14 as calculated by finding the charge on the *outer* conductor agrees with equation 5.85.

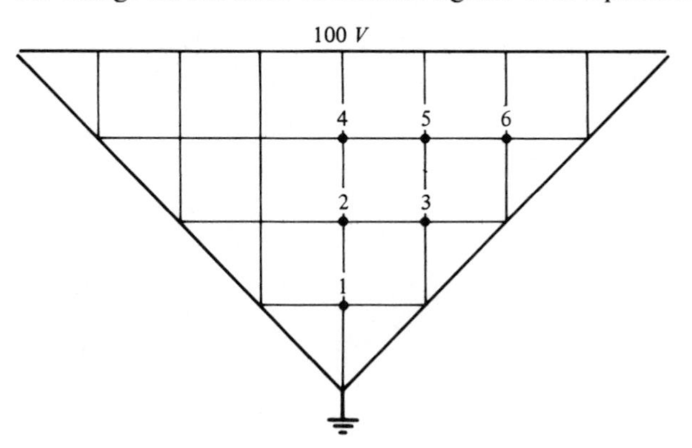

Figure 5.25. A two-dimensional boundary value problem.

24. (a) Use an iterative scheme to find the approximate potential at the six points labeled in Figure 5.25 below. Start with $\Phi_1 = 6$ V, $\Phi_2 = 24$ V, $\Phi_3 = 20$ V, $\Phi_4 = 60$ V, $\Phi_5 = 55$ V, and $\Phi_6 = 40$ V. These voltages may have been obtained from voltage measurements on a resistance sheet analog. Stop iterating when the answers are within 0.1 V of a final answer ($\varepsilon = \varepsilon_0$).
    (b) If the trough if filled to points 2 and 3 with water ($\varepsilon_R = 80$), repeat (a).
25. Show that $\mathbf{\nabla} \cdot \varepsilon \mathbf{\nabla} \Phi = \varepsilon \nabla^2 \Phi + \mathbf{\nabla} \Phi \cdot \mathbf{\nabla} \varepsilon$ using Cartesian coordinates.
26. Suppose that you have a strong prejudice toward even functions. Describe how Example 4 could be solved in terms of cosine functions rather than sine functions.

Chapter 6
Steady Electric Currents

In this chapter, the concept of electric charges in motion will be investigated explicitly. We will find that an electric current results from charge motion due to an applied electric field whenever the charges are not constrained so that they cannot move. This current may assume many forms, but we will be primarily interested in conduction currents and convection currents. Heating occurs in conductors, and, so, Joule's law is derived. The second of the three important circuit parameters, resistance, is defined and determined for some simple geometries. A brief, simplified picture of the electrical properties of conductors and semiconductors will be presented. The concepts of electric current will be of much use to us in the next chapter, where we will find that a new force field arises because of the charge motion.

6.1 ELECTRIC CURRENT DISTRIBUTIONS

In circuit theory we frequently speak of the current in a branch or through an element of the circuit. The current is measured in amperes or coulombs per second, indicating that it is the rate of movement of charge passing some reference point. In field theory we are usually interested in point forms, and thus the concept of current density (which may vary over the cross section of a conductor) is more useful. Charge, in the form of a volume density, ρ_v, in uniform motion with velocity $\mathbf{u}$ constitutes a vector current density $\mathbf{J}$ according to

$$\boxed{\mathbf{J} = \rho_v \mathbf{u}} \quad (\text{A/m}^2) \tag{6.1}$$

Notice that so far as equation 6.1 is concerned, we have said nothing about conductors. In other words, a current density, as described by equation 6.1 can exist in a region free of conductors, although we usually think of conductors when we speak of electric current. A current density, such as that between the cathode and plate of a vacuum diode, or that between the cathode and screen of a cathode ray tube in an oscilloscope, may be called a *convection current density*.

Let us now consider the various mathematical and physical distributions that an electric current density may assume in special cases. We found that line charge densities, ρ_l, and surface charge densities, ρ_s, were special cases of the general volume charge density, ρ_v. If the differential elements of charge making up these various distributions are moving with velocity, $\mathbf{u}$, and the distributions are otherwise unchanged, then we have line (filamentary) currents, I (in amperes), surface current densities, $\mathbf{J}_s$ (in amperes per meter), and the general current density, $\mathbf{J}$ (in amperes per square meter) of equation 6.1. It will be helpful here, and particularly so in the following chapters, if we utilize the idea of a *differential current element*. We will find in the next chapter that these current elements act as sources for a new field (force) quantity. Consider a filamentary current I (a moving thread of charge with negligible cross section) which exists over a differential length dl. Current, I, in amperes, is not a vector (although we commonly speak of the direction of the current in an electric circuit), but the differential current element is a vector, being given by $I\,d\mathbf{l}$. The vector $d\mathbf{l}$ gives the direction of the current. Notice that $I\,d\mathbf{l}$ corresponds to the differential line charge element, $\rho_l\,dl$, where the charges are in motion. If the current is in the form of a surface current density, then the differential element is $\mathbf{J}_s\,ds$. If the current is in the form of a general current density, $\mathbf{J}$, then the differential element is $\mathbf{J}\,dv$. We may replace ρ_v with dQ/dv in equation 6.1 so that $\mathbf{J} = \mathbf{u}\,dQ/dv$, or $\mathbf{J}\,dv = \mathbf{u}\,dQ$. The current elements then are

$$I\,d\mathbf{l},\ \mathbf{J}_s\,ds,\ \mathbf{J}\,dv,\ \mathbf{u}\,dQ \quad (\text{A}\cdot\text{m}) \tag{6.2}$$

These current elements, elementary sources, or basic "building blocks," will be useful to us in the material in future sections.

The total current flowing in a filamentary path is obviously I, and if we desire the *total* current in a nonfilamentary path, we merely need to integrate the normal component of the current density, $\mathbf{J}$, over the given cross section.

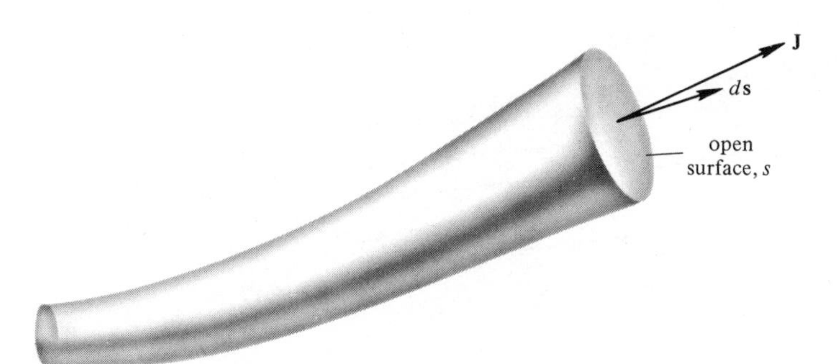

Figure 6.1. A general current density **J** over the cross section s.

As shown in Figure 6.1,

$$\boxed{I = \iint_s \mathbf{J} \cdot d\mathbf{s}} \quad \text{(A)} \tag{6.3}$$

is the *definition* of the total current in amperes. If the current is a surface current density, $\mathbf{J}_s$, the total current flowing can be found using equation 6.3 if we express the surface current density as a general current density by way of the impulse function. For example, if we have a z-directed surface current density flowing on a circular cylinder of radius a, we may use the infinite $z = 0$ plane surface with $\mathbf{J} = J_{sz}\,\delta(\rho - a)\mathbf{a}_z$, and equation 6.3 gives

$$I = \int_0^{\infty} \int_0^{2\pi} J_{sz}\,\delta(\rho - a)\mathbf{a}_z \cdot \mathbf{a}_z \rho \, d\rho \, d\phi$$

or

$$I = a \int_0^{2\pi} J_{sz}\, d\phi$$

If, in addition, J_{sz} is independent of ϕ, then

$$I = J_{sz}(2\pi a) \quad \text{(surface current times width)}$$

A general surface current density is shown in Figure 6.2. It is a current per unit width.

6.2 CONSERVATION OF CHARGE

The principle of conservation of charge simply states that charge can be neither created nor destroyed. *Equal* amounts of *positive* and *negative* charge may, of course, be simultaneously created and destroyed. Consider a

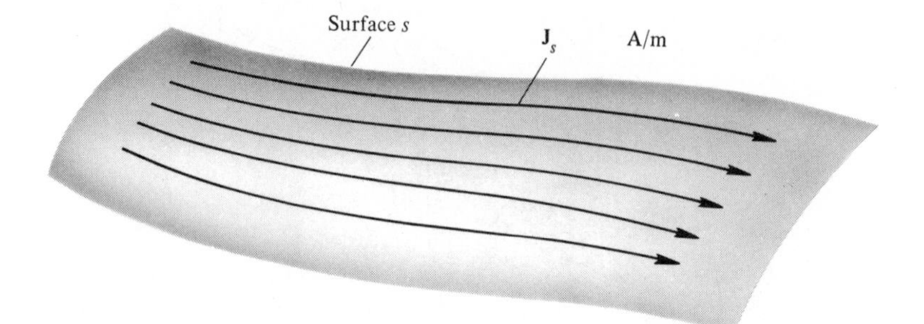

Figure 6.2. A surface current density, $\mathbf{J}_s$, on a surface s.

closed surface s. Conservation of charge requires that the net, steady (non-time varying) current passing through this surface be zero. Thus,

$$\oiint_s \mathbf{J} \cdot d\mathbf{s} = 0 = \iiint_{vol} \nabla \cdot \mathbf{J} \, dv \tag{6.4}$$

by means of the divergence theorem. Since the volume in question is completely arbitrary, we have the conservation of charge equation for *steady* currents,

$$\boxed{\nabla \cdot \mathbf{J} = 0} \tag{6.5}$$

which may be recognized as the point form of Kirchhoff's current law of circuit theory. This result may also be called the *continuity equation* for *steady* currents. We will find that the right side of equation 6.5 *is not* zero for currents that vary with time! Equation 6.5 simply states that the steady current diverging from a small volume per unit volume is zero. This statement follows from the definition of the divergence as a limit.[1]

6.3 CONDUCTION CURRENT

We have previously stated that a conductor is a material that allows charge (electron) motion under the influence of an applied field. These electrons are called *conduction*, *valence*, or *free* electrons, *but are not excess* electrons since the conductor is charge neutral. The force involved is easy to calculate from the definition of electric field.

$$\mathbf{F} = Q\mathbf{E} \quad \text{(N)}$$

Since

$$Q = -1.602 \times 10^{-19} \equiv -e \quad \text{(C)} \tag{6.6}$$

we have

$$\mathbf{F} = -e\mathbf{E} \quad \text{(N)} \tag{6.7}$$

[1] See equation 2.20.

for an electron. An electron in free space would accelerate unimpeded, but an electron in a conducting material cannot do this. Instead, an electron attains a constant *average* velocity called the *drift* velocity, $\mathbf{u}_d$. The ability of an electron to traverse the obstacle course created by the thermally excited crystalline lattice structure of the conductor is called the *mobility*, μ_e, of the electron in the material. Thus, due to collisions and the resulting loss of energy, an electron reaches an average velocity. The equation that relates drift velocity to mobility and applied field is

$$\mathbf{u}_d = -\mu_e \mathbf{E} \quad (\text{m/s}) \tag{6.8}$$

Mobility is measured in $\text{m}^2/(\text{V} \cdot \text{s})$. The drift velocity for good conductors is surprisingly small, being of the order of 10^{-2} m/s or less. The mobility of aluminum is approximately 1.4×10^{-4}, while for copper and silver it has values of about 3.2×10^{-3} and 5.2×10^{-3}, respectively.

Current density, **J**, in a conductor is then equal to the density of the conduction or valence electrons times their average velocity, so

$$\mathbf{J} = \rho_{ve}(-\mu_e \mathbf{E})$$

or

$$\mathbf{J} = -\rho_{ve} \mu_e \mathbf{E} \tag{6.9}$$

where ρ_{ve} is the density of valence electrons in the conductor. It is more common to relate current density and applied field through the parameter conductivity, σ, so that *conduction current density* is given by *Ohm's law* in point form,

$$\boxed{\mathbf{J} = \sigma \mathbf{E}} \tag{6.10}$$

so that **J** is linearly related to **E** through σ. Conductivity is measured in $\text{A}/(\text{V} \cdot \text{m})$, or $(\Omega \cdot \text{m})^{-1}$, or mho/m, since a volt per ampere is an ohm. For our purposes, we may consider conductivity to be a scalar constant. Values for some common conductors are listed in Appendix B for room temperature. Temperature is bound to affect conductivity, for a higher temperature gives rise to a more excited crystalline lattice structure and thus lower drift velocity. These effects are more clearly seen if we solve equation 6.9 and 6.10 for σ. That is,

$$\sigma = -\rho_{ve} \mu_e \tag{6.11}$$

a positive quantity, since μ_e is positive and ρ_{ve} is negative (electrons). Equation 6.10 is called the *point form* of Ohm's law, because Ohm's law of circuit theory is only a special case of equation 6.10.

6.4 JOULE'S LAW

It was mentioned in the previous section that (macroscopically) an average or drift velocity is reached by valence electrons in the conduction process. This occurs because of the inevitable collision between these electrons and the atoms of the conductor. Thus, in accelerating and decelerating between collisions the electrons give up part of their energy. If the steady current is to be maintained in the conductor, it is necessary to continually supply energy to the charges which will ultimately appear as *heat*.

Suppose that inside a conductor there are N valence or conduction electrons available per cubic meter. Thus, equation 6.1 may be written

$$\mathbf{J} = N(-e)\mathbf{u} \tag{6.12}$$

According to equation 6.7, the force on each electron is $\mathbf{F} = -e\mathbf{E}$ if $\mathbf{E}$ is the electric field in the conductor. Since *work* is the dot product of *force* and *distance*, and *distance* is the product of *velocity* and *time*, the differential work done by the electric forces in the differential time, dt, on a single electron is

$$dW_e = -e\mathbf{E} \cdot \mathbf{u}\, dt$$

Then, the differential work done in moving all the electrons in a small volume, dv, is

$$dW = -e\mathbf{E} \cdot \mathbf{u}\, dt\, N\, dv = N(-e)\mathbf{u} \cdot \mathbf{E}\, dv\, dt$$

or, with equation 6.12,

$$dW = \mathbf{J} \cdot \mathbf{E}\, dv\, dt \tag{6.13}$$

Equation 6.13 gives the energy transformed into heat within a small volume over a small time interval. The *rate* at which energy is converted into heat per unit volume is the power dissipated per unit volume and is

$$\frac{dP}{dv} = \frac{dW/dt}{dv} = \mathbf{J} \cdot \mathbf{E} \quad (\text{W/m}^3) \tag{6.14}$$

Equation 6.14 is the point form of Joule's law. The integral form of Joule's law is obtained by simply integrating the differential power density over whatever volume we are interested in. This is usually the volume occupied by the conductor. Thus,

$$\boxed{P = \iiint_{\text{vol}} \mathbf{J} \cdot \mathbf{E}\, dv} \quad (\text{W}) \tag{6.15}$$

It should be mentioned here that a term exactly like that given by equation 6.15 will appear when Poynting's theorem is considered in Chapter 9.

6.5 RESISTANCE

It is convenient at this point to define the electric circuit parameter resistance, R, for steady currents as

$$\boxed{R \equiv \frac{\Phi_{ab}}{I} = \frac{1}{G}} \quad (\text{ohms}, \Omega) \tag{6.16}$$

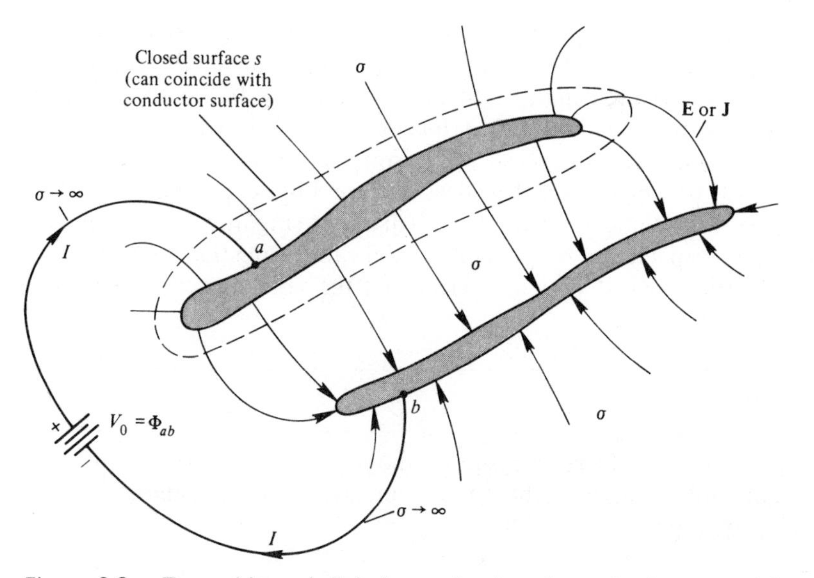

Figure 6.3. Two arbitrary infinitely conducting electrodes in a material (σ) for calculating resistance or conductance.

where the source current I enters the higher potential terminal of the device we wish to call a resistor. This is shown in Figure 6.3. G is called the conductance. Notice that we assume that the conductivity of the connecting wires and electrodes is infinite, but that of the intervening material is finite. It is still possible to use $\mathbf{E} = -\nabla\Phi$, because the charges are in *uniform* motion and the charge distribution is the same (macroscopically) at any point at any instant of time. Thus, with equations 6.3 and 6.10, we have

$$G = \frac{\oiint_s \sigma \mathbf{E} \cdot d\mathbf{s}}{\Phi_{ab}} = \frac{-\oiint_s \sigma \nabla\Phi \cdot d\mathbf{s}}{\Phi_{ab}} \quad (\text{mhos}, \mho) \tag{6.17}$$

Notice that a *closed* surface integral is used to ensure that all electric field (or current density) lines *leaving* the positive electrode are included. The current entering the closed surface s by way of the interconnecting wire is to be excluded because, according to equation 6.4, the *net* current out of s is zero. Resistance and conductance as defined here are positive quantities, and will depend only on geometry and σ.

Comparing equation 6.17 with equation 4.37, and also comparing Figure 6.3 with Figure 4.10, shows that if σ and ε have the same dependence on position, then, for the same geometries, C will be proportional to G. In particular, if σ and ε are constant,

$$\boxed{\frac{C}{G} = \frac{\varepsilon}{\sigma}} \tag{6.18}$$

and the electrostatic capacitance problem is *dual* to the steady electric current conductance problem. Several of the problems at the end of this chapter are designed to demonstrate this. It should also be obvious that lines of **J** (current density) and lines of **D** (flux density) are the same.

We now see why the mesh of resistors in Example 5 of Chapter 5 (Figure 5.13) becomes a resistive sheet (Figure 5.14) as the number of field points becomes infinite. In Example 6 of Chapter 5 (Figure 5.15) we found the approximate capacitance of a square coaxial cable using the finite difference result, equation 5.84. It should be clear that the same numerical technique can be employed here.

EXAMPLE 1

What is the finite difference result for the shunt resistance of the square coaxial cable of Figure 5.14 if the conductivity of the material is σ? Using equations 6.18 and 5.85,

$$R = \frac{1}{G} = \frac{\varepsilon}{\sigma C} = \frac{1}{2.26 l \sigma}$$

Notice that the resistance is inversely proportional to σ. It is also inversely proportional to l because the current density ($\mathbf{J} = \sigma \mathbf{E}$) is perpendicular to the lengthwise dimension l at the surface of the conductors.

For a given geometry, R is relatively easy to measure, and if this is done, the capacitance for the same geometry can easily be calculated using equation 6.18. Much use is made of resistance analogs in problems of a practical nature. See Problem 21.

With the aid of equation 4.37 we can easily show that $R = d/\sigma s$ for a resistive material between parallel plates. If the resistive material is confined to the region *between* plates with air or some other material with negligible conductivity outside, then the current will be confined ($\mathbf{J} = \sigma \mathbf{E}$) to the resistive material and leakage current is negligible. Furthermore, since $I = Js$ and $V_0 = Ed = (J/\sigma)d$ in this simple example, we have $V_0 = IR$, which is Ohm's law of circuit theory. This result should not be surprising since it is incorporated in the definition, equation 6.16. The power being dissipated in the resistor is also easy to calculate. Using equation 6.15,

$$P = \iiint_{\text{vol}} \mathbf{J} \cdot \mathbf{E} \, dv = JE \iiint_{\text{vol}} dv = JEls$$

or

$$P = \frac{V^2}{R} = I^2 R = VI \quad \text{(W)}$$

6.6 SOURCES FOR CURRENTS

We have said very little up to this point about the sources that supply the forces to maintain a current flow. In the case of conduction current density, Ohm's law relates the current density and the electric field intensity *in the conductor*, but where did this electric field come from? In other words, a source of energy is required somewhere if the conduction process is to be continual. A conductor, as we have seen, possesses a resistance. It also normally has two ends. Into one of these ends charges are being supplied, while at the same time equal amounts of charge are being removed from the other end. Between the two ends of the conductor there must exist a source. These energy sources must supply nonelectric, or independent, forces to propel the charges. It is customary to refer to these nonelectric sources as suppliers or generators of electromotive force (emf).

Generators of emf may take many different forms. For example, Faraday's law (Chapter 9) states that a time dependent magnetic field is a source of emf. Perhaps the most common source (and certainly the first of practical use) of emf is the chemical cell or battery. The ordinary rotating dc generator is another example of an emf source. Free electric charges at any point may be acted upon by two distinct forces: the Coulomb electric forces due to the distributed electric charges themselves, and the nonelectric forces, which are not due in any way to the charges. For this reason the nonelectric forces are said to be *independent*, *external*, or *impressed* forces. The emf being produced by a generator may be defined as the line integral of the *impressed* electric field of the generator from the negative end (terminal) to the positive end *through the generator*. That is,

$$\text{emf} = \int_{(-)}^{(+)} \mathbf{E}_i \cdot d\mathbf{l} \quad \text{(volts, taken through source)} \tag{6.19}$$

The impressed electric field is fictitious, because it is not due to charges, but it does give the correct force on charges when it is multiplied by the charge ($\mathbf{F}_i = Q\mathbf{E}_i$).

EXAMPLE 2

Consider a battery with no external load (i.e., open circuited) as shown in Figure 6.4. Then, from equation 6.19

$$\text{emf} = \int_{\substack{a \\ l_1}}^{b} \mathbf{E}_i \cdot d\mathbf{l}$$

But $\mathbf{E}_i = -\mathbf{E}$: that is, the impressed field is the negative of the field due to the stationary charges since no current is flowing. Therefore,

$$\text{emf} = -\int_{\substack{a \\ l_1}}^{b} \mathbf{E} \cdot d\mathbf{l}$$

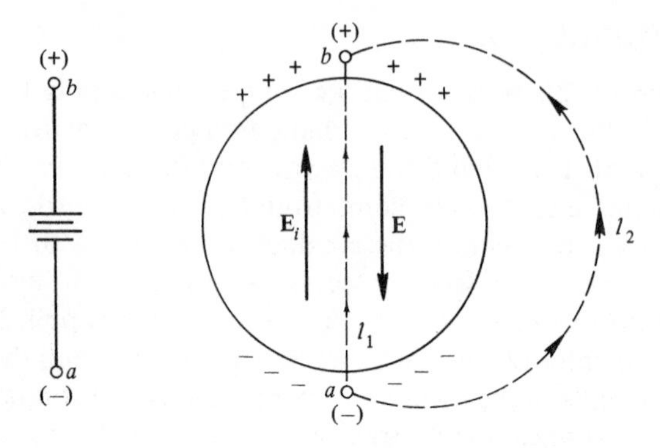

Figure 6.4. An open-circuit ($I = 0$) battery, $\mathbf{E}_i = -\mathbf{E}$.

Now, **E** is a conservative field, so its line integral is independent of the path, and

$$\text{emf} = -\int_{\substack{a \\ l_1}}^{b} \mathbf{E} \cdot d\mathbf{l} = -\int_{\substack{a \\ l_2}}^{b} \mathbf{E} \cdot d\mathbf{l} = -\int_{\substack{a \\ l_2}}^{b} -\nabla\Phi \cdot d\mathbf{l}$$

or

$$\text{emf} = \Phi_b - \Phi_a = \Phi_{ba} = V \quad \text{(V)} \tag{6.20}$$

Thus, we have the well-known circuit theory result that the emf of a generator is the potential difference between its open-circuited terminals. Notice that this result agrees with that given in Figure 6.4.

In order to gain more insight into the role of nonelectric forces, let us consider a problem in which there will be a current flow ($\mathbf{E}_i \neq -\mathbf{E}$).

EXAMPLE 3

A large conducting slab (σ, ε_0) is separated into three layers as shown in Figure 6.5. On the two outer sides of the slab are placed very thin sheets of a highly conducting material. These two thin sheets are connected by a wire of negligible resistance (short-circuit) so that a current flows when a uniform electric field, $\mathbf{E}_i$, is impressed in the middle section between the two outer sections. We want to find **J**, **E**, and ρ_s everywhere. First of all, we will neglect fringing, as we usually do for problems of this type. The problem is (therefore) one-dimensional. The current density, **J**, must be the same everywhere in the slab (Why?), so

$$J_z = \sigma E_{z2} = \sigma(E_i + E_{z1})$$

where $\mathbf{E}_1 = \mathbf{a}_z E_{z1}$, $\mathbf{E}_2 = \mathbf{a}_z E_{z2}$, and $\mathbf{E}_i = \mathbf{a}_z E_i$ are as shown in their respective regions in Figure 6.5. Therefore,

$$E_{z2} = E_i + E_{z1}$$

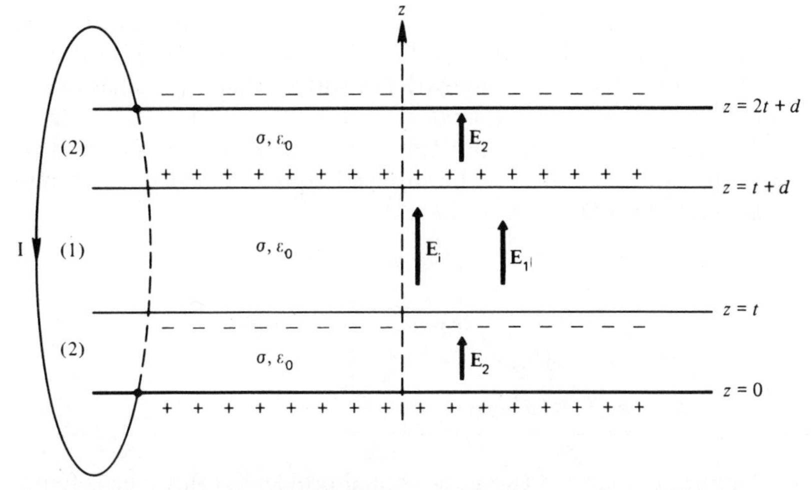

Figure 6.5. A large conducting slab (σ, ε_0) for $0 \leq z \leq 2t + d$. Thin sheets of a highly conducting material exist at $z = 0$ and $z = 2t + d$. $\mathbf{E}_i$ exists for $t < z < t + d$.

Now, even though there is a current flow and the charges are continually in motion, the same charge distribution will exist at any point at any given instant of time! The same situation exists for the vacuum diode of this chapter and also for Example 2. The charge distribution is thus the same as a *static* distribution of charge (which is conservative), and the circulation of **E** (not including the nonelectric $\mathbf{E}_i$) is zero, or

$$\oint \mathbf{E} \cdot d\mathbf{l} = 0 = E_{z2}t + E_{z1}d + E_{z2}t + 0$$

The closed path used is a straight line from $z = 0$ to $z = 2t + d$ parallel to the z axis and around through the shorting wire (where **E** is essentially zero). The last equation gives

$$E_{z1} = -E_{z2}\frac{2t}{d}$$

We now have two equations and two unknowns, so

$$E_{z1} = -E_i\frac{2t}{2t + d}$$

and

$$E_{z2} = E_i\frac{d}{2t + d}$$

Also,

$$J_z = \frac{\sigma}{2t + d}E_i d$$

This result could have been written down initially because $E_i d$ is the emf of the impressed field in the middle section and $\sigma/(2t + d)$ is the conductance of the slab per unit area! The surface charge densities can be found from the (static) boundary condition $\rho_s = \mathbf{a}_n \cdot (\mathbf{D}_1|_s - \mathbf{D}_2|_s)$ which is simply equation 4.33b. Refer to Figure 4.8, where it is shown that $\mathbf{a}_n$ is unit vector normal to the surface in question and pointing into region (1). With $\mathbf{D} = \varepsilon_0 \mathbf{E}$ we have

$$\rho_s = \varepsilon_0 E_{z2}, \qquad z = 0$$

$$\rho_s = \varepsilon_0(E_{z1} - E_{z2}) = -\varepsilon_0 E_i, \qquad z = t$$

$$\rho_s = \varepsilon_0(E_{z2} - E_{z1}) = \varepsilon_0 E_i, \qquad z = t + d$$

$$\rho_s = -\varepsilon_0 E_{z2}, \qquad z = 2t + d$$

A more complicated two-dimensional problem of this type is found as Problem 8 at the end of this chapter.

In this section we have been primarily concerned with the sources for conduction currents. What about the sources for convection currents, or currents in a vacuum or in a rarefied gas? In a vacuum there are no conductors, obviously, and the source of the energy to move the charges usually comes from the electric field of some stationary distribution of charge. The vacuum diode is an example. The impressed electric field from a nonelectric emf is normally not present. Also, we do not find a linear relation between convection current density and the electric field intensity such as existed in the form of Ohm's law for conductors. The Langmuir–Child law for vacuum diodes shows this very well. This will be examined very shortly.

Convection currents and conduction currents have other differences, besides those attributed to their respective energy sources. In the former case the charges can accelerate and acquire considerable kinetic energy, while in the latter case this is not so. In the former case, an excess of charge exists, while in the latter case, the conductor is charge neutral. They both can be described, however, by charges in motion.

6.7 INTRINSIC SEMICONDUCTORS

In order to complete this chapter on steady current, a brief discussion of the behavior of intrinsic semiconductors will be included. According to the concepts of *quantum theory*, those electrons in semiconductors which have enough energy will exist in the conduction band. To exist in the conduction band these electrons must have crossed the forbidden-band energy gap which is about 1 eV (1.6×10^{-19} J) for semiconductors. The surprising thing about this behavior is the fact that the *vacancies* left below the forbidden band by these conduction electrons behave as if they were real *positive* charges with charge equal to e. The mobility of this vacancy or *hole*, as it is logically called, is less than that for the electron whose absence created it. In the semiconductor, then, we have two current carriers available for conduction. The currents resulting from the application of an electric field to a semiconductor will be in the same direction for both types of carriers (Why?). The conductivity of semiconductors then is simply

$$\sigma = -\rho_{ve}\mu_e + \rho_{vh}\mu_h \tag{6.21}$$

where μ_h is the hole mobility. This explanation of semiconductor behavior is rather fictitious, but, nevertheless, gives good answers.

Germanium and silicon are the two most useful semiconductors. Intrinsic germanium has electron and hole mobilities of 0.36 and 0.17, respectively. For intrinsic silicon these values are 0.12 and 0.025.

It may seem surprising that the mobilities in semiconductors are one or two orders of magnitude larger than mobilities for good conductors. The charge densities are, however, very much less in semiconductors, so that the conductivity of good conductors is something like 10^8 times that of semiconductors and 10^{25} times that of good *insulators*. Semiconductors possess conductivities that depend strongly on temperature (a distinct difficulty). This dependence is due to a drastic increase of charge densities with increasing temperature. Thus, the conductivity of semiconductors increases with temperature. The opposite effect was noted for conductors. Typical values of conductivity for germanium and silicon, the most useful semiconductors, are 1.7 ℧/m and 0.7×10^{-3} ℧/m, respectively, at 300°K.

6.8 THE SEMICONDUCTOR JUNCTION DIODE AND THE VACUUM DIODE

Now that electric current has been formally introduced and semiconductors discussed, we can proceed, as promised, to investigate the semiconductor junction diode and the vacuum diode in terms of Poisson's equation.

EXAMPLE 4

As another example of the solution of Poisson's equation, let us consider a symmetrical *P-N* semiconductor junction as shown in Figure 6.6. We assume that for $x < 0$ the material is *P* type. That is, this

Figure 6.6. *P-N* semiconductor junction.

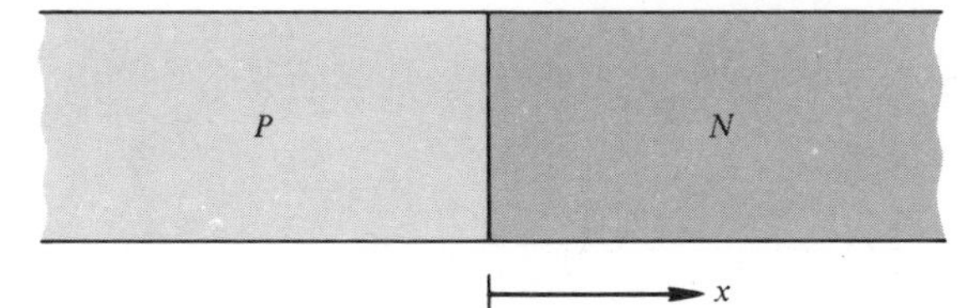

material has been doped so that there is an excess of *holes* (equivalent positive charges). For $x > 0$ the material has the same degree of doping as for $x < 0$, but has been doped so that there is an excess of electrons (negative charges). Equilibrium is established when the holes diffuse to the right and the electrons diffuse to the left in such a manner that an electric field is established which will reduce this diffusion current to zero (neglecting drift currents). To prevent holes from moving to the right and electrons to the left, a negative electric field, E_x, is required. This, in turn, requires that there exist a net positive charge for $x > 0$ and a net negative charge (with the same distribution) for $x < 0$. Thus, the charge distribution is an odd function, and consists of holes and *donor* ions for $x > 0$ and electrons and *acceptor* ions for $x < 0$. The proper charge distribution may be approximated by many mathematical functions. A reasonable model of the junction is realized if we *assume*

$$\rho_v(x) = \begin{cases} A_0(e^{-\alpha_1 x} - e^{-\alpha_2 x}), & x > 0 \\ -A_0(e^{\alpha_1 x} - e^{\alpha_2 x}), & x < 0 \end{cases} \tag{6.22}$$

where $\alpha_2 > \alpha_1$.

Our problem here is to determine the potential and electric field, given Poisson's equation. As in other similar situations, we have assumed uniformity, so that the only variable here is x. Poisson's equation is

$$\frac{d^2\Phi}{dx^2} = -\frac{\rho_v}{\varepsilon} \tag{6.23}$$

Some labor is avoided if we work only with the region $x > 0$, results for $x < 0$ being deduced from symmetry considerations. Substitution of equation 6.22 into equation 6.23 gives

$$\frac{d^2\Phi}{dx^2} = -\frac{A_0}{\varepsilon}(e^{-\alpha_1 x} - e^{-\alpha_2 x}), \qquad x > 0 \tag{6.24}$$

Integrating once,

$$\frac{d\Phi}{dx} = \frac{A_0}{\varepsilon}\left(\frac{e^{-\alpha_1 x}}{\alpha_1} - \frac{e^{-\alpha_2 x}}{\alpha_2}\right) + C_1 \tag{6.25}$$

Then, since $\mathbf{E} = -\nabla\Phi = -\mathbf{a}_x(\partial\Phi/\partial x)$, we have

$$E_x = -\frac{A_0}{\varepsilon}\left(\frac{e^{-\alpha_1 x}}{\alpha_1} - \frac{e^{-\alpha_2 x}}{\alpha_2}\right) - C_1, \qquad x > 0 \tag{6.26}$$

The electric field is zero for large x, so in the limit as x tends to infinity we have $C_1 = 0$. Then,

$$\frac{d\Phi}{dx} = \frac{A_0}{\varepsilon}\left(\frac{e^{-\alpha_1 x}}{\alpha_1} - \frac{e^{-\alpha_2 x}}{\alpha_2}\right), \qquad x > 0 \tag{6.27}$$

Integrating again,

$$\Phi(x) = -\frac{A_0}{\varepsilon}\left(\frac{e^{-\alpha_1 x}}{\alpha_1^2} - \frac{e^{-\alpha_2 x}}{\alpha_2^2}\right) + C_2 \tag{6.28}$$

Now, we must choose a zero reference for potential. A convenient choice is the junction itself. Thus, $\Phi = 0$ when $x = 0$, and from equation 6.28

$$C_2 = \frac{A_0}{\varepsilon}\left(\frac{1}{\alpha_1^2} - \frac{1}{\alpha_2^2}\right) \tag{6.29}$$

Finally,

$$\Phi(x) = \frac{A_0}{\varepsilon}\left(\frac{1 - e^{-\alpha_1 x}}{\alpha_1^2} - \frac{1 - e^{-\alpha_2 x}}{\alpha_2^2}\right), \qquad x > 0 \tag{6.30}$$

A sketch of $\rho_v(x)$, $\Phi(x)$, and $E_x(x)$ is given in Figure 6.7. Values for

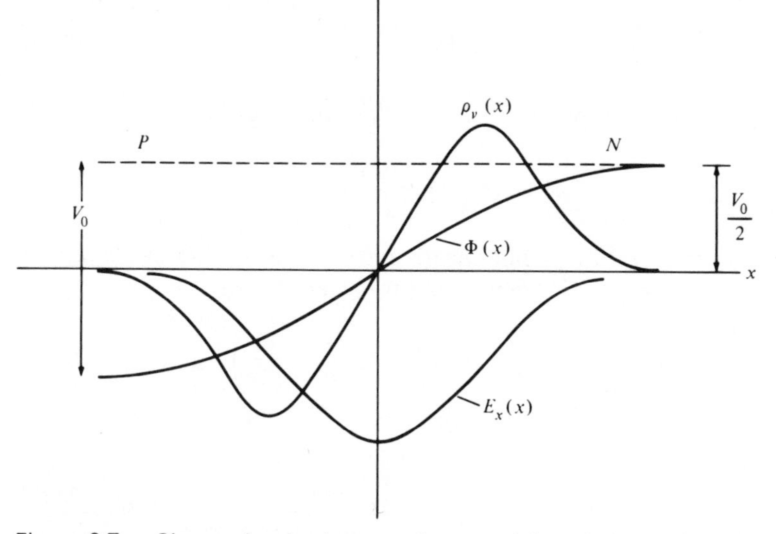

Figure 6.7. Charge density (assumed), potential, and electric field for a semiconductor junction.

$x < 0$ may be obtained from values for $x > 0$ by simply noting that $\rho_v(x)$ is an *odd* function, $\Phi(x)$ is an *odd* function and $E_x(x)$ is an *even* function. The parameters, A_0, α_1, and α_2 are arbitrary in Figure 6.7.

The potential difference, V_0, from one end of the device to the other may be obtained from equation 6.30. The highest potential occurs for *large* x and is $V_0/2$, so

$$\frac{V_0}{2} = \frac{A_0}{\varepsilon}\left(\frac{1}{\alpha_1^2} - \frac{1}{\alpha_2^2}\right)$$

or

$$V_0 = \frac{2A_0}{\varepsilon}\left(\frac{1}{\alpha_1^2} - \frac{1}{\alpha_2^2}\right) \quad \text{(V)} \tag{6.31}$$

The total positive charge for $x > 0$ and a cross-sectional area of s is

$$Q = \iiint_{\text{vol}} \rho_v \, dv = s \int_0^\infty A_0(e^{-\alpha_1 x} - e^{-\alpha_2 x}) \, dx$$

or

$$Q = sA_0\left(\frac{1}{\alpha_1} - \frac{1}{\alpha_2}\right) \quad \text{(C)} \tag{6.32}$$

with $\alpha_2 > \alpha_1$ we have the *approximate* relations.

$$V_0 \approx \frac{2A_0}{\varepsilon\alpha_1^2} \tag{6.33}$$

and

$$Q \approx \frac{sA_0}{\alpha_1} \tag{6.34}$$

Eliminating α_1 in equation 6.34 by using equation 6.33 gives

$$Q = s\left(\frac{A_0 \varepsilon V_0}{2}\right)^{1/2} \tag{6.35}$$

Since Q and V_0 are not linearly related, a new definition of capacitance is necessary. This definition (from circuit theory) is

$$C \equiv \frac{dQ}{dV_0} \qquad \left(I = C\frac{dV_0}{dt} = \frac{dQ}{dt}\right) \tag{6.36}$$

Therefore,

$$C = s\left(\frac{A_0 \varepsilon}{8V_0}\right)^{1/2} \quad \text{(F)} \tag{6.37}$$

The potential difference, V_0, is the *sum* of any external voltage source and the unbiased barrier voltage. If V_0 is increased, then C decreases. Another equation for the capacitance (from equations 6.33 and 6.37) is

$$C = \frac{s\varepsilon\alpha_1}{4} \quad \text{(F)} \tag{6.38}$$

which is similar to that for an ordinary parallel plate capacitor if we note that α_1 has the dimension per meter (m^{-1}).

This simplified picture of the behavior of the junction diode demonstrates that it has some interesting practical applications besides those attributed to its well-known rectifying ability. One of these is its use as a

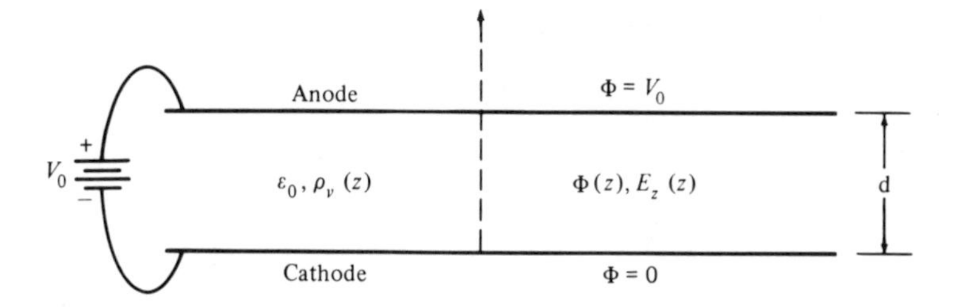

Figure 6.8. Geometry for a planar, vacuum diode.

voltage controlled capacitor as shown by equation 6.37. If the diode is part of a resonant circuit, for example, then the resonant frequency can be varied over a small range by an external voltage. Automatic frequency control (AFC) systems use the diode in this manner.

The vacuum diode is another device that still finds uses today in spite of the many advances in semiconductor technology. The planar diode is like a parallel plate capacitor except that the lower plate at $z = 0$, called the *cathode*, is a very special conducting material and is usually heated so as to readily emit electrons. It is shown in Figure 6.8. We again assume that d is small compared to the plate (electrode) dimensions so that fringing is safely neglected and we have a one-dimensional problem. The plate at $z = d$ is called the *anode* and is raised to a potential V_0 with respect to the cathode. In the same way, the coaxial version has a cathode at $\rho = a$ and an anode at $\rho = b$. The current in the diode is *time-independent*, thus, at any point in the cathode-anode region there will always be as much charge arriving as there is leaving, so the amount of charge at the point remains constant. In moving through the region the identity of the charges in a volume dv is continually changing, but $\rho_v dv$ is not. Thus, Poisson's equation (a *static* form) can be used to find Φ. Our immediate goal is to show that the current depends on the three-halves power of the potential for *any* geometry.

An electron starting from the cathode (at zero potential) has zero initial energy since $W = Q\Phi$ and $\Phi = 0$ when $z = 0$. If the electrons just barely manage to leave the cathode because of the force of repulsion offered by the electron cloud already present in the interspace, then the initial velocity of the electrons is essentially zero. Hence, the kinetic energy, which is proportional to the square of the velocity, is also zero initially. The total energy is constant and zero initially, so it is always zero, or

$$0 = Q\Phi + \tfrac{1}{2}mu^2$$

or

$$u = \left(\frac{-2Q}{m}\right)^{1/2} \Phi^{1/2} \tag{6.39}$$

We also have the equation for convection current density,

$$\mathbf{J} = \rho_v \mathbf{u} \tag{6.40}$$

conservation of charge,

$$\nabla \cdot \mathbf{J} = 0 \tag{6.41}$$

and Poisson's equation

$$\nabla^2 \Phi = \frac{-\rho_v}{\varepsilon_0} \tag{6.42}$$

Let us assume that we are dealing with a one-dimensional problem of unspecified geometry (like the planar diode, for example) in which the current density has only one component, and the current density vector **J** lies in the direction of the one variable. The current density will be perpendicular to, and, constant on, the equipotential surfaces. Calling the independent variable η, and substituting equations 6.39 and 6.40 into equation 6.42 gives

$$\Phi^{1/2} \nabla^2 \Phi = C J_\eta \tag{6.43}$$

where

$$\Phi = \Phi(\eta)$$

and

$$C = -\frac{1}{\varepsilon_0} \left(\frac{m}{-2Q} \right)^{1/2} = -0.191 \times 10^6 \tag{6.44}$$

Integrating both sides of equation 6.43 over an equipotential surface of area s, where η is constant, gives

$$\Phi^{1/2} \nabla^2 \Phi \iint_s ds = C \iint_s J_\eta \, ds$$

or

$$\Phi^{1/2} \nabla^2 \Phi s = C I_\eta$$

where I_η is the current. Thus,

$$I_\eta = \Phi^{1/2} \nabla^2 \Phi s / C \tag{6.45}$$

If we assume for the moment that a solution to equation 6.45 has been found,

$$\Phi(\eta) = A f(\eta) \quad \text{or} \quad A = \frac{\Phi(\eta)}{f(\eta)} \tag{6.46}$$

then equation 6.45 becomes

$$I_\eta = A^{3/2} [f(\eta)]^{1/2} \nabla^2 f(\eta) s(\eta) / C$$

Eliminating A,

$$I_\eta = \frac{s(\eta)}{C} \Phi^{3/2} \frac{\nabla^2 f(\eta)}{f(n)} \tag{6.47}$$

At the anode, where $\eta = \eta_a$ and $s = s_a$, the potential is V_0, so

$$I_\eta = \frac{s_a}{C} V_0^{3/2} \left[\frac{\nabla^2 f(\eta)}{f(\eta)}\right]_{\eta=\eta_a} \tag{6.48}$$

and the current depends on the three-halves power of the potential for *any* geometry. This result is known as the *Child–Langmuir law* or *three-halves power law*. Equation 6.45 must be solved for a particular geometry if all the important quantities are desired.

EXAMPLE 5

For the *planar* diode of Figure 6.8 the area (in equation 6.45) is the same for any equipotential (that is, $s = s_a$), $\eta = z$, and equation 6.45 can be written

$$\Phi^{1/2}\nabla^2\Phi = \frac{C}{s_a} I_z \quad \text{(constant)} \tag{6.49}$$

where

$$\Phi = \Phi(z)$$

Since the potential is proportional to z for the planar capacitor, it is reasonable to try $\Phi(z) = Az^\alpha$, or $f(z) = z^\alpha$, in equation 6.46. It is easy to show (Problem 10) that $\alpha = \frac{4}{3}$. Equation 6.48 then gives

$$I_z = \frac{s_a}{C} V_0^{3/2} \left[\frac{(d^2/dz^2)(z^{4/3})}{z^{4/3}}\right]_{z=d},$$

or

$$I_z = -2.33 \times 10^{-6} s_a \frac{V_0^{3/2}}{d^2} \tag{6.50}$$

It is relatively easy to go back now and show that[2]

$$u(z) = 5.93 \times 10^5 V_0^{1/2} \left(\frac{z}{d}\right)^{2/3} \tag{6.51}$$

$$\rho_v(z) = -3.93 \times 10^{-12} V_0 (zd^2)^{-2/3} \tag{6.52}$$

and

$$\Phi(z) = V_0 \left(\frac{z}{d}\right)^{4/3} \tag{6.53}$$

[2] See Problem 11.

for the planar diode. The results for this geometry are shown in Figure 6.9. Notice that both boundary conditions are satisfied by equation 6.53.

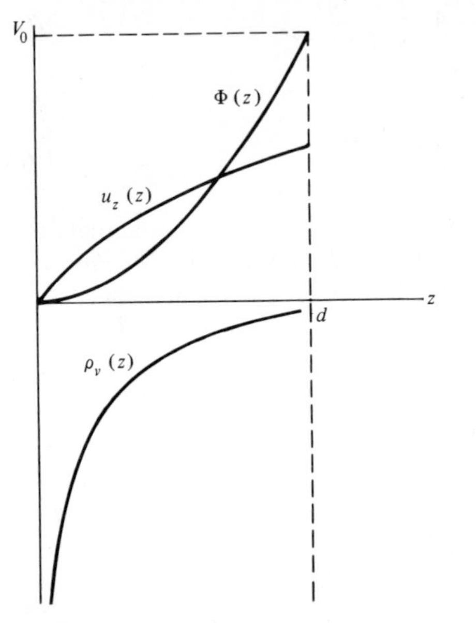

Figure 6.9. Potential, electron velocity, and charge density versus distance for the planar, vacuum diode.

Notice also that if the cathode does not emit electrons, the potential is simply that of a parallel plate capacitor,

$$\Phi(z) = V_0 \frac{z}{d}.$$

The nonlinear relation between current and potential in the vacuum diode enables it to operate as a signal detector in certain applications.

6.9 CONCLUSIONS

In this chapter the various forms of current density were discussed and related to the total current in a given region. The continuity of current equation, which is of fundamental importance, was derived. A brief discussion of conductors showed that an important parameter, the conductivity, varies over wide limits for insulators to semiconductors to ordinary conductors (at ordinary temperatures). For some conductors, the conductivity becomes infinite if the temperature is lowered enough. These materials are called *superconductors*. In the next chapter we will show that current, or moving charge, is the source of an additional force field superimposed on that due to fixed charges and described by Coulomb's law.

The second circuit parameter, resistance, and its reciprocal, conductance were defined. In the steady current case, conductance is just as easy (or difficult!) to calculate as capacitance because for similar geometries they are simply related. This relation will be used when transmission lines are discussed.

Having introduced current, we returned to Poisson's equation and determined the volt-ampere relationship for the vacuum diode. We were able to do this using static concepts, even though the charges were in motion.

REFERENCES

Hayt, W. H. (see references for Chapter 2). A slightly different treatment of the *P-N* junction starts on page 213.

Rao, Nannapaneni N. *Basic Electromagnetics with Applications*. Englewood Cliffs, N.J.: Prentice-Hall, 1972. The vacuum diode is treated on page 350.

PROBLEMS

1. A conduction current density, **J**, flows in a long, circular conductor coaxial with the z axis. Find **E** everywhere.
2. If $\mathbf{J} = \mathbf{a}_z/(\rho^2 + a^2)[u(\rho) - u(\rho - a)]$:

   (a) Find $\nabla \cdot \mathbf{J}$ everywhere.
   (b) Find the total current in the $\mathbf{a}_z$ direction.
   (c) Find the total current out of *any* closed surface.

3. Find the total current crossing the $z = 0$ plane in the $+\mathbf{a}_z$ direction, if it is in the form of an electron beam with charge density $-10^{-9}e^{-\rho}$ and velocity $-2 \times 10^7\mathbf{a}_z$.
4. If $\mathbf{J} = a^2/r^2\mathbf{a}_r$, find $\nabla \cdot \mathbf{J}$ everywhere. Explain what happens at $r = 0$.
5. A truncated cone is concentric with the z axis and has a conductivity $\sigma = 3 \times 10^6$. The two faces have radii 3×10^{-3} m and 10^{-4} m and are 0.5 m apart. Assuming $J_z = I/\pi\rho^2$, find the resistance between the end faces. The cone apex is at (0, 0, 0).
6. A 1-V potential difference is applied between faces 2×10^{-2} m apart for intrinsic germanium. It has a uniform cross-sectional area, 10^{-3} m^2. If $\sigma = 4.35 \times 10^6\, e^{-4350/T}$, find:

   (a) E, σ, J, and I for electrons at 300°K.
   (b) Repeat (a) for 360°K.

7. Assume that a certain round wire has a diameter of 10^{-1} m and can carry a maximum current of 20 A. At this current level, what is the current density? Explain what happens if the maximum current rating is exceeded.

8. A uniform electric field, $\mathbf{E}_i = a_x \mathbf{E}_i$, is impressed in the cylindrical region (only), $\rho < a$, of a large homogeneous conducting material (σ, ε_0).

   (a) Show that a solution for the potential of the static distribution of charges compatible with the boundary conditions is

$$\Phi(\rho, \phi) = \begin{cases} \dfrac{E_i}{2} \rho \cos \phi, & \rho \leq a \\ \dfrac{E_i a^2}{2\rho} \cos \phi, & \rho \geq a \end{cases}$$

   (b) Find E, J, and ρ_s everywhere.

9. (a) Find the time required for an electron to leave the cathode and reach the plate for the planar diode of Example 5 (transit-time). Let $V_0 = 100$, $d = 10^{-3}$.
   (b) Estimate the highest usable frequency for this device.
10. Derive equation 6.50 starting with equation 6.49.
11. Derive equation 6.51, 6.52, and 6.53.
12. Calculate the *shunt* conductance per unit length for a coaxial cable with inner conductor radius a, outer conductor radius b, and a lossy dielectric with conductivity σ_d.
13. Calculate the *series* resistance per unit length for the coax in Problem 12 if the conductors have conductivity σ and thickness δ.
14. Repeat Problem 12 for a hollow conducting sphere with inner radius a and outer radius b and $\mathbf{E} = (1/r^2)\mathbf{a}_r$.
15. Calculate the conductance of a parallel plate capacitor.
16. If $\varepsilon_R = 4$ and $\sigma = 10^{-5}$ for the dielectric in a parallel plate capacitor, and if a 100-V battery draws 1 mA when connected across it, find C.
17. A two-wire line is embedded in a material with $\varepsilon = 4\varepsilon_0$ and $\sigma = 10^{-6}$. If the wires are spaced 2×10^{-2} m and have diameters 2×10^{-3} m, find the conductance (G) per unit length.
18. It is possible to construct an electric circuit with a piece of paper and a pencil. Assuming that graphite has a conductivity of 7×10^4, how "thick" would a 10^3-Ω resistor need to be if it is 2×10^{-2} m long and 10^{-3} m wide?
19. Find the electric field intensity in a plane halfway between the anode and cathode of a planar diode if $d = 10^{-3}$ m and $V_0 = 100$ V.
20. A semiconductor diode is used to alter the resonant frequency of a high Q parallel resonant circuit.

   (a) Show that a 1% increase in diode voltage produces a $\frac{1}{2}$% decrease in diode capacitance.

(b) Show that a 1% decrease in total capacitance produces a $\frac{1}{2}$% increase in the resonant frequency.

(c) If the diode capacitance is 10% of the total capacitance and it decreases by 1%, what percentage increase in resonant frequency can be expected?

21. The resistance measured between opposite edges of a 10-cm^2 piece of resistance paper is 500 Ω. What is the film thickness if the conductivity is 10^2 ℧/m?

22. (a) What differential equation must be satisfied by a coaxial vacuum diode?

    (b) What are the boundary conditions?

    (c) Show that $\Phi(\rho) = A\rho^{2/3}$ is a solution if the cathode is a directly heated filamentary wire.

23. Suppose an analog for Figure 5.5 is constructed out of resistance paper as in Figure 5.13. If a sharp knife is used to make a vertical cut in the paper from the upper electrode to the lower electrode, and if this cut is made midway between the vertical electrodes, what happens to the current measured by an ammeter in series with the source supplying the potential V_0?

Chapter 7
The Steady Magnetic Field

In the preceding chapters we found that most of the laws governing electrostatics result from Coulomb's experimental law. This law gives the force between fixed point charges. If we calculate the force between *moving* charges, as in Appendix F, we not only find the previously mentioned Coulomb force that gave us the definition of electric field, but also a relativistic correction to the Coulomb force. A new force thus appears, and a *magnetic field* can be defined in terms of this force. This introduction of the magnetic field is very acceptable to those who prefer a general approach. The magnetic field is also given in equation form by the Biot–Savart law for those who prefer a direct approach. The Biot–Savart law, however, is an experimental law and makes no mention of force by itself, and is therefore not directly related to the material in the preceding chapters of this text.

The Biot–Savart law is closely related to another famous experimental law called Ampere's law. Given the integral form of Ampere's law, we will be able to solve a limited number of problems that have a high degree of symmetry. The precedure is much like that used with Gauss's law of electrostatics. The differential form of Ampere's law is easy to obtain and is more commonly listed as another of Maxwell's (static) equations.

The relationship between magnetic field intensity, magnetic flux density, and magnetic flux is next investigated. These quantities are closely analogous to their counterparts of the electrostatic field, but there is one notable exception. There are no magnetic charges in nature, and, instead, the sources for the magnetic field are moving electric charges. A magnetic scalar potential will be derived, but it will be found that this potential function is of limited use. A much more useful magnetic *vector* potential will be derived to aid in solving magnetic field problems.

Analogous to the electrostatic dipole, there is a magnetostatic dipole, which is simply an electric current loop. The fields produced by this loop are similar to those produced by the electrostatic dipole, and we can use the knowledge of the behavior of the magnetic dipole when under the influence of an external magnetic field to explain the magnetization of magnetic materials. The magnetic dipole serves as a model for an electron orbiting around the nucleus in an atom, so there are magnetic dipoles in materials. The qualitative description of magnetization is presented in the next chapter.

Forces on current elements or moving charges are next considered. These forces are not only important in explaining the behavior of magnetic materials, but are also important in understanding the basic principles behind the operation of such devices as electric motors. The torque equations for a magnetic dipole are again very similar in general to those for the torque on an electric dipole.

7.1 THE BIOT–SAVART LAW[1]

A calculation is performed in Appendix F for the force on a charge Q_2 detected by a fixed observer when charges Q_1 and Q_2 are in uniform motion with velocities $\mathbf{u}_1$ and $\mathbf{u}_2$, respectively. The equation for this force is equation F.28 and is repeated here for convenience. A subscript (2) is added to the symbol for this force to indicate that it is the force on Q_2.

$$\mathbf{F}_2 = \mathbf{a}_r \frac{Q_1 Q_2}{4\pi\varepsilon r^2} + \frac{Q_1 Q_2}{4\pi\varepsilon c^2 r^2} \mathbf{u}_2 \times (\mathbf{u}_1 \times \mathbf{a}_r) \qquad \text{(F.28)}$$

if $u_1 \ll c$, where c is the velocity of light in the medium. See Figure 7.1. We now define (for the time being)

$$\boxed{\mu \equiv \frac{1}{\varepsilon c^2}} \qquad \text{(F.29)}$$

Then,

$$\mathbf{F}_2 = Q_2 \frac{Q_1 \mathbf{a}_r}{4\pi\varepsilon r^2} + Q_2 \mathbf{u}_2 \times \left(\mu \mathbf{u}_1 \times \frac{Q_1 \mathbf{a}_r}{4\pi r^2}\right)$$

[1] The description of the Biot–Savart law begins below equation 7.7. It is possible to begin there.

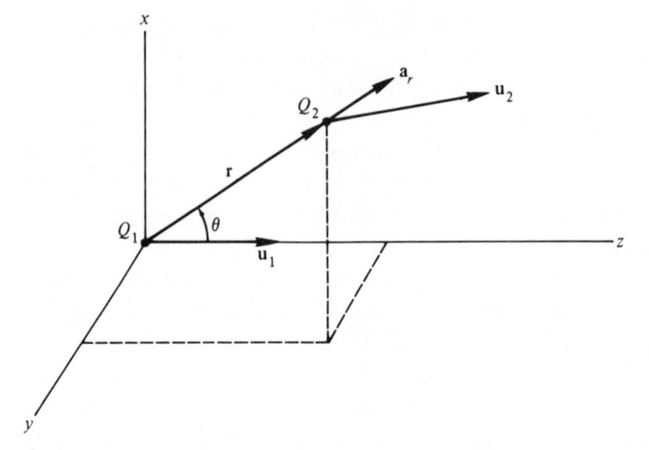

Figure 7.1. Geometry for the interaction force between two moving charges.

or, recognizing the electric field intensity, **E**, and the electric flux density, **D**, due to Q_1,

$$\mathbf{F}_2 = Q_2\mathbf{E} + Q_2\mathbf{u}_2 \times (\mu\mathbf{u}_1 \times \mathbf{D})$$

Subscripts (1) could be added to **E** and **D** to indicate that these terms arise from Q_1, but this has not been done here. The first term in the preceding equation is easily recognized, while the second term arises because of the charge motion. We can think of the second term as a correction to the ordinary Coulomb force. The term in parentheses is a vector such that when its cross product is taken with the current element ($Q_2\mathbf{u}_2$), a force results. The term in parentheses is therefore a force per unit current element much like the electric field is the force per unit charge. A definition is in order.

$$\mathbf{B} \equiv \mu\mathbf{H} = \mu\mathbf{u}_1 \times \mathbf{D} \tag{F.30}$$

or

$$\mathbf{H} = \mathbf{u}_1 \times \mathbf{D}$$

so that

$$\mathbf{F}_2 = Q_2(\mathbf{E} + \mathbf{u}_2 \times \mathbf{B}) \tag{F.31}$$

Equation F.31 is called the *Lorentz-force* equation and will be mentioned again. Notice that the electric and magnetic parts are easy to identify. It is also apparent from equations F.28 and F.29 that the magnetic force is normally much less than the electric force ($u_1, u_2 \ll c \approx 3 \times 10^8$ m/s for free space). **B** is called the magnetic flux density (webers per square meter), and **H** is called the magnetic field intensity (amperes per meter).

In much the same way that an electrostatic field is related to its sources (stationary charges) by Coulomb's law, a magnetic field is related to its sources by an experimental law called the *Biot–Savart law*. Before describing

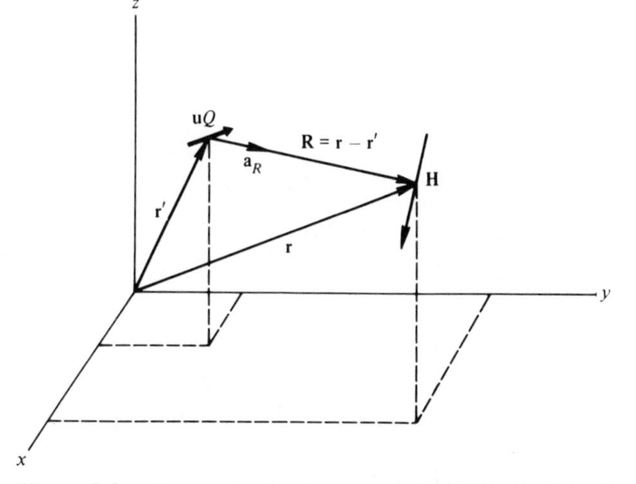

Figure 7.2. Geometry for the magnetic field strength, **H**, of a moving charge Q.

this law as an experimental law, we can easily obtain its mathematical form from the definition of **H** given in the preceding paragraph.

$$\mathbf{H} = \mathbf{u}_1 \times \mathbf{D} = \mathbf{u}_1 \times \frac{Q_1}{4\pi r^2}\mathbf{a}_r = \mathbf{u}_1 Q_1 \times \frac{\mathbf{a}_r}{4\pi r^2} \tag{7.1}$$

If the charge Q_1 is at r' rather than the origin, then as shown in Figure 7.2,

$$\mathbf{H} = \frac{Q}{4\pi}\mathbf{u} \times \frac{\mathbf{a}_R}{R^2} \quad \text{(A/m)} \tag{7.2}$$

where $\mathbf{a}_R$ and **R** have the usual meaning, and we have dropped the subscripts on **u** and Q since they are no longer necessary. It is very simple to obtain superposition integrals for the magnetic field intensity in terms of the various current sources. As discussed in Chapter 6, a current element can be described variously as

$$\mathbf{u}\,dQ \quad \text{or} \quad \mathbf{u}\rho_v\,dv \quad \text{or} \quad \mathbf{J}\,dv \quad \text{or} \quad \mathbf{J}_s\,ds \quad \text{or} \quad I\,d\mathbf{l} \tag{7.3}$$

The unit impulse response is obtained directly from equation 7.2 by setting $Q\mathbf{u} = \mathbf{u}/u = \mathbf{a}_u$, a *unit* vector in the direction of the charge velocity. It is

$$\mathbf{h}(\mathbf{r} - \mathbf{r}') = \frac{1}{4\pi}\left(\mathbf{a}_u \times \frac{\mathbf{a}_R}{R^2}\right) \tag{7.4}$$

Since the most general form for a current element is $\mathbf{J}\,dv$, we first obtain a superposition integral for **H** in terms of **J**. It is simply

$$\boxed{\mathbf{H}(\mathbf{r}) = \frac{1}{4\pi}\iiint_{\text{vol}'} \mathbf{J}(\mathbf{r}') \times \frac{\mathbf{a}_R}{R^2}\,dv'} \tag{7.5}$$

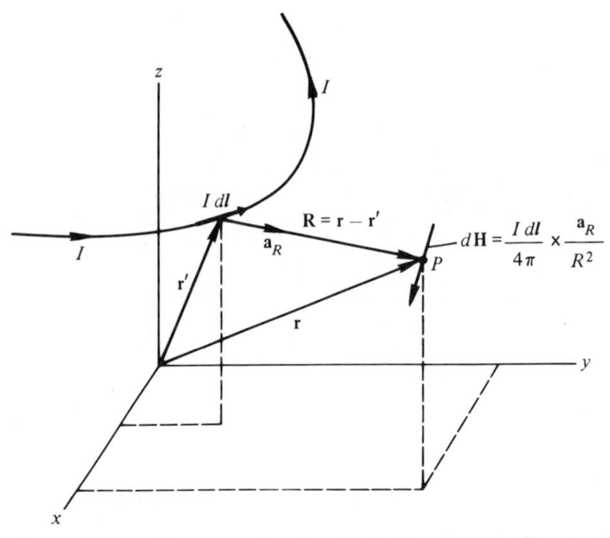

Figure 7.3. Geometry for the Biot–Savart law, filamentary current I.

Notice that, once again, knowing the unit impulse response enables us to immediately write down the response for *any* source. If the current source is a surface current density, $\mathbf{J}_s$,

$$\mathbf{H}(\mathbf{r}) = \frac{1}{4\pi} \iint_{s'} \mathbf{J}_s(\mathbf{r}') \times \frac{\mathbf{a}_R}{R^2}\, ds' \tag{7.6}$$

and if the current source is a filamentary current, I,

$$\mathbf{H}(\mathbf{r}) = \frac{1}{4\pi} \int_{l'} I\, d\mathbf{l}(\mathbf{r}') \times \frac{\mathbf{a}_R}{R^2} \tag{7.7}$$

The Biot–Savart law can now be described as an experimental law applying to a current element arbitrarily located. Consider a differential element of filamentary current, $I\,d\mathbf{l}$, as shown in Figure 7.3. In a manner consistent with previous formulations, this current element is located by $\mathbf{r}'$ since it is a source. Notice again that the current is not a vector, but its differential path is. According to the Biot–Savart law, a differential amount of magnetic field strength, $d\mathbf{H}$, is produced at some point P, which is located by $\mathbf{r}$. This law states that the field is inversely proportional to the square of the distance from the current filament to the point where the field is being evaluated, directly proportional to the current, directly proportional to the differential path length, and finally, directly proportional to the sine of the angle between the current filament and the line joining the current filament to the field point. The *direction* of the magnetic field is perpendicular to the plane formed by the current filament and the line joining the current filament to the field point. Of the two possible perpendicular directions, the correct one is obtained by the direction of progress of a right-hand-threaded screw turned from the current filament through the smaller angle to the line joining the current filament to the field point. For the MKSA system of units, the constant of proportionality is $1/4\pi$.

In terms of Figure 7.3, it is very simple to write this law down as a mathematical equation. It is

$$d\mathbf{H} = \frac{I\,d\mathbf{l}}{4\pi} \times \frac{\mathbf{a}_R}{R^2} \quad \text{(A/m)} \tag{7.8}$$

We now see that this is merely the differential form of equation 7.7, which in turn is a special case of equation 7.5.

7.2 THE INFINITE FILAMENTARY WIRE

EXAMPLE 1

As an example of the use of the Biot–Savart law, consider the infinitely long filamentary wire lying on the z axis as shown in Figure 7.4. We must

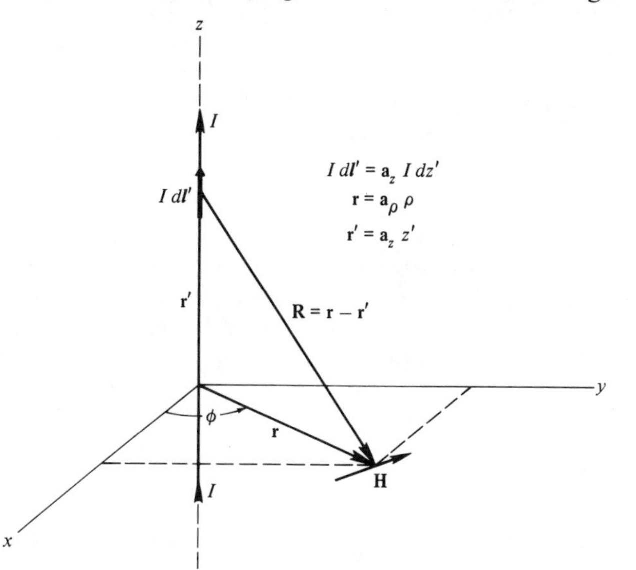

Figure 7.4. Geometry for finding **H** around an infinite filamentary current on the z axis.

keep in mind that there must be a return current somewhere. In the present example we consider the return current to be so far removed that it will not affect our answer. Since the filament is of infinite extent, our answer will not be affected if the point at which we choose to evaluate **H** is placed in the $z = 0$ plane. Symmetry considerations tell us that the field must also be independent of the azimuth angle, ϕ. Using equation 7.7, we have (in cylindrical coordinates)

$$\mathbf{H} = \int_{-\infty}^{\infty} \frac{I\,dz'\,\mathbf{a}_z}{4\pi} \times \frac{\mathbf{a}_\rho \rho - \mathbf{a}_z z'}{[\rho^2 + (z')^2]^{3/2}}$$

or

$$\mathbf{H} = \frac{I}{4\pi} \int_{-\infty}^{\infty} \frac{\rho \mathbf{a}_\phi}{[\rho^2 + (z')^2]^{3/2}}\,dz'$$

In the last equation the only variable is z', in which case ρ and $\mathbf{a}_\phi$ are *both constant*, and so

$$\mathbf{H} = \frac{I\rho\mathbf{a}_\phi}{4\pi} \int_{-\infty}^{\infty} \frac{dz'}{[\rho^2 + (z')^2]^{3/2}}$$

or simply

$$H_\phi(\rho) = \frac{I\rho}{4\pi} \int_{-\infty}^{\infty} \frac{dz'}{[\rho^2 + (z')^2]^{3/2}}$$

The required integral can be found in the tables, and so

$$H_\phi(\rho) = \frac{I}{2\pi\rho} \text{ (A/m)} \left[\text{or } H_\phi = \frac{I}{2\pi\rho}\,u(\rho)\right] \tag{7.9}$$

This result bears a striking resemblance to that for the electrostatic field of an infinite line charge which is repeated here for convenience.

$$E_\rho(\rho) = \frac{\rho_l}{2\pi\varepsilon\rho} \text{ (V/m)} \tag{7.10}$$

In amplitude equations 7.9 and 7.10 behave the same way (inverse distance). The electric field begins with the line charge and is radially directed. The magnetic field, however, is azimuthally directed such that *field lines close on themselves*. This would seem to indicate that the magnetic field *does not diverge* ($\nabla \cdot \mathbf{H} = 0$).

EXAMPLE 2

As a second example, suppose we imagine an infinite cylindrical shell of radius a carrying a uniform surface current density.

$$\mathbf{J}_s = \mathbf{a}_z \frac{I}{2\pi a}$$

where I is the total current according to equation 6.3. Particular care must be taken in setting up the equation for $\mathbf{H}$ for there are many inherent pitfalls in it. The geometry for calculating $\mathbf{H}$ is shown in Figure 7.5. Notice particularly that $\mathbf{a}_\rho \neq \mathbf{a}_{\rho'}$, for $\mathbf{a}_\rho$ is fixed in the up-

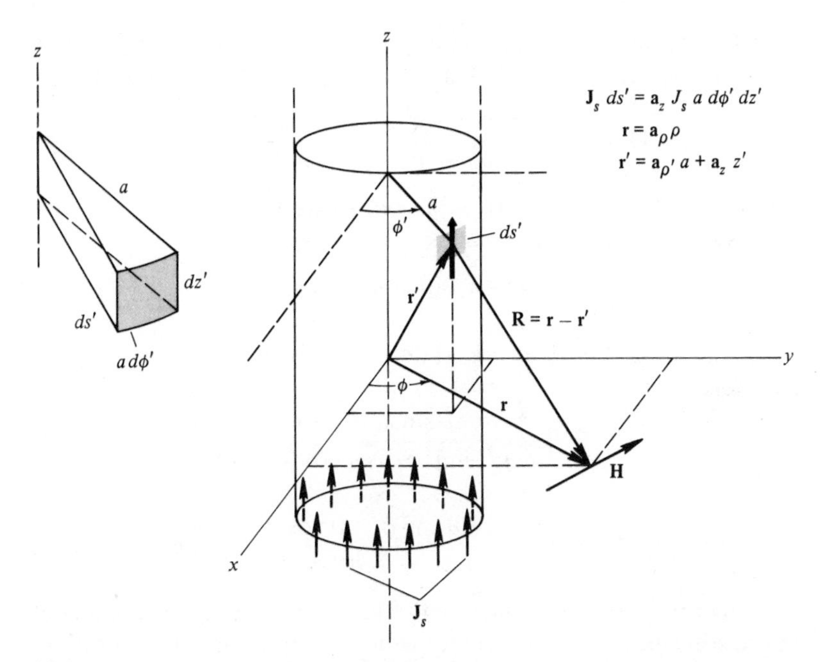

Figure 7.5. Geometry for finding **H** around an infinite uniform surface current density $\mathbf{J}_s$ on a cylinder of radius *a*.

coming integration, while $\mathbf{a}_{\rho'}$ has a direction which changes with ϕ'! We then have

$$\mathbf{R} = \mathbf{r} - \mathbf{r}' = \mathbf{a}_\rho \rho - \mathbf{a}_{\rho'} a - \mathbf{a}_z z'$$

and from equation 1.32b

$$R = |\mathbf{r} - \mathbf{r}'| = [\rho^2 + a^2 - 2\rho a \cos(\phi - \phi') + (z')^2]^{1/2}$$

Notice that **H** may be calculated in the $z = 0$ plane without loss of generality. Now using equation 7.6,

$$\mathbf{H} = \int_0^{2\pi} \int_{-\infty}^{\infty} \frac{I\mathbf{a}_z}{2\pi a(4\pi)} \times \frac{\mathbf{a}_\rho \rho - \mathbf{a}_{\rho'} a - \mathbf{a}_z z'}{[\rho^2 + a^2 - 2\rho a \cos(\phi - \phi') + (z')^2]^{3/2}} a \, d\phi' \, dz'$$

or

$$\mathbf{H} = \frac{I}{8\pi^2} \int_0^{2\pi} \int_{-\infty}^{\infty} \frac{(\mathbf{a}_\phi \rho - \mathbf{a}_{\phi'} a) \, d\phi' \, dz'}{[\rho^2 + a^2 - 2\rho a \cos(\phi - \phi') + (z')^2]^{3/2}}$$

since $\mathbf{a}_z \times \mathbf{a}_\rho = \mathbf{a}_\phi$ and $\mathbf{a}_z \times \mathbf{a}_{\rho'} = \mathbf{a}_{\phi'}$. The unit vector $\mathbf{a}_\phi$ is constant in the integration, while the unit vector $\mathbf{a}_{\phi'}$ depends on ϕ' but not on z'.

Therefore, we may immediately integrate on z', and this integration is of the same form as that for the filamentary current (Why?). We have

$$\mathbf{H} = \frac{I}{4\pi^2} \int_0^{2\pi} \frac{(\mathbf{a}_\phi \rho - \mathbf{a}_{\phi'} a)\, d\phi'}{\rho^2 + a^2 - 2\rho a \cos(\phi - \phi')}$$

It is now necessary to work in a coordinate system where the unit vectors are constant, and so we use

$$\mathbf{a}_\phi = -\mathbf{a}_x \sin\phi + \mathbf{a}_y \cos\phi$$

and

$$\mathbf{a}_{\phi'} = -\mathbf{a}_x \sin\phi' + \mathbf{a}_y \cos\phi'$$

Therefore,

$$H_x = \frac{-I}{4\pi^2} \int_0^{2\pi} \frac{\rho \sin\phi - a \sin\phi'}{\rho^2 + a^2 - 2\rho a \cos(\phi - \phi')}\, d\phi' \qquad (7.11)$$

and

$$H_y = \frac{I}{4\pi^2} \int_0^{2\pi} \frac{\rho \cos\phi - a \cos\phi'}{\rho^2 + a^2 - 2\rho a \cos(\phi - \phi')}\, d\phi' \qquad (7.12)$$

All of the subtle difficulties in this problem have been exposed, but we are still faced with some rather tedious integration. Equations 7.11 and 7.12 can be evaluated in terms of infinite series, but this would defeat the real purpose of this example. Instead, let us determine the answer for $\rho \gg a$. In this case,

$$H_x = \frac{-I \sin\phi}{4\pi^2 \rho} \int_0^{2\pi} d\phi' = -\frac{I}{2\pi\rho} \sin\phi \qquad (7.13)$$

$$H_y = \frac{I \cos\phi}{4\pi^2 \rho} \int_0^{2\pi} d\phi' = \frac{I}{2\pi\rho} \cos\phi \qquad (7.14)$$

In terms of H_x and H_y, we have

$$H_\rho = H_x \cos\phi + H_y \sin\phi$$

$$H_\phi = -H_x \sin\phi + H_y \cos\phi$$

and

$$H_z = 0$$

from Appendix A.2. So, finally we have

$$\left.\begin{aligned} H_\rho &= 0 \\ H_\phi &= \frac{I}{2\pi\rho} \\ H_z &= 0 \end{aligned}\right\} \rho \gg a \qquad (7.15)$$

This result agrees with that for the infinite filamentary current. We could have predicted this (Why?). We will return to this problem in the next section where we will find a simpler method to solve it exactly.

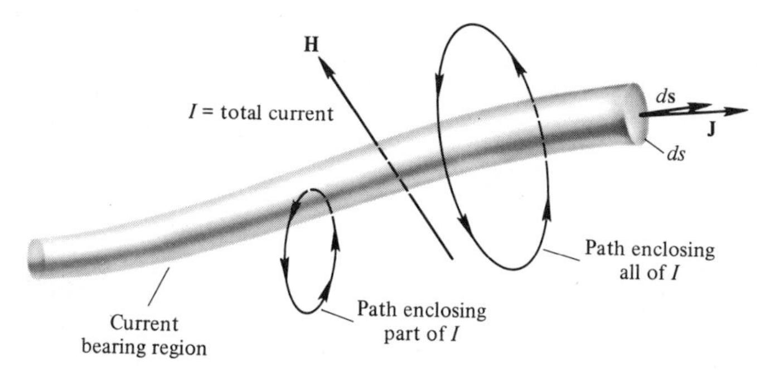

Figure 7.6. Direction of integration and current for $\oint \mathbf{H} \cdot d\mathbf{l}$ (right-hand rule).

7.3 AMPERE'S LAW

Ampere's law (or perhaps more correctly, Ampere's circuital law) simply states that the line integral of **H** around *any* closed path (or the circulation of **H**) is equal to the current enclosed by that path. The path is completely arbitrary. The direction of the current is that of an advancing right-hand threaded screw turned in the direction in which the closed path is traversed (right-hand rule). Thus, as seen in Figure 7.6, Ampere's law is simply

$$\boxed{\oint \mathbf{H} \cdot d\mathbf{l} = I_{\text{enc}} = \iint_s \mathbf{J} \cdot d\mathbf{s}} \tag{7.16}$$

Ampere's law can be derived directly from the Biot–Savart law, but this derivation is not essential at this point in the development. It is included in Appendix G for the interested reader. Instead, we will show at an appropriate time that Ampere's law is consistent with the Biot–Savart law. It is worth mentioning that a very useful instrument, the "clamp-on" ammeter, is based on the principle behind Ampere's law.

EXAMPLE 3

Just as Gauss's law enabled us to find electrostatic fields from charges (when the proper symmetry was present), Ampere's law enables us to determine magnetic fields when certain symmetry conditions are met. As a first example let us consider the infinite filamentary current on the z axis. This problem was solved in Section 7.2 by means of the Biot–Savart law. It is obvious from the symmetry in Figure 7.7 that *whatever* field is present must be independent of ϕ and z as far as its magnitude is concerned. The elementary form of the Biot–Savart law tells us that the magnetic field is ϕ-directed. Hence, we know that $H_\phi = H_\phi(\rho)$. In order to obtain any advantage from this method, we must choose as our path of integration one for which the magnitude of **H** is constant and to

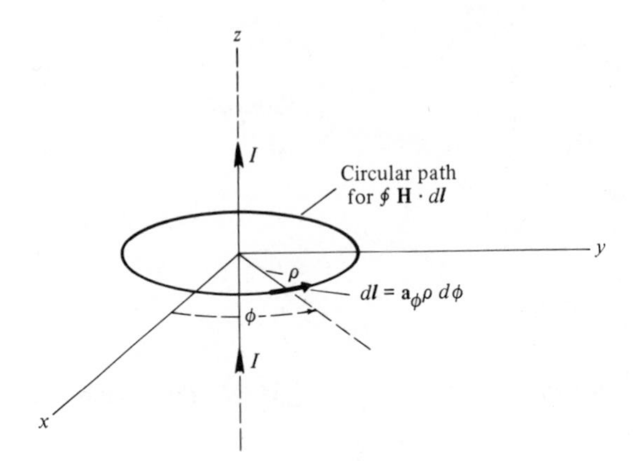

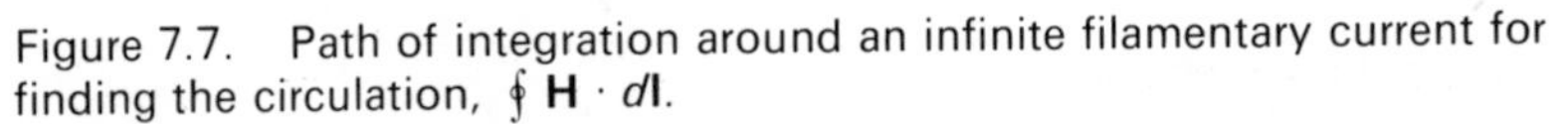
Figure 7.7. Path of integration around an infinite filamentary current for finding the circulation, $\oint \mathbf{H} \cdot d\mathbf{l}$.

which $\mathbf{H}$ is either perpendicular or tangent. In Figure 7.7 we have chosen a path to which $\mathbf{H}$ is tangent and for which $|\mathbf{H}|$ is constant: namely a circle of radius ρ. Hence,

$$I = \oint \mathbf{H} \cdot d\mathbf{l} = \int_0^{2\pi} \mathbf{a}_\phi H_\phi \cdot \mathbf{a}_\phi \rho \, d\phi$$

or

$$I = \rho H_\phi \int_0^{2\pi} d\phi = 2\pi \rho H_\phi$$

or

$$H_\phi(\rho) = \frac{I}{2\pi\rho} \quad \text{(A/m)} \tag{7.17}$$

which agrees with equation 7.9, and was obtained with much less effort.

EXAMPLE 4

Let us next take another look at the infinite cylindrical sheet of surface current, which was the second example of Section 7.2. As seen in Figure 7.8, symmetrically located current elements produce (acting together) only a y component of field. Since symmetry arguments tell us that the field is independent of ϕ and z, the point chosen to evaluate $d\mathbf{H}$ can be placed anywhere. Thus, we have $H_\phi = H_\phi(\rho)$ only. We now apply Ampere's law, and

$$H_\phi(\rho) = \frac{I}{2\pi\rho}, \qquad \rho > a$$

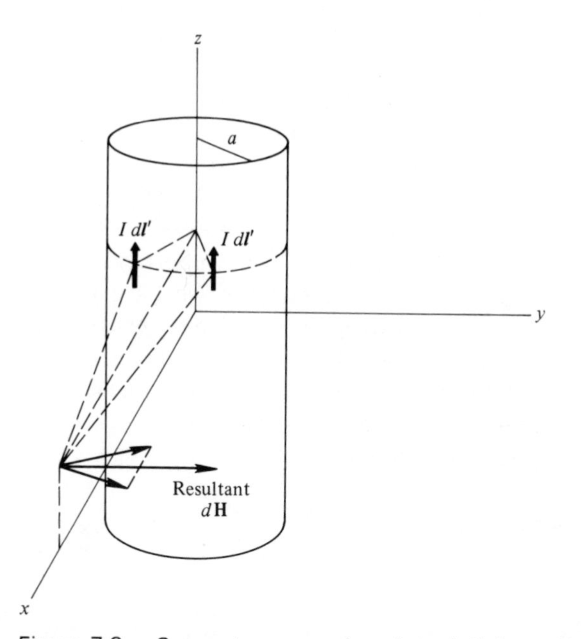

Figure 7.8. Symmetry properties of the infinite uniform surface current density, $\mathbf{J}_s$, on a cylinder of radius a.

since the procedure is identical to that in the preceding problem. This answer is *exact*! Since no current is enclosed when the radius of the path of integration is less than the radius of the current sheet ($\rho < a$), we have $H_\phi(\rho) = 0, \rho < a$, and

$$H_\phi(\rho) = \frac{I}{2\pi\rho} u(\rho - a) \tag{7.18}$$

The preceding example has done at least two things. First, it shows how simple Ampere's law really is when we can pick a path along which $|\mathbf{H}|$ is constant (or zero) and to which $\mathbf{H}$ is tangent. Secondly, it establishes two mathematical identities because the x and y components of equation 7.18 must equal equations 7.11 and 7.12, respectively. Since

$$H_x = -H_\phi \sin\phi = -\frac{I}{2\pi\rho}\sin\phi \tag{7.19}$$

and

$$H_y = H_\phi \cos\phi = \frac{I}{2\pi\rho}\cos\phi \tag{7.20}$$

we have (for $\rho > a$)

$$2\pi \sin\phi = \rho \int_0^{2\pi} \frac{\rho\sin\phi - a\sin\phi'}{\rho^2 + a^2 - 2\rho a\cos(\phi - \phi')} d\phi' \tag{7.21}$$

and

$$2\pi \cos \phi = \rho \int_0^{2\pi} \frac{\rho \cos \phi - a \cos \phi'}{\rho^2 + a^2 - 2\rho a \cos (\phi - \phi')} d\phi' \tag{7.22}$$

EXAMPLE 5

Another example, of considerably more practical importance, is the coaxial cable, whose capacitance per unit length we have already determined. A total current of $+IA$ with uniform density exists in the center conductor (radius $= a$), and a total (return) current of $-IA$ with uniform density exists in the outer conductor (shield) whose inner radius is b and outer radius is c. This geometry is shown in Figure 7.9. The cable has infinite length so that end effects are eliminated.

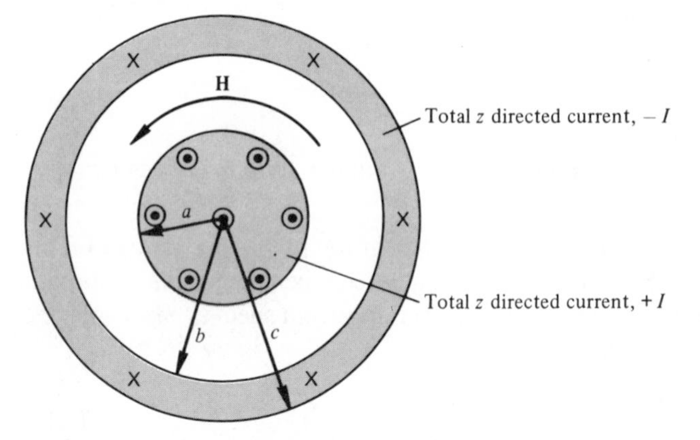

Figure 7.9. Coaxial cable with uniform current densities.

Using exactly the same arguments as in the two preceding examples, we can show that we only have $H_\phi = H_\phi(\rho)$. We also know from the preceding examples that

$$H_\phi = \frac{I}{2\pi\rho}, \qquad a < \rho < b \tag{7.23}$$

For $\rho < a$ we choose a path of integration which is a circle of radius ρ. The current *enclosed* by such a path is, from equation 6.3,

$$I_{\text{enc}} = \iint_s \mathbf{J} \cdot d\mathbf{s} = \int_0^{\rho}\int_0^{2\pi} J_z \mathbf{a}_z \cdot \mathbf{a}_z \rho \, d\rho \, d\phi$$

Now, for the inner conductor

$$J_z = \frac{I}{\pi a^2} \quad (\text{A/m}^2)$$

since the current density is uniformly distributed. Therefore,

$$I_{\text{enc}} = \frac{I}{\pi a^2} \int_0^{\rho} \int_0^{2\pi} \rho \, d\rho \, d\phi = I \frac{\rho^2}{a^2}$$

Applying Ampere's law,

$$\int_0^{2\pi} H_\phi \rho \, d\phi = I_{\text{enc}} = I \frac{\rho^2}{a^2} \qquad \rho < a$$

or

$$H_\phi \rho \int_0^{2\pi} d\phi = I \frac{\rho^2}{a^2}$$

so,

$$H_\phi = \frac{I\rho}{2\pi a^2}, \qquad \rho < a \tag{7.24}$$

For $b < \rho < c$, we have

$$I_{\text{enc}} = I + \int_b^{\rho} \int_0^{2\pi} J_z \mathbf{a}_z \cdot \mathbf{a}_z \rho \, d\rho \, d\phi$$

but for the outer conductor

$$J_z = \frac{-I}{\pi(c^2 - b^2)}$$

Then,

$$I_{\text{enc}} = I - \frac{I}{\pi(c^2 - b^2)} \int_b^{\rho} \int_0^{2\pi} \rho \, d\rho \, d\phi$$

or

$$I_{\text{enc}} = I - I \frac{\rho^2 - b^2}{c^2 - b^2} = I \frac{c^2 - \rho^2}{c^2 - b^2}$$

Ampere's law gives

$$H_\phi = \frac{I}{2\pi\rho} \frac{c^2 - \rho^2}{c^2 - b^2}, \qquad b < \rho < c \tag{7.25}$$

For $\rho > c$, $H_\phi = 0$, so that the system is "shielded." A plot of H_ϕ against ρ shows that H_ϕ is continuous. Such a plot is sketched in Figure 7.10.

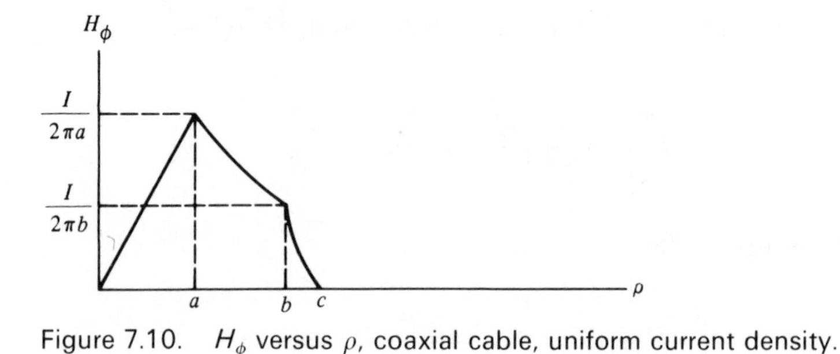

Figure 7.10. H_ϕ versus ρ, coaxial cable, uniform current density.

EXAMPLE 6

The concept of a uniform surface current density flowing on a plane surface will be extremely useful to us in introducing transmission lines later on. Consider such a z-directed surface current shown in Figure 7.11.

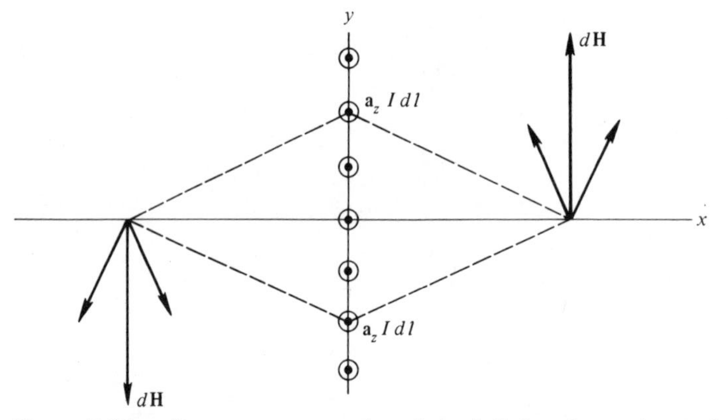

Figure 7.11. Symmetry properties of the infinite planar sheet of z-directed current.

It should be clear from the symmetry and a comparison with previous examples that the only component of magnetic field present is a positive y component for $x > 0$ and a negative y component for $x < 0$. There are several ways to determine what this field is. Perhaps the simplest is the use of Ampere's law. Consider Figure 7.12. Surface s_1 defines a path (c_1) of integration around its periphery which lies symmetrically in the $z = 0$ plane. Applying Ampere's law, we have

$$\oint_{c_1} \mathbf{H} \cdot d\mathbf{l} = \int_{y_0}^{y_0+L} \mathbf{a}_y H_{y1} \cdot \mathbf{a}_y\, dy + \int_{a}^{-a} 0 \cdot \mathbf{a}_x\, dx + \int_{y_0+L}^{y_0} -\mathbf{a}_y H_{y1} \cdot \mathbf{a}_y\, dy$$

$$+ \int_{-a}^{a} 0 \cdot \mathbf{a}_x\, dx$$

$$= H_{y1}[y_0 + L - y_0] - H_{y1}[y_0 - y_0 - L]$$

$$= H_{y1}L + H_{y1}L = J_{sz}L = I_{\text{enc}}$$

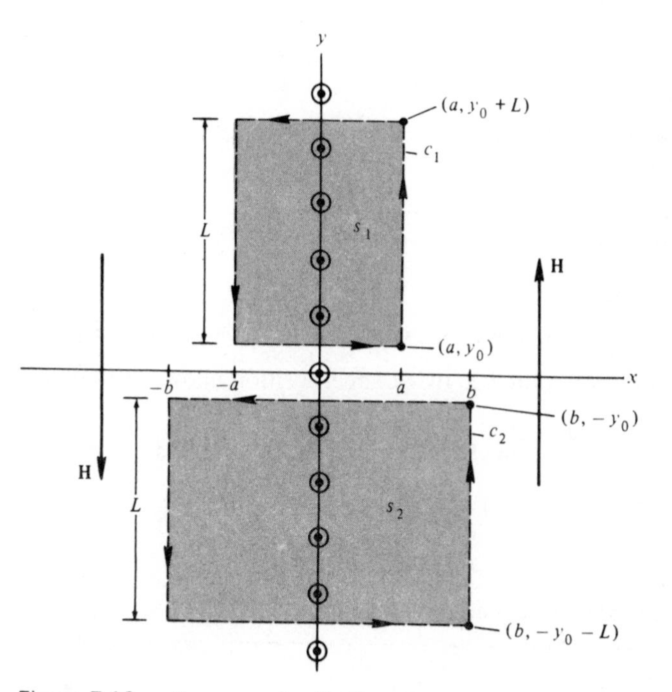

Figure 7.12. Geometry for finding **H** for the infinite sheet of z-directed (planar) current, $\mathbf{a}_z J_{sz}$.

or

$$H_{y1} = \frac{J_{sz}}{2} \quad (x > 0)$$

The *same* answer is obtained for path c_2, so **H** is independent of x, y, and z. This result, as it stands is correct only for $x > 0$, because we used the fact that H_y was negative for $x < 0$ to derive it. A form that is correct for any x is

$$H_y = \frac{J_{sz}}{2}[2u(x) - 1] \tag{7.26}$$

A result something like this could have been predicted almost immediately because we already know the electrostatic field of an infinite sheet of surface charge density.

EXAMPLE 7

In order to utilize this result in any practical configuration, such as a transmission line, a return surface current must be included. Consider Figure 7.13, where a pair of parallel oppositely directed surface currents

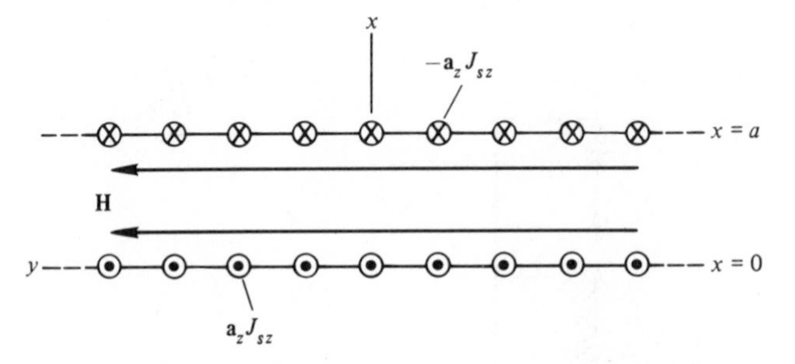

Figure 7.13. Parallel oppositely directed planar surface current densities.

is shown. Using equation 7.26 twice, superposition allows us to obtain

$$H_y = \frac{J_{sz}}{2}[2u(x) - 1] - \frac{J_{sz}}{2}[2u(x - a) - 1]$$

or

$$H_y = J_{sz}[u(x) - u(x - a)] \tag{7.27}$$

which again represents a shielded system.

7.4 AMPERE'S LAW IN POINT FORM

Stokes theorem,[2] equation 3.31, may be applied to Ampere's law, giving

$$\oint_c \mathbf{H} \cdot d\mathbf{l} \equiv \iint_{s_1} \nabla \times \mathbf{H} \cdot d\mathbf{s} = I \tag{7.28}$$

where s_1 is any one of the possible *open* surfaces defined by the path of integration used for the line integral. See Figure 7.14. We also have equation 6.3,

$$I = \iint_{s_2} \mathbf{J} \cdot d\mathbf{s} \tag{7.29}$$

Combining equations 7.28 and 7.29,

$$\iint_{s_1} \nabla \times \mathbf{H} \cdot d\mathbf{s} = \iint_{s_2} \mathbf{J} \cdot d\mathbf{s} \tag{7.30}$$

Since the surfaces s_1 and s_2 are arbitrary, they may be made identical, but still arbitrary. In this case, we have

$$\iint_{s} \nabla \times \mathbf{H} \cdot d\mathbf{s} = \iint_{s} \mathbf{J} \cdot d\mathbf{s} \tag{7.31}$$

[2] Stoke's theorem is derived in Appendix G.

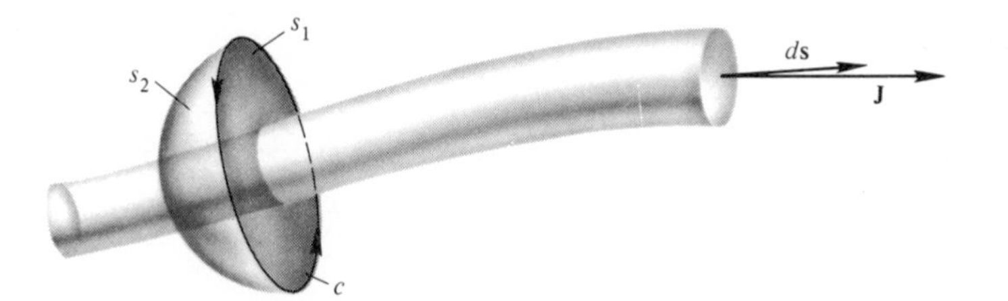

Figure 7.14. Geometry for deriving the point form of Ampere's law (Maxwell's equation).

where s is arbitrary. Since s is still arbitrary, we arrive at the result (also, see Appendix G),

$$\boxed{\nabla \times \mathbf{H} = \mathbf{J}} \tag{7.32}$$

This is the point form of Ampere's law, and is also one of Maxwell's equations for *steady* currents. The modification to this equation for time dependent currents was probably Maxwell's most important contribution. Recall that the electrostatic field had no curl ($\nabla \times \mathbf{E} = 0$) and was conservative. The magnetic field is thus nonconservative.

Let us next take the divergence of both sides of equation 7.32. We have

$$\nabla \cdot \nabla \times \mathbf{H} \equiv 0 = \nabla \cdot \mathbf{J} \tag{7.33}$$

since the divergence of the curl of *any* vector is identically zero. Equation 7.33 verifies equation 6.5, which we derived on the basis of conservation of charge.

EXAMPLE 8

In order to get a check on equation 7.32, let us take as an example an infinite cylinder of radius a carrying a uniform current density, $\mathbf{J} = \mathbf{a}_z(I/\pi a^2)u(a - \rho)$. From the coaxial cable example, we have

$$H_\phi = \frac{I}{2\pi\rho}, \qquad \rho > a$$

and

$$H_\phi = \frac{I\rho}{2\pi a^2}, \qquad \rho < a$$

or

$$H_\phi = \frac{I\rho}{2\pi a^2} u(a - \rho) + \frac{I}{2\pi\rho} u(\rho - a) \qquad 0 \leq \rho \leq \infty$$

so,

$$\nabla \times \mathbf{H} = \mathbf{a}_z \frac{1}{\rho} \frac{\partial}{\partial \rho} (\rho H_\phi)$$

$$= \mathbf{a}_z \frac{1}{\rho} \left[-\frac{I\rho^2}{2\pi a^2} \delta(\rho - a) + \frac{I\rho}{\pi a^2} u(a - \rho) + \frac{I}{2\pi} \delta(\rho - a) \right]$$

or

$$\nabla \times \mathbf{H} = \mathbf{a}_z \frac{I}{\pi a^2} u(a - \rho)$$

or

$$\nabla \times \mathbf{H} = \mathbf{J}, \qquad 0 \leq \rho \leq \infty$$

7.5 MAGNETIC FLUX AND FLUX DENSITY

In free space we may define the magnetic flux density as

$$\boxed{\mathbf{B} = \mu_0 \mathbf{H}} \quad (\text{webers/m}^2, \text{Wb/m}^2) \tag{7.34}$$

where μ_0 is the *permeability* of free space and is given by $4\pi \times 10^{-7}$ henrys per meter (H/m). The units of **B** are webers per square meter. The flux, as expected, may be found by integrating the normal component of the flux density over whatever surface we are interested in. In other words, the flux through the surface s is

$$\boxed{\psi_m = \iint_s \mathbf{B} \cdot d\mathbf{s}} \quad (\text{Wb}) \tag{7.35}$$

We have already noted in previous examples that magnetic field lines (**H**) close on themselves. Compared to electric field lines, this behavior is easy to explain. Magnetic charges do not exist in nature, and so there are no sources or sinks for magnetic lines to begin or terminate on (as opposed to the electrostatic field). In this case, for a *closed* surface,

$$\psi_m = \oiint_s \mathbf{B} \cdot d\mathbf{s} = 0 \tag{7.36}$$

It also follows (using the divergence theorem) that

$$\oiint_s \mathbf{B} \cdot d\mathbf{s} = \iiint_{\text{vol}} \nabla \cdot \mathbf{B} \, dv = 0$$

Since the surface and volume are arbitrary, we have

$$\boxed{\nabla \cdot \mathbf{B} = 0} \tag{7.37}$$

The comparison between magnetic and electric fields in terms of sources and sinks is even more striking now, for Gauss's law gives

$$\nabla \cdot \mathbf{D} = \rho_v$$

A field such as **B**, or **J**, or **H**, which has no divergence is a *solenoidal* field. A more elegant proof of the validity of equation 7.37 is obtained if we take the divergence of $\mu_0 \mathbf{H}$ with **H** given by the Biot–Savart law. (See Problem 4.)

7.6 MAGNETIC SCALAR AND VECTOR POTENTIAL

In a region for which the current density is zero, we have

$$\nabla \times \mathbf{H} = 0 \tag{7.38}$$

from equation 7.32. In this case we may use the same procedures as for electrostatics, where it was found that

$$\nabla \times \mathbf{E} = 0 \quad \text{and} \quad \mathbf{E} = -\nabla\Phi$$

In other words, we may write

$$\mathbf{H} = -\nabla\Phi_m \tag{7.39}$$

where Φ_m is the scalar *magnetic* potential in amperes. It also follows that the differential equation for Φ_m is Laplace's equation,

$$\nabla^2\Phi_m = 0 \tag{7.40}$$

since $\mathbf{B} = \mu_0 \mathbf{H}$ and $\nabla \cdot \mathbf{H} = 0$, or

$$0 = \nabla \cdot \mathbf{H} = \nabla \cdot (-\nabla\Phi_m) = -\nabla^2\Phi_m$$

and we are only considering vacua here. Since we have spent considerable time discussing the methods available for solving problems of this type (Laplace's equation), we will not solve any examples here.

A difficulty arises in using the magnetic scalar potential which never occurs when using the electric scalar potential. The magnetic scalar potential Φ_m is *not single* valued. To put it in another way, in electrostatics we *always* have

$$\oint \mathbf{E} \cdot d\mathbf{l} = 0$$

whereas, for steady magnetic fields we have

$$\oint \mathbf{H} \cdot d\mathbf{l} = I$$

if I is enclosed by the path. Thus, $\mathbf{H}$ is not *in general* a conservative field ($\nabla \times \mathbf{H} \neq 0$). Everytime we encircle the current in calculating the line integral we pick up a contribution of I. If we choose a particular path for magnetic potential difference and agree not to completely encircle the current, then a single-valued magnetic scalar potential difference may be defined as

$$\boxed{\Phi_{m,\,ab} = -\int_b^a \mathbf{H} \cdot d\mathbf{l}} \tag{7.41}$$

Our most important use of the scalar magnetic potential will occur when we consider magnetic materials and magnetic circuits. We will use equation 7.41 at that time.

We next return to equation 7.37,

$$\nabla \cdot \mathbf{B} \equiv 0$$

Since the divergence of the curl of *any* vector is identically zero, it is certainly true that

$$\boxed{\mathbf{B} = \nabla \times \mathbf{A}} \tag{7.42}$$

or

$$\nabla \cdot \nabla \times \mathbf{A} \equiv 0$$

Equation 7.42 does not completely define $\mathbf{A}$, for the gradient of any scalar ($\nabla\alpha$) added to $\mathbf{A}$ will give a new function that still satisfies equation 7.37. Why is this so? $\mathbf{A}$ has the dimensions of webers per meter (Wb/m). This potential is a vector, called the *magnetic vector potential*. It does not have the same physical significance (it is not easily measured) as the scalar potential, but can be very useful to us.

From equations 7.42 and 7.34, we have

$$\mathbf{H} = \frac{1}{\mu_0} \nabla \times \mathbf{A} \tag{7.43}$$

With this equation and equation 7.32, we have

$$\nabla \times \nabla \times \mathbf{A} = +\mu_0 \mathbf{J} \tag{7.44}$$

The left side of equation 7.44 may be expanded by using the vector identity (see Appendix A, A.10.)

$$\nabla \times \nabla \times \mathbf{A} \equiv \nabla(\nabla \cdot \mathbf{A}) - \nabla^2 \mathbf{A} = \mu_0 \mathbf{J} \tag{7.45}$$

In equation 7.45, $\nabla^2 \mathbf{A}$ is the Laplacian of a vector. That is,

$$\nabla^2 \mathbf{A} = \mathbf{a}_x \nabla^2 A_x + \mathbf{a}_y \nabla^2 A_y + \mathbf{a}_z \nabla^2 A_z \tag{7.46}$$

(only) in rectangular coordinates. In other coordinate systems (see Appendix

A), a result as simple as equation 7.46 does not occur! This point should be kept in mind.

In order to completely define **A**, we need to find its divergence, $\nabla \cdot \mathbf{A}$. We can specify $\nabla \cdot \mathbf{A}$ any way we wish, but an examination of equation 7.45 reveals that if $\nabla \cdot \mathbf{A} \equiv 0$, we obtain a vector equation that looks like Poisson's and whose rectangular components (equation 7.46) are precisely those in Poisson's equation! For the time being then, we will simply define $\nabla \cdot \mathbf{A}$ as being *identically* zero.

$$\boxed{\nabla \cdot \mathbf{A} \equiv 0} \tag{7.47}$$

In this case, **A** is said to be in the *Coulomb gauge*. This point will be discussed again at an appropriate time. Combining equations 7.47 and 7.45, we finally have

$$\boxed{\nabla^2 \mathbf{A} = -\mu_0 \mathbf{J}} \tag{7.48}$$

This is the differential equation that **A** must satisfy. Recalling that the solution to Poisson's equation,

$$\nabla^2 \Phi = -\frac{\rho_v}{\varepsilon_0}$$

is the scalar Helmholtz superposition integral,

$$\Phi(\mathbf{r}) = \frac{1}{4\pi\varepsilon_0} \iiint_{\text{vol}'} \frac{\rho_v(\mathbf{r}')}{R}\, dv'$$

we suspect that the solution of equation 7.48 may be a vector Helmholtz integral,

$$\boxed{\mathbf{A}(\mathbf{r}) = \frac{\mu_0}{4\pi} \iiint_{\text{vol}'} \frac{\mathbf{J}(\mathbf{r}')}{R}\, dv'} \tag{7.49}$$

In fact, we *know* that rectangular components of **A**, given by equation 7.46 and used in equation 7.49, will satisfy equation 7.48. This follows from the principle of *duality*. That is, if

$$\phi(\mathbf{r}) = \frac{1}{4\pi\varepsilon_0} \iiint_{\text{vol}'} \frac{\rho_v(\mathbf{r}')}{R}\, dv' \quad \text{satisfies} \quad \nabla^2 \phi(\mathbf{r}) = -\frac{\rho_v(\mathbf{r})}{\varepsilon_0}$$

then it is certainly true for A_x, for example, that

$$A_x(\mathbf{r}) = \frac{\mu_0}{4\pi} \iiint_{\text{vol}'} \frac{J_x(\mathbf{r}')}{R}\, dv' \quad \text{satisfies} \quad \nabla^2 A_x(\mathbf{r}) = -\mu_0 J_x(\mathbf{r}) \tag{7.50}$$

because the only difference in the two sets of equations is a change of symbols!

It is not necessary for the present treatment to prove that equation 7.49 is a general solution to equation 7.48, but this proof is supplied in Appendix G. It is important to recognize that **A** must be a continuous function if we are to avoid impulses of magnetic field. Recall that a similar situation exists for the scalar electric potential Φ.

If equation 7.49 is correct, it must be consistent with equation 7.5.

$$\mathbf{H}(\mathbf{r}) = \frac{1}{4\pi} \iiint_{\text{vol}'} \mathbf{J}(\mathbf{r}') \times \frac{\mathbf{a}_R}{R^2}\, dv'$$

This can be verified by showing that equation 7.5 results from equation 7.49 by way of $\mathbf{H} = (1/\mu_0)\nabla \times \mathbf{A}$, or

$$\mathbf{H}(\mathbf{r}) = \frac{1}{\mu_0} \nabla \times \frac{\mu_0}{4\pi} \iiint_{\text{vol}'} \frac{\mathbf{J}(\mathbf{r}')}{R}\, dv' \tag{7.51}$$

or

$$\mathbf{H}(\mathbf{r}) = \frac{1}{4\pi} \iiint_{\text{vol}'} \nabla \times \frac{\mathbf{J}(\mathbf{r}')}{R}\, dv' \tag{7.52}$$

since ∇ operates on the unprimed coordinates. As usual, we have need of a vector identity.

$$\nabla \times (\alpha \mathbf{C}) \equiv \nabla\alpha \times \mathbf{C} + \alpha(\nabla \times \mathbf{C})$$

as found in Appendix A. In the present problem

$$\nabla \times \mathbf{C} = \nabla \times \mathbf{J}(\mathbf{r}') \equiv 0$$

since $\mathbf{J}(\mathbf{r}')$ depends on primed quantities only. Then equation 7.52 becomes

$$\mathbf{H}(\mathbf{r}) = \frac{1}{4\pi} \iiint_{\text{vol}'} \nabla\left(\frac{1}{R}\right) \times \mathbf{J}(\mathbf{r}')\, dv' \tag{7.53}$$

From equation 3.21, $\nabla(1/R) = -(\mathbf{a}_R/R^2)$, so equation 7.53 becomes

$$\mathbf{H}(\mathbf{r}) = \frac{1}{4\pi} \iiint_{\text{vol}'} -\frac{\mathbf{a}_R}{R^2} \times \mathbf{J}(\mathbf{r}')\, dv'$$

which is the same as equation 7.5.

Is Ampere's circuital law consistent with **A**? We have

$$\oint_l \mathbf{H} \cdot d\mathbf{l} = \iint_s \nabla \times \mathbf{H} \cdot d\mathbf{s} = \frac{1}{\mu_0} \iint_s \nabla \times (\nabla \times \mathbf{A}) \cdot d\mathbf{s} = \frac{1}{\mu_0} \iint_s -\nabla^2 \mathbf{A} \cdot d\mathbf{s}$$

$$= \frac{1}{\mu_0} \iint_s \mu_0 \mathbf{J} \cdot d\mathbf{s} = \iint_s \mathbf{J} \cdot d\mathbf{s} \equiv I$$

Each step in this sequence of operations should be verified.

When the term $\nabla \cdot \mathbf{A}$ arose in equation 7.45, we simply chose $\nabla \cdot \mathbf{A} \equiv 0$, and this choice gave us equation 7.48, which we knew could be solved by equation 7.49 (at least in rectangular coordinates). If we now take the divergence of **A**, when **A** is given by equation 7.49, will we obtain zero?

$$\nabla \cdot \mathbf{A} = \frac{\mu_0}{4\pi} \iiint\limits_{\text{vol}'} \nabla \cdot \frac{\mathbf{J}(\mathbf{r}')}{R}\, dv' = \frac{\mu_0}{4\pi} \iiint\limits_{\text{vol}'} \nabla \cdot \left\{\left(\frac{1}{R}\right)[\mathbf{J}(\mathbf{r}')]\right\} dv'$$

but

$$\nabla \cdot (\alpha \mathbf{C}) \equiv \mathbf{C} \cdot \nabla \alpha + \alpha \nabla \cdot \mathbf{C}$$

and

$$\nabla \cdot \mathbf{C} = \nabla \cdot \mathbf{J}(\mathbf{r}') \equiv 0 \quad \text{(Why?)}$$

So,

$$\nabla \cdot \mathbf{A} = \frac{\mu_0}{4\pi} \iiint\limits_{\text{vol}'} \mathbf{J}(\mathbf{r}') \cdot \nabla\left(\frac{1}{R}\right) dv'$$

or

$$\nabla \cdot \mathbf{A} = -\frac{\mu_0}{4\pi} \iiint\limits_{\text{vol}'} \mathbf{J}(\mathbf{r}') \cdot \nabla'\left(\frac{1}{R}\right) dv'$$

using equation 3.21 again. We now use the vector identity once more, but this time with ∇'.

$$\mathbf{C} \cdot \nabla' \alpha \equiv \nabla' \cdot (\alpha \mathbf{C}) - \alpha \nabla' \cdot \mathbf{C}$$

But,

$$\nabla' \cdot \mathbf{C} = \nabla' \cdot \mathbf{J}(\mathbf{r}') \equiv 0$$

because of conservation of charge, equation 6.5. Therefore,

$$\nabla \cdot \mathbf{A} = -\frac{\mu_0}{4\pi} \iiint\limits_{\text{vol}'} \nabla' \cdot \frac{\mathbf{J}(\mathbf{r}')}{R}\, dv'$$

Now the original Helmholtz integral for **A** must be over a volume (vol′) large enough to contain all of the current density, **J**. It can be *larger* than the minimum volume required to contain **J** and all of our equations are still correct. This is true for all of our superposition-type volume integrals. This being the case, let the volume for the last equation be that within a sphere whose radius is approaching infinity. When the divergence theorem is applied to the last equation, we obtain

$$\nabla \cdot \mathbf{A} = -\frac{\mu_0}{4\pi} \iint\limits_{s'} \frac{\mathbf{J}(\mathbf{r}')}{R} \cdot d\mathbf{s}'$$

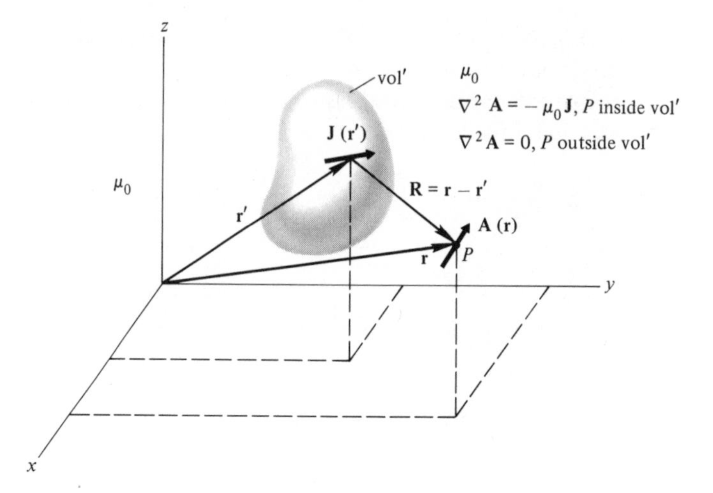

Figure 7.15. Geometry for finding the magnetic vector potential **A** from its source **J**.

Now s' is the surface of a sphere whose radius is approaching infinity, and on this surface $\mathbf{J}(\mathbf{r}') = 0$ (except for some nonrealistic examples where we may have sources at infinity). Then,

$$\nabla \cdot \mathbf{A} = 0$$

Thus, our equations for $\nabla \times \mathbf{A}$ and $\nabla \cdot \mathbf{A}$ are consistent with the Biot–Savart law and the continuity equation if **A** is given by the vector Helmholtz integral.

Before looking at examples, let us picture the general system where equation 7.49, the Helmholtz integral, applies. Consider Figure 7.15 where a volume (vol′) contains a general current density, **J**, and free space conditions exist outside this volume. Equation 7.49 is a solution to $\nabla^2 \mathbf{A} = \mu_0 \mathbf{J}$, if the point where **A** is to be found is within the volume containing the current. If the point where **A** is to be found is not within this volume, then equation 7.49 satisfies the equation, $\nabla^2 \mathbf{A} = 0$.

If the current sources are surface current densities, $\mathbf{J}_s$, or filamentary line currents, I, then

$$\mathbf{A}(\mathbf{r}) = \frac{\mu_0}{4\pi} \iint_{s'} \frac{\mathbf{J}_s(r')}{R} \, ds' \tag{7.54}$$

and

$$\mathbf{A}(\mathbf{r}) = \frac{\mu_0}{4\pi} \int_{l'} \frac{I \, d\mathbf{l}'}{R} \tag{7.55}$$

respectively, as special cases of equation 7.49.

7.7 THE ELECTRIC CURRENT LOOP OR MAGNETIC DIPOLE

EXAMPLE 9

As a first attempt to use **A**, consider the circular loop of radius a carrying a filamentary current I and lying in the $z = 0$ plane. For small a, this source is called a *magnetic dipole* because its field is dual to that of the *electric dipole* (which we have already considered) if the distance to the field point is large compared to the dipole dimensions. In other words, the magnetic field of the loop is in the same mathematical form as the electric field of the electric dipole for large distances. As a start, let us attempt to calculate the magnetic field of the loop shown in Figure 7.16

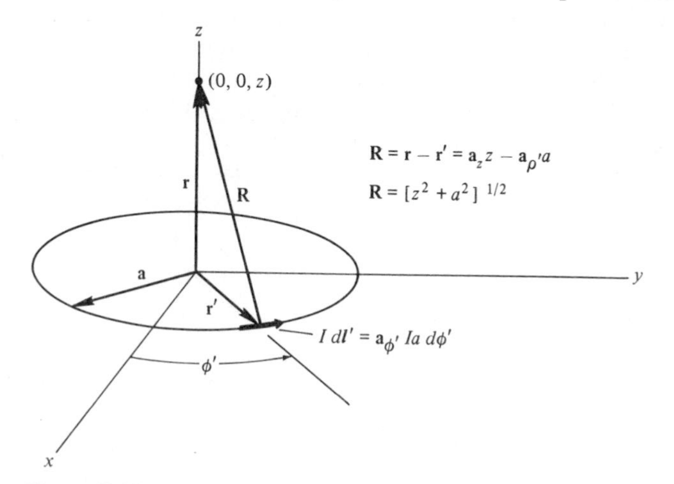

Figure 7.16. Geometry for finding the magnetic field on the axis of an electric current loop (magnetic dipole).

at a point on the z axis using the vector potential. Recalling that $\mathbf{a}_{\phi'}$ is not a constant vector, we have ($\mathbf{a}_{\phi'} = -\mathbf{a}_x \sin\phi' + \mathbf{a}_y \cos\phi'$)

$$\mathbf{A}(0, 0, z) = \frac{\mu_0}{4\pi}\int_0^{2\pi} \frac{Ia\mathbf{a}_{\phi'}\, d\phi'}{(a^2 + z^2)^{1/2}}$$

$$= \frac{\mu_0 Ia}{4\pi(a^2 + z^2)^{1/2}}\int_0^{2\pi} (-\mathbf{a}_x \sin\phi' + \mathbf{a}_y \cos\phi')\, d\phi'$$

or

$$\mathbf{A}(0, 0, z) = 0$$

This result is obvious from the symmetry, but it does not tell us that $\mathbf{B}(0, 0, z) = 0$! Just because a function itself is zero at a point is no guarantee that its derivative ($\mathbf{B} = \nabla \times \mathbf{A}$) is zero at that point. We now have the choice of evaluating **A** at some point off of the z axis where **A** is not zero, or evaluating **B** by some other method. The Biot–Savart law gives

$$\mathbf{B}(0, 0, z) = \frac{\mu_0 I a^2 \mathbf{a}_z}{2(a^2 + z^2)^{3/2}}$$

In the general case, we know that **B** must be independent of azimuth angle but dependent on r and θ (or ρ and z).

EXAMPLE 10

We will now calculate the magnetic field produced by an electric current loop or magnetic dipole. Rather than attempting the most general case, we will allow the loop to have arbitrary shape, but constrain it to lie in the $z = 0$ plane. That is, it is a *planar loop.* We are only interested in finding the field at distances that are large compared to the loop dimensions. The loop is shown in Figure 7.17. Our goal is to show that this

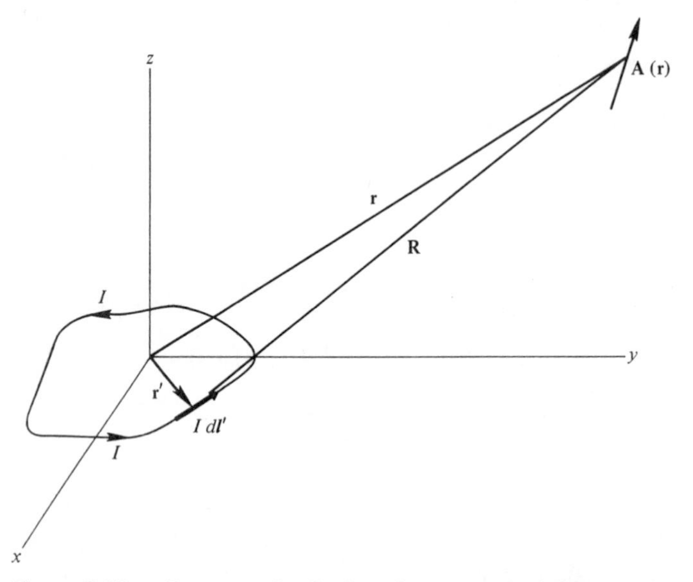

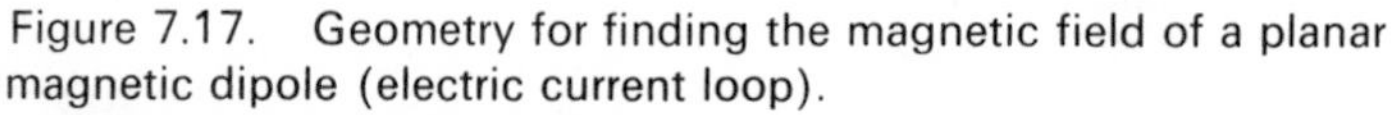

Figure 7.17. Geometry for finding the magnetic field of a planar magnetic dipole (electric current loop).

field depends on the *area* of the loop, not its shape.

The magnetic vector potential is given by equation 7.55 as

$$\mathbf{A} = \frac{\mu_0 I}{4\pi} \oint_{l'} \frac{1}{R}\, d\mathbf{l}' \tag{7.56}$$

Using the last identity in Appendix A.10, equation 7.56 can be written as

$$\mathbf{A} = \frac{\mu_0 I}{4\pi} \iint_{s'} \mathbf{a}_z \times \nabla'\left(\frac{1}{R}\right) ds' \tag{7.57}$$

since $\mathbf{a}_n = \mathbf{a}_z$ when the loop lies in the $z = 0$ plane. Now, using equation 3.21, equation 7.57 becomes

$$\mathbf{A} = \frac{\mu_0 I}{4\pi} \iint_{s'} \mathbf{a}_z \times \frac{\mathbf{a}_R}{R^2} ds' \tag{7.58}$$

It can be seen in Figure 7.17 that for large distances $R \approx r$ and $\mathbf{a}_r \approx \mathbf{a}_R$, so that equation 7.58 can be written

$$\mathbf{A} \approx \frac{\mu_0 I}{4\pi} \iint_{s'} \mathbf{a}_z \times \frac{\mathbf{a}_r}{r^2} ds' = \frac{\mu_0 I}{4\pi r^2} \iint_{s'} \mathbf{a}_z \times \mathbf{a}_r \, ds' \tag{7.59}$$

But

$$\mathbf{a}_r = \mathbf{r}/r = (\mathbf{a}_x x + \mathbf{a}_y y = \mathbf{a}_z z)/r$$

$$\mathbf{a}_r = (\mathbf{a}_x r \sin\theta \cos\phi + \mathbf{a}_y r \sin\theta \sin\phi + \mathbf{a}_z r \cos\theta)/r$$

so, taking the cross product

$$A_x = \frac{-\mu_0 I y}{4\pi r^3} \iint ds' = \frac{-\mu_0 I s}{4\pi r^2} \sin\theta \cos\phi \tag{7.60}$$

and

$$A_y = \frac{\mu_0 I x}{4\pi r^3} \iint ds' = \frac{\mu_0 I s}{4\pi r^2} \sin\theta \sin\phi \tag{7.61}$$

where s is the area of the loop.

Now using Appendix A.2, for finding the spherical components of $\mathbf{A}$ in terms of A_x and A_y, we have

$$\left.\begin{aligned} A_r &= 0 \\ A_\theta &= 0 \\ A_\phi &= \frac{\mu_0 I s}{4\pi r^2} \sin\theta \end{aligned}\right\} \tag{7.62}$$

Taking the curl of $\mathbf{A}$ in spherical coordinates gives the desired result,

$$\left.\begin{aligned} B_r &= \frac{\mu_0 I s}{2\pi r^3} \cos\theta \\ B_\theta &= \frac{\mu_0 I s}{4\pi r^3} \sin\theta \\ B_\phi &= 0 \end{aligned}\right\} \tag{7.63}$$

The magnetic field intensity, $\mathbf{H} = \mathbf{B}/\mu_0$, is

$$\mathbf{H} = \frac{Is}{4\pi r^3}(\mathbf{a}_r 2\cos\theta + \mathbf{a}_\theta \sin\theta) \tag{7.64}$$

If equation 7.64 is compared to equation 4.12, we see that the magnetic field for the magnetic dipole is like the electric field of the electric dipole. They are numerically the same if

$$\frac{Qd}{\varepsilon_0} = Is \tag{7.65}$$

The magnetic dipole moment is defined as

$$\boxed{\mathbf{m} = I\mathbf{s}} \tag{7.66}$$

Based on equation 7.62, an equation that is independent of coordinate system can now be written for **A** in terms of **m** for the magnetic dipole. This equation is

$$\boxed{\mathbf{A} = \frac{\mu_0}{4\pi}\frac{\mathbf{m} \times \mathbf{a}_R}{R^2}} \tag{7.67}$$

and we will have need of it in the next chapter when we discuss magnetic materials.

7.8 THE INFINITE FILAMENTARY CURRENT (AGAIN)

EXAMPLE 11

As another example of the use of the magnetic vector potential let us recalculate the magnetic field **H** of an infinite z-directed current filament of strength I. The differential equation for **A** is

$$\nabla^2\mathbf{A} = -\mu_0\mathbf{J} = 0, \qquad \rho > 0$$

Since the current is z-directed, this equation becomes

$$\nabla^2 A_z = 0, \qquad \rho > 0$$

or

$$\frac{1}{\rho}\frac{\partial}{\partial\rho}\left(\rho\frac{\partial A_z}{\partial\rho}\right) + \frac{1}{\rho^2}\frac{\partial^2 A_z}{\partial\phi^2} + \frac{\partial^2 A_z}{\partial z^2} = 0$$

from Appendix A. The last two terms are zero because of the symmetry and the infinite filament length, respectively, so finally we have

$$\frac{1}{\rho}\frac{d}{d\rho}\left(\rho\frac{dA_z}{d\rho}\right) = 0$$

or

$$A_z = C_1 \ln \rho + C_2$$

Now,

$$\mathbf{H} = \frac{1}{\mu_0}\nabla \times \mathbf{A}$$

or,

$$\mathbf{H}_\phi = \frac{-1}{\mu_0}\frac{dA_z}{d\rho} = -\frac{C_1}{\mu_0 \rho}$$

Ampere's law enables us to find the constant C_1. That is,

$$I = \oint \mathbf{H}\cdot d\mathbf{l} = \int_0^{2\pi} -\frac{C_1}{\mu_0 \rho}\mathbf{a}_\phi \cdot \mathbf{a}_\phi \rho \, d\phi = -\frac{2\pi C_1}{\mu_0}$$

so that

$$C_1 = -\frac{\mu_0 I}{2\pi}$$

or

$$H_\phi = \frac{I}{2\pi\rho}$$

and

$$A_z = \frac{-\mu_0 I}{2\pi}\ln \rho + C_2 \tag{7.68}$$

Notice that C_2 is unimportant in finding **H**, but we have the same difficulties in establishing a zero reference for A_z here that we had in establishing a zero reference for Φ in the case of the infinite line charge density in Section 3.5.

7.9 FORCE

It was stated in the introduction to this chapter (and derived in Appendix F) that the magnetic field represents a relativistic correction to the force on a moving charge in the presence of another moving charge. This added force is (see equation F.31)

$$\mathbf{F} = Q\mathbf{u} \times \mathbf{B} \quad \text{(N)} \tag{7.69}$$

and this is the force on a charge Q moving with velocity $\mathbf{u}$ in the presence of an external magnetic flux density $\mathbf{B}$. The force on a charge due to an external electric field is

$$\mathbf{F} = Q\mathbf{E} \quad \text{(N)} \tag{7.70}$$

that is, the force is in the same direction as the external electric field. This latter relation holds, as we have seen, even if the charge is in motion. On the other hand, the force resulting from a magnetic field acting on a moving charge is always at right angles to the charge velocity. In this case, the *magnitude* of the charge velocity is unchanged because the *acceleration* vector ($\mathbf{F} = m\mathbf{a}$) is perpendicular to the velocity vector. The kinetic energy ($\frac{1}{2}mu^2$) is constant. The external magnetic field cannot change the kinetic energy of the charge.

It is worthwhile at this point to consider Newton's second law which governs particle motion under the influence of a force field. In terms of an equation, this law is

$$\mathbf{F} = \frac{d}{dt}(m\mathbf{u})$$

or, in other words, the applied force is equal to the time rate of change of momentum. If all velocity components are small compared to the speed of light, relativistic corrections are not necessary, and the mass, m, is constant, or

$$\mathbf{F} = m\frac{d\mathbf{u}}{dt}$$

By definition, the velocity vector $\mathbf{u}$ is the time rate of change of the position vector $\mathbf{r}$ which locates the particle in some coordinate system. Thus,

$$\mathbf{u} = \frac{d\mathbf{r}}{dt}$$

In rectangular coordinates Newton's law is simply

$$\mathbf{F} = m\left(\mathbf{a}_x\frac{du_x}{dt} + \mathbf{a}_y\frac{du_y}{dt} + \mathbf{a}_z\frac{du_z}{dt}\right)$$

but in cylindrical coordinates it becomes

$$\mathbf{F} = \mathbf{a}_\rho\left(m\frac{d^2\rho}{dt^2} - \rho m\omega^2\right) + \mathbf{a}_\phi\frac{m}{\rho}\frac{d}{dt}(\rho^2\omega) + \mathbf{a}_z m\frac{d^2z}{dt^2}$$

where

$$\omega = \frac{d\phi}{dt} \quad \text{(angular velocity)}$$

$$u_\rho = \frac{d\rho}{dt}, \qquad u_\phi = \rho\frac{d\phi}{dt} = \rho\omega, \qquad u_z = \frac{dz}{dt}$$

Newton's law is quite obviously much more complicated looking in circular cylindrical coordinates.

EXAMPLE 12

Consider a simple classic example. Suppose the force is provided by a uniform magnetic flux density, $\mathbf{B} = \mathbf{a}_z B_0$, so that for an electron

$$\begin{aligned}\mathbf{F} &= -e\mathbf{u} \times \mathbf{B} = -e(\mathbf{a}_\rho u_\rho + \mathbf{a}_\phi u_\phi + \mathbf{a}_z u_z) \times \mathbf{a}_z B_0 \\ &= eB_0 u_\rho \mathbf{a}_\phi - eB_0 u_\phi \mathbf{a}_\rho \\ &= -eB_0 \rho \frac{d\phi}{dt} \mathbf{a}_\rho + eB_0 \frac{d\rho}{dt} \mathbf{a}_\phi\end{aligned}$$

Equating ρ and ϕ components of force,

$$m \frac{d^2\rho}{dt^2} - m\rho\left(\frac{d\phi}{dt}\right)^2 + eB_0 \rho \frac{d\phi}{dt} = 0$$

and

$$\frac{m}{\rho} \frac{d}{dt}\left(\rho^2 \frac{d\phi}{dt}\right) - eB_0 \frac{d\rho}{dt} = 0$$

The simplest solution to this set of coupled equations is obtained if $\rho = \rho_0$ (constant), for in this case they reduce to

$$\frac{d\phi}{dt} = \frac{eB_0}{m} = \omega$$

Thus,

$$\omega_c = \frac{eB_0}{m} \tag{7.71}$$

is the *cyclotron frequency* or angular velocity of an electron in a circular orbit with a radius of $\rho_0 = u_\phi/\omega_c$. Another solution to this problem is asked for in Problem 10 (magnetostatic deflection).

If a charged particle in motion is subjected to *both* an external electric and magnetic field we have the *Lorentz* force equation,

$$\boxed{\mathbf{F} = Q(\mathbf{E} + \mathbf{u} \times \mathbf{B})} \quad \text{(N)} \tag{7.72}$$

which has already been mentioned.

In differential form, equation 7.69 may be written

$$d\mathbf{F} = dQ\mathbf{u} \times \mathbf{B} \tag{7.73}$$

The term $dQ\mathbf{u}$ is dimensionally a differential current element and may be thought of, equivalently, as a current element, $I\, d\mathbf{l}$. That is, we may replace $dQ\mathbf{u}$ by $I\, d\mathbf{l}$ in equation 7.73 giving

$$d\mathbf{F} = I\, d\mathbf{l} \times \mathbf{B} \quad \text{(N)} \tag{7.74}$$

Equation 7.74 states that if we have a differential current element in an external magnetic field it will experience a force. This force is perpendicular to $I\, d\mathbf{l}$ and $\mathbf{B}$ (using the right-hand rule), and its magnitude is equal to the product of $|I\, d\mathbf{l}|$, $|\mathbf{B}|$ and the sine of the angle between $I\, d\mathbf{l}$ and $\mathbf{B}$. In a sense, we are now *defining* the magnetic flux density, $\mathbf{B}$, as the force on a unit current element, $I\, d\mathbf{l}$, if the directions of the vector quantities are taken into account properly.

In a *conductor* carrying a current, the force, just described, is actually a force on the electrons that are moving (rather slowly) to constitute the current. We might expect that this magnetic force would merely shift the electrons with respect to the total conductor. There are, however, very strong Coulomb forces between the electrons and positive ions in a conductor, so any electron displacement due to the magnetic force is very small. Thus, the magnetic force is transferred to the *total conductor* according to equation 7.74.

Any filamentary current path must ultimately be closed, so equation 7.74 results in an equation that can be verified experimentally.

$$\boxed{\mathbf{F} = \oint_l I\, d\mathbf{l} \times \mathbf{B}} \quad \text{(N)} \tag{7.75}$$

We may extend this result, as we have previously done in similar situations, to include surface currents and general current densities,

$$\mathbf{F} = \iint_s \mathbf{J}_s \times \mathbf{B}\, ds \tag{7.76}$$

and

$$\mathbf{F} = \iiint_{\text{vol}} \mathbf{J} \times \mathbf{B}\, dv \tag{7.77}$$

One fact should be mentioned at this point. The magnetic field *produced* by a current in a closed circuit will *not* produce a net force on the circuit (itself) which would cause the circuit (itself) to move. (Why not?) It *will* create a force of tension in the circuit. For example, a large current in a very flexible square loop would create a force of tension which would tend to make the loop circular in shape.

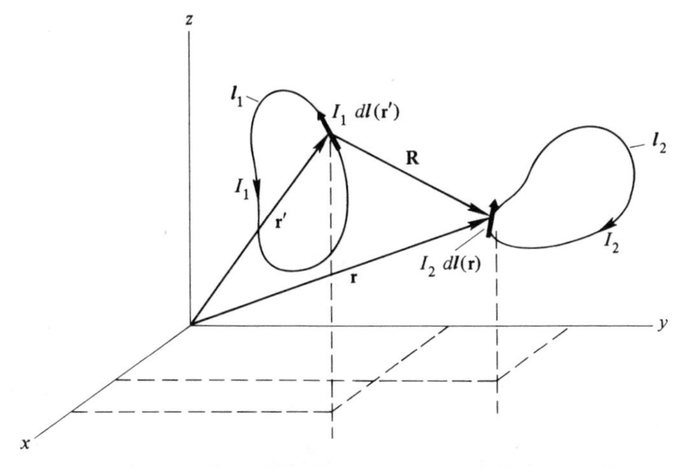

Figure 7.18. Geometry for finding the force between filamentary current elements.

7.10 INTERACTION BETWEEN FILAMENTARY CURRENTS

Consider two differential filamentary current elements as shown in Figure 7.18.

Suppose we want to calculate the force on I_2 due to I_1. From the Biot–Savart law, the field due to I_1 is

$$\mathbf{B}(\mathbf{r}) = \frac{\mu_0}{4\pi} \int_{l_1} I_1\ d\mathbf{l}(\mathbf{r}') \times \frac{\mathbf{a}_R}{R^2} \tag{7.78}$$

So, from equation 7.75,

$$\mathbf{F}(\mathbf{r}) = \frac{\mu_0 I_1 I_2}{4\pi} \int_{l_2} d\mathbf{l}(\mathbf{r}) \times \int_{l_1} d\mathbf{l}(\mathbf{r}') \times \frac{\mathbf{a}_R}{R^2} \tag{7.79}$$

Two interchanges of cross-products in equation 7.79 give

$$\mathbf{F}(\mathbf{r}) = \frac{\mu I_1 I_2}{4\pi} \int_{l_2} \left[\int_{l_1} \frac{\mathbf{a}_R}{R^2} \times d\mathbf{l}(\mathbf{r}') \right] \times d\mathbf{l}(\mathbf{r}) \tag{7.80}$$

Equations 7.79 and 7.80 can easily be modified to accommodate surface current densities or general current densities.

EXAMPLE 13

Suppose we wish to calculate the force per unit length repelling two parallel infinite filamentary currents as shown in Figure 7.19. Let us calculate the force per unit length on the upper conductor. The field

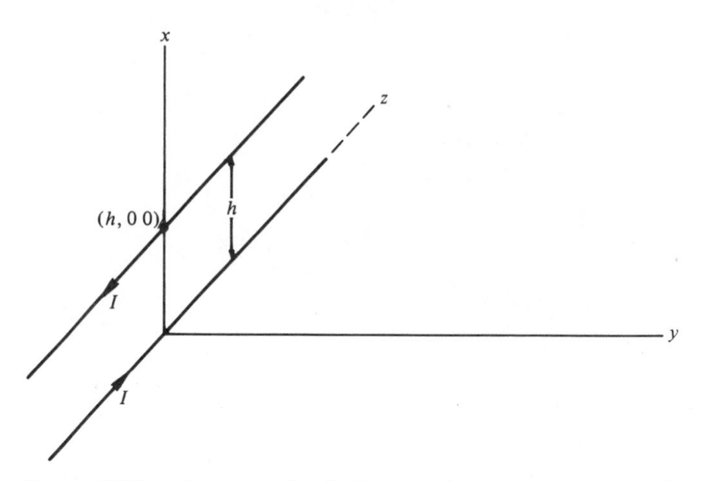

Figure 7.19. Geometry for finding the force per unit length between parallel current filaments.

at the upper conductor due to the current in the lower conductor acting *alone* is

$$\mathbf{H} = \mathbf{a}_y \frac{I}{2\pi h}$$

from equation 7.9. This term is the negative of $I_1/4\pi$ times the bracketed term of equation 7.80. Hence, equation 7.80 gives

$$\mathbf{F} = \mu_0 I \int_0^1 (-\mathbf{a}_y) \frac{I}{2\pi h} \times (-\mathbf{a}_z)\, dz$$

or

$$\mathbf{F} = \frac{\mu_0 I^2}{2\pi h} \mathbf{a}_x \quad \text{(N/m)} \tag{7.81}$$

7.11 TORQUE

Perhaps the easiest way to introduce torque is to reconsider the electric dipole. Let us assume we have a dipole such that the two charges are located at the ends of a dielectric rod. A uniform external field **E** is applied as shown in Figure 7.20. It is immediately obvious that a force exists on the positive charge and a force exists on the negative charge which would tend to make the dipole rotate and align its own field opposite to that of the applied field, or to align the dipole moment with the applied field. This situation may be described by saying that there is a *torque* on the dipole. The torque is

$$\mathbf{T} = \mathbf{R}_0 \times \mathbf{F} \quad (\text{N} \cdot \text{m}) \tag{7.82}$$

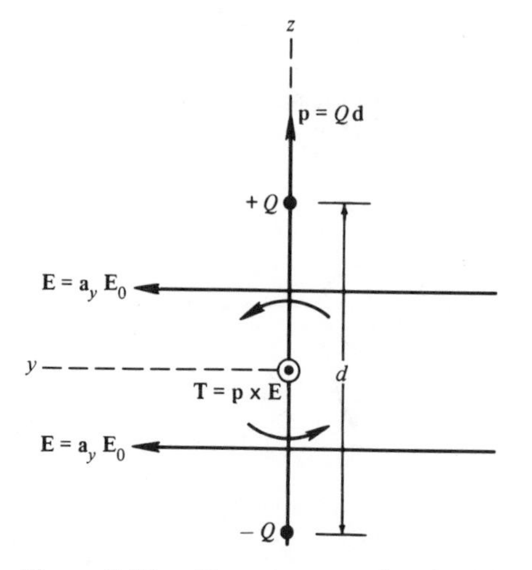

Figure 7.20. Torque on an electric dipole.

where $\mathbf{R}_0$ is the vector lever-arm, and $\mathbf{F}$ is the force acting at the end of the lever-arm. For the electric dipole,

$$\mathbf{T} = \mathbf{a}_z \frac{d}{2} \times Q\mathbf{E} + (-\mathbf{a}_z)\frac{d}{2} \times (-Q)\mathbf{E}$$

or

$$\mathbf{T} = \mathbf{a}_z \frac{d}{2} \times \mathbf{a}_y QE_0 + \mathbf{a}_z \frac{d}{2} \times \mathbf{a}_y \mathbf{E}_0$$

or

$$\mathbf{T} = -Q\, dE_0 \mathbf{a}_x \tag{7.83}$$

The direction of the torque is along the axis about which the dipole would tend to rotate (using the right-hand rule). Another general form is possible. This is

$$\boxed{\mathbf{T} = \mathbf{p} \times \mathbf{E}} \quad (\text{with } \mathbf{p} = Q\mathbf{d}) \tag{7.84}$$

which gives the same result. Notice that the *total force* on the dipole is zero!

In the same way, the *total* force on a closed filamentary current from a uniform external magnetic field is zero. That is,

$$\mathbf{F} = \oint I\, d\mathbf{l} \times \mathbf{B} = -\oint \mathbf{B} \times I\, d\mathbf{l}$$

or

$$\mathbf{F} = -\mathbf{B} \times \oint I\, d\mathbf{l} = -I\mathbf{B} \times \oint d\mathbf{l} = 0$$

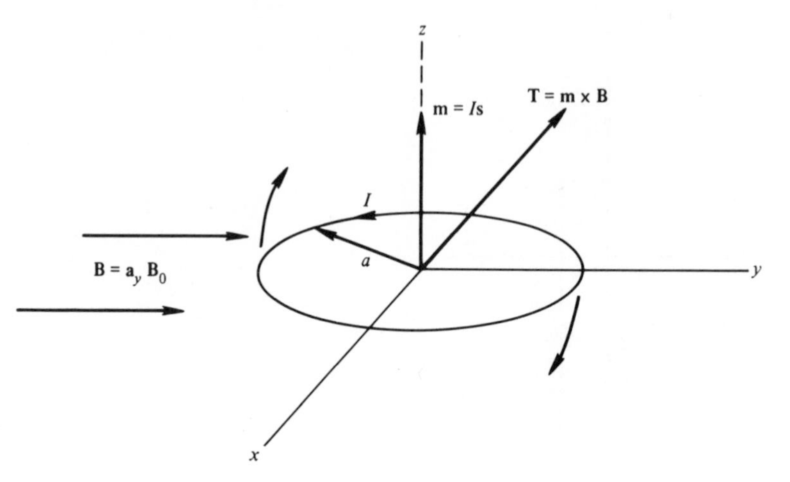

Figure 7.21. Torque on a magnetic dipole.

since $\oint d\mathbf{l} = 0$. Consider the circular loop (magnetic dipole) shown in Figure 7.21. An external uniform magnetic flux density **B** is applied as shown. A z-directed force is created on the left ($y < 0$) side of the loop, while a negative z-directed force is created on the right ($y > 0$) side of the loop. Thus, the loop will attempt to rotate and align its field and moment with the external field when under the influence of a torque. The axis of rotation is the x axis, and the torque is in the $-x$ direction from the right-hand rule. Calculating the torque may be difficult, but we should be able to predict that the answer will be $\mathbf{m} \times \mathbf{B}$ by analogy with the electric dipole case.

Consider Figure 7.22. We have

$$\begin{aligned} d\mathbf{F} &= I\, d\mathbf{l} \times \mathbf{B} = Ia\, d\phi\, \mathbf{a}_\phi \times \mathbf{a}_y \mathbf{B}_0 \\ &= Ia\, d\phi(-\mathbf{a}_x \sin\phi + \mathbf{a}_y \cos\phi) \times \mathbf{a}_y \mathbf{B}_0 \end{aligned}$$

or

$$d\mathbf{F} = -\mathbf{a}_z Ia \sin\phi\, d\phi\, B_0$$

Then,

$$\begin{aligned} d\mathbf{T} &= \mathbf{R}_0 \times d\mathbf{F} \\ &= Ia^2 B_0\, d\phi(\mathbf{a}_y \sin\phi \cos\phi - \mathbf{a}_x \sin^2\phi) \end{aligned}$$

or

$$\mathbf{T} = Ia^2 B_0 \int_0^{2\pi} (\mathbf{a}_y \sin\phi \cos\phi - \mathbf{a}_x \sin^2\phi)\, \mathrm{d}\phi$$

or

$$\mathbf{T} = I\pi a^2 B_0(-\mathbf{a}_x)$$

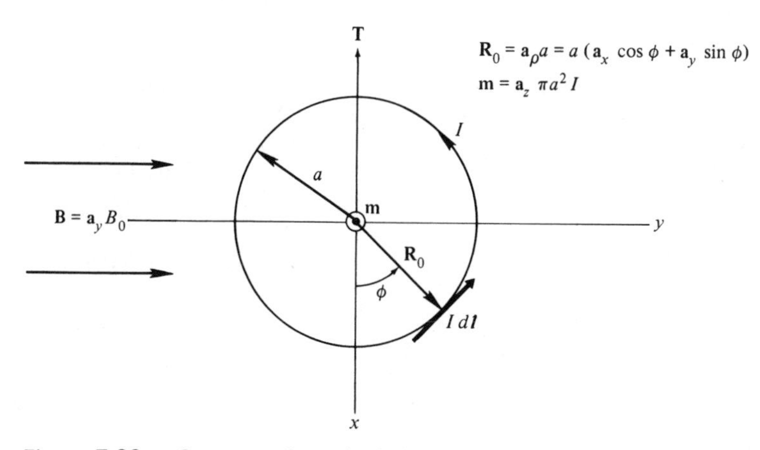

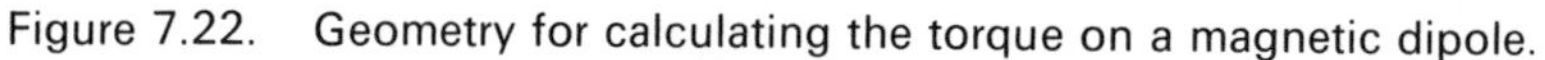

Figure 7.22. Geometry for calculating the torque on a magnetic dipole.

By inspection a general result is as predicted.

$$\boxed{\mathbf{T} = \mathbf{m} \times \mathbf{B}} \tag{7.85}$$

where

$$\mathbf{m} = I\mathbf{s}$$

as previously defined.

These results are not only very important as the basis for motor operation, but are useful in explaining the behavior of magnetic materials.

7.12 CONCLUSIONS

The magnetic field intensity, **H**, was introduced in this chapter as merely a term to be added to the Coulomb force when two charges are in relative motion. The magnetic field intensity is then the force on a current element in somewhat the same way that the electric field intensity, **E**, is the force on a charge. It is more satisfying to introduce the magnetic field in terms of a force than to simply introduce the Biot–Savart law. With no mention of force, the Biot–Savart law introduces a rather abstract quantity.

The Biot–Savart law relates the magnetic field to the source producing it. Ampere's circuital law was introduced and shown to be the integral form of one of Maxwell's equations. A very useful magnetic vector potential, **A**, was introduced: its existence being due to the solenoidal nature of magnetic fields. The formulations of the Biot–Savart law, Ampere's law, and the magnetic vector potential were shown to be consistent. Finally, the force and torque on a magnetic dipole were found. In Chapter 8, we will investigate the modifications that must be made in order to use these results when the medium is not a vacuum.

REFERENCES

Lorrain, P. and Corson, D. R. *Electromagnetic Fields and Waves.* San Francisco: Freeman, 1970. Relativity is discussed in Chapter 5.

Page, L. "A Derivation of the Fundamental Relations of Electrodynamics from Those of Electrostatics." *Am. J. Sci.* 34 (1912): 57–68. This is apparently the original work whereby the magnetostatic field is obtained from the electrostatic field of moving charges.

Plonus, M. A. *Applied Electromagnetics.* New York: McGraw-Hill, 1978. Relativity is discussed in Chapter 12.

Shedd, P. C. *Fundamentals of Electric Waves.* New York: Prentice-Hall, 1954. The electromagnetic field is derived from Coulomb's law in Chapter 18.

Silvester (see References for Chapter 5). A relativistic preamble is used.

PROBLEMS

1. Find the magnetic field strength produced by a filamentary current, $I[u(z+h) - u(z-h)]$, on the z axis. Take a circular path of radius a in the $z = 0$ plane and find $\oint \mathbf{H} \cdot d\mathbf{l}$. Explain the answer.
2. Find the magnetic field on the z axis from a square filamentary loop having sides l and located in the $z = 0$ plane. Assume the current is counterclockwise from above. The loop center is at (0, 0, 0). See Figure 7.23.

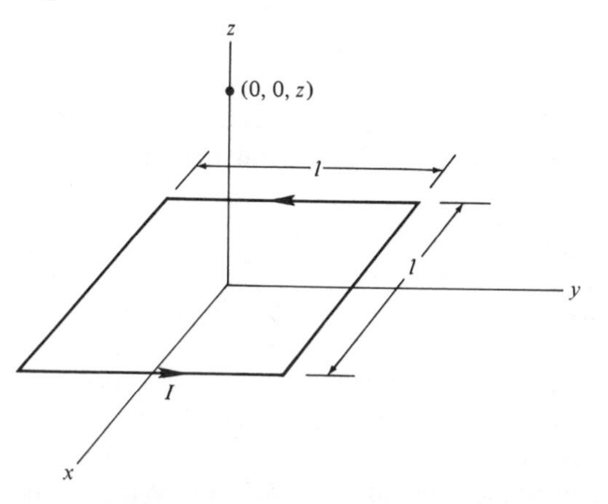

Figure 7.23. A square loop located symmetrically in the $z = 0$ plane.

3. An infinite sheet of surface current, $\mathbf{J}_s = J_0 \mathbf{a}_z$, is located in the $x = 0$ plane. Find $\int \mathbf{H} \cdot d\mathbf{l}$, if the path is a circle of radius a in the $z = 0$ plane.
4. Take the divergence of $\mathbf{B} = \mu_0 \mathbf{H}$, where $\mathbf{H}$ is given by the Biot–Savart law, and show explicitly that $\nabla \cdot \mathbf{B} = 0$.
5. Find $\mathbf{A}$ for an infinite cylinder of radius a carrying a uniform current density in the $+z$ direction. Use Poisson's equation. Find the magnetic field strength. $[A_z(a) = 0.]$

6. Find the force of repulsion per unit length between the two conductors of a planar transmission line. The two conductors are plane strips, of width b and separation d, carrying equal and opposite currents. Assume $b \gg d$.
7. A constant external magnetic field, B_0, is in the direction of $\mathbf{r}_0 = 10\mathbf{a}_x + 5\mathbf{a}_y - 2\mathbf{a}_z$. What torque is produced on a loop of area 10 m^2 carrying a current of 10 A? The loop lies in the $z = 0$ plane and the current is counterclockwise when viewed from above. Describe the final position of the loop (if allowed to rotate) in terms of the equation of a plane.
8. A very flexible conductor, in the form of an arbitrarily shaped loop, has a rather large current applied to it. What shape will the loop assume? See Problem 9.
9. A circular loop of moment, $\mathbf{m} = I\mathbf{s}$, is exposed to a uniform field, $\mathbf{B} = B_0\mathbf{a}_m$, where $\mathbf{a}_m$ is in the direction of $\mathbf{m}$, and $\mathbf{B}$ is produced partially by the current I. Find the force of tension which is attempting to break the loop apart.
10. A uniform magnetic field, $\mathbf{B} = -B_0\mathbf{a}_y$, exists approximately in the region, $-d/2 \le x \le d/2$, $-\infty \le y \le \infty$, $0 \le z \le l$. An electron enters this field at (0, 0, 0) with an initial velocity $u_{z0}\mathbf{a}_z$. Find the equations of motion for the electron (*magnetostatic deflection system*). See Figure 7.24.

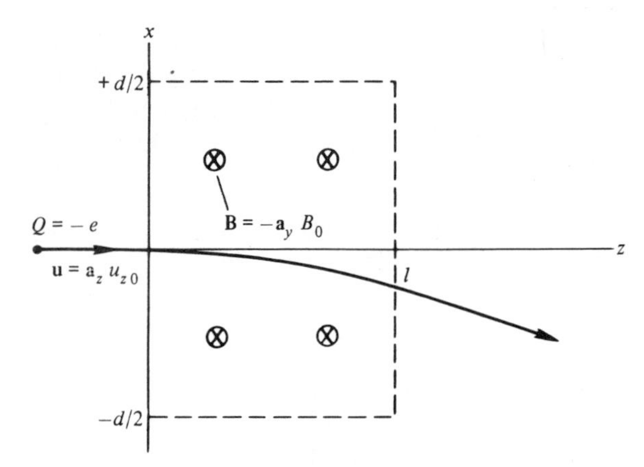

Figure 7.24. Geometry for a magnetostatic deflection system.

11. (a) If the filamentary current $I[u(z + l/2) - u(z - l/2)]$ is located symmetrically on the z axis, show that $A_z = \mu Il/4\pi r$ if $r \gg l$.
    (b) Show that the same answer is obtained if $\mathbf{J} = Il\,\delta(x)\,\delta(y)\,\delta(z)\mathbf{a}_z$.

12. A magnetic dipole (center at 0, 0, 0) is located in the $z = 0$ plane with current in the $\mathbf{a}_\phi$ direction. A magnetic flux density, $\mathbf{B} = -\mathbf{a}_y B_0$, is present. Find the energy required to rotate the dipole to a stable position. Hint: See Problem 19, Chapter 4.

13. As an example to demonstrate the importance of symmetry, consider an infinite filamentary current which was on the z axis but is now rotated α degrees in the $y = 0$ plane. Ignore the fact that the answer is known (use it as a check), and use the Biot–Savart law to find **H** at a point in the $y = 0$ plane on a line passing through the origin. See Figure 7.25.

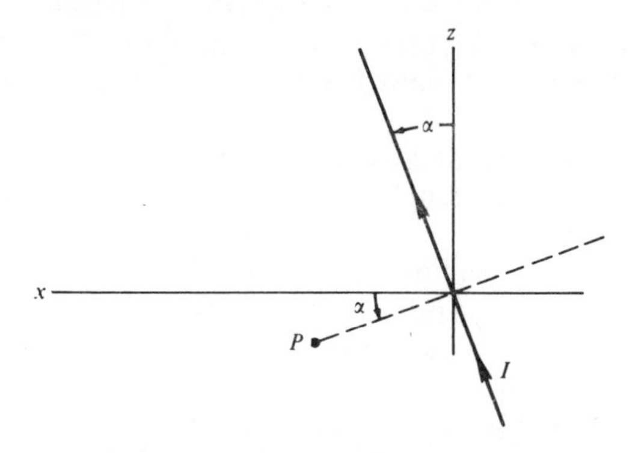

Figure 7.25. Geometry for Problem 13.

14. In Figure 7.19 let the conductor that passes through $(h, 0, 0)$ be rotated in the $x = h$ plane about $(h, 0, 0)$ so that it makes an angle α with respect to the $y = 0$ plane. Find the *total* force on this conductor.
15. A filamentary current I_1 is located on the z axis while another filamentary current I_2 in the form of a square loop 1 m on a side lies in the $x = 0$ plane with its sides parallel to the y and z axes and its center at $(0, 1.5, 0)$. I_2 flows in the $+z$ direction at $y = 2$.

    (a) Find the total vector force on the loop.
    (b) Find the total vector torque on the loop about an axis through the center of the loop. See Figure 7.26.

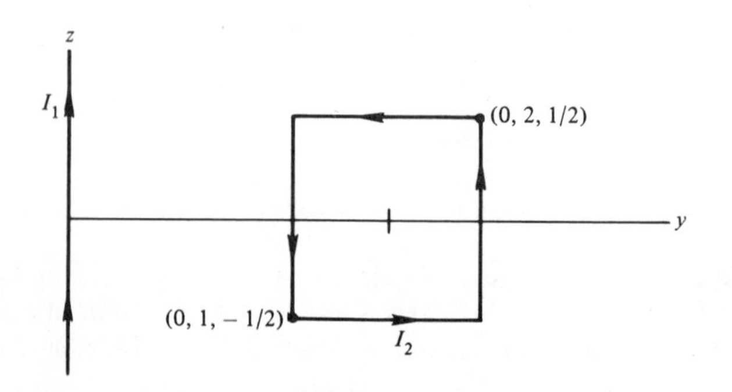

Figure 7.26. Geometry for Problem 15.

16. Repeat Problem 15 if the loop is rotated 90° such that the leg that was at $y = 2$ is now at $x = -\frac{1}{2}$, $y = \frac{3}{2}$. See Figure 7.27.

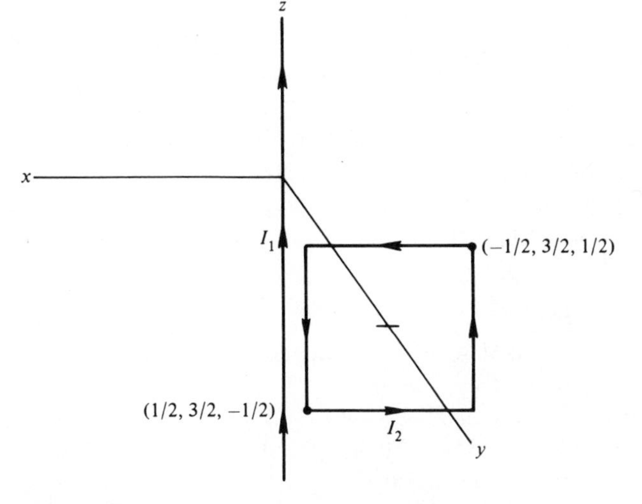

Figure 7.27. Geometry for Problem 16.

17. The cylinder $\rho = 0.05$, $0 \leq z \leq 0.20$ carries a uniform surface current density $\mathbf{J}_s = \mathbf{a}_z 10^3$. A magnetic flux density $\mathbf{B} = \mathbf{a}_\rho$ exists at the cylinder surface (*idealized motor*).

    (a) Find the total torque on the cylinder.
    (b) If the cylinder rotates at 1000 rpm, what power is provided? See Figure 7.28.

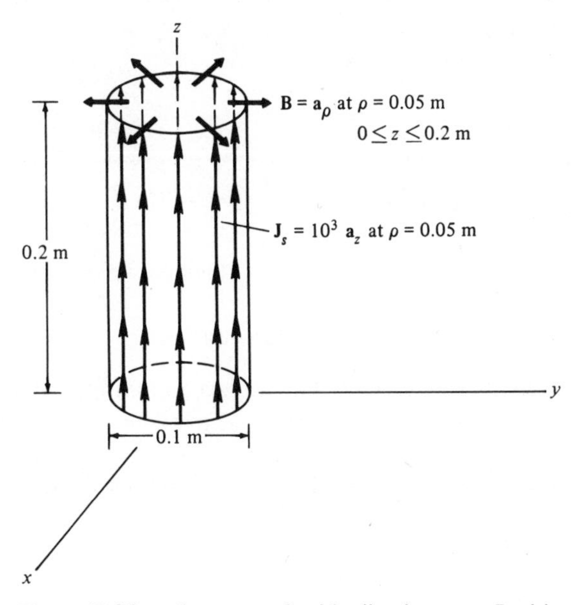

Figure 7.28. Geometry for idealized motor, Problem 17.

18. Find B at (0, 0, 0) from a pair of circular loops in the $z = \pm 1$ m planes, radii 1 m and centers at $z = \pm 1$ m. The current in the lower loop is 1 A in the $\mathbf{a}_\phi$ direction, while the current in the upper loop is 1 A in the $\mathbf{a}_\phi$ direction.
19. An idealized solenoid can be thought of as an infinite number of the loops of Problem 18 stacked on each other. This ultimately results in a current density of $J_{s\phi}\,\delta(\rho - a)\mathbf{a}_\phi$, where a is the solenoid radius. In the next chapter (Figure 8.3) it is shown by means of Ampere's law that

$$H_z(\rho) = J_{s\phi}u(a - \rho).$$

(a) Show that $\nabla^2\mathbf{A} = -\mu_0\mathbf{J}$ for the idealized solenoid.
(b) Show that $\mathbf{B} = \nabla \times \mathbf{A}$.
Hint: Try $A_\phi = (\mu_0/2)J_{s\phi}[\rho u(a - \rho) + (a^2/\rho)u(\rho - a)]$.
20. An idealized toroid can be thought of as a finite length ideal solenoid bent around to close on itself to form a doughnut shape. If the surface current density at $\rho = \rho_a - a$ is J_{sz} as shown in Figure 7.29,

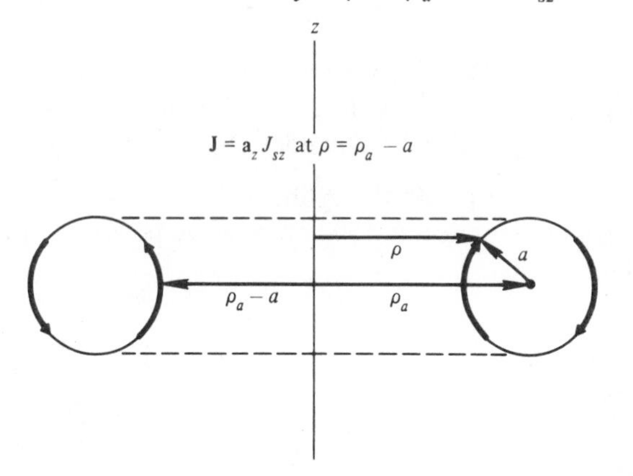

Figure 7.29. Ideal toroid.

then it can be shown that

$$\mathbf{H} = \begin{cases} J_{sz}\dfrac{\rho_a - a}{\rho}\mathbf{a}_\phi, & \text{inside toroid} \\ 0, & \text{outside toroid} \end{cases}$$

Find the circulation of $\mathbf{H}$ for a circular path of radius ρ (in the $z = 0$ plane) if (a) $0 < \rho < \rho_a - a$, (b) $\rho_a - a < \rho < \rho_a + a$, and (c) $\rho > \rho_a + a$.
21. Show explicitly that $\nabla \times \mathbf{H} = \mathbf{J}$ for an infinite filamentary current on the z axis.
22. Show that $\nabla \cdot \nabla \times \mathbf{A} \equiv 0$ using Cartesian coordinates.
23. Show that $\nabla \times (\alpha\mathbf{C}) \equiv \nabla\alpha \times \mathbf{C} + \alpha(\nabla \times \mathbf{C})$ using Cartesian coordinates.

Chapter 8
Materials and Material Discontinuities with the Steady Magnetic Field

Having investigated the properties of the magnetic dipole in the preceding chapter, we are now in a position to investigate the magnetic properties of materials. The procedure is almost identical to that used in examining the behavior of dielectric materials. Since we have established a number of basic equations governing the behavior of magnetic fields in a vacuum, we would hope that these results are not severely modified by the addition of other materials. In most cases they are not. Once material properties are understood we can establish boundary conditions that must exist at the interface between different magnetic materials. Simple magnetic circuits, which have much practical use, can be investigated. We will also define the third circuit parameter, inductance, which is usually much more complicated than capacitance and more difficult to calculate. Finally, the energy associated with the steady magnetic field will be found.

8.1 MAGNETIC MATERIALS

As in our discussion of the dielectric properties of materials, we will avoid a detailed discussion of quantum theory in arriving at the magnetic properties of materials. The reader is referred to the many excellent texts on solid-state physics for the quantum theory approach to material behavior. This is a study in itself, and so we must use a simplified picture of material behavior in the presence of applied fields. In the simplest picture of an atom, an electron or electrons orbit around the nucleus. As such, these electrons are effectively current loops, or magnetic dipoles. In the previous chapter, we investigated the behavior of magnetic dipoles under the influence of a uniform external magnetic field. We should, then, be able to develop a fairly accurate theory of material behavior, at least for present purposes.

There are other vector moments present in an atom besides those attributed to the magnetic dipole (orbiting electron). For example, *electron spin* and *nuclear spin* also contribute moments. It is the combination of all the moments which determines the magnetic properties of materials. For an *isolated* current loop or magnetic dipole, the field of the loop itself always acts to add to the external field giving an *increased* field in the neighborhood of the loop. In a material, the internal magnetic field may increase or decrease, depending on *all* of the moments. Let us examine briefly some classes of magnetic materials.

1. In *diamagnetic* materials (bismuth) spin moments tend to be dominant and produce fields that *oppose* the external field. Thus, the internal magnetic field is reduced slightly compared to the external field.
2. If the dipole moments dominate slightly, then the internal field is increased slightly over the external field and the material is *paramagnetic* (tungsten).
3. Large dipole moments are produced in certain regions or *domains* for the *ferromagnetic* materials (iron). A random domain alignment exists for virgin ferromagnetic material. When an external field is applied and then removed, a net alignment occurs giving *permanent magnetization* and a *hysteresis* effect. Alloys of some of the ferromagnetic materials are also ferromagnetic (alnico).
4. In *ferrimagnetic* materials, adjacent atoms develop unequal, but oppositely directed moments, allowing a rather larger response to external fields. From the point of view of engineering applications, the *ferrites* are very important ferrimagnetic materials. Ferrites possess a very high resistance, and hence give very little eddy current loss at higher frequencies when used as transformer cores.
5. The magnetic tape used for audio and video recording is a superparamagnetic material and is composed of an array of small ferromagnetic particles.

8.2 MAGNETIZATION AND BOUNDARY CONDITIONS

Magnetization, perhaps more aptly called magnetic polarization, is a vector quantity that is quite analogous to the polarization vector used to explain dielectric behavior. The development here is very similar to that used for dielectrics in Chapter 4.

We will attempt to show that the magnetic dipoles act as sources for the magnetic field. The current that gives rise to the magnetic field will be due to the motion of *bound* charge. (This should not be unexpected, for it was the action of the *bound* charge in a dielectric which caused it to be polarized.) The material will then be magnetized, or magnetically polarized.

The magnetic dipole moment may be described in terms of equation 7.67 as $\mathbf{m} = I\mathbf{s}$. If we now simply recognize that the magnetic dipole is a source of strength or moment $I\mathbf{s}$, then a unit dipole would have a moment $\mathbf{a}_m$, a unit vector directed along $\mathbf{s}$. Then the (macroscopic) unit impulse response (for $\mathbf{A}$) for the dipole is obtained directly from equation 7.67. It is

$$\mathbf{h}(\mathbf{r} - \mathbf{r}') = \frac{\mu_0}{4\pi} \frac{\mathbf{a}_m \times \mathbf{a}_R}{R^2} \tag{8.1}$$

The magnetic dipole is being treated here as a point source. If we now define a density function, *which must be dipole moment per unit volume*, we can obtain a general solution for $\mathbf{A}$ by simply writing down the superposition integral. We take the definition of the *magnetization vector*, $\mathbf{M}$, to be the magnetic dipole moment per unit volume with dimension A/m. It then follows that

$$\mathbf{A}(\mathbf{r}) = \frac{\mu_0}{4\pi} \iiint_{\text{vol}'} \frac{\mathbf{M}(\mathbf{r}') \times \mathbf{a}_R}{R^2} \, dv' \tag{8.2}$$

where we are still using μ_0 because we are including the magnetic effects in the molecular magnetic dipoles. We next replace $\mathbf{a}_R/R^2$ with $\nabla'(1/R)$, obtaining

$$\mathbf{A}(\mathbf{r}) = \frac{\mu_0}{4\pi} \iiint_{\text{vol}'} \mathbf{M}(\mathbf{r}') \times \nabla'\left(\frac{1}{R}\right) dv' \tag{8.3}$$

We have previously used the vector identity.

$$\mathbf{C} \times \nabla'\alpha \equiv \alpha\nabla' \times \mathbf{C} - \nabla' \times (\alpha\mathbf{C})$$

In equation 8.3 this identity gives

$$\mathbf{A}(\mathbf{r}) = \frac{\mu_0}{4\pi} \iiint_{\text{vol}'} \frac{1}{R} \nabla' \times \mathbf{M}(\mathbf{r}') \, dv' - \frac{\mu_0}{4\pi} \iiint_{\text{vol}'} \nabla' \times \left[\frac{\mathbf{M}(\mathbf{r}')}{R}\right] dv' \tag{8.4}$$

We now need a vector identity not previously encountered, but it can be found in Appendix A. It is

$$\iiint_{\text{vol}} (\nabla \times \mathbf{C}) \, dv \equiv -\oiint_s \mathbf{C} \times d\mathbf{s} = -\oiint_s \mathbf{C} \times \mathbf{a}_n \, ds \qquad (\mathbf{a}_n \, ds = d\mathbf{s})$$

With this identity equation 8.4 becomes

$$\mathbf{A}(\mathbf{r}) = \frac{\mu_0}{4\pi} \iiint_{\text{vol}'} \frac{1}{R} \nabla' \times \mathbf{M}(\mathbf{r}')\, dv' + \frac{\mu_0}{4\pi} \iint_{s'} \frac{M(\mathbf{r}')}{R} \times \mathbf{a}_n\, ds' \tag{8.5}$$

We next calculate the magnetic vector potential due to *ordinary* general current density, **J**, and surface current density, $\mathbf{J}_s$, flowing in and on an ordinary nonmagnetic (μ_0) conductor. This is just the sum of equations 7.49 and 7.54 by superposition. The result is

$$\mathbf{A}(\mathbf{r}) = \frac{\mu_0}{4\pi} \iiint_{\text{vol}'} \frac{\mathbf{J}(\mathbf{r}')}{R}\, dv' + \frac{\mu_0}{4\pi} \iint_{s'} \frac{\mathbf{J}_s(\mathbf{r}')}{R}\, ds' \tag{8.6}$$

The currents in equation 8.6 are due to *conduction* electrons. A comparison of equations 8.5 and 8.6 reveals that $\nabla \times \mathbf{M}(\mathbf{r})$ has the effect of a general current density, while $\mathbf{M}(\mathbf{r})$ has the effect of a surface current density! We are then justified in writing

$$\nabla \times \mathbf{M} = \mathbf{J}^b \tag{8.7}$$

and

$$\mathbf{M} \times \mathbf{a}_n = \mathbf{J}_s^b \tag{8.8}$$

These are the *bound* currents, or currents due to bound charges, which were mentioned earlier.

We now have a situation where the permeability is that of free space, μ_0, and where *both* bound current and ordinary current may exist. Let us write Maxwell's equation in terms of **B** instead of **H**. The reason for this is that we intend to obtain a new, more general, relation between **B** and **H** (other than the original $\mathbf{B} = \mu_0 \mathbf{H}$). We have

$$\nabla \times \frac{\mathbf{B}}{\mu_0} = \text{total vector current density} \tag{8.9}$$

This equation is correct, because the permeability is μ_0, and *all* of the current is considered. Equation 8.9 may be written

$$\nabla \times \frac{\mathbf{B}}{\mu_0} = \mathbf{J} + \mathbf{J}^b$$

With equation 8.7, we have

$$\nabla \times \frac{\mathbf{B}}{\mu_0} = \mathbf{J} + \nabla \times \mathbf{M}$$

or

$$\nabla \times \left(\frac{\mathbf{B}}{\mu_0} - \mathbf{M}\right) = \mathbf{J} \tag{8.10}$$

The right side of equation 8.10 is the current density due to ordinary processes and is not bound current. We now have a way of avoiding the use of bound current if we simply accept Maxwell's equation (or Ampere's law),

$$\nabla \times \mathbf{H} = \mathbf{J} \tag{8.11}$$

as being correct. A comparison of equations 8.10 and 8.11 gives us the new relation between **B** and **H**.

$$\mathbf{H} = \frac{\mathbf{B}}{\mu_0} - \mathbf{M} \tag{8.12}$$

or

$$\boxed{\mathbf{B} = \mu_0(\mathbf{H} + \mathbf{M})} \tag{8.13}$$

Notice that the right side of equation 8.11 is a current density due to conduction electrons, while the effect of the bound charges is included on the left side in **H**.

We next need a relation between **M** and **H** to simplify equation 8.13. For linear isotropic materials, a simple relationship exists.

$$\mathbf{M} = \chi_m \mathbf{H} \tag{8.14}$$

where χ_m is the *magnetic susceptibility*. In this case, we have

$$\mathbf{B} = \mu_0(\mathbf{H} + \chi_m \mathbf{H}) = \mu_0(1 + \chi_m)\mathbf{H}$$

or simply

$$\boxed{\mathbf{B} = \mu_0 \mu_R \mathbf{H} = \mu \mathbf{H}} \tag{8.15}$$

That is,

$$\boxed{\mu = \mu_0 \mu_R} \tag{8.16}$$

is the permeability of the material, and μ_R is its *relative permeability*. If the material is anisotropic, μ is *not* a scalar, but is a tensor (as was ε under the same conditions). The relation between **B** and **H** is still equation 8.15 for anisotropic material, but **B** and **H** may not be parallel.

In most of the problems we will be concerned with here, we simply replace μ_0 by μ if magnetic material is present, and then use the equations in their original form (for free space). This is a great step forward. It is rewarding to return to Chapter 4 to see that the development for dielectric effects there exactly parallels the development here. As a matter of fact, we could have invoked the principle of duality and written down the results of this section after examining the results of Chapter 4.

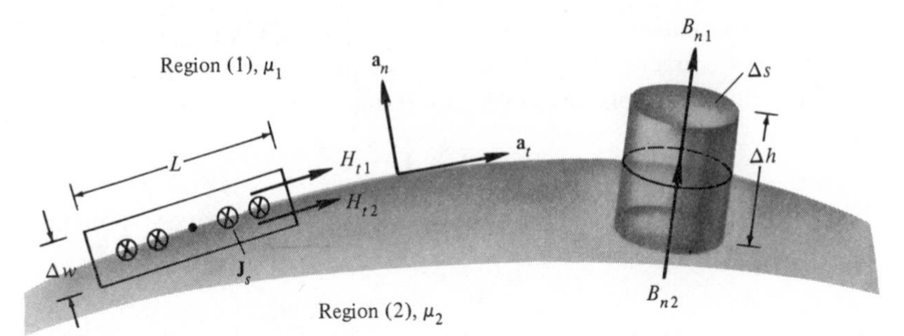

Figure 8.1. Geometry for determining boundary conditions at the interface between simple magnetic media.

It is a very simple matter to obtain the boundary conditions at the interface between two linear, isotropic, and homogeneous magnetic materials. Consider Figure 8.1 which is similar to Figure 4.8 (which was used for the analogous dielectric case). The development here parallels that for the dielectric case, and the details will be omitted. Applying Ampere's law to the rectangle for vanishing Δw gives

$$H_{t1} - H_{t2} = J_s, \qquad \boxed{\mathbf{a}_n \times (\mathbf{H}_1 - \mathbf{H}_2) = \mathbf{J}_s} \tag{8.17}$$

So, the tangential component of magnetic field strength is discontinuous by an amount equal to the surface current density. Using equation 8.15 yields

$$\frac{B_{t1}}{\mu_1} - \frac{B_{t2}}{\mu_2} = J_s \tag{8.18}$$

Applying the flux equation,

$$\oiint \mathbf{B} \cdot d\mathbf{s} = 0$$

to the "pill box" in Figure 8.1 for vanishing Δh gives

$$B_{n1} = B_{n2}, \qquad \boxed{\mathbf{a}_n \cdot (\mathbf{B}_1 - \mathbf{B}_2) = 0} \tag{8.19}$$

So the normal component of the magnetic flux density is continuous. It immediately follows that

$$\frac{H_{n1}}{\mu_2} = \frac{H_{n2}}{\mu_1} \tag{8.20}$$

EXAMPLE 1

An example is in order. Suppose the $z = 0$ plane is the locus of an interface for which $z > 0$ has $\mu_1 = 5\mu_0$ and $z < 0$ has $\mu_2 = 3\mu_0$. If it is known that no surface current is flowing, and $\mathbf{H}_2 = 10\mathbf{a}_x + 20\mathbf{a}_z$, find $\mathbf{B}_2$, $\mathbf{B}_1$, and $\mathbf{H}_1$. We immediately obtain

$$\mathbf{B}_2 = \mu_2 \mathbf{H}_2 = 3\mu_0 \mathbf{H}_2 = \mu_0(30\mathbf{a}_x + 60\mathbf{a}_z)$$

Since $\mathbf{J}_s = 0$, tangential $\mathbf{H}$ and normal $\mathbf{B}$ are continuous, so

$$H_{x1} = H_{t1} = H_{t2} = 10 \quad \text{and} \quad B_{z1} = B_{n1} = B_{n2} = 60\ \mu_0$$

Also, from equation 8.18 and 8.20,

$$B_{x1} = \frac{\mu_1}{\mu_2} B_{x2} = \frac{5}{3}(30\mu_0) = 50\mu_0$$

and

$$H_{z1} = \frac{\mu_2}{\mu_1} H_{z2} = \frac{3}{5}(20) = 12$$

Thus,

$$\mathbf{B}_1 = \mu_0(50\mathbf{a}_x + 60\mathbf{a}_z)$$

and

$$\mathbf{H}_1 = 10\mathbf{a}_x + 12\mathbf{a}_z$$

This result is shown in Figure 8.2.

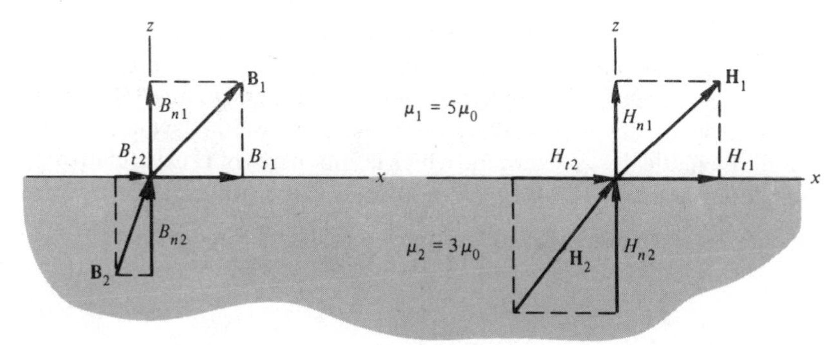

Figure 8.2. Normal and tangential components of **B** and **H** at a plane interface for $\mu_1 = \frac{5}{3}\mu_2$.

8.3 SIMPLE MAGNETIC CIRCUITS

EXAMPLE 2

In order to define what we mean by a *magnetic* circuit, let us consider the field of a very long *solenoid*. A solenoid is usually made by winding many closely spaced turns on a circular cylindrical form. In other words, it is a coil. For simplicity, let us consider first the case where the *core* (the interior of the solenoid) is air ($\mu = \mu_0$) and the current is filamentary. Since the windings carry a steady current, I, circumferentially around the axis of the solenoid, we may ideally consider the current to be an azimuthal sheet of surface current, $\mathbf{J}_s = \mathbf{a}_\phi J_{s\phi}$ as shown in Figure 8.3.

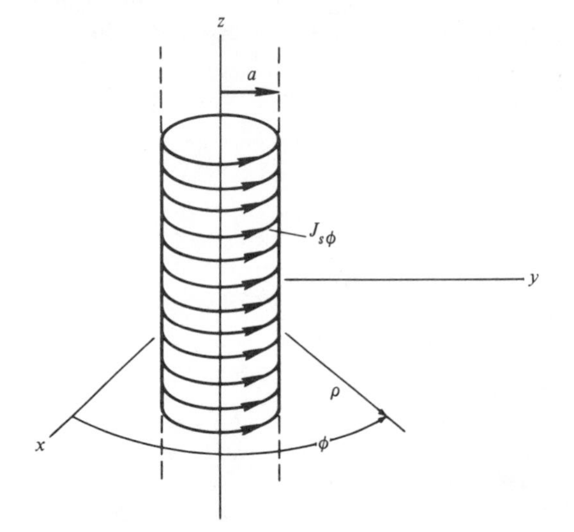

Figure 8.3. Idealized solenoid.

We desired to find $\mathbf{H}$ for $\rho < a$ and $\rho > a$. Since the solenoid is very long, there are no variations with z. There are also no variations with ϕ because of the complete azimuthal symmetry. Thus, $|\mathbf{H}|$ can at most depend only on ρ. At this point we can argue either symmetrically or mathematically to determine which components of $\mathbf{H}$ exist, choosing the latter course to avoid making several rather difficult sketches to

Figure 8.4. The Biot-Savart law applied to the idealized solenoid.

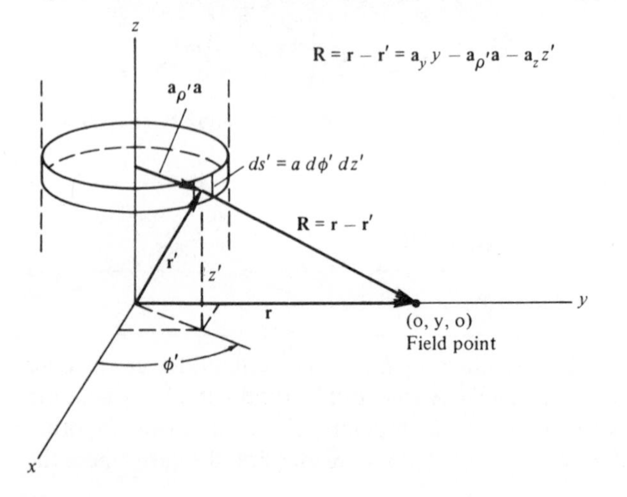

exploit the symmetry, we proceed to Figure 8.4 and apply the Biot–Savart law.

The Biot–Savart law gives

$$\mathbf{H} = \frac{\mu_0 a J_{s\phi}}{4\pi} \int_0^{2\pi} \int_{-\infty}^{\infty} \mathbf{a}_{\phi'} \times \frac{\mathbf{r} - \mathbf{r}'}{|\mathbf{r} - \mathbf{r}'|^3} \, d\phi' \, dz'$$

For a field point on the y axis, which is general because of the symmetry,

$$\mathbf{r} - \mathbf{r}' = \mathbf{a}_y y - \mathbf{a}_{\rho'} a - \mathbf{a}_z z' = -\mathbf{a}_x a \cos \phi' - \mathbf{a}_y(a \sin \phi' - y) - \mathbf{a}_z z'$$

Since $\mathbf{a}_{\phi'} = -\mathbf{a}_x \sin \phi' + \mathbf{a}_y \cos \phi'$, we have

$$\mathbf{a}_{\phi'} \times (\mathbf{r} - \mathbf{r}') = -\mathbf{a}_x z' \cos \phi' - \mathbf{a}_y z' \sin \phi' - \mathbf{a}_z(y \sin \phi' - a)$$

after some trigonometric simplification. Also,

$$|\mathbf{r} - \mathbf{r}'|^3 = [y^2 + a^2 - 2ya \sin \phi' + (z')^2]^{3/2}$$

Thus, the integrands in the integral above, for the x and y components of $\mathbf{H}$, are *odd* functions of z' being integrated over symmetric limits $(-\infty, +\infty)$. These components are thus both identically zero, leaving

$$H_z = \frac{\mu_0 J_{s\phi} a}{4\pi} \int_0^{2\pi} \int_{-\infty}^{\infty} \frac{(a - y \sin \phi') \, d\phi' \, dz'}{[y^2 + a^2 - 2ya \sin \phi' + (z')^2]^{3/2}}$$

Symmetry arguments would have led to the same results. We have now shown that $\mathbf{H} = \mathbf{a}_z H_z(y)$, or more generally, $\mathbf{H} = \mathbf{a}_z H_z(\rho)$, without integrating.

It is easy, however, to show that H_z is *not* even a function of ρ for $\rho < a$ or $\rho > a$! Consider Figure 8.5. We see that $H_{z1} = H_{z2}$ in order for $\oint \mathbf{H} \cdot d\mathbf{l} = 0$ for the rectangular path $\rho > a$. Therefore, H_z is at best a constant for $\rho > a$. For large ρ, H_z must vanish (Why?), so this constant is zero, and $H_z = 0$, $\rho > a$.

The circulation around the other path (for $\rho < a$), or, alternatively, use of equation 8.17, shows that $H_z = J_{s\phi}$ for $\rho = a$. Therefore, since H_z is constant, we have[1]

$$H_z(\rho) = J_{s\phi} u(a - \rho) \quad \text{(A/m)} \tag{8.21}$$

Thus, we actually have evaluated the integral above! It is worth mentioning that if the solenoid current is not entirely ϕ-directed, that is, if there is a z component of current from a coil with a finite pitch, however small, then Ampere's law applied to a circular path lying in the $z = 0$ plane with radius $\rho > a$ gives $\oint \mathbf{H} \cdot d\mathbf{l} = I$, and $H_\phi \neq 0$ for $\rho > a$. Thus, a real solenoid will always produce leakage flux.

[1] See Problem 19, Chapter 7.

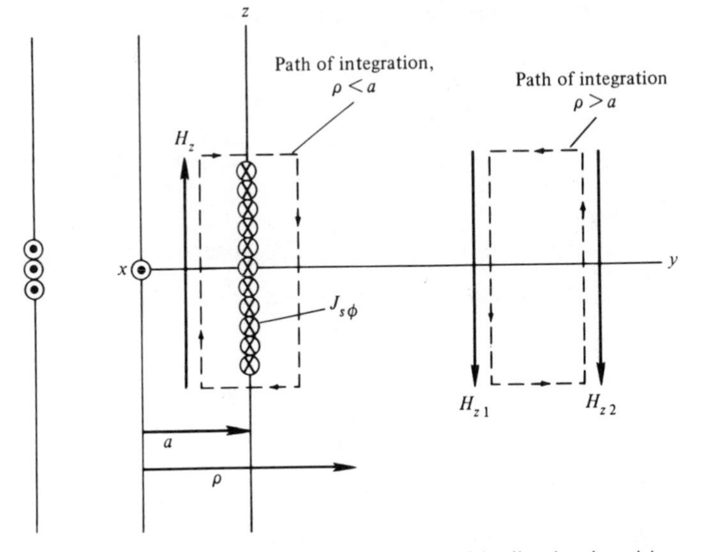

Figure 8.5. Ampere's law applied to the idealized solenoid.

It follows from equation 8.21 that

$$B_z = \mu_0 H_z = \mu_0 J_{s\phi} \quad (\text{Wb/m}^2) \tag{8.22}$$

and

$$\psi_m = \mu_0 J_{s\phi} \pi a^2 = \mu_0 J_{s\phi} s \quad (\text{Wb}) \tag{8.23}$$

where s is the cross-sectional area of the solenoid. It should be clear now that the *magnetic circuit* in this example is the interior of the solenoid, where **B** is uniform. Fortunately, this kind of situation exists, approximately, for many practical magnetic circuits. The ideal solenoid is treated as a boundary value problem in Problem 19, Chapter 7. That method is perhaps shorter, but more mathematical and less physical.

Suppose next that the surface current in the previous example is being produced approximately by a coil of n turns/m carrying I A. Then, ignoring leakage flux, equation 8.23 gives

$$\psi_m = \mu_0 n I s \tag{8.24}$$

since $nI \approx J_{s\phi}$. If the solenoid is l m long ($l \gg a$) with a total of N turns, the magnetic flux is approximately

$$\psi_m \approx \frac{\mu_0 N I s}{l}$$

or

$$\psi_m = \frac{NI}{l/\mu_0 s} \quad (\text{Wb}) \tag{8.25}$$

In the analogous problem of a uniform current carrying conductor, it is well known that

$$I = \frac{V}{R} = \frac{V}{l/\sigma s} \tag{8.26}$$

Comparing equations 8.25 and 8.26 shows that magnetic flux is analogous to electric current. NI is like a forcing function, analogous to an ideal voltage source or supplier of emf and is thus called magnetomotive force, mmf. The denominator of equation 8.25 is analogous to electric resistance and is called the *reluctance* of the magnetic circuit (the core of the solenoid). That is,

$$\mathscr{R} = \frac{l}{\mu_0 s} \quad (\mathrm{H}^{-1}) \tag{8.27}$$

If the magnetic material is not air, but is still linear, isotropic, and homogeneous, then equation 8.27 becomes

$$\boxed{\mathscr{R} = \frac{l}{\mu s}} \quad (\mathrm{H}^{-1}) \tag{8.28}$$

If $\mu \gg \mu_0$, then any leakage flux occurring for $\mu = \mu_0$ will now become smaller, and almost all of the magnetic flux will be confined to the magnetic material; that is, the core. This simple picture of magnetic circuits will serve us well in the examples to follow, because we can lean heavily on simple electric circuit theory.

Let us next consider the behavior of ferromagnetic materials graphically since the relation between B and H is nonlinear and μ_R for a given sample is not unique. This graph is called the B-H curve or hysteresis curve. If an mmf is applied to a virgin (unmagnetized) sample of ferromagnetic material with B and H both zero (point a), a magnetization curve such as that shown in Figure 8.6 is established (from point a to b). As H is increased, for example, by increasing the current in a solenoid around the sample, B increases, slowly at first, then more rapidly, then more slowly to point b. Further increase in H gives little increase in B, and finally B will not increase at all, indicating magnetic saturation (point c). The curve a-b-c can be called the magnetization curve.

If, in the process of magnetization, point b is reached and then H is reduced, the original curve is not retraced, but instead, the B-H relation follows a curve such as b-d, so that at point d a remnant *flux density*, B_r exists even though H is zero. This is *permanent* magnetization. Now if H is decreased (negative), the B-H relation follows the curve d-e, and at point e B is zero, but, H has the value H_c, *the coercive mmf.* Further decrease of H can lead to saturation again, but if, instead, H is now increased at point f (which is symmetrical to b), then the B-H curve will follow the path f-g-h-b. The path b-d-e-f-g-h-b is called the hysteresis loop.

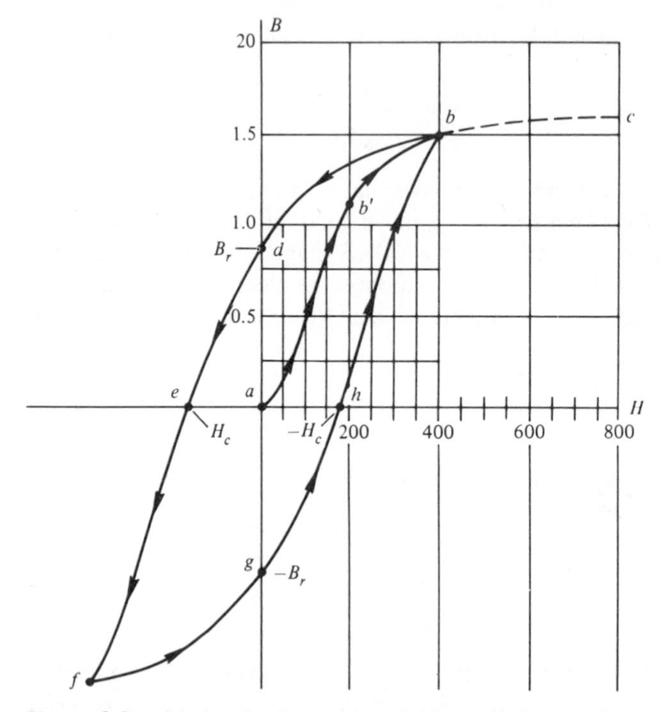

Figure 8.6. Magnetization curve and hysteresis loop for a ferromagnetic sample.

If, when the sample was first being magnetized, point b' had been reached and then H had been decreased, then a smaller hysteresis loop could be traced out. In this way a sample can be demagnetized. That is, if an ac current is applied to the solenoid around the sample, and if the amplitude of this ac signal is gradually reduced, the hysteresis loops will become smaller and eventually degenerate to point a where B is zero.

EXAMPLE 3

Let us next solve, approximately, a simple problem involving a magnetic circuit composed of a ferromagnetic core with an air gap. Let the mmf be applied by means of a toroidal coil as shown in Figure 8.7. Let the air gap be 1 mm, N be 1000 turns, and the *mean* radius of the toroid be 22.5 cm. We would like to find the coil current required to produce $B = 0.5$ Wb/m^2 in the core. Some approximations must be made if we are to accomplish this. First of all we assume that B is uniform in the ferromagnetic core. This is not the case (see Problem 20, Chapter 7), but since we intend to use the mean radius of the toroid in our calculations, and this radius is not very different from the minimum (21.25 cm) and maximum (23.75 cm) radii, B is almost uniform. Secondly, the air gap is very

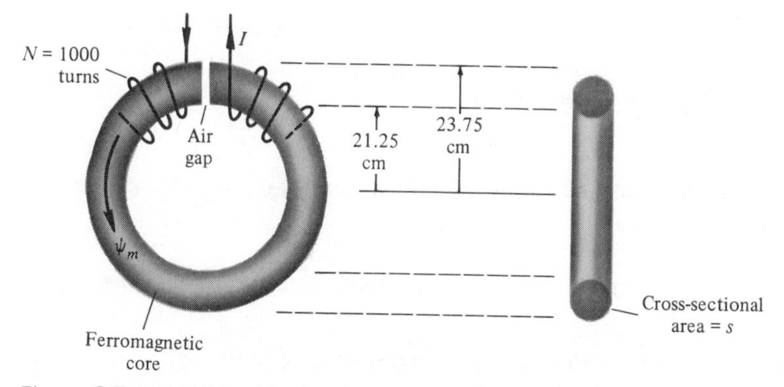

Figure 8.7. Toroid with air gap.

small compared to the diameter of the core, so the flux density is essentially *normal* to the interface between the air and the ferromagnetic core and must therefore be continuous (equation 8.19).

The small leakage flux around the periphery of the air gap is ignored. There will also be some leakage flux between the turns of the coil, and in some cases flux lines may even extend across the interior of the toroid through the air. These effects are small and are neglected, but it should be recognized that magnetic flux lines "prefer" the ferromagnetic material over air by a factor of only about 10^3, whereas in the equivalent electric problem the electric flux lines "prefer" good conductors over air by a factor of about 10^{15}. That is, the permeability of the ferromagnetic material will be roughly 10^3 times that of air, but the conductivity of a good conductor is about 10^{15} times that of air. Thus, the magnetic problem leads to a much more crude (but still useful) model than the electric problem.

We now assume that the virgin magnetization curve *a-b* of Figure 8.6 applies to our problem, so that if $B = 0.5$, then $H = 100$. The assumptions we have made lead to the equivalent electric circuit shown in Figure 8.8. For the air gap

$$\mathcal{R}_{ag} = \frac{l}{\mu_0 s} = \frac{10^{-3}}{4\pi \times 10^{-7}[\pi(2.5 \times 10^{-2})^2]/4} = 1.621 \times 10^6$$

Figure 8.8. Equivalent circuit of the idealized toroid. $\mathcal{R}_{ag}$ = linear reluctance of air gap. $\mathcal{R}_{fc}$ = nonlinear reluctance of ferromagnetic core.

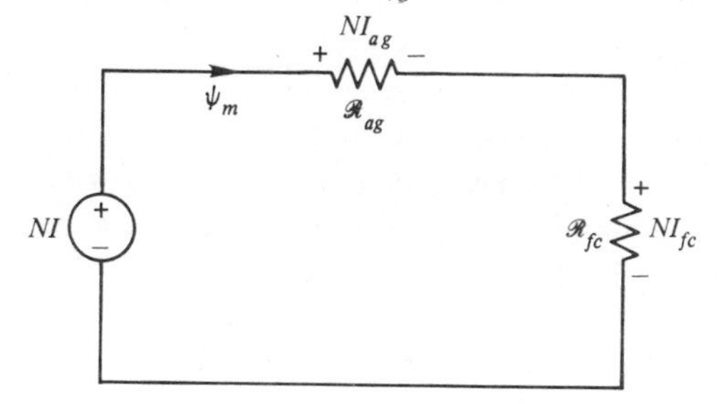

and (everywhere)

$$\psi_m = Bs = 0.5 \frac{\pi(2.5 \times 10^{-2})^2}{4} = 2.454 \times 10^{-4}$$

Therefore the mmf "drop" across the air gap is $\psi_m \mathscr{R}_{ag}$, neglecting flux fringing at the gap; so

$$(NI)_{ag} = \psi_m \mathscr{R}_{ag} = (1.621 \times 10^6)(2.454 \times 10^{-4}) = 398 \quad (\text{A} \cdot \text{t})$$

Producing 0.5 Wb/m^2 in the ferromagnetic core requires $H_{fc} = (NI)_{fc}/l = 100 \quad (\text{A} \cdot \text{t/m})$ (Figure 8.6), so

$$(NI)_{fc} = 100l = 100 \times 2\pi \times 0.225 = 141.4 \quad (\text{A} \cdot \text{t})$$

using the mean radius for finding l. Therefore,

$$NI = (NI)_{ag} + (NI)_{fc} = 539$$

or

$$I = \frac{539}{1000} = 0.539 \quad (\text{A})$$

The solution to this problem was simple for two related reasons. First, we neglected flux leakage and assumed a uniform flux density in the core and gap, and second, because the flux density was uniform and also a *given* quantity, we were able to avoid difficulties with the nonlinear B-H curve. We did not need to calculate $\mathscr{R}_{fc}$.

EXAMPLE 4

Instead of requiring the $B = 0.5$, suppose we specify that $I = 0.5$ A and find the resulting flux density. A trial and error technique will be followed to find B. The answer to the preceding problem gives us a first trial.

1. We know the smaller current will produce a B smaller than 0.5, so we try $B = 0.45$. An enlarged (and linearized) version of the magnetization curve of Figure 8.6 is given in Figure 8.9 showing details in the neighborhood of $B = 0.5$. From this curve, we see that $H_{fc} = 90$ for $B = 0.45$. Then,

$$(NI)_{fc} = H_{fc} l = (90)(2\pi \times 0.225) = 127.2$$

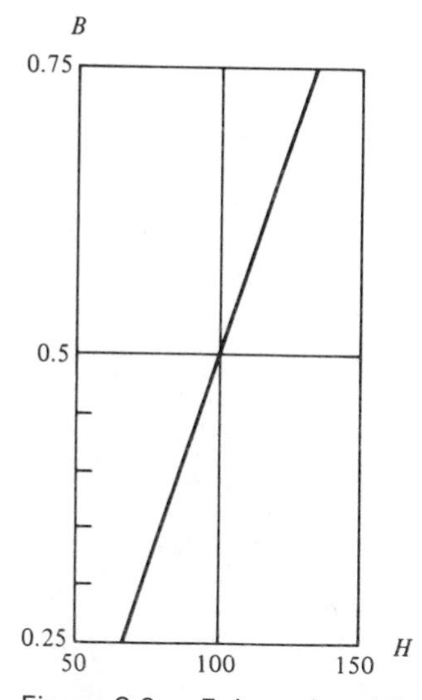

Figure 8.9. Enlarged and linearized portion of Figure 8.6 around $B = 0.5$.

Now,

$$\psi_m = Bs = 0.45 \times \frac{\pi(2.5 \times 10^{-2})^2}{4} = 2.21 \times 10^{-4}$$

so

$$(NI)_{ag} = \psi_m \mathcal{R}_{ag} = (2.21 \times 10^{-4})(1.621 \times 10^6) = 358 \quad (\mathrm{A \cdot t})$$

Thus, $NI = 485.2$ or

$$I = \frac{485.2}{1000} = 0.485 \neq 0.5 \quad (\mathrm{A})$$

2. Try $B = 0.46$. $H_{fc} = 92$ from Figure 8.9. Then,

$$(NI)_{fc} = 130, \qquad \psi_m = 2.26 \times 10^{-4}, \qquad (NI)_{ag} = 366,$$
$$NI = 496, \qquad \text{or} \qquad I = 0.496 \quad (\mathrm{A})$$

This last result is probably close enough, so we conclude that $B = 0.46$ is produced by $I = 0.5$ A.

One last remark is in order before passing on to a new topic. Figure 8.8 shows that a constant-"current" (flux) generator results if $\mathcal{R}_{ag} \gg \mathcal{R}_{fc}$. That is, if the air gap reluctance is very large compared to that of the ferromagnetic core, then the flux is simply

$$\psi_m \approx \frac{NI}{\mathcal{R}_{ag}} \tag{8.29}$$

and the ferromagnetic material core has little effect.

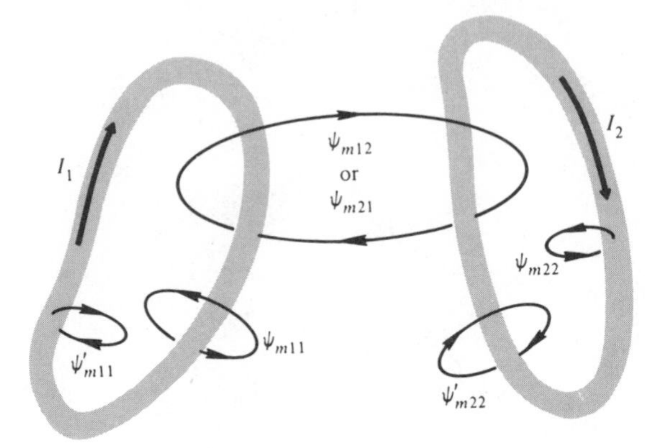

Figure 8.10. Magnetically coupled circuits showing flux linkages.

8.4 INDUCTANCE

Inductance, L, is the third parameter to be considered which is useful in circuit theory. An inductor stores magnetic energy in the region surrounding it. Recall that resistance, R, depended on conductivity and the geometry of the conductor. Capacitance, C, depended on permittivity and the capacitor geometry. In the same way, we expect that inductance will depend on permeability and the inductor geometry. Consider Figure 8.10, which shows two circuits magnetically coupled. ψ'_{m11} is a flux line due to current I_1 linking partially with I_1 by passing through the conductor. This type *flux linkage* gives rise to an *internal self inductance*, L'_{11}. Those flux lines like ψ_{m11}, due to I_1, which link I_1 externally, give rise to an *external* self-inductance, L_{11}. The same things can be said about circuit 2. Flux line ψ_{m12} is produced by I_1 and links I_2, giving rise to a *mutual inductance*, L_{12}. ψ_{m12} could just as well be called ψ_{m21} because it could also be considered to be due to I_2 and linking I_1. We suspect that $L_{12} = L_{21}$.

Inductance will be defined as the ratio of magnetic flux linkage to the current producing the flux linkage, or

$$\boxed{L = \frac{\psi_{ml}}{I}} \quad \text{(H)} \tag{8.30}$$

This definition can accommodate any of the various types of inductance mentioned in the previous paragraph. It is obvious from equation 8.30 that the flux linkage must be proportional to the current if L is to be independent of I. This, in turn, requires that the medium be linear so that μ is constant. In this way, L, as defined in equation 8.30, will depend only on permeability and the geometry. Inductance is a characteristic possessed by one or more *closed* circuits.

EXAMPLE 5

Consider a long straight filamentary current, I, flowing in the positive z direction, surrounded by a medium with permeability μ. Ampere's law gives $B_\phi = \mu I/2\pi\rho$. *All* of the flux links all of the current, so for a length l,

$$L_{\text{ext}} = \frac{\psi_{ml}}{I} = \frac{\displaystyle\int_0^\infty \int_0^l \frac{\mu I}{2\pi\rho} \mathbf{a}_\phi \cdot \mathbf{a}_\phi \, d\rho \, dz}{I},$$

or

$$L_{\text{ext}} = \frac{\mu l}{2\pi} \int_0^\infty \frac{d\rho}{\rho} \rightarrow \infty + \infty$$

This answer should not be surprising, for, first of all, filamentary currents are nonphysical. Secondly, and more importantly, there is no return current. In other words, we have not considered a *closed* circuit.

EXAMPLE 6

Next, consider a long cylinder of radius a carrying an assumed uniform current density, $\mathbf{J} = \mathbf{a}_z(I/\pi a^2)$. In this case, we have internal flux linkage and would expect an internal inductance. We must exercise care in finding the flux linkage. Perhaps the best way to do this is to find the flux linkage on a differential basis and then integrate (sum) over the interior of the conductor to find the total flux linkage. Actually, this gives us an equation that is more general and more useful than equation 8.30. The differential flux linkage will be the differential flux times the fraction of the total current which is actually enclosed. That is,

$$L = \frac{1}{I} \iint_s d\psi_{ml} = \frac{1}{I} \iint_s \left[\frac{I_{\text{enc}}}{I}\right] d\psi_m \quad \text{(H)} \tag{8.31}$$

The geometry is shown in Figure 8.11. From Section 7.3, where Ampere's

Figure 8.11. Geometry for finding the internal flux linkages for a long straight wire of radius a.

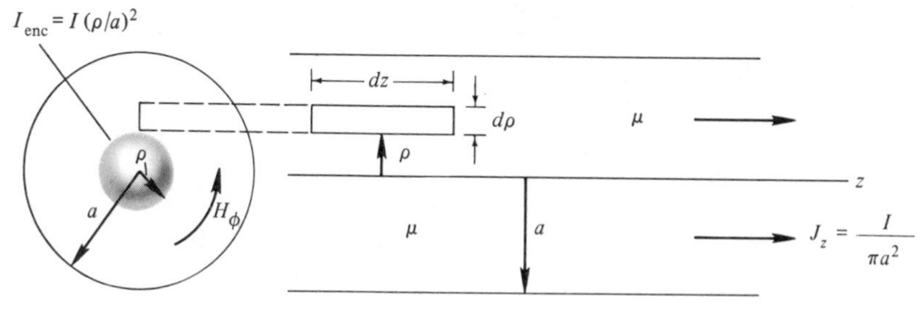

law was used, we have

$$I_{\text{enc}} = I\frac{\rho^2}{a^2} \quad \text{and} \quad H_\phi = \frac{I\rho}{2\pi a^2}, \qquad \rho < a$$

For a section l m long, equation 8.31 gives

$$L_{\text{int}} = \frac{1}{I}\int_0^a\int_0^l \left[\frac{I(\rho/a)^2}{I}\right]\left[\frac{\mu I\rho}{2\pi a^2}\, d\rho\, dz\right]$$

if the conductor permeability is μ. This reduces easily to

$$L_{\text{int}} = \frac{\mu l}{8\pi} \tag{8.32}$$

and is the *internal* self-inductance. The *external* self-inductance for a length l is still infinite!

EXAMPLE 7

As another example, consider the coaxial cable shown in Figure 8.12. Suppose (initially) there is a total current, I_1, flowing in the positive z direction in the inner conductor and a total current, I_2, flowing in the

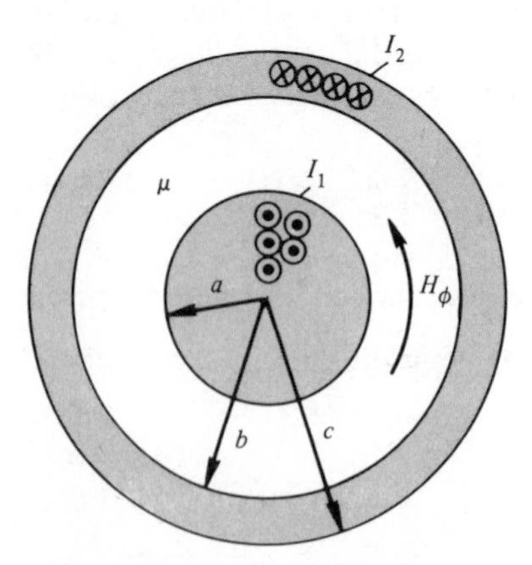

Figure 8.12. The coaxial cable.

negative z direction in the outer conductor. These currents may not be equal if they are due to *different* forcing functions. In other words, we may be dealing with parts of two *different* circuits. In this case, it is not proper to speak of the inductance of the coaxial cable, but we may speak of the self-inductances of the two conductors separately and the mutual inductance between them. The external inductance per unit length of both inner and outer conductor will be infinite as in the previous paragraph. The mutual inductances per unit length will also be infinite. At the risk of being repetitious, we simply emphasize that we are still not dealing with one or two closed circuits, but merely two partial circuits. The currents I_1 and I_2 must have return paths, and when the flux due to the currents in these return paths are considered, previous statements about infinite inductances per unit length must be modified.

It is informative to calculate the mutual inductances per unit length in spite of the nonphysical aspects of this problem. If the current densities are assumed uniform, we have

$$L_{12} = \frac{1}{I_1} \int_0^\infty \int_0^l \left[\frac{I_2(\rho^2 - b^2)/(c^2 - b^2)}{I_2} \right] \left[\frac{\mu I_1}{2\pi\rho} d\rho \, dz \right]$$

from Section 7.3. The first bracketed term is the fraction of I_2 enclosed, while the second bracketed term is the differential flux linkage due to I_1. In the same way

$$L_{21} = \frac{1}{I_2} \int_b^\infty \int_0^l \left[\frac{I_1}{I_1} \right] \left[\frac{\mu I_2}{2\pi\rho} \frac{\rho^2 - b^2}{c^2 - b^2} d\rho \, dz \right]$$

The first bracketed term is the fraction of I_1 enclosed, while the second bracketed term is the differential flux linkage due to I_2. Since the integrals are identical, we suspect very strongly that $L_{12} = L_{21}$! We may not, however, consider this as any sort of formal proof since both L_{12} and L_{21} are infinite.

EXAMPLE 8

In normal use, the coaxial cable is a closed circuit when the end effects are included. That is, there is only *one* current involved. The philosophy here is quite different from that of the previous paragraph. In this case, the magnetic field is zero for $\rho > c$ (see Section 7.3). It is true that the flux links *both* the current in the inner and outer conductors, but we do not interpret this as mutual inductance. The coaxial cable is now an entity, or single closed circuit, and the next calculation will give the *self-*inductance of a coaxial cable for a length l. We *assume* that the current densities are uniform. Using equation 8.31 and the results of Section 7.3, we have

$$L = \frac{1}{I}\int_0^a \int_0^l \left[\frac{I\rho^2/a^2}{I}\right]\left[\frac{\mu I\rho}{2\pi a^2}\,d\rho\,dz\right]$$

$$+\frac{1}{I}\int_a^b \int_0^l \left[\frac{I}{I}\right]\left[\frac{\mu I}{2\pi\rho}\,d\rho\,dz\right]$$

$$+\frac{1}{I}\int_b^c \int_0^l \left[\frac{I(c^2-\rho^2)/(c^2-b^2)}{I}\right]\left[\frac{\mu I}{2\pi\rho}\frac{c^2-\rho^2}{c^2-b^2}\,d\rho\,dz\right] \quad (8.33)$$

assuming μ is the same everywhere. The bracketed terms are easy to identify as before. Equation 8.33 simplifies to

$$L = \frac{\mu l}{8\pi} + \frac{\mu l}{2\pi}\ln\frac{b}{a} + \frac{\mu l}{2\pi}\int_b^c \left[\frac{c^2-\rho^2}{c^2-b^2}\right]^2 \frac{d\rho}{\rho} \quad (8.34)$$

or

$$L = \frac{\mu l}{8\pi} + \frac{\mu l}{2\pi}\ln\frac{b}{a} + \frac{\mu l}{2\pi}\left[\frac{c^2}{c^2-b^2}\right]^2 \ln\frac{c}{b}$$

$$-\frac{\mu l}{2\pi}\left[\frac{c^2}{c^2-b^2}\right] + \frac{\mu l}{8\pi}\left[\frac{c^2+b^2}{c^2-b^2}\right] \quad \text{(H)} \quad (8.35)$$

Since this is the inductance of *the coax*, it is not desirable, nor is it necessary, to attempt to associate various parts of equation 8.35 with the inner and outer conductors.

EXAMPLE 9

If the currents flow in the thin hollow cylinders, they are essentially the same as surface currents. Assume these surface currents flow on the outside of the inner conductor and on the inside of the outer conductor. They will be uniformly distributed because of the symmetry, and equation 8.31 gives

$$L_{\text{ext}} = \frac{\mu l}{2\pi}\ln\frac{b}{a} \quad \text{(H)} \quad (8.36)$$

a very useful result. This is all of the inductance because there are no internal fields! This result is also identified in equation 8.35.

EXAMPLE 10

Another practical geometry is that of the two-wire line shown in Figure 8.13. Finding the inductance of this circuit is not easy because

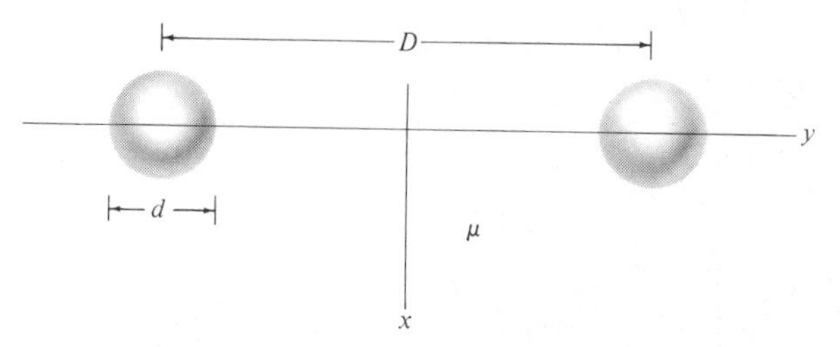

Figure 8.13. The two-wire line.

it does not fit any simple coordinate system. Instead of attempting an exact solution for internal currents, let us find an approximate solution. If we align the z axis with the axis of one of the conductors, and find the flux it produces and links with itself (neglecting all the flux from the near edge of the other conductor to infinity) then,

$$L \approx \frac{1}{I}\int_0^{d/2}\int_0^{l}\left[\frac{I\rho^2/a^2}{I}\right]\left[\frac{\mu I\rho}{2\pi a^2}\right]d\rho\, dz + \frac{1}{I}\int_{d/2}^{D-d/2}\int_0^{l}\left[\frac{I}{I}\right]\left[\frac{\mu I}{2\pi\rho}\right]d\rho\, dz \quad (8.37)$$

for uniform current densities. The other conductor produces the same flux linkage, so the total inductance is twice that given above, or

$$L \approx \frac{\mu l}{4\pi} + \frac{\mu l}{\pi}\ln\frac{D - d/2}{d/2} \quad \text{(H)} \qquad (8.38)$$

This approximation is obviously good only if $D \gg d$; so

$$L \approx \frac{\mu l}{4\pi} + \frac{\mu l}{\pi}\ln\frac{2D}{d} \quad \text{(H)} \qquad (8.39)$$

Even though the method used to obtain the equation is not very satisfying, equation 8.39 is accurate for $D \gg d$. It is emphasized that the current density in this example was assumed to be uniform in this and previous examples. In most practical cases (nondirect current), the current density will not be uniform.

We shall next show that the two-dimensional problem of finding the *external* inductance per unit length of a uniform two-conductor transmission line is equivalent to the electrostatic problem of determining the capacitance per unit length of the same geometry. We have already shown that the shunt conductance can be determined from the capacitance (see equation 6.18). In order to avoid internal fields and internal inductance consider a pair of

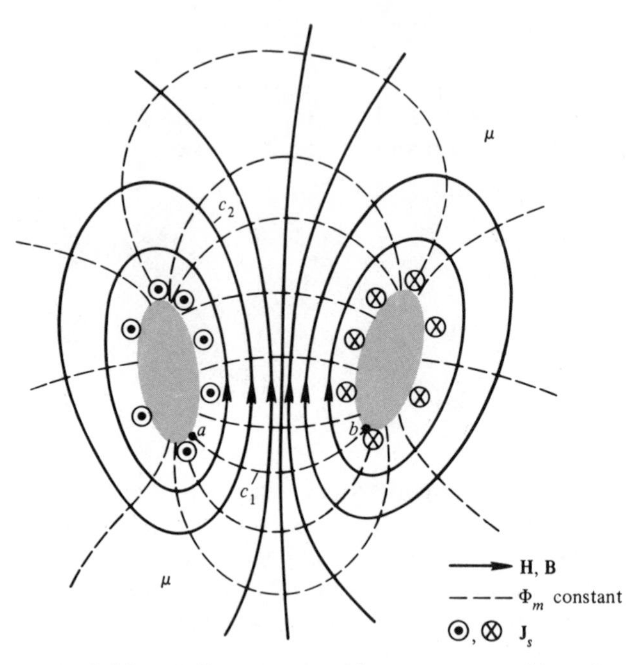

Figure 8.14. Uniform two-conductor system with surface currents showing **H** lines and contours of constant magnetic scalar potential, Φ_m.

(infinitesimally) thin wall conductors that are parallel with an arbitrary, but uniform, cross section. The conductors are embedded in a medium with a scalar constant permeability as shown in Figure 8.14. The capacitance problem with the same geometry and scalar constant permittivity is shown in Figure 8.15.

We have $\mathbf{J} = 0$ in the region outside the conductors, so we also have $\nabla \times \mathbf{H} = 0$ and $\nabla \cdot \mathbf{H} = 0$. Thus,

$$\nabla \times (\nabla \times \mathbf{H}) \equiv \nabla(\nabla \cdot \mathbf{H}) - \nabla^2 \mathbf{H} = 0 \tag{8.40}$$

or

$$\nabla^2 \mathbf{H} = 0 \tag{8.41}$$

For the capacitor problem we have $\rho_v = 0$ outside the conductors, so we also have $\nabla \times \mathbf{E} = 0$ and $\nabla \cdot \mathbf{E} = 0$, leading (as above) to

$$\nabla^2 \mathbf{E} = 0 \tag{8.42}$$

Thus, the differential equations for **E** and **H** are the same except for symbols (i.e., dual). The boundary conditions at the conductor surface are $H_{\text{norm}} = 0$ and $H_{\tan} = J_s$, while for the capacitance problem they are $E_{\text{norm}} = \rho_s/\varepsilon$ and $E_{\tan} = 0$. Thus, the boundary conditions between normal and tangential components are *reversed*. A comparison of superposition integrals, equations 1.42 and 7.6, shows essentially the same thing. Furthermore, since these

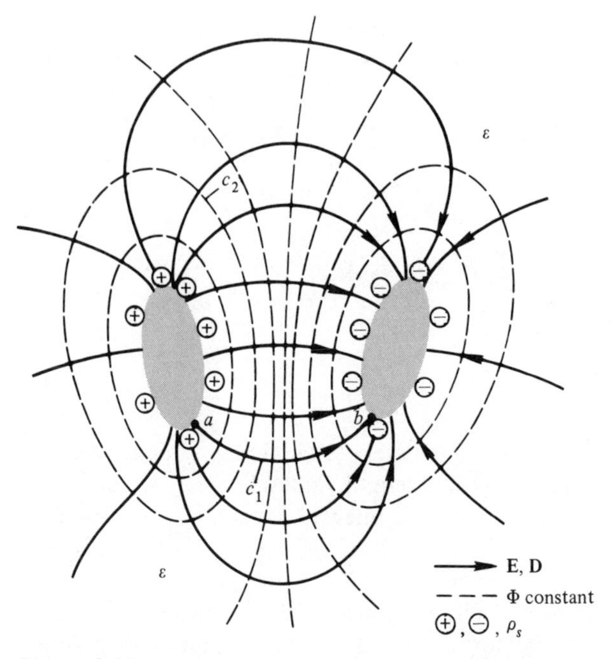

Figure 8.15. Uniform two-conductor system (identical to that in Figure 8.14) with surface charges showing **E** lines and contours of constant electric scalar potential, Φ.

integrals are dual equations in the present two-dimensional problem, except for the cross-product in equation 7.6, and since they can be solved as integral equations in order to determine the exact *distribution* of ρ_s and J_s around the conductors, we conclude that ρ_s and J_s are distributed in exactly the same way. In other words ρ_s is proportional to J_s. Therefore $|\mathbf{E}|$ is proportional to $|\mathbf{H}|$ and **E** and **H** are everywhere normal. This, in turn, means that **H** lines in the inductance problem lie along the equipotentials in the capacitance problem.

From equation 4.36,

$$C = \frac{\varepsilon \oiint \mathbf{E} \cdot d\mathbf{s}}{\int_{a,\, c_1}^{b} \mathbf{E} \cdot d\mathbf{l}_1} = \frac{\varepsilon \oint_{c_2} E_n dl_2}{\int_{a,\, c_1}^{b} E_n dl_1} \quad \text{(F/m)} \tag{8.43}$$

where we have chosen a 1-m length, c_2 lies along an equipotential, and c_1 lies along an **E** line. Notice that these choices mean that the subscript n means *normal* to c_2. From equation 8.30,

$$L_{\text{ext}} = \frac{\mu \oiint \mathbf{H} \cdot d\mathbf{s}}{\oint_{c_2} \mathbf{H} \cdot d\mathbf{l}_2} = \frac{\mu \int_{a,\, c_1}^{b} H_t\, dl_1}{\oint_{c_2} H_t\, dl_2} \quad \text{(H/m)} \tag{8.44}$$

where we have chosen a 1-m length, c_2 lies along an **H** line (or an equipotential in the other problem), and c_1 lies along an **E** line in the other problem. The subscript t means *tangent* to c_2. Multiplying the last two equations and using $E_n = KH_t$ gives

$$L_{\text{ext}} C = \frac{\mu \int_{a,\, c_1}^{b} H_t\, dl_1}{\oint_{c_2} H_t\, dl_2} \, \frac{\varepsilon K \oint_{c_2} H_t\, dl_2}{K \int_{a,\, c_1}^{b} H_t\, dl_1}$$

Therefore,

$$\boxed{L_{\text{ext}} C = \mu\varepsilon} \tag{8.45}$$

and if we already know C, we can immediately calculate L_{ext}. Remember that the values of L_{ext} and C here are per unit length values. We shall see later that for high-frequency transmission lines the internal inductance is negligible, and all the inductance is essentially L_{ext}.

Equation 8.45 is an important general result. It tells us that we need not solve two separate problems in finding L_{ext} and C for problems of this type. It is usually easier to find C first and then find L_{ext} from equation 8.45. As an example, equation 8.36 gives the external inductance for a coaxial cable with surface currents. Substituting equation 4.40, with $L = 1$, into equation 8.45 gives the identical answer as equation 8.36

EXAMPLE 11

We now return to the two-wire line of Figure 8.13. Equation 4.51, with $l = 1$, gives the capacitance per unit length for this geometry. Substituting equation 4.51 into equation 8.45 gives

$$L_{\text{ext}} = \frac{\mu}{\pi} \cosh^{-1} \frac{D}{d} \quad \text{(H/m)} \tag{8.46}$$

or

$$L_{\text{ext}} \approx \frac{\mu}{\pi} \ln \frac{2D}{d} \quad \text{(H/m)} \tag{8.47}$$

It is also possible to find the surface current distribution. As has already been noted, this distribution is the same as that of the surface charge in the capacitance problem. This charge distribution is the same as the electric flux density distribution at the conductor surface. The latter quantity can be found because we know the electric scalar potential. Knowledge of the surface currents would enable us to find the magnetic field, the magnetic flux and, finally, the external inductance which we already know.

EXAMPLE 12

As a final example, consider the idealized N turn solenoid of Figure 8.3. For this case the flux links all N turns, or in other words, the flux links the current N times with *no leakage flux*. Then,

$$\psi_{ml} = N\psi_m = \frac{N\mu NIs}{l} = \frac{\mu N^2 Is}{l}$$

and

$$L = \frac{\mu N^2 s}{l} \quad \text{(H)} \tag{8.48}$$

For practical solenoids with leakage flux, end effects, and finite diameter conductors, we can use equation 8.48 as a first guess, or simply use empirical formulas found in engineering handbooks.

8.5 INDUCTANCE AND MAGNETOSTATIC ENERGY

Our discussion of inductance will continue in this section, for it provides a means of introducing magnetostatic energy. We cannot introduce magnetostatic energy in a direct way, as we did for electrostatic energy, for the simple reason that we do not have any *point magnetic charges* to move in a magnetic field! Equation 8.31 gives us a convenient starting point for the development of magnetostatic energy relationships. We have

$$L = \frac{1}{I} \iint_s \frac{I_{\text{enc}}}{I} \, d\psi_m \tag{8.31}$$

It has also been accepted that

$$I_{\text{enc}} = \oint_l \mathbf{H} \cdot d\mathbf{l} \quad \text{(Ampere's law)}$$

and

$$d\psi_m = \mathbf{B} \cdot d\mathbf{s}$$

so that

$$L = \frac{1}{I^2} \iint_s \left(\oint_l \mathbf{H} \cdot d\mathbf{l} \right) \mathbf{B} \cdot d\mathbf{s} \tag{8.49}$$

If we choose a path, l, for the line integral coincident with $\mathbf{H}$ and perpendicular to $d\mathbf{s}$, then it is possible to write ($\mathbf{B} = \mu\mathbf{H}$, μ a scalar)

$$L = \frac{1}{I^2} \iint BH \int dl \, ds$$

or

$$L = \frac{1}{I^2} \iiint_{\text{vol}} BH \, dv$$

or

$$\boxed{L = \frac{1}{I^2} \iiint_{\text{vol}} \mathbf{B} \cdot \mathbf{H} \, dv} \quad \text{(H)} \tag{8.50}$$

Equation 8.50 requires that we integrate $\mathbf{B} \cdot \mathbf{H} = \mu H^2 = B^2/\mu$ over all space to find L.

In Chapter 4, we found that

$$C = \frac{2W_E}{V_0^2} = \frac{1}{V_0^2} \iiint_{\text{vol}} \mathbf{D} \cdot \mathbf{E} \, dv$$

where V_0 is the potential difference between the conductors. Comparing the last equation with equation 8.50, and also recalling circuit theory results, suggests strongly that

$$\boxed{W_H = \frac{1}{2} LI^2 = \frac{1}{2} \iiint_{\text{vol}} \mathbf{B} \cdot \mathbf{H} \, dv} \quad \text{(J)} \tag{8.51}$$

For the present, we will accept equation 8.51 as correctly giving the magnetostatic energy stored in a region of linear media. A term like the right side of equation 8.51 appears when we consider *Poynting's theorem* in a later chapter. We will consider

$$w_H = \tfrac{1}{2}\mathbf{B} \cdot \mathbf{H} \quad (\text{J/m}^3) \tag{8.52}$$

to represent magnetostatic energy density at a point even though, strictly speaking, we have no justification for doing so.

EXAMPLE 13

A practical example of the use of equation 8.51 can now be considered. Suppose that we have an ideal N turn solenoid whose ferromagnetic core consists of two identical halves that are touching. We apply a mechanical force to separate the two halves, creating an air gap. The required work appears as magnetostatic energy stored in the (linear media) air gap. This is a practical problem which must be answered in designing solenoid operated devices (e.g., relays). Equation 8.51 may be written

$$W_H = \frac{1}{2} \iiint_{\text{vol}} \frac{B^2}{\mu_0}\, dv$$

Since B is the same in the ferromagnetic core and the air gap (and is constant by the assumption of an ideal solenoid), we have in the air gap

$$W_H = \frac{B^2 s l_{ag}}{2\mu_0}$$

or

$$dW_H = \frac{B^2 s}{2\mu_0}\, dl_{ag} = F\, dl_{ag}$$

where F is the mechanical force required to create the air gap of length l_{ag}, Therefore,

$$F = \frac{B^2 s}{2\mu_0} \quad \text{(N)} \tag{8.53}$$

We next return to further discussion of inductance. Equation 8.50 has been obtained as an alternate expression for finding inductance. It is worthwhile to test this formula on the coaxial cable example. For the case of uniform current densities and equal but opposite currents, equation 8.50 gives

$$L = \frac{\mu}{I^2} \iiint_{\text{vol}} H^2\, dv$$

$$= \frac{\mu}{I^2} \int_0^a \int_0^l \int_0^{2\pi} \left(\frac{I\rho}{2\pi a^2}\right)^2 \rho\, d\rho\, d\phi\, dz + \frac{\mu}{I^2} \int_a^b \int_0^l \int_0^{2\pi} \left(\frac{I}{2\pi\rho}\right)^2 \rho\, d\rho\, d\phi\, dz$$

$$+ \frac{\mu}{I^2} \int_b^c \int_0^l \int_0^{2\pi} \left(\frac{I}{2\pi\rho}\frac{c^2-\rho^2}{c^2-b^2}\right)^2 \rho\, d\rho\, d\phi\, dz$$

or

$$L = \frac{\mu l}{8\pi} + \frac{\mu l}{2\pi} \ln\frac{b}{a} + \frac{\mu l}{2\pi} \int_b^c \left(\frac{c^2-\rho^2}{c^2-b^2}\right)^2 \frac{d\rho}{\rho} \tag{8.54}$$

Equation 8.54 is identical with equation 8.34 which was obtained from equation 8.31.

If $\mathbf{B}$ is replaced by $\nabla \times \mathbf{A}$ in equation 8.50, we obtain

$$L = \frac{1}{I^2} \iiint_{\text{vol}} (\nabla \times \mathbf{A}) \cdot \mathbf{H} \, dv \tag{8.55}$$

The integrand of equation 8.55 may be replaced using a vector identity[2]

$$\nabla \times \mathbf{A} \cdot \mathbf{H} \equiv \nabla \cdot (\mathbf{A} \times \mathbf{H}) + \mathbf{A} \cdot (\nabla \times \mathbf{H})$$

Then,

$$L = \frac{1}{I^2} \iiint_{\text{vol}} \nabla \cdot (\mathbf{A} \times \mathbf{H}) \, dv + \frac{1}{I^2} \iiint_{\text{vol}} \mathbf{A} \cdot (\nabla \times \mathbf{H}) \, dv \tag{8.56}$$

We next apply the divergence theorem to the first integral of equation 8.56 and replace $\nabla \times \mathbf{H}$ by $\mathbf{J}$ in the second integral. This gives

$$L = \frac{1}{I^2} \oiint_{s} (\mathbf{A} \times \mathbf{H}) \cdot d\mathbf{s} + \frac{1}{I^2} \iiint_{\text{vol}} \mathbf{A} \cdot \mathbf{J} \, dv \tag{8.57}$$

The closed surface s of the first integral in equation 8.57 must enclose all of the magnetostatic energy if the *total* L is desired, and on this surface $\mathbf{H}$ is zero *if the surface extends to infinity*. (See Problem 17.) Therefore,

$$L = \frac{1}{I^2} \iiint_{\text{vol}} \mathbf{A} \cdot \mathbf{J} \, dv \tag{8.58}$$

or

$$\boxed{L = \frac{1}{I^2} \iiint_{\text{cond}} \mathbf{A} \cdot \mathbf{J} \, dv} \tag{8.59}$$

since $\mathbf{J} = 0$ outside the volume containing the conductor. The vector potential $\mathbf{A}$ is that due to $\mathbf{J}$. If a vector potential exists inside the volume, and is being produced by a current in some *other* circuit, then it leads to a mutual inductance and is ignored here. It is left as an exercise to verify that equation 8.59 or 8.57 leads to equation 8.54 for the coaxial cable example.

Equation 7.49 with $\mu = \mu_0$ gives

$$\mathbf{A}(\mathbf{r}) = \frac{\mu}{4\pi} \iiint_{\text{vol}'} \frac{\mathbf{J}(\mathbf{r}')}{R} \, dv'$$

so equation 8.59 can be written

$$L = \frac{1}{I^2} \iiint_{\text{vol}} \left[\frac{\mu}{4\pi} \iiint_{\text{vol}'} \frac{\mathbf{J}(\mathbf{r}')}{R} \, dv' \right] \cdot \mathbf{J}(\mathbf{r}) \, dv \tag{8.60}$$

[2] See Appendix A.10.

Equation 8.60 is not exactly inviting to say the least, but can be reduced to two single integrals in case the current densities are essentially filamentary.

It is convenient now to define the mutual inductance between closed circuits 1 and 2. Since we have referred to mutual inductance at several points in previous paragraphs, its definition should be rather obvious. It is

$$\boxed{L_{12} \equiv \frac{\psi_{ml,12}}{I_1} \quad \text{(H)}} \tag{8.61}$$

where $\psi_{ml,12}$ is the magnetic flux produced by I_1 which links I_2. As in the case of self-inductance, it is convenient to write equation 8.61 as

$$L_{12} = \frac{1}{I_1} \iint_s \frac{I_{2\text{enc}}}{I_2} \, d\psi_{ml,12} \tag{8.62}$$

Following exactly the same procedure as that used to obtain equation 8.50, we find that

$$L_{12} = \frac{1}{I_1 I_2} \iiint_{\text{vol}} \mathbf{B}_1 \cdot \mathbf{H}_2 \, dv \tag{8.63}$$

where $\mathbf{H}_1 = \mathbf{B}_1/\mu$ is the field due to I_1 alone ($I_2 = 0$), and $\mathbf{H}_2$ is the field due to I_2 alone ($I_1 = 0$). Defining L_{21} as in equation 8.61 and following the same procedure as above, we have

$$L_{21} = \frac{1}{I_2 I_1} \iiint_{\text{vol}} \mathbf{B}_2 \cdot \mathbf{H}_1 \, dv \tag{8.64}$$

Since equations 8.63 and 8.64 are identical ($\mathbf{B} = \mu\mathbf{H}$, μ a scalar), we have formally shown that

$$L_{12} = L_{21} \tag{8.65}$$

Beginning with equation 8.50 and following the same procedure as that used to obtain equation 8.59, we have

$$L_{12} = \frac{1}{I_1 I_2} \iiint_{\text{vol}} \mathbf{A}_1 \cdot \mathbf{J}_2 \, dv \tag{8.66}$$

where $\mathbf{A}_1$ is the vector potential due to I_1 alone but existing within the volume of $\mathbf{J}_2$. A more explicit form for equation 8.66 is

$$L_{12} = \frac{1}{I_1 I_2} \iiint_{\text{vol 2}} \left[\frac{\mu}{4\pi} \iiint_{\text{vol 1}} \frac{\mathbf{J}_1(\mathbf{r}_1)}{R_{12}} \, dv_1 \right] \cdot \mathbf{J}_2(\mathbf{r}_2) \, dv_2 \tag{8.67}$$

where

$$R_{12} = R_{21} = |\mathbf{r}_2 - \mathbf{r}_1|$$

It is apparent that all of the equations we have developed for self- and mutual inductance are rather formidable, and only those geometries with considerable symmetry will be easy to handle. We next consider two such cases.

EXAMPLE 14

Consider first the magnetically coupled circuits shown in Figure 8.16.

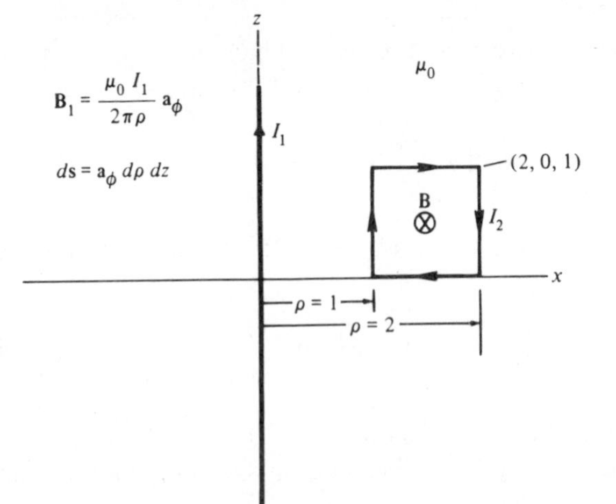

Figure 8.16. An infinite filamentary current coupled to a filamentary current loop.

The currents are filamentary, so the labor required to find the mutual inductance should be much less than for the more general case. Using equation 8.62, we have[3]

$$L_{12} = \frac{1}{I_1}\int_1^2\int_0^1 \frac{I_2}{I_2}\frac{\mu_0 I_1}{2\pi\rho}\,\mathbf{a}_\phi \cdot \mathbf{a}_\phi\, d\rho\, dz$$

or

$$L_{12} = \frac{\mu_0}{2\pi}\ln 2 = 0.139 \quad (\mu\text{H})$$

EXAMPLE 15

Finally, consider the somewhat more practical case of ideal air core coaxial solenoids. Let the inner solenoid have N_1 turns, I_1 A, and a radius a, while the outer solenoid has N_2 turns, I_2 A, and a radius b.

[3] The reader is welcome to calculate L_{21}.

From equation 8.25, we have

$$\psi_m = \frac{\mu_0 N I s}{l} \tag{8.25}$$

The geometry is shown in Figure 8.17. No integration is necessary since

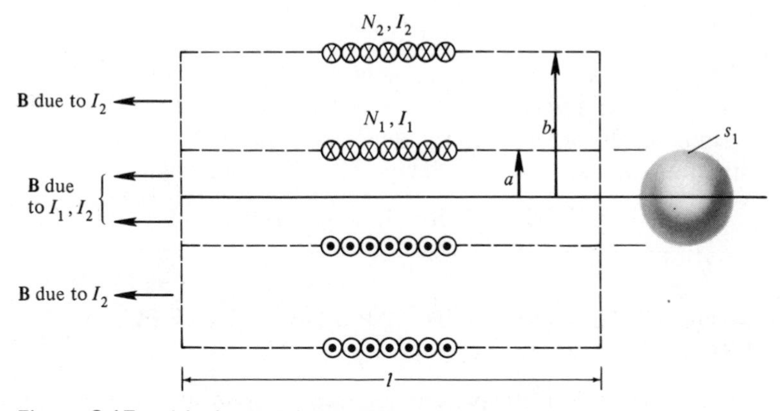

Figure 8.17. Ideal coaxial solenoids.

ψ_m is uniform. Therefore, equation 8.61 gives

$$L_{12} = \frac{1}{I_1}\left[\frac{N_2 I_2}{I_2}\right]\left[\frac{\mu_0 N_1 I_1 s_1}{l}\right]$$

The first bracketed term is the fraction of I_2 linked by the flux produced by I_1, while the second bracketed term is the flux produced by I_1 which links I_2 (N_2 times). Thus,

$$L_{12} = \frac{\mu_0 N_1 N_2 s_1}{l} \quad \text{(H)} \tag{8.68}$$

In the same way,

$$L_{21} = \frac{1}{I_2}\left[\frac{N_1 I_1}{I_1}\right]\left[\frac{\mu_0 N_2 I_2 s_1}{l}\right]$$

or

$$L_{21} = \frac{\mu_0 N_1 N_2 s_1}{l} \quad \text{(H)}$$

so

$$L_{12} = L_{21}$$

From equation 8.48, we know that

$$L_1 = \frac{\mu_0 N_1^2 s_1}{l} \quad \text{and} \quad L_2 = \frac{\mu_0 N_2^2 s_2}{l}$$

A coefficient of magnetic coupling can be defined as

$$k_{mc} = \frac{L_{12}}{\sqrt{L_1 L_2}} \tag{8.69}$$

and gives

$$k_{mc} = (s_1/s_2)^{1/2} \tag{8.70}$$

for the present example. This parameter is a measure of the leakage flux. In the present example, for instance, not all of the flux produced by I_2 links I_1. In order to eliminate leakage flux in this example, we must require that $s_1 = s_2$, which gives the maximum value of k_{mc},

$$k_{mc} = 1 \quad \text{(no leakage flux)} \tag{8.71}$$

8.6 A NUMERICAL METHOD FOR MAGNETIC FIELD PROBLEMS

Consider a region that is isotropic, but not necessarily linear or homogeneous. Equations 7.32 and 7.43 give

$$\nabla \times \left(\frac{1}{\mu} \nabla \times \mathbf{A}\right) = \mathbf{J} \tag{8.72}$$

Expanding equation 8.72 by means of an identity in Appendix A.10, we have

$$\frac{1}{\mu} \nabla(\nabla \cdot \mathbf{A}) - \frac{1}{\mu} \nabla^2 \mathbf{A} - (\nabla \times \mathbf{A}) \times \nabla\left(\frac{1}{\mu}\right) = \mathbf{J} \tag{8.73}$$

If $\nabla \cdot \mathbf{A} \equiv 0$, then equation 8.73 reduces to

$$\frac{1}{\mu} \nabla^2 \mathbf{A} + (\nabla \times \mathbf{A}) \times \nabla\left(\frac{1}{\mu}\right) = -\mathbf{J} \tag{8.74}$$

We now specialize to the case $\mathbf{A} = \mathbf{a}_z A_z$ so that

$$\frac{1}{\mu} \nabla^2(\mathbf{a}_z A_z) + (\nabla \times \mathbf{a}_z A_z) \times \nabla\left(\frac{1}{\mu}\right) = -\mathbf{J}$$

which, in Cartesian coordinates, gives three scalar equations.

$$\left(\frac{\partial A_z}{\partial x}\right)\left(\frac{\partial\, 1/\mu}{\partial z}\right) = J_x$$

$$\left(\frac{\partial A_z}{\partial y}\right)\left(\frac{\partial\, 1/\mu}{\partial z}\right) = J_y$$

$$\frac{1}{\mu} \nabla^2 A_z + \left(\frac{\partial A_z}{\partial x}\right)\left(\frac{\partial\, 1/\mu}{\partial x}\right) + \left(\frac{\partial A_z}{\partial y}\right)\left(\frac{\partial\, 1/\mu}{\partial y}\right) = -J_z$$

Simplifying one more time to the two-dimensional case where $\partial/\partial z \equiv 0$, gives $J_x = J_y = 0$ and

$$\frac{1}{\mu}\frac{\partial^2 A_z}{\partial x^2} + \frac{1}{\mu}\frac{\partial^2 A_z}{\partial y^2} + \left(\frac{\partial A_z}{\partial x}\right)\left(\frac{\partial\, 1/\mu}{\partial x}\right) + \left(\frac{\partial A_z}{\partial y}\right)\left(\frac{\partial\, 1/\mu}{\partial y}\right) = -J_z \quad (8.75)$$

Returning to the electrostatic case where the region is isotropic, but perhaps nonlinear and inhomogeneous, equation 5.74 is

$$\varepsilon\nabla^2\Phi + \nabla\Phi \cdot \nabla\varepsilon = -\rho_v$$

For a two-dimensional case with $\partial/\partial z \equiv 0$, the last equation becomes

$$\varepsilon\frac{\partial^2\Phi}{\partial x^2} + \varepsilon\frac{\partial^2\Phi}{\partial y^2} + \left(\frac{\partial\Phi}{\partial x}\right)\left(\frac{\partial\varepsilon}{\partial x}\right) + \left(\frac{\partial\Phi}{\partial y}\right)\left(\frac{\partial\varepsilon}{\partial y}\right) = -\rho_v \quad (8.76)$$

which is dual to equation 8.75. A difference equation solution to equation 8.76 was found in Section 5.4 to be equation 5.82.

$$\Phi_0 = \frac{\rho_0 d^2 + \sum_{n=1}^{4} \dfrac{\Phi_n}{1/2\varepsilon_0 + 1/2\varepsilon_n}}{\sum_{n=1}^{4} \dfrac{1}{1/2\varepsilon_0 + 1/2\varepsilon_n}} \quad (8.77)$$

Comparing equations 8.75, 8.76, and 8.77 gives

$$A_0 = \frac{J_0 d^2 + \sum_{n=1}^{4} \dfrac{A_n}{\mu_0/2 + \mu_n/2}}{\sum_{n=1}^{4} \dfrac{1}{\mu_0/2 + \mu_n/2}} \quad (8.78)$$

Notice that equation 8.78 leads to a simple electric circuit analog as shown in Figure 8.18, and we have found a method for obtaining a difference equation solution to equation 8.72 in two-dimensional form. The method can be expanded to handle a three-dimensional case, but, in the interest of simplicity, that course is not pursued here. Notice also that equation 8.78 is nothing more than a solution for the integral form of Ampere's law.

$$\oint \frac{\nabla \times \mathbf{A}}{\mu} \cdot d\mathbf{l} = I$$

A_0 is the vector potential at the center of a $(d \times d)$ square cell in terms of the current $(J_0 d^2)$ in the square and the vector potential and permeability

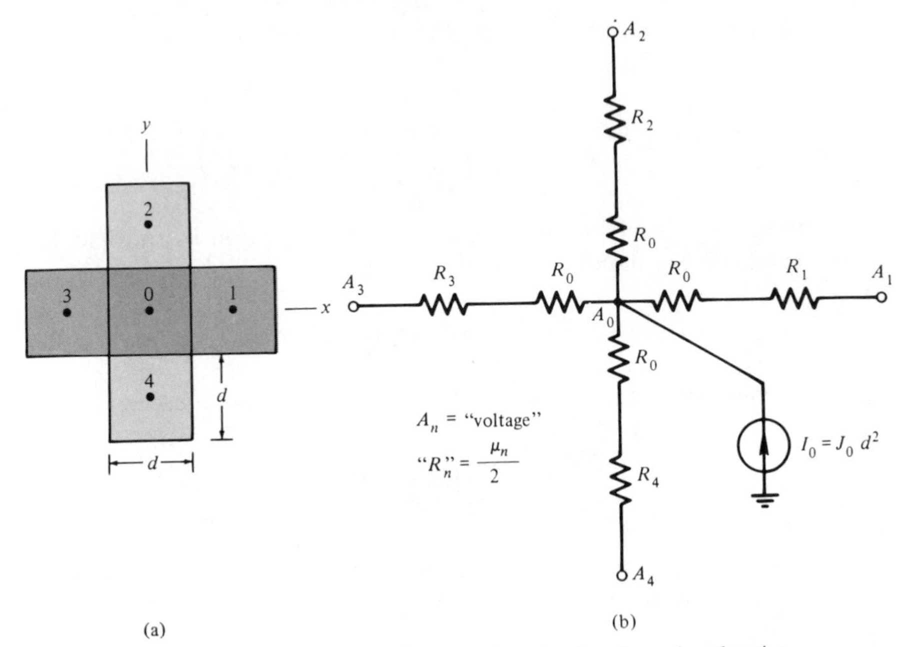

Figure 8.18. (a) Cells. (b) Finite element electric circuit analog for the two-dimensional solution of $\nabla \times [(\nabla \times \mathbf{A})/\mu] = \mathbf{J}$ or $\oint (\nabla \times \mathbf{A})/\mu \cdot d\mathbf{l} = I$.

in the four adjacent cells. The boundary conditions on **H** and **B** in general form are given by equations 8.17 and 8.19, where $\mathbf{B} = \nabla \times \mathbf{A}$, or

$$B_x = \frac{\partial A_z}{\partial y} \tag{8.79}$$

and

$$B_y = -\frac{\partial A_z}{\partial x} \tag{8.80}$$

The derivatives in equations 8.79 and 8.80 are to be approximated by difference forms using Figure 8.18 for the geometry.

The iterative scheme outlined in Section 5.4 can of course be employed here. Problems where the permeability depends on a *B-H* curve (in ferromagnetic materials, for example) can be treated as in Section 5.4. It should be mentioned that contours of constant A_z are magnetic flux lines.

For one-dimensional problems equation 8.78 easily reduces to

$$A_0 = \frac{J_0 d^2 + \sum_{n=1}^{2} \frac{A_n}{\mu_0/2 + \mu_n/2}}{\sum_{n=1}^{2} \frac{1}{\mu_0/2 + \mu_n/2}} \tag{8.81}$$

It should be remembered that we also have a second numerical method to use in case it is necessary to place field points on interfaces. Using equations 5.87 and 5.88 we have

$$A_0 = \frac{d^2 J_0 + \sum_{n=1}^{4} A_n \left\langle \frac{1}{\mu_{0n}} \right\rangle}{\sum_{n=1}^{4} \left\langle \frac{1}{\mu_{0n}} \right\rangle} \tag{8.82}$$

for the two-dimensional case, where $\langle 1/\mu_{0n} \rangle$ is the average of the reciprocal of the permeability of the two quadrants sharing the line 0-n, and

$$A_0 = \frac{d^2 J_0 + \sum_{n=1}^{2} A_n \frac{1}{\mu_{0n}}}{\sum_{n=1}^{2} \frac{1}{\mu_{0n}}} \tag{8.83}$$

for the one-dimensional case, where $1/\mu_{0n}$ is the reciprocal of the permeability on the line 0-n.

EXAMPLE 16

As a numerical example consider the boundary value problem of determining the inductance of a pair of oppositely directed parallel surface current densities (or stripline) with two different magnetic materials as shown in Figure 8.19. Assume that $\partial/\partial x = \partial/\partial z = 0$

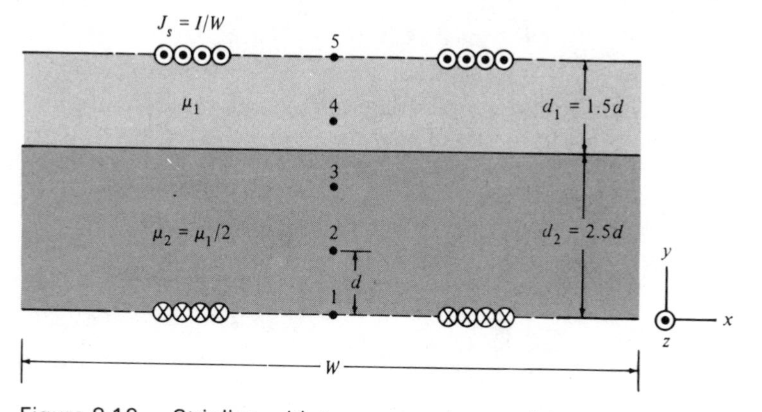

Figure 8.19. Stripline with two magnetic materials.

so that the problem is one-dimensional (and relatively easy to solve by analytical methods). Notice that the interface has been chosen to be at $y = 2.5d$ so that it lies midway between the field points 2 and 3. The depth in the z direction is l. Choose boundary conditions such that $A_1 = 0$ and $A_5 = 10$. The three unknowns, A_2, A_3, and A_4, are determined from three simultaneous equations generated by equation 8.81 with only two terms per sum. The equations and solutions are shown below.

$$A_2 = \tfrac{1}{2}A_3 \qquad A_2 = \tfrac{20}{11}$$

$$A_3 = A_2 + \tfrac{2}{3}A_4 \qquad A_3 = \tfrac{40}{11}$$

$$A_4 = \tfrac{4}{7}A_3 + \tfrac{30}{7} \qquad A_4 = \tfrac{70}{11}$$

These are exact answers for this problem (Why?), and the vector potential is linear (with different slopes) in each region and continuous at the interface. The inductance can be calculated by several methods. For example, the definition, equation 8.30, gives

$$L = \frac{B_1 d_1 l + B_2 d_2 l}{J_s W} = \frac{\mu_1 H_1 d_1 l + \mu_2 H_1 d_2 l}{J_s W}$$

Notice that the fields are uniform in their respective regions since A_z is linear and $H_1 = H_2 = J_s$. Therefore,

$$L = \frac{\mu_1 d_1 l}{W} + \frac{\mu_2 d_2 l}{W} \tag{8.84}$$

which is equivalent to two inductors in series. (Can you explain this?) In order to show that the inductance can be calculated directly from the numerical work, consider equation 8.59, which gives

$$L = \frac{1}{I^2} A_5 J_s W l = \frac{l A_5}{I}$$

There is no contribution at the other surface current since $A_1 = 0$. The remaining problem is that of relating A_5 to I. This relation is

$$J_s = H_5 = \frac{B_5}{\mu_1} = \frac{1}{\mu_1} \frac{\partial A_z}{\partial y}\bigg|_5 \approx \frac{1}{\mu_1} \frac{A_5 - A_4}{d}$$

or

$$J_s = \frac{1}{\mu} \frac{\frac{4}{11}}{d} A_5 \quad \text{or} \quad A_5 = \frac{2.75 \mu_1 I d}{W}$$

Therefore,

$$L = \frac{2.75 \mu_1 \, dl}{W} \tag{8.85}$$

Since $\mu_1 d_1 + \mu_2 d_2 = \mu_1 (d_1 + d_2/2) = 2.75 \mu_1 d$, equation 8.85 agrees with equation 8.84.

8.7 CONCLUSIONS

In this chapter, we considered the general behavior of magnetic materials and found that for simple magnetic materials most of the previously developed equations still apply. A set of boundary conditions for magnetic materials was developed. Only the simplest cases were investigated, and for these cases, we developed a simple theory of magnetic circuits. Most of this chapter concerned the development of equations for finding the electric circuit parameter, inductance. Magnetostatic energy was defined and related to inductance and the magnetic field for simple media.

A numerical technique for solving magnetic problems with steady electric currents was developed. It is applicable to all cases except those for which the magnetic material is anisotropic, but only one- and two-dimensional cases were considered. This technique is not only useful for solving many practical problems, but gives much insight into the behavior of magnetic materials as governed by Ampere's law. It was based on the similarity between the magnetostatic and electrostatic cases, and led to a very similar resistive electric circuit analog.

REFERENCES

Matsch. (See references for Chapter 4.)

Popović. (See references for Chapter 3.) Magnetic fields and materials are extensively treated.

Ramo. (See references for Chapter 4.) Magnetization is discussed on page 131.

Silvester. (See references for Chapter 5.) Some interesting examples of the numerical solution of magnetic problems are given.

PROBLEMS

1. For the region $z > 0$, $\mu_R = 4$, while for $z < 0$, $\mu_R = 1$. If **B** is uniform for $z > 0$ and in the direction $\theta = 60°$, $\phi = 45°$ with a magnitude of 1 Wb/m^2, find **B** and **H** for $z < 0$.
2. Show that the boundary conditions for **A** at the interface between two simple magnetic materials, μ_1 and μ_2, are $A_{t1} = A_{t2}$ and $A_{n1} = A_{n2}$.
3. Find the reluctance of a simple magnetic material which is 0.1 m long, 10^{-2} m^2 in cross-sectional area, and for which $H = 500\,B$.
4. For the magnetic core shown in Figure 8.20, assume that the lengths

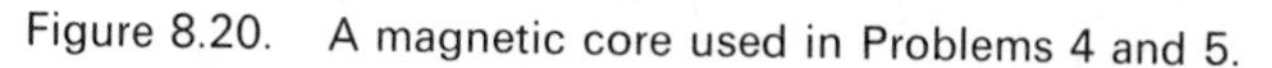
Figure 8.20. A magnetic core used in Problems 4 and 5.

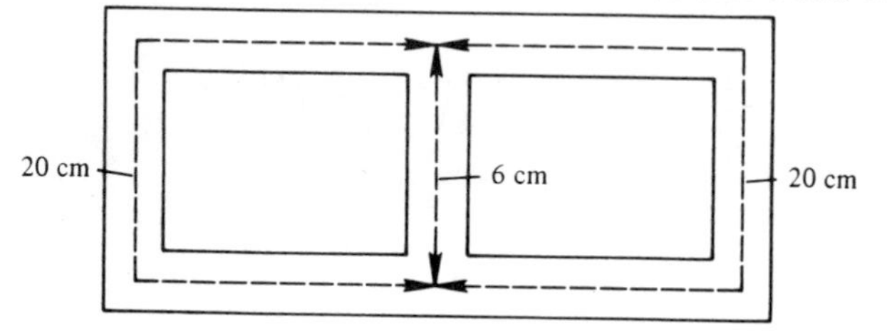

of the outer legs are 20 cm and the length of the center leg is 6 cm. The cross-sectional areas are 4 cm^2. For this material, we assume $H = 500\,B$. If a 1000 turn coil carrying 50 mA is placed on the center leg, make reasonable assumptions and find B in each leg.

5. If a 10^{-4} m air gap is introduced in the center leg of the magnetic circuit of Problem 4, find B in each leg.
6. Find the inductance of a 400 turn solenoid of 4 cm^2 cross-sectional area and length 10 cm, if it is producing 0.05 Wb/m^2 with 0.1 A.
7. Find the inductance of the solenoid of Problem 6, if a 10^{-3} m air gap is present and the current is readjusted so that $B = 400\mu_0 H$. B remains 0.05 Wb/m^2.
8. Find the external inductance per unit length for a planar transmission line consisting of parallel planes w m wide spaced d m such that $w \gg d$.
9. What maximum (retaining) spring force can be used in a solenoid for which $B = 1$ Wb/m^2 and $s = 10^{-4}$ m^2?
10. Find the vector potential in the $z = 0$ plane from a loop of radius a carrying a ϕ-directed current I. Assume the current is filamentary. Leave the answer in integral form. See Figure 8.21.

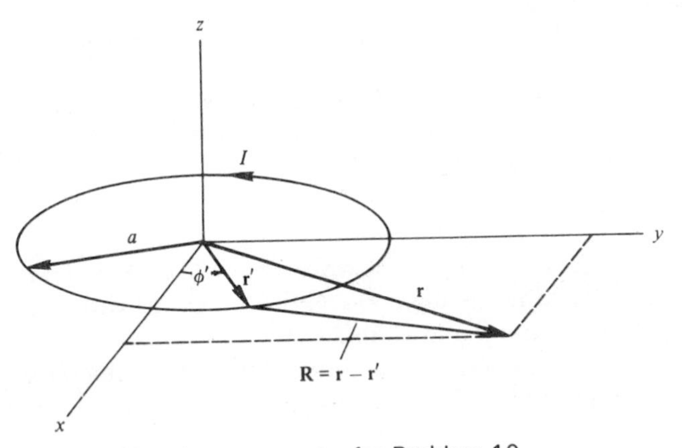

Figure 8.21. Loop geometry for Problem 10.

11. Find the *external* inductance of the loop of Problem 10 by using $L = 1/I \oint \mathbf{A} \cdot d\mathbf{l}$ where the path is one of radius $a - b$. b is the wire radius. Hint:

$$L_{\text{ext}} = \mu(2a - b)\left[\left(1 - \frac{V^2}{2}\right)K(V) - E(V)\right]$$

where

$$V^2 = \frac{4a(a-b)}{(2a-b)^2}$$

and

$$E(V) = \int_0^{\pi/2} \sqrt{1 - V^2 \sin^2 \alpha}\, d\alpha$$

$$= \text{complete elliptic integral (first kind)}$$

$$K(V) = \int_0^{\pi/2} \frac{d\alpha}{\sqrt{1 - V^2 \sin^2 \alpha}}$$

$$= \text{complete elliptic integral (second kind)}$$

12. Show that the external inductance of the loop of Problem 11 is

$$L_{\text{ext}} \approx \mu a \left(\ln \frac{8a}{b} - 2 \right) \quad \text{(H)}$$

if $a \gg b$. Hint: If $V \simeq 1$,

$$K(V) \approx \ln \frac{4}{\sqrt{1 - V^2}}, \qquad E(V) \approx 1$$

13. Show that the external inductance of a square loop of sides L and wire radius b is

$$L_{\text{ext}} \approx \frac{2\mu L}{\pi} \left(-2 + \ln 3.41 \frac{L}{b} \right) \quad \text{(H)}$$

if $L \gg b$. Hint: Start with the results of Problem 1, Chapter 7.

14. Compare the results of Problems 12 and 13 in terms of C_0, the loop circumferences. Could Problem 13 be solved with the results of Problem 12 with reasonable accuracy?
15. Repeat Problem 13 using the capacitance of a two-wire line and equation 8.45. Comment on the results.
16. Find the mutual inductance between filamentary, concentric, and coplanar (sides parallel) square loops of sides l_1 and l_2 ($l_2 > l_1$). See Figure 8.22.
17. Can equation 8.59 be used to calculate the *internal* inductance per unit length of a long round wire carrying a uniform current density? Why? Repeat for equation 8.57. See Problem 5, Chapter 7.
18. (a) What are the differential equations for **A** and the boundary conditions for **H** for the problem shown in Figure 8.23(a)?
    (b) Show that this problem is equivalent to that shown in Figure 8.23(b) for $z > 0$ and to that shown in Figure 8.23(c) for $z < 0$ if

$$I' = I \frac{\mu_2 - \mu_1}{\mu_2 + \mu_1} \quad \text{and} \quad I'' = I - I \frac{\mu_2 - \mu_1}{\mu_2 + \mu_1}$$

This is another example of the use of the *image principle.* See Problem 15, Chapter 5, and Section 5.6.

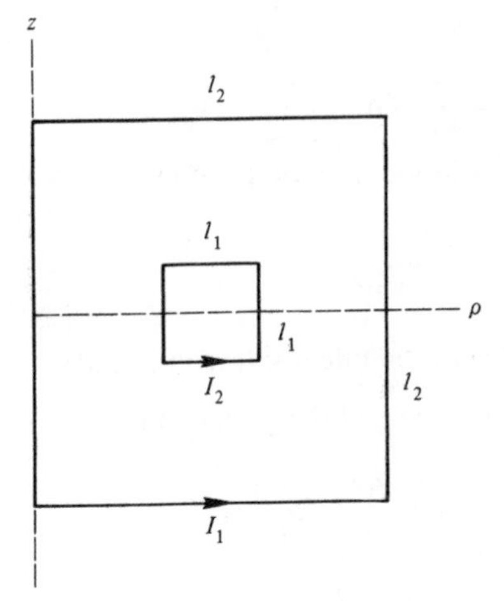

Figure 8.22. Geometry for Problem 16.

19. Derive the formula (given as a hint in Problem 19, Chapter 7) for the vector potential for the ideal solenoid.
20. A single layer solenoid is 10 cm long with a radius of 0.5 cm and 100 turns. Coaxial with this solenoid is a one turn circular loop near the middle of the solenoid ($\mu = \mu_0$).

    (a) Find L_{12} if the radius of the loop is 0.25 cm.
    (b) Find L_{12} if the radius of the loop is 1 cm.

21. A magnetic circuit is excited with 100 A · t and is closed except for a 0.214-mm air gap. Its mean length is 15 cm. Start with the equation $\psi_m = V_m/(\mathscr{R}_{ag} + \mathscr{R}_{fc}) = Bs$ and eliminate all unknowns except B and H. Use Figure 8.6 to find B.
22. Find the mutual inductance between an infinite filamentary wire on the z axis and a triangular loop whose corners are located at (0.5, 0, 0), (1, 0, 0.5) and (1, 0, −0.5).

Figure 8.23. (a) Filamentary current and a plane interface between two magnetic materials. (b) Equivalent problem for $z > 0$. (c) Equivalent problem for $z < 0$.

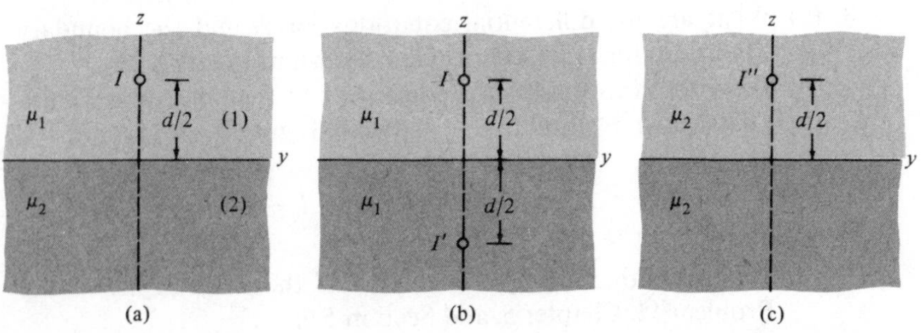

23. Two arbitrarily shaped pieces of copper sheet are placed (no contact resistance) on a large sheet of paper which has been uniformly covered with a material for which $\sigma = 10^2$ to a depth of 0.1 mm. A 1.5-V source connected across the pieces of copper produces a current of 15 mA. See Figure 8.24.

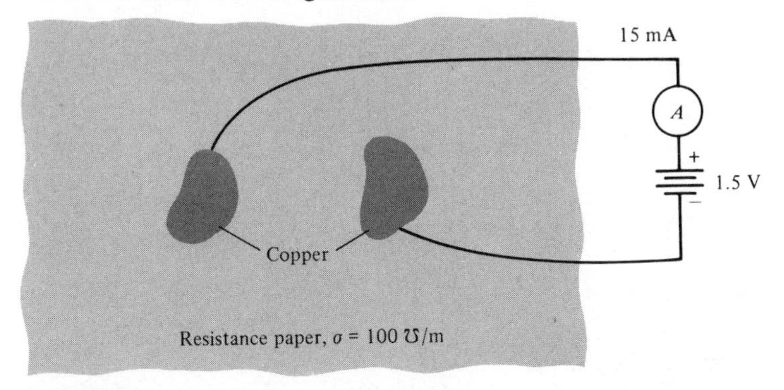

Figure 8.24. Geometry for Problem 23.

(a) What is the conductance, G?
(b) What capacitance per unit length would a pair of long conductors with the same cross section and spacing as the copper sheets possess if placed in air?
(c) Repeat (b) for external inductance per unit length.
(d) If the pieces of copper in (a) are American pennies, what is their center-to-center spacing?

24. If a convention is adopted whereby a dot is placed at a terminal of one of the windings where an *entering* current produces a flux that is *adding* to the flux being produced by the other winding, where should the dots be placed for the transformer in Figure 8.25?

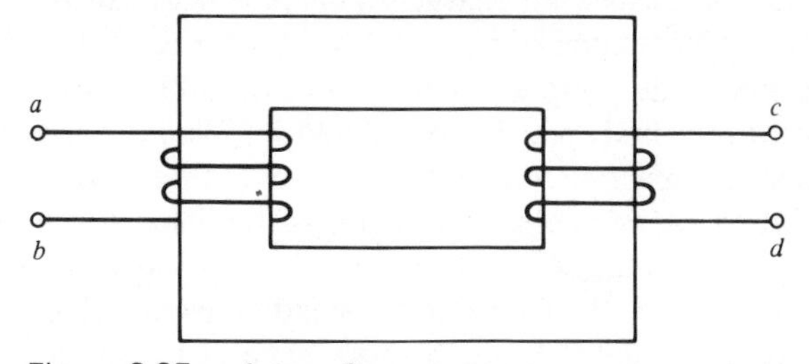

Figure 8.25. A transformer.

25. An ideal air core solenoid has N_1 turns, length l, and area s. A second such solenoid has N_2 turns, length l, area s, and is coaxial with the first solenoid and is connected in series with it such that the fluxes are additive. Show that

$$L_{eq} = \frac{\mu s}{l}(N_1 + N_2)^2 = L_1 + L_2 + 2\sqrt{L_1 L_2}$$

Chapter 9
Maxwell's Equations

Electrostatic relationships were initiated with Coulomb's experimental law, which we accepted, along with Gauss's law, as being true. We discovered nothing to indicate that we had been misled. We introduced magnetostatics with the magnetic field of moving charges. This new field can be found by the Biot–Savart law. The analog of Gauss's law was Ampere's law. Nothing presented in the first eight chapters would indicate difficulties with these experimental laws. In the first part, we considered *stationary* charge distributions, while in the second part we considered uniform charge motion (constant velocity). The relation between electric and magnetic fields which occurred was one resulting from charges in relative motion. We also found that a steady electric field in a conductor forced a steady current which, in turn, led to a steady magnetic field. A much more general case remains to be investigated. If a charge is somehow *accelerated*, a time dependent magnetic field would be expected to result. Can this occur without the existence of a time dependent electric field? The answer to this question will be found in another experimental law, called Faraday's law, which we will also accept as being correct.

In this chapter we will begin with Faraday's law and obtain a set of equations pertaining to the dynamic case. These equations will give the same static results we have already obtained. The consolidation of this set of equations, plus the development of the concept of displacement current density, was due primarily to J. C. Maxwell. Therefore, this set of equations bears his name. One may accurately state that a study of "field theory" is a study of Maxwell's equations.

Equations for the auxiliary potential functions will be rederived for the dynamic case. These equations and their integral solutions will be in forms that reduce to those already derived for the static case. Whether or not these potential functions are actually used to solve a problem will depend to some extent on the problem itself. There are certain problems, which we will encounter in later chapters, which can be solved very easily using potential functions. Other problems may be solved more quickly by using Maxwell's equations directly.

The boundary conditions need to be reexamined, and this will be done. A general conservation of energy relationship, called Poynting's theorem will be derived. It is comforting to find that the energy dissipated (Joule's law) plus energies stored in the electric and magnetic fields given by this theorem agree with the forms used in previous chapters. As an example of the use of Maxwell's equations, and to demonstrate their fundamental and primary nature, a simple series electric circuit is investigated by beginning with Maxwell's equations. The final result is the well-known circuit equation for this geometry.

9.1 FARADAY'S LAW

In 1831, Faraday was successful in demonstrating that a *time changing* magnetic field could produce an electric current. It would perhaps be more accurate to say that what Faraday actually discovered was the following. When the magnetic flux linking a closed circuit is *altered*, a voltage, or *electromotive force* (emf), is induced which may produce a current in this circuit. Faraday's law is usually written mathematically as

$$\boxed{\text{emf} = -\frac{d\psi_m}{dt}} \quad \text{(V or Wb/s)} \tag{9.1}$$

where ψ_m is the magnetic flux passing through any open surface bounded by the circuit (closed path l). The flux that would be produced by the resultant current *opposes* the original variation of the flux. The last sentence is a statement of *Lenz's law* and accounts for the minus sign in equation 9.1.

Electromotive force, emf, is a voltage due to some form of energy other than electric (e.g., batteries and generators also produce emf). We shall define emf as

$$\boxed{\text{emf} \equiv \oint_l \mathbf{E} \cdot d\mathbf{l}} \quad \text{(V)} \tag{9.2}$$

Equation 9.2 implies a *particular* closed path l and if some other closed path is chosen, the emf will, in general, change! In electrostatics, we spoke of voltage and potential difference interchangeably, and this is permissible. In the present situation, we will refrain from mentioning scalar potential, or potential difference, until it is necessary or desirable to do so. As we shall eventually see, voltage and potential difference *are usually not* equivalent except in electrostatics. One fact concerning equation 9.2 is obvious. **E** cannot be the gradient of a simple scalar potential alone, or else the right side of equation 9.2 would be identically zero. This means, equivalently, that in the general dynamic case **E** *is nonconservative*!

It is still true that

$$\psi_m = \iint_{s_1} \mathbf{B} \cdot d\mathbf{s} \quad \text{(Wb)} \tag{9.3}$$

where s_1 is any open surface bounded by the closed path in equation 9.2. Combining equations 9.1, 9.2, and 9.3, we have

$$\oint_l \mathbf{E} \cdot d\mathbf{l} = -\frac{d}{dt} \iint_{s_1} \mathbf{B} \cdot d\mathbf{s} \tag{9.4}$$

We follow the conventional right-hand rule (as in Ampere's law) for determining the positive sense of circulation with respect to positive flow through the surface. This is demonstrated in Figure 9.1. Applying Stoke's theorem to the left side of equation 9.4 gives

$$\iint_{s_2} \nabla \times \mathbf{E} \cdot d\mathbf{s} = -\frac{d}{dt} \iint_{s_1} \mathbf{B} \cdot d\mathbf{s} \tag{9.5}$$

where s_2, like s_1, is any open surface bounded by the closed path of equation 9.2. Notice that s_1 and s_2 are not necessarily the same, but their *limits* are the same.

If the closed path is fixed or stationary, then s_1 and s_2 are not time dependent. In this case, the limits of the integrals of equation 9.5 are fixed, and we may differentiate **B** inside the integral sign partially with time. That is,

$$\iint_{s_2} \nabla \times \mathbf{E} \cdot d\mathbf{s} = -\frac{d}{dt} \iint_{s_1} \mathbf{B} \cdot d\mathbf{s} = -\iint_{s_1} \frac{\partial \mathbf{B}}{\partial t} \cdot d\mathbf{s} \tag{9.6}$$

Equation 9.6 is valid regardless of s_1 and s_2 (and the limits); so if s_1 and s_2 are identical, the left side and the right side of equation 9.6 can only be equal if the *integrands are equal.* Therefore,

$$\boxed{\nabla \times \mathbf{E} = -\frac{\partial \mathbf{B}}{\partial t}} \tag{9.7}$$

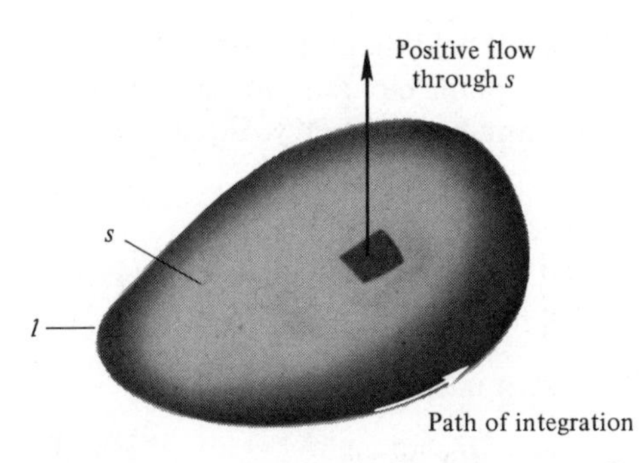

Figure 9.1. Right-hand rule applied to Faraday's law.

and **E** indeed is nonconservative ($\nabla \times \mathbf{E} \neq 0$) as predicted. This is the differential form of one of Maxwell's equations. The integral form is obtained from equation 9.4 with s fixed. It is

$$\text{emf} = \oint_l \mathbf{E} \cdot d\mathbf{l} = \iint_s \frac{\partial \mathbf{B}}{\partial t} \cdot d\mathbf{s} \tag{9.8}$$

Notice particularly that if there are no time variations, we immediately obtain

$$\oint \mathbf{E} \cdot d\mathbf{l} = 0 \qquad \left(\frac{\partial}{\partial t} = 0\right)$$

and

$$\nabla \times \mathbf{E} = 0 \qquad \left(\frac{\partial}{\partial t} = 0\right)$$

which agree with our electrostatic results.

EXAMPLE 1

An example is in order. Suppose a magnetic flux density,

$$\mathbf{B} = B_0 \cos\left(\omega t - \frac{2\pi x}{\lambda}\right)\mathbf{a}_z,$$

is being produced by some external source. ω and λ are constants having dimensions of radians per second and meters, respectively. For present purposes, we are interested in a confined region where $\lambda \gg x$. In this case, we may use $\mathbf{B} = (B_0 \cos \omega t)\mathbf{a}_z$ for the given flux density. The original given form of **B** will satisfy *all* of Maxwell's equations, while the new approximate form will not. This, nevertheless, is a standard type of approximation for ordinary ac circuit theory. We will elaborate on this point at an appropriate time.

With $\mathbf{B} = (B_0 \cos \omega t)\mathbf{a}_z$, we have a $\mathbf{B}$ field that is uniform in space. We would like to find the emf around a plane circular path of radius ρ in the $z = 0$ plane where E_ϕ is constant by symmetry. Equation 9.8 gives

$$\text{emf} = \int_0^{2\pi} E_\phi \mathbf{a}_\phi \cdot \mathbf{a}_\phi \rho \, d\phi = -\int_0^{2\pi}\int_0^{\rho} \frac{\partial}{\partial t}(B_0 \cos \omega t)\mathbf{a}_z \cdot \mathbf{a}_z \rho \, d\rho \, d\phi$$

following the right-hand rule. The last equation reduces to

$$\text{emf} = 2\pi\rho E_\phi = \pi\omega\rho^2 B_0 \sin \omega t \tag{9.9}$$

or

$$E_\phi = \frac{\omega\rho}{2} B_0 \sin \omega t \quad \text{(V/m)} \tag{9.10}$$

On the other hand, equation 9.7 gives

$$(\nabla \times \mathbf{E})_z = \frac{1}{\rho}\frac{\partial}{\partial\rho}(\rho E_\phi) = -\frac{\partial}{\partial t}(B_0 \cos \omega t) \tag{9.11}$$

when we equate z components. Performing the indicated differentiation on the right side gives

$$\frac{\partial}{\partial\rho}(\rho E_\phi) = \omega\rho B_0 \sin \omega t$$

Integrating with t held constant,

$$\rho E_\phi = \frac{\omega\rho^2}{2} B_0 \sin \omega t + C$$

where C is a constant insofar as ρ is concerned. We now have

$$E_\phi = \frac{\omega\rho}{2} B_0 \sin \omega t + \frac{C}{\rho} \tag{9.12}$$

For the static case ($\partial/\partial t = 0$), equation 9.11 gives

$$E_\phi = \frac{C}{\rho}$$

so we may say that the second term on the right side of equation 9.12 is a static solution and does not concern us here. Actually, we are handicapped by a lack of *all* of Maxwell's equations. Later on we will be able to make more definite statements about the constant in equation 9.12. For our purposes here, equation 9.12 reduces to

$$E_\phi = \frac{\omega\rho}{2} B_0 \sin \omega t \tag{9.13}$$

which is the same as equation 9.10.

It is interesting to verify Lenz's law in this example. For $0 < \omega t < \pi/2$ and $B_0 > 0$, we have **B** *decreasing* with time in the positive z direction while the emf from equation 9.9 is *increasing* with time in the positive ϕ direction. Then, a filamentary circular loop of finite conductivity in the $z = 0$ plane would have a current in the positive ϕ direction *increasing* with time because of the emf. The flux produced by this current would be increasing with time in the positive z direction, opposing the original applied flux.

In the preceding paragraph, we have seen how an emf can be induced in a closed *stationary* path by means of a time-changing magnetic flux. This may be called *transformer* action. It is also possible to produce an emf with a *steady* magnetic field if the path or circuit is changing in time. This may be called *generator* action. Equation 9.1, Faraday's law, indicates clearly that such a thing is possible. In field theory, we are usually interested in the former mechanism (interaction between time dependent electric and magnetic fields), but an investigation of generator action is also very useful.

Generator action, or motional emf, is perhaps best introduced by means of a simple example. Consider a conductor moving in a steady magnetic field as shown in Figure 9.2. Equation 7.72,

$$\mathbf{F} = Q\mathbf{u} \times \mathbf{B} \quad \text{(N)} \tag{7.72}$$

gives the Lorentz force on a charge moving with velocity **u** in a magnetic field **B**. This force is experienced by both the positive and negative charges in the conductor and is capable of producing a current because we have acting a force per unit charge or electric field given by

$$\mathbf{E} = \frac{\mathbf{F}}{Q} = \mathbf{u} \times \mathbf{B} \quad \text{(V/m)} \tag{9.14}$$

The emf produced by this electric field is called the motional emf and is given by

$$\text{emf} = \oint_l \mathbf{E} \cdot d\mathbf{l} = \oint_l \mathbf{u} \times \mathbf{B} \cdot d\mathbf{l} \tag{9.15}$$

A more elegant derivation which produces the results in *both* equations 9.8 and 9.15 can be obtained by considering a more general case. Let us return to the general form of Faraday's law as given by equation 9.4,

$$\oint_l \mathbf{E} \cdot d\mathbf{l} = -\frac{d}{dt} \iint_s \mathbf{B} \cdot d\mathbf{s}$$

and consider a moving loop or path. In this case, the limits on the integrals are explicitly time dependent and direct differentiation inside the integral is not permissible. We can, however, handle this situation by using coordinates

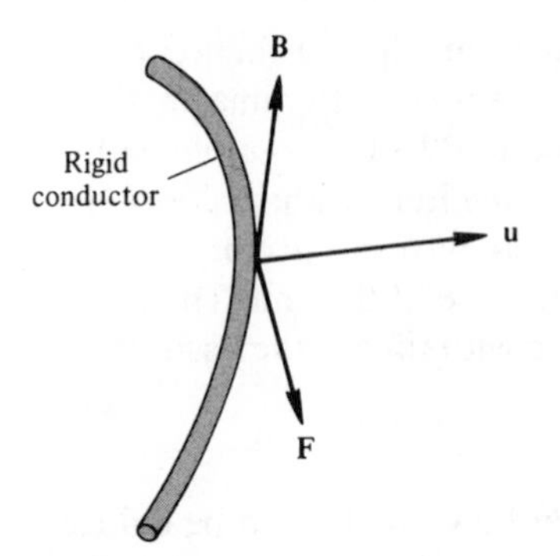

Figure 9.2. Lorentz force producing a motional emf in a rigid conductor moving with a uniform velocity in a uniform **B** field.

that are *stationary in the loop.* Instead of attempting the most general case of rotation *and* translation, we will only consider translation. We are seeking a transformation of d/dt to a frame of reference (the loop) moving with velocity **u** relative to the laboratory coordinates. This transformation is[1]

$$\frac{d}{dt} = \mathbf{u} \cdot \nabla + \frac{\partial}{\partial t} \tag{9.16}$$

In the rest frame of the loop, the induced emf is emf′, and

$$\text{emf}' = \oint_{l'} \mathbf{E}' \cdot d\mathbf{l}' = -\frac{d}{dt} \iint_{s'} \mathbf{B} \cdot d\mathbf{s}' \tag{9.17}$$

where the primes denote variables stationary in the rest frame of the loop. Using equation 9.16, we have

$$\oint_{l'} \mathbf{E}' \cdot d\mathbf{l}' = -\iint_{s'} (\mathbf{u} \cdot \nabla)\mathbf{B} \cdot d\mathbf{s}' - \iint_{s'} \frac{\partial \mathbf{B}}{\partial t} \cdot d\mathbf{s}' \tag{9.18}$$

Now it can be shown, using a straightforward, *but lengthy*, expansion in rectangular coordinates, that

$$\nabla \times (\mathbf{u} \times \mathbf{B}) \equiv (\nabla \cdot \mathbf{B})\mathbf{u} - (\nabla \cdot \mathbf{u})\mathbf{B} + (\mathbf{B} \cdot \nabla)\mathbf{u} - (\mathbf{u} \cdot \nabla)\mathbf{B} \tag{9.19}$$

We know, however, that there are no magnetic charges, so $\nabla \cdot \mathbf{B} = 0$ in *any* case. In the present problem we have only translation, or *constant* velocity **u**, so $\nabla \cdot \mathbf{u} = 0$ and $(\mathbf{B} \cdot \nabla)\mathbf{u} = 0$. Thus, equation 9.19 reduces to

$$(\mathbf{u} \cdot \nabla)\mathbf{B} = -\nabla \times (\mathbf{u} \times \mathbf{B}) \tag{9.20}$$

[1] See Owen in the references at the end of the chapter.

$$\mathbf{u} \cdot \nabla = u_x \frac{\partial}{\partial x} + u_y \frac{\partial}{\partial y} + u_z \frac{\partial}{\partial z}$$

Then, equation 9.18 may be written

$$\oint_{l'} \mathbf{E}' \cdot d\mathbf{l}' = \iint_{s'} \nabla \times (\mathbf{u} \times \mathbf{B}) \cdot d\mathbf{s}' - \iint_{s'} \frac{\partial \mathbf{B}}{\partial t}\, d\mathbf{s}' \tag{9.21}$$

or, with the aid of Stoke's theorem,

$$\boxed{\oint_{l'} \mathbf{E}' \cdot d\mathbf{l}' = \oint_{l'} \mathbf{u} \times \mathbf{B} \cdot d\mathbf{l}' - \iint_{s'} \frac{\partial \mathbf{B}}{\partial t} \cdot d\mathbf{s}'} \tag{9.22}$$

Equation 9.22 gives the general result, showing quite clearly the two terms that may contribute to the emf.

EXAMPLE 2

Let us next consider an example that has been used by many authors. Figure 9.3 shows a rectangular conducting loop, one arm of which has

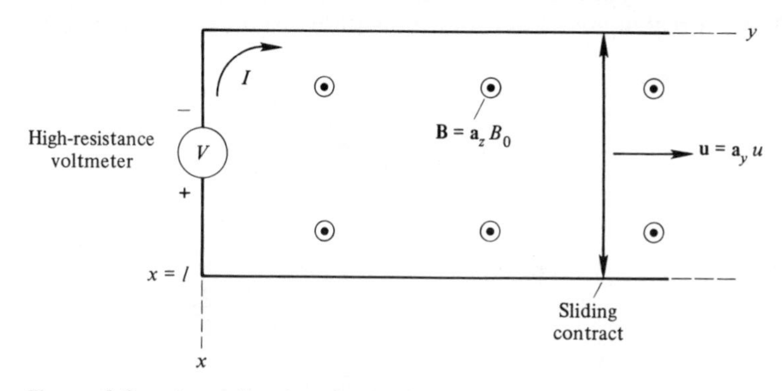

Figure 9.3. A motional emf calculation.

sliding contacts. This arm could also be a rolling bar. The external flux density $\mathbf{B} = \mathbf{a}_z \mathbf{B}_0$ is uniform in space and constant in time. This, then, is a case of motion only, and so we need to use equation 9.15.

$$\text{emf} = \oint \mathbf{E} \cdot d\mathbf{l} = \oint \mathbf{u} \times \mathbf{B} \cdot d\mathbf{l}$$

Where does the emf appear? We consider all the arms to have negligible resistance compared to that of the voltmeter, so the entire emf must appear across the voltmeter terminals. Furthermore, the polarity at the voltmeter must be as shown, and the very small current must be in the direction shown in order to provide an opposing flux (Lenz's law). If this polarity is correct, then the electric field at the voltmeter must be in the direction of a unit vector that points from the (+) to the (−)

sign. Thus, in integrating around the loop in a counterclockwise (positive) direction we must obtain a *negative* emf from the (only) contribution at the voltmeter. The only velocity is that of the (flux cutting) moving arm, so

$$\text{emf} = \int_l^0 (u\mathbf{a}_y \times B_0\mathbf{a}_z) \cdot \mathbf{a}_x \, dx = -B_0 lu \quad \text{(V)} \tag{9.23}$$

as predicted. Remember that $d\mathbf{l} = \mathbf{a}_x\, dx + \mathbf{a}_y\, dy + \mathbf{a}_z\, dz$. Equation 9.4 gives

$$\text{emf} = -\frac{d}{dt}\int_0^l\int_0^y B_0\mathbf{a}_z \cdot \mathbf{a}_z \, dx\, dy = -B_0\frac{d}{dt}(ly) = -B_0 lu$$

which is the same result.

EXAMPLE 3

A more practical example, the ac generator, is considered next. Figure 9.4 shows a rectangular loop rotating with constant angular velocity

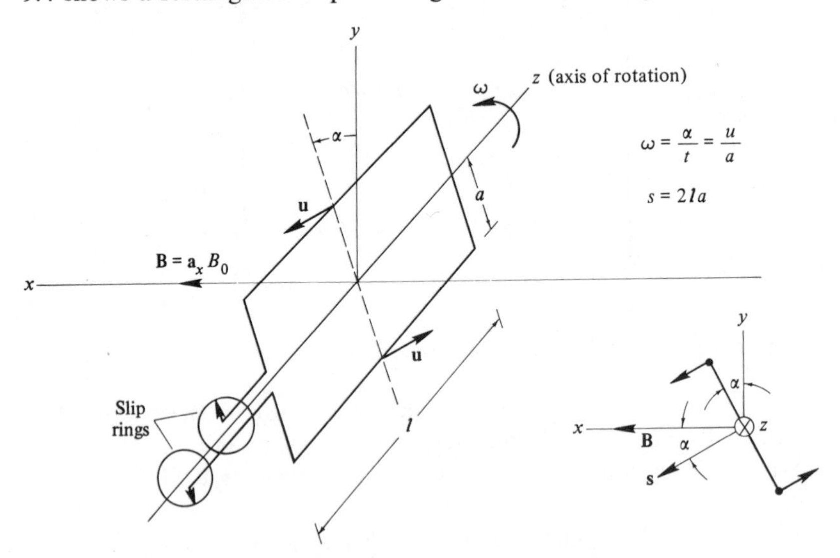

Figure 9.4. A simple ac generator.

$\omega = \alpha/t$. The emf ultimately appears at the terminals connected to the slip rings. This again is a case of motion only, if $\mathbf{B}$ is constant. Two legs of the loop enter the calculation, the quantity $\mathbf{u} \times \mathbf{B} \cdot d\mathbf{l}$ being zero for the legs perpendicular to the axis of rotation. Then,

$$\text{emf} = \int_{-l/2}^{l/2} uB_0 \sin\alpha(+\mathbf{a}_z) \cdot \mathbf{a}_z \, dz + \int_{l/2}^{-l/2} uB_0 \sin\alpha(-\mathbf{a}_z) \cdot \mathbf{a}_z \, dz$$

But $\alpha = \omega t$ and $u = \omega a$, so

$$\text{emf} = 2\omega B_0 la \sin \omega t = \omega B_0 s \sin \omega t \tag{9.24}$$

Application of equation 9.4 to this problem gives equation 9.24 almost immediately. It is left as an exercise to verify that Lenz's law is obeyed. Finally, if **B** is not constant in time, but is still uniform in space, both types of emf are present and must be included.

EXAMPLE 4

Suppose $\mathbf{B} = (B_0 \cos \omega t)\mathbf{a}_x$. Using equation 9.22, we have

$$\text{emf} = \oint_l \mathbf{u} \times \mathbf{B} \cdot d\mathbf{l} - \iint_s \frac{\partial \mathbf{B}}{\partial t} \cdot d\mathbf{s}$$

$$= \int_{-l/2}^{+l/2} uB_0 \cos \omega t \sin \alpha \mathbf{a}_z \cdot \mathbf{a}_z \, dz$$

$$+ \int_{+l/2}^{-l/2} uB_0 \cos \omega t \sin \alpha(-\mathbf{a}_z) \cdot \mathbf{a}_z \, dz$$

$$+ \omega B_0 \int_{-l/2}^{l/2} \int_{-a}^{a} \sin \omega t \mathbf{a}_x \cdot (\mathbf{a}_x \cos \alpha - \mathbf{a}_y \sin \alpha) \, dz \, dv$$

where $-l/2 \leq z \leq l/2$ and $-a \leq v \leq a$. That is,

$$d\mathbf{s} = (\mathbf{a}_x \cos \alpha - \mathbf{a}_y \sin \alpha) \, dz \, dv$$

Therefore

$$\text{emf} = uB_0 l \sin \omega t \cos \omega t + uB_0 l \sin \omega t \cos \omega t + 2\omega B_0 la \sin \omega t \cos \omega t$$

or

$$\text{emf} = 2\omega B_0 s \sin \omega t \cos \omega t$$

or

$$\text{emf} = \omega B_0 s \sin 2\omega t \quad \text{(V)} \tag{9.25}$$

In this example, we have a second harmonic generator.

9.2 CONSERVATION OF CHARGE (AGAIN)

In Section 6.2 we considered conservation of charge (or continuity of current) for steady currents and found that the net current through a closed surface was zero, or

$$\nabla \cdot \mathbf{J} = 0$$

We now inquire as to what conservation of charge requires in the time-varying case. In words, the principle of conservation of charge is always the same: charges are neither created nor destroyed. Let us begin with our original definition of current,

$$I = \iint_s \mathbf{J} \cdot d\mathbf{s}$$

or, in other words, the current out of a closed surface is

$$I = \oint\!\!\oint_s \mathbf{J} \cdot d\mathbf{s} \quad \text{(A)} \tag{9.26}$$

and is zero for the steady current case. In general, though, this outward flow of *positive* charge must be accompanied by a *decrease* in time of positive charge inside the volume, $-dQ/dt$. Thus, we have simply

$$I = \oint\!\!\oint_s \mathbf{J} \cdot d\mathbf{s} = -\frac{dQ}{dt} = -\frac{d}{dt}\iiint_{\text{vol}} \rho_v \, dv \tag{9.27}$$

If the volume in question is held constant in time, then we may differentiate partially with time inside the integral sign (as we did in Faraday's law). In this case,

$$\oint\!\!\oint_s \mathbf{J} \cdot d\mathbf{s} = -\iiint_{\text{vol}} \frac{\partial \rho_v}{\partial t} \, dv \tag{9.28}$$

or, with the divergence theorem,

$$\iiint_{\text{vol}} \nabla \cdot \mathbf{J} \, dv = -\iiint_{\text{vol}} \frac{\partial \rho_v}{\partial t} \, dv \tag{9.29}$$

Since the limits are arbitrary, the integrands must be the same, so

$$\boxed{\nabla \cdot \mathbf{J} = -\frac{\partial \rho_v}{\partial t}} \tag{9.30}$$

Equation 9.30 is the general form for the conservation of charge and agrees with the steady current form ($\partial/\partial t = 0$, $\nabla \cdot \mathbf{J} = 0$).

9.3 MAXWELL'S SECOND EQUATION

Faraday's law (Maxwell's first equation, dynamic case) tells us that a time dependent magnetic field results in, or is accompanied by an electric field. It is only natural to suspect that a time dependent electric field is accompanied by a magnetic field, or, in other words, perhaps time dependent electric and magnetic fields coexist. Maxwell's second equation for magnetostatics is equation 7.32,

$$\nabla \times \mathbf{H} = \mathbf{J} \tag{7.32}$$

Since the divergence of the curl of any vector is identically zero,

$$\nabla \cdot (\nabla \times \mathbf{H}) \equiv 0 = \nabla \cdot \mathbf{J} \tag{9.31}$$

Equation 9.31 is correct for magnetostatics, but, in light of equation 9.30, cannot be correct for the time-varying case. This implies, in turn, that equation 7.32 is incorrect for the general case also. A *correct* equation for the general case is equation 9.31 with $\partial \rho_v / \partial t$ added to the right side, since this would give

$$\nabla \cdot (\nabla \times \mathbf{H}) \equiv 0 = \nabla \cdot \mathbf{J} + \frac{\partial \rho_v}{\partial t} \qquad \left(\nabla \cdot \mathbf{J} = -\frac{\partial \rho_v}{\partial t}\right) \tag{9.32}$$

If it is still true that $\nabla \cdot \mathbf{D} = \rho_v$, then

$$\nabla \cdot (\nabla \times \mathbf{H}) \equiv 0 = \nabla \cdot \mathbf{J} + \frac{\partial}{\partial t}(\nabla \cdot \mathbf{D}) \tag{9.33}$$

If **D** *and its spatial and time derivatives are continuous*, then

$$\frac{\partial}{\partial t}(\nabla \cdot \mathbf{D}) = \nabla \cdot \frac{\partial \mathbf{D}}{\partial t}$$

and

$$\nabla \cdot \nabla \times \mathbf{H} \equiv 0 = \nabla \cdot \left(\mathbf{J} + \frac{\partial \mathbf{D}}{\partial t}\right) \tag{9.34}$$

Equation 9.34 implies that Maxwell's equation is, in general,

$$\boxed{\nabla \times \mathbf{H} = \mathbf{J} + \frac{\partial \mathbf{D}}{\partial t}} \tag{9.35}$$

This result is consistent with all previous equations.

The added term, $\partial \mathbf{D} / \partial t$, was Maxwell's primary contribution, and because of this contribution, his name is associated with the whole set of equations. This term is obviously a current density (A/m^2), and so Maxwell named it *displacement current density* (time derivative of electric flux density). It is probably easier for us to recognize the need for this added term than it was for Maxwell. Hindsight always seems easy. We know that the ammeter in Figure 9.5 will give a continuous reading, indicating a continuous current. For closed surface s_1 it is obvious that the divergence of the *total* current density is zero, because only a conduction current density is involved. For closed surface s_2, however, we suddenly lose our conducting path. It is still true that the divergence of the *total* current density is zero, or

$$\nabla \cdot \left(\mathbf{J} + \frac{\partial \mathbf{D}}{\partial t}\right) = 0$$

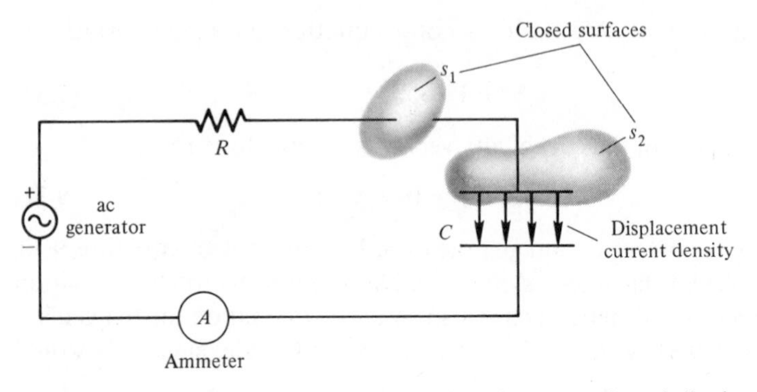

Figure 9.5. A simple ac circuit showing the necessity of displacement current density.

as in equation 9.34. In other words, the current density is continuous because of the added time derivative of electric flux density between the capacitor plates. In this simple (to us) example, the necessity of displacement current is obvious.

In many low-frequency applications, the displacement current term is not important and is usually neglected. This is one reason why its presence was not easy to verify or detect. With the advent of higher frequency sources, the displacement current term became more important in relation to conduction current or convection current. This statement will be verified in the chapters to follow. Problem 17 demonstrates the relation between conduction current and displacement current and how a material behaves as a conductor and dielectric simultaneously.

9.4 A SUMMARY OF MAXWELL'S EQUATIONS

The story has been told about the professor who, at the first meeting of a class for a course in electromagnetics, wrote the set of equations called Maxwell's equations on the board and left the room. The implication is that these equations tell the whole story of electromagnetic theory. This is essentially true. In this section, we will do (more or less) the same thing.

We have found that

$$\boxed{\nabla \times \mathbf{E} = -\frac{\partial \mathbf{B}}{\partial t}} \tag{9.36}$$

and

$$\boxed{\nabla \times \mathbf{H} = \mathbf{J} + \frac{\partial \mathbf{D}}{\partial t}} \tag{9.37}$$

Two other equations give us the set of four equations most frequently called Maxwell's equations. These two equations are unchanged from their static forms.

$$\boxed{\nabla \cdot \mathbf{D} = \rho_v} \tag{9.38}$$

$$\boxed{\nabla \cdot \mathbf{B} = 0} \tag{9.39}$$

These last two equations relate flux lines to sources or sinks. It is emphasized, once again, that the magnetic flux density is solenoidal, having no source or sink. The *constitutive* relations are

$$\boxed{\mathbf{D} = \varepsilon \mathbf{E}} \tag{9.40}$$

and

$$\boxed{\mathbf{B} = \mu \mathbf{H}} \tag{9.41}$$

where the parameters μ and ε depend on the material present. The force on a charge is of fundamental importance because we have seen that this is one way to define the electric and magnetic field. The Lorentz force is

$$\boxed{\mathbf{F} = Q(\mathbf{E} + \mathbf{u} \times \mathbf{B})} \tag{9.42}$$

Finally, conduction current density and convection current density have been defined by

$$\boxed{\mathbf{J} = \sigma \mathbf{E}} \tag{9.43}$$

and

$$\boxed{\mathbf{J} = \rho_v \mathbf{u}} \tag{9.44}$$

respectively.

Maxwell's equations, as given in equations 9.36 through 9.39, are in point form. The integral forms of Maxwell's equations are related directly to the experimental laws we accepted and used to "derive" equations 9.36 through 9.39. Faraday's law is obtained by integrating both sides of equation 9.36 over a fixed open surface and then applying Stoke's theorem to the left side. We have

$$\boxed{\oint_l \mathbf{E} \cdot d\mathbf{l} = -\iint_s \frac{\partial \mathbf{B}}{\partial t} \cdot d\mathbf{s}} \quad \text{(Faraday's law)} \tag{9.45}$$

The same process applied to equation 9.37 gives Ampere's (generalized) law,

$$\oint_l \mathbf{H} \cdot d\mathbf{l} = I + \iint_s \frac{\partial \mathbf{D}}{\partial t} \cdot d\mathbf{s} \quad \text{(Ampere's law)} \tag{9.46}$$

If we integrate both sides of equation 9.38 throughout a volume and apply the divergence theorem to the left side, we have Gauss's law,

$$\oint\!\!\oint_s \mathbf{D} \cdot d\mathbf{s} = \iiint_{\text{vol}} \rho_v \, dv \quad \text{(Gauss's law)} \tag{9.47}$$

The same process applied to equation (9.39) gives

$$\oint\!\!\oint_s \mathbf{B} \cdot d\mathbf{s} \equiv 0 \tag{9.48}$$

One special case needs to be given special attention because of its importance. The most important time-varying case, from an engineering applications viewpoint, is the time-harmonic case. For this case, we need to assume sinusoidal time variations and steady-state conditions. If this is done, all the methods of Fourier analysis (superposition of different sinusoids) are available to us. In particular, all the advantages of *phasor* forms (complex algebra) can be utilized. If we merely replace $\partial/\partial t$ by $j\omega$, then Maxwell's equations become

$$\nabla \times \mathbf{E} = -j\omega\mathbf{B} \tag{9.49}$$

$$\nabla \times \mathbf{H} = \mathbf{J} + j\omega\mathbf{D} \tag{9.50}$$

$$\nabla \cdot \mathbf{D} = \rho_v \tag{9.51}$$

and

$$\nabla \cdot \mathbf{B} = 0 \tag{9.52}$$

It should be noticed that the same notation is being employed for both phasors and real time dependent quantities, whether scalar or vector. This should cause no difficulties, as we will state explicitly what form we are working with, if it is not otherwise obvious from the context. The price of a new set of symbols seems too high to pay.

If we wish to examine the behavior of a function versus time, and we are employing phasor notation, we simply multiply the phasor by $e^{j\omega t}$ and take the real part of the resultant. As an example, suppose we have the phasor electric field,

$$\mathbf{E}(\omega) = \mathbf{E}_r + j\mathbf{E}_i$$

where $\mathbf{E}_r$ and $\mathbf{E}_i$ are *real* vectors. Then, in the real time domain

$$\mathbf{E}(t) = \text{Re}[(\mathbf{E}_r + j\mathbf{E}_i)e^{j\omega t}]$$

or

$$\mathbf{E}(t) = \mathbf{E}_r \cos \omega t - \mathbf{E}_i \sin \omega t.$$

Since $\cos \omega t \rightarrow 1$ and $\sin \omega t \rightarrow -j$, when going from the time domain to the frequency domain (phasor), the phasor form for $\mathbf{E}(t)$ above is

$$\mathbf{E}(\omega) = \mathbf{E}_r - (-j)\mathbf{E}_i = \mathbf{E}_r + j\mathbf{E}_i$$

Notice that in the preceding equations an ω was inserted to show that the phasor $\mathbf{E}(\omega)$ was a function of radian frequency. This is not necessary, strictly speaking, because the use of phasors implies *single* frequency excitation. It is advantageous, however, to employ this notation because quite often we want to find a time domain response when the excitation is not sinusoidal. That is, it is easy to find the response to any excitation if the phasor response is known. This is accomplished by means of the inverse Fourier (or Laplace) transform. Consider the case of a simple R-L series circuit driven by an ideal voltage source at a single frequency, ω. The response is defined as a current. Then,

$$I(\omega) = \frac{V(\omega)}{Z(\omega)} = \frac{V(\omega)}{R + j\omega L}$$

A very simple *phasor* result. Now suppose the excitation is not sinusoidal, but is an arbitrary function of time, $v(t)$. We now consider $V(\omega)$ to be the Fourier transform of $v(t)$ and so the response, $i(t)$, is the inverse Fourier transform of $I(\omega)$. That is,

$$i(t) = \mathscr{F}^{-1}\{I(\omega)\} = \mathscr{F}^{-1}\left\{\frac{V(\omega)}{R + j\omega L}\right\}$$

Ordinarily, we find such inverse transforms by means of tables of transform pairs. Such a table is included in Appendix E. Suppose $v(t) = \delta(t)$. (Then we are requesting the unit impulse response which was used in the example following equation 1.39.) From the table, $V(\omega) = 1$ and so

$$i(t) = \mathscr{F}^{-1}\left\{\frac{1}{R + j\omega L}\right\} = \frac{1}{L}\mathscr{F}^{-1}\left\{\frac{1}{R/L + j\omega}\right\}$$

Using the table once more,

$$i(t) = u(t)\frac{e^{-Rt/L}}{L}$$

9.5 POTENTIALS

In electrostatics and magnetostatics, we found that certain problems could be solved by first solving for auxiliary potential functions, Φ or $\mathbf{A}$. The desired field quantity, $\mathbf{E}$ or $\mathbf{B}$, could then be found from the potential by simple differentiation. We would like now to find a set of potentials from which time dependent fields may be derived and which are also consistent with those potential functions already found for statics.

From equation 9.39, there is an $\mathbf{A}$ such that

$$\mathbf{B} = \nabla \times \mathbf{A} \tag{9.53}$$

since the divergence of the curl of *any* vector is always zero. Equation 9.53, however, does not completely define $\mathbf{A}$, for if $\nabla\alpha$ (where α is any scalar field) is added to $\mathbf{A}$, equations 9.39 and 9.53 still hold. α is called a *gauge function.* Since equation 9.53 is the same as equation 7.42 for magnetostatics, let us try to utilize $\mathbf{A}$ here. Substituting equation 9.53 into equation 9.36 gives

$$\nabla \times \mathbf{E} = -\frac{\partial}{\partial t}(\nabla \times \mathbf{A})$$

Now, *if* $\mathbf{A}$ *and its derivatives are continuous*, we have

$$\nabla \times \mathbf{E} = -\nabla \times \frac{\partial \mathbf{A}}{\partial t}$$

or

$$\nabla \times \left(\mathbf{E} + \frac{\partial \mathbf{A}}{\partial t}\right) = 0 \tag{9.54}$$

This means that $\mathbf{E} + \partial\mathbf{A}/\partial t$ is a lamellar or conservative (zero curl) field and a scalar Φ exists such that

$$\mathbf{E} + \frac{\partial \mathbf{A}}{\partial t} = -\nabla\Phi \tag{9.55}$$

Note also that

$$\int_a^b \left(\mathbf{E} + \frac{\partial \mathbf{A}}{\partial t}\right) \cdot d\mathbf{l}$$

is independent of the path. $\mathbf{E}$ is (in general) nonconservative, but $\mathbf{E} + \partial\mathbf{A}/\partial t$ is conservative. Equation 9.55 may be written

$$\boxed{\mathbf{E} = -\nabla\Phi - \frac{\partial \mathbf{A}}{\partial t}} \tag{9.56}$$

and agrees with electrostatic result ($\mathbf{E} = -\boldsymbol{\nabla}\Phi$). Equation 9.56 is not exactly inviting for determining $\mathbf{E}$, because it contains both a scalar and a vector potential.

A pair of coupled equations for Φ and $\mathbf{A}$ may be obtained in the following way. Substituting equation 9.40 into equation 9.38 and then, in turn, substituting equations 9.40 and 9.41 into 9.37, (μ, ε scalars) gives

$$\boldsymbol{\nabla} \cdot \mathbf{E} = \frac{\rho_v}{\varepsilon} \tag{9.57}$$

and

$$\boldsymbol{\nabla} \times \mathbf{B} - \mu\varepsilon \frac{\partial \mathbf{E}}{\partial t} = \mu \mathbf{J} \tag{9.58}$$

Equations 9.57 and 9.58 are inhomogeneous in that they have *sources* on the right sides. We next eliminate $\mathbf{B}$ with equation 9.53 and $\mathbf{E}$ with equation 9.56. We then have

$$\boldsymbol{\nabla} \cdot \left(-\boldsymbol{\nabla}\Phi - \frac{\partial \mathbf{A}}{\partial t} \right) = \frac{\rho_v}{\varepsilon} \tag{9.59}$$

and

$$\boldsymbol{\nabla} \times (\boldsymbol{\nabla} \times \mathbf{A}) - \mu\varepsilon \frac{\partial}{\partial t} \left(-\boldsymbol{\nabla}\Phi - \frac{\partial \mathbf{A}}{\partial t} \right) = \mu \mathbf{J} \tag{9.60}$$

a pair of coupled equations. If Φ and its derivatives are continuous, equations 9.59 and 9.60 may be written

$$\nabla^2 \Phi + \frac{\partial}{\partial t} (\boldsymbol{\nabla} \cdot \mathbf{A}) = -\frac{\rho_v}{\varepsilon} \tag{9.61}$$

and

$$\boldsymbol{\nabla} \times (\boldsymbol{\nabla} \times \mathbf{A}) + \mu\varepsilon \frac{\partial^2 \mathbf{A}}{\partial t^2} + \mu\varepsilon \boldsymbol{\nabla} \frac{\partial \Phi}{\partial t} = \mu \mathbf{J} \tag{9.62}$$

Now,

$$\boldsymbol{\nabla} \times (\boldsymbol{\nabla} \times \mathbf{A}) \equiv \boldsymbol{\nabla}(\boldsymbol{\nabla} \cdot \mathbf{A}) - \nabla^2 \mathbf{A}$$

by vector identity (see Appendix A); so the pair of coupled equations becomes

$$\nabla^2 \Phi + \frac{\partial}{\partial t} (\boldsymbol{\nabla} \cdot \mathbf{A}) = -\frac{\rho_v}{\varepsilon} \tag{9.63}$$

and

$$\nabla^2 \mathbf{A} - \boldsymbol{\nabla}\left(\boldsymbol{\nabla} \cdot \mathbf{A} + \mu\varepsilon \frac{\partial \Phi}{\partial t} \right) - \mu\varepsilon \frac{\partial^2 \mathbf{A}}{\partial t^2} = -\mu \mathbf{J} \tag{9.64}$$

We have already specified the curl of $\mathbf{A}$ in equation 9.53. In order to completely define $\mathbf{A}$, we must also specify its divergence. Equation 9.64 suggests very strongly that we choose

$$\boxed{\nabla \cdot \mathbf{A} = -\mu\varepsilon \frac{\partial \Phi}{\partial t}} \tag{9.65}$$

called the *Lorentz condition.* If this is done, then $\mathbf{A}$ is unique in the Lorentz gauge and equations 9.63 and 9.64 are uncoupled, for then we have

$$\boxed{\nabla^2 \Phi - \mu\varepsilon \frac{\partial^2 \Phi}{\partial t^2} = -\frac{\rho_v}{\varepsilon}} \tag{9.66}$$

and

$$\boxed{\nabla^2 \mathbf{A} - \mu\varepsilon \frac{\partial^2 \mathbf{A}}{\partial t^2} = -\mu \mathbf{J}} \tag{9.67}$$

Notice that equations 9.65, 9.66, and 9.67 are consistent with their static counterparts,

$$\nabla \cdot \mathbf{A} = 0$$

$$\nabla^2 \Phi = -\frac{\rho_v}{\varepsilon}$$

and

$$\nabla^2 \mathbf{A} = -\mu \mathbf{J}$$

respectively. Equations 9.66 and 9.67 are called the inhomogeneous scalar and vector Helmholtz *wave equations*, respectively.

The static solutions for Φ and $\mathbf{A}$ were the Helmholtz integrals,

$$\Phi(\mathbf{r}) = \frac{1}{4\pi\varepsilon} \iiint\limits_{\text{vol}'} \frac{\rho_v(\mathbf{r}')}{R}\, dv'$$

and

$$\mathbf{A}(\mathbf{r}) = \frac{\mu}{4\pi} \iiint\limits_{\text{vol}'} \frac{\mathbf{J}(\mathbf{r}')}{R}\, dv'$$

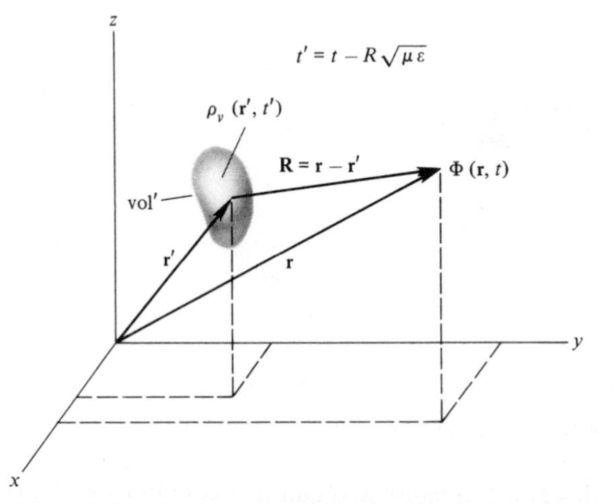

Figure 9.6. Geometry for the calculation of scalar potential, $\Phi(\mathbf{r}, t)$.

Solutions to equations 9.66 and 9.67 are similar as might be expected. Without proof,[2] the solutions are

$$\Phi(\mathbf{r}, t) = \frac{1}{4\pi\varepsilon} \iiint_{\text{vol}'} \frac{\rho_v(\mathbf{r}', t - R\sqrt{\mu\varepsilon})}{R} \, dv' \tag{9.68}$$

and

$$\mathbf{A}(\mathbf{r}, t) = \frac{\mu}{4\pi} \iiint_{\text{vol}'} \frac{\mathbf{J}(\mathbf{r}', t - R\sqrt{\mu\varepsilon})}{R} \, dv' \tag{9.69}$$

That is, to evaluate Φ at $\mathbf{r}$ and time t, the value of ρ_v at $\mathbf{r}'$ and time $t' = t - R\sqrt{\mu\varepsilon}$, or *retarded* time, should be used in the integrand. In the same way, to evaluate $\mathbf{A}$ at $\mathbf{r}$ and time t, the value of $\mathbf{J}$ at $\mathbf{r}'$ and $t - R\sqrt{\mu\varepsilon}$ should be used. The potentials given by equations 9.68 and 9.69 are thus called *retarded* potentials. In other words, a change in the source cannot be observed at the field point until a later time. Apparently, the effect propagates at a velocity given by $(\mu\varepsilon)^{-1/2}$. Figures 9.6 and 9.7 demonstrate the geometry. If the region of interest does not include the source (x, y, z not inside the region labeled vol'), then equations 9.68 and 9.69 (still) are solutions to the *homogeneous* differential equations (wave equations)

$$\nabla^2\Phi - \mu\varepsilon \frac{\partial^2 \Phi}{\partial t^2} = 0 \tag{9.70}$$

and

$$\nabla^2\mathbf{A} - \mu\varepsilon \frac{\partial^2 \mathbf{A}}{\partial t^2} = 0 \tag{9.71}$$

[2] See Appendix G.

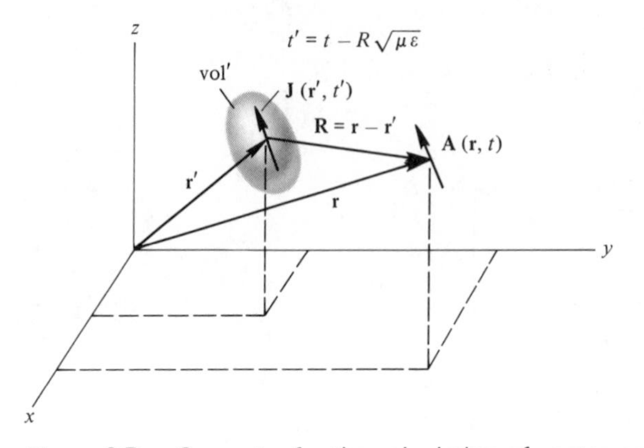

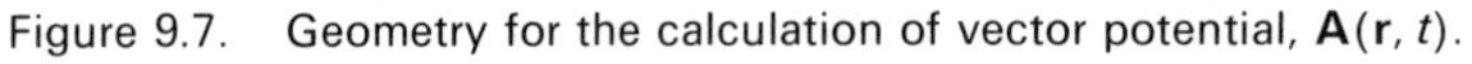
Figure 9.7. Geometry for the calculation of vector potential, $\mathbf{A}(\mathbf{r}, t)$.

It is worthwhile at this point to pause and consider equations 9.68 and 9.69 as superposition integrals. What are the unit impulse responses? A unit point charge (impulse) at $\mathbf{r}'$ and $t' = t - R\sqrt{\mu\varepsilon}$ produces a unit impulse response at $\mathbf{r}$ and t, so the unit impulse response for equation 9.68 and for rectangular components of equation 9.69 must be

$$h(\mathbf{r} - \mathbf{r}', t - t') = \frac{1}{4\pi\varepsilon R}$$

and

$$h(\mathbf{r} - \mathbf{r}', t - t') = \frac{\mu}{4\pi R}$$

For the special case of time dependent surface charge and current densities, convolution gives

$$\Phi(\mathbf{r}, t) = \frac{1}{4\pi\varepsilon} \iint_{s'} \frac{\rho_s(\mathbf{r}', t')}{R}\, ds \tag{9.72}$$

and

$$\mathbf{A}(\mathbf{r}, t) = \frac{\mu}{4\pi} \iint_{s'} \frac{\mathbf{J}_s(\mathbf{r}', t')}{R}\, ds' \tag{9.73}$$

In the same way, for filamentary charges and currents

$$\Phi(\mathbf{r}, t) = \frac{1}{4\pi\varepsilon} \int_{l'} \frac{\rho_l(\mathbf{r}', t')}{R}\, dl' \tag{9.74}$$

and

$$\mathbf{A}(\mathbf{r}, t) = \frac{\mu}{4\pi} \int_{l'} \frac{I(\mathbf{r}', t')}{R}\, d\mathbf{l}' \tag{9.75}$$

It is, of course, not necessary to work with both potentials. Suppose we want to work with **A** only. Then equation 9.53 gives us **B** if we have found **A**. We may then find **E** from Maxwell's equation. Alternatively, we may use equation 9.56, eliminating Φ by means of equation 9.65. It is simpler to show this in phasor form. There is no loss in generality in this approach because if the response to a general sinusoidal (phasor) excitation is known, then the response to any excitation can be found by the Fourier integral theorem. What is the phasor form for equation 9.69, or put a different way, what is the Fourier transform ($t \leftrightarrow \omega$) of equation 9.69? This is easy to answer because a time shift in the time domain is a phase shift in the frequency domain. Therefore, the phasor form is

$$\mathbf{A}(\mathbf{r}, \omega) = \frac{\mu}{4\pi} \iiint_{\text{vol}'} \frac{\mathbf{J}(\mathbf{r}', \omega)}{R} e^{-j\omega\sqrt{\mu\varepsilon}R} \, dv' \tag{9.76}$$

which satisfies the phasor form of equation 9.67; namely,

$$\nabla^2 \mathbf{A} + \omega^2 \mu\varepsilon \mathbf{A} = -\mu \mathbf{J}$$

The impulse response for scalar components of equation 9.76 is

$$h(\mathbf{r} - \mathbf{r}', \omega) = \frac{\mu}{4\pi} \frac{e^{-j\omega\sqrt{\mu\varepsilon}R}}{R}$$

In terms of this **A**,

$$\boxed{\mathbf{B} = \nabla \times \mathbf{A}} \tag{9.77}$$

and

$$\boxed{\mathbf{E} = \frac{1}{j\omega\mu\varepsilon} \nabla(\nabla \cdot \mathbf{A}) - j\omega\mathbf{A}} \tag{9.78}$$

where

$$\boxed{\Phi = -\frac{1}{j\omega\mu\varepsilon} (\nabla \cdot \mathbf{A})} \tag{9.79}$$

It is worthwhile at this point to emphasize that the use of potential functions is not necessary. Problem 18 sheds more light on this statement. Many times, however, considerable labor is saved by the use of **A** and Φ.

The *voltage* between two points may be defined *in general* as the negative of the line integral of the electric field taken along a *specific* path from point 1 to point 2. That is, the voltage between point 2 and point 1 is

$$V_{21} \equiv -\int_1^2 \mathbf{E} \cdot d\mathbf{l} = \int_2^1 \mathbf{E} \cdot d\mathbf{l} = -V_{12} \quad \text{(V)} \tag{9.80}$$

Substituting equation 9.56 into equation 9.80 gives

$$V_{21} = +\int_1^2 \left(\nabla\Phi + \frac{\partial \mathbf{A}}{\partial t}\right) \cdot d\mathbf{l}$$

or

$$V_{21} = \int_1^2 \left(\frac{\partial \Phi}{\partial x}\, dx + \frac{\partial \Phi}{\partial y}\, dy + \frac{\partial \Phi}{\partial z}\, dz\right) + \int_1^2 \frac{\partial \mathbf{A}}{\partial t} \cdot d\mathbf{l}$$

or

$$V_{21} = \int_1^2 d\Phi + \int_1^2 \frac{\partial \mathbf{A}}{\partial t} \cdot d\mathbf{l} \tag{9.81}$$

or

$$\boxed{V_{21} = \Phi_2 - \Phi_1 + \int_1^2 \frac{\partial \mathbf{A}}{\partial t} \cdot d\mathbf{l}} \tag{9.82}$$

The last term on the right of equation 9.82 is zero in *statics*, and, thus, the voltage and potential difference are the same in statics. For the general dynamic case, however, there is an added path dependent (in general) term; and so *voltage and potential difference are not generally the same.*[3]

9.6 POYNTING'S THEOREM

An identity of vector analysis from Appendix A is

$$\nabla \cdot \mathbf{E} \times \mathbf{H} \equiv \mathbf{H} \cdot \nabla \times \mathbf{E} - \mathbf{E} \cdot \nabla \times \mathbf{H} \tag{9.83}$$

for any **E** and **H**. Substituting Maxwell's equations, equations 9.36 and 9.37, into the right side of equation 9.83 gives

$$\nabla \cdot \mathbf{E} \times \mathbf{H} = \mathbf{H} \cdot \left(-\frac{\partial \mathbf{B}}{\partial t}\right) - \mathbf{E} \cdot \left(\mathbf{J} + \frac{\partial \mathbf{D}}{\partial t}\right)$$

or

$$\nabla \cdot \mathbf{E} \times \mathbf{H} = -\mathbf{H} \cdot \frac{\partial \mathbf{B}}{\partial t} - \mathbf{E} \cdot \mathbf{J} - \mathbf{E} \cdot \frac{\partial \mathbf{D}}{\partial t} \tag{9.84}$$

Equation 9.84 may be integrated throughout some region of interest, giving

$$-\iiint_{\text{vol}} \nabla \cdot (\mathbf{E} \times \mathbf{H})\, dv = \iiint_{\text{vol}} \left(\mathbf{H} \cdot \frac{\partial \mathbf{B}}{\partial t} + \mathbf{E} \cdot \frac{\partial \mathbf{D}}{\partial t} + \mathbf{E} \cdot \mathbf{J}\right) dv$$

or

$$-\oint\!\!\oint_{s} (\mathbf{E} \times \mathbf{H}) \cdot d\mathbf{s} = \iiint_{\text{vol}} \left(\mathbf{H} \cdot \frac{\partial \mathbf{B}}{\partial t} + \mathbf{E} \cdot \frac{\partial \mathbf{D}}{\partial t} + \mathbf{E} \cdot \mathbf{J}\right) dv \tag{9.85}$$

[3] If **A** is zero along the path of integration, then $V_{21} = \Phi_{21}$. See Figure 8.14, for example.

Now, if μ and ε are time independent scalars, then

$$\mathbf{E} \cdot \frac{\partial \mathbf{D}}{\partial t} = \frac{1}{2} \frac{\partial}{\partial t} (\mathbf{D} \cdot \mathbf{E})$$

$$\mathbf{H} \cdot \frac{\partial \mathbf{B}}{\partial t} = \frac{1}{2} \frac{\partial}{\partial t} (\mathbf{B} \cdot \mathbf{H})$$

and

$$-\oint\!\!\!\oint_s (\mathbf{E} \times \mathbf{H}) \cdot d\mathbf{s} = \iiint_{\text{vol}} \left[\frac{\partial}{\partial t} \left(\frac{\mathbf{D} \cdot \mathbf{E}}{2} + \frac{\mathbf{B} \cdot \mathbf{H}}{2} \right) + \mathbf{E} \cdot \mathbf{J} \right] dv \qquad (9.86)$$

Equation 9.86 may be rewritten

$$-\oint\!\!\!\oint_s (\mathbf{E} \times \mathbf{H}) \cdot d\mathbf{s} = \frac{\partial}{\partial t} \iiint_{\text{vol}} \left(\frac{\mathbf{D} \cdot \mathbf{E}}{2} + \frac{\mathbf{B} \cdot \mathbf{H}}{2} \right) dv + \iiint_{\text{vol}} \mathbf{E} \cdot \mathbf{J} \, dv \qquad (9.87)$$

if the volume limits are fixed in time. In Chapter 3, we showed that

$$\iiint_{\text{vol}} \frac{\mathbf{D} \cdot \mathbf{E}}{2} \, dv$$

represented the electrostatic energy stored in a volume. In the same way,

$$\iiint_{\text{vol}} \frac{\mathbf{B} \cdot \mathbf{H}}{2} \, dv$$

was taken as the magnetostatic energy stored in a volume. The interpretation of these terms is now extended to cover dynamics as well as statics. The first term on the right of equation 9.87 is the time rate of increase of the stored energy in the electric field, while the second term is the time rate of increase of the stored energy in the magnetic field. The last term represents energy dissipated (Joule's law) in heat per unit time or energy to accelerate isolated charges per unit time, depending on whether $\mathbf{J}$ is $\sigma\mathbf{E}$ or $\rho_v \mathbf{u}$, respectively. Normally we will use this term to calculate power lost or dissipated (P_d) in a material with conductivity. If there are *sources* inside the volume, $\mathbf{E} \cdot \mathbf{J}$ will be of opposite sign, and will represent the power density added to the system by the sources. The left side of equation 9.87 must then represent the energy flow *into* the volume per unit time. The energy flow *out* of the volume per unit time is therefore

$$\begin{aligned} P_f &= \oint\!\!\!\oint_s (\mathbf{E} \times \mathbf{H}) \cdot d\mathbf{s} \\ &= -\iiint_{\text{vol}} \frac{\partial}{\partial t} \left(\frac{\mathbf{D} \cdot \mathbf{E}}{2} + \frac{\mathbf{B} \cdot \mathbf{H}}{2} \right) dv - \iiint_{\text{vol}} \mathbf{E} \cdot \mathbf{J} \, dv \end{aligned} \qquad \text{(W)} \quad (9.88)$$

We *interpret* the vector $\mathbf{E} \times \mathbf{H}$ as the vector giving the direction and magnitude of power density at any point. This interpretation is a matter of convenience and does not follow directly from Poynting's theorem. The Poynting vector is

$$\boxed{\mathbf{S} = \mathbf{E} \times \mathbf{H}} \quad (\text{W/m}^2) \tag{9.89}$$

In the time-harmonic case, the time average Poynting vector is (in phasor notation)

$$\boxed{\langle \mathbf{S} \rangle = \tfrac{1}{2} \operatorname{Re} \{\mathbf{E} \times \mathbf{H}^*\}} \tag{9.90}$$

analogous to $\frac{1}{2} \operatorname{Re} \{VI^*\}$ in circuit theory. $\mathbf{H}^*$ is the complex conjugate of $\mathbf{H}$.

EXAMPLE 5

Let us close this section with a simple example using the Poynting vector. Suppose we have a cylindrical conductor of circular cross section whose axis is the z axis. Using previous results for $\mathbf{E}$ and $\mathbf{H}$ (steady current, uniform density)

$$P_f = \oint\!\!\oint_s \mathbf{E} \times \mathbf{H} \cdot d\mathbf{s} = \int_0^1 \int_0^{2\pi} \frac{J}{\sigma}\Big|_{\rho=a} (\mathbf{a}_z \times H_\phi\big|_{\rho=a} \mathbf{a}_\phi) \cdot \mathbf{a}_\rho a \, d\phi \, dz$$

for a section 1 m long. The integration is over the conductor surface, where $H_\phi = I/2\pi a$. Thus,

$$P_f = -\frac{J_z}{\sigma} \frac{I}{2\pi a} \int_0^1 \int_0^{2\pi} a \, d\phi \, dz = -\frac{J_z I}{\sigma}$$

But,

$$R = \frac{1}{\sigma \pi a^2} \quad \text{and} \quad J_z = \frac{I}{\pi a^2}$$

so

$$P_f = -I^2 R$$

Thus, we have a power flow of I^2R W *into* the conductor. If we accept $\mathbf{S}$ as correctly giving the power density flow at every point, then this concept tells us here that a battery (or some other source of emf) sets up fields such that energy flows through the fields and into the conductor through its surface. A circuit theory interpretation may differ in detail, but will correctly give I^2R as the power being dissipated. Regardless of the interpretation, the Poynting theorem, equation 9.88, will always give the correct total power balance for a given region. Other examples will be considered in a later chapter.

9.7 BOUNDARY CONDITIONS

We have already seen in electrostatics and magnetostatics that it is necessary to have a complete set of boundary conditions in order to solve many problems, especially when dealing with boundary value problems explicitly. The same thing is true in the more general dynamic case. In fact, the boundary conditions are almost the same for statics and dynamics. The required boundary conditions are most easily determined from Maxwell's equations in integral form, equations 9.45 through 9.48,

$$\oint_l \mathbf{E} \cdot d\mathbf{l} = -\iint_s \frac{\partial \mathbf{B}}{\partial t} \cdot d\mathbf{s} \tag{9.45}$$

$$\oint_l \mathbf{H} \cdot d\mathbf{l} = I + \iint_s \frac{\partial \mathbf{D}}{\partial t} \cdot d\mathbf{s} \tag{9.46}$$

$$\oint\!\!\oint \mathbf{D} \cdot d\mathbf{s} = \iiint_{\text{vol}} \rho_v \, dv \tag{9.47}$$

and

$$\oint\!\!\oint_s \mathbf{B} \cdot d\mathbf{s} = 0 \tag{9.48}$$

Applying equation 9.45 to the rectangular path in Figure 9.8, leads to

$$E_{t1} = E_{t2} \quad \text{or} \quad \boxed{\mathbf{a}_n \times (\mathbf{E}_1 - \mathbf{E}_2) = 0} \tag{9.91}$$

as long as $\partial \mathbf{B}/\partial t$ is finite, and we allow Δw to approach zero. Equation 9.91 simply states that between any two *physically realizable media* the tangential components of **E** are continuous. An important special case, to be considered shortly, is the unrealizable (but very useful) perfect conductor.

Applying equation 9.46 to the same path leads (no surface currents) to

$$H_{t1} = H_{t2} \quad \text{or} \quad \boxed{\mathbf{a}_n \times (\mathbf{H}_1 - \mathbf{H}_2) = 0} \tag{9.92}$$

as long as **J** and $\partial \mathbf{D}/\partial t$ are finite. (Perfect conductors are again an exception.) Equation 9.92 states in words that the tangential components of **H** are continuous at the interface between two physically realizable media.

Applying equation 9.47 to the cylindrical can of Figure 9.9 (for $\Delta w \to 0$) gives the same result as in electrostatics, namely,

$$D_{n1} - D_{n2} = \rho_s \quad \text{or} \quad \boxed{\mathbf{a}_n \cdot (\mathbf{D}_1 - \mathbf{D}_2) = \rho_s} \tag{9.93}$$

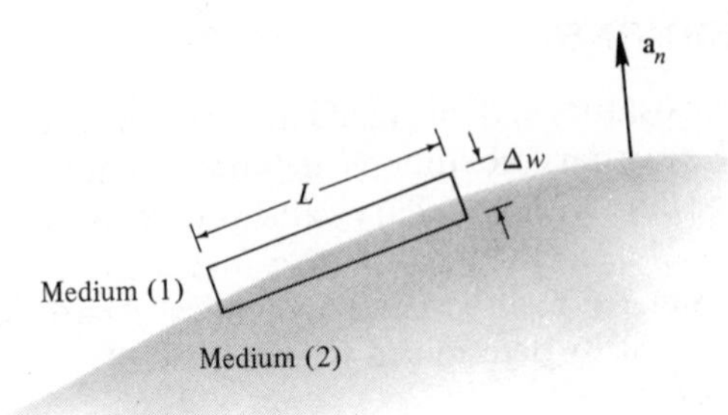

Figure 9.8. Rectangular path for evaluating $\oint \mathbf{E} \cdot d\mathbf{l}$ or $\oint \mathbf{H} \cdot d\mathbf{l}$.

Equation 9.48, applied to this surface, gives

$$B_{n1} = B_{n2} \quad \text{or} \quad \boxed{\mathbf{a}_n \cdot (\mathbf{B}_1 - \mathbf{B}_2) = 0} \tag{9.94}$$

Next, suppose that region 2 in Figure 9.8 and Figure 9.9 is a perfect conductor. In a perfect conductor the conductivity, σ, is infinite. Insofar as external fields are concerned, it is often convenient to treat good conductors, like copper or silver, as if they were perfect conductors, neglecting the small errors introduced. Ohm's law, $\mathbf{J} = \sigma\mathbf{E}$, requires that $\mathbf{J}$ be infinite unless $\mathbf{E}$ is zero. In other words, to keep $\mathbf{J}$ finite in the perfect conductor $\mathbf{E}$ must be identically zero. Then from equation 9.36, $\mathbf{H}$ must be identically zero inside the perfect conductor. It then follows from equation 9.37 that $\mathbf{J} = 0$ inside the perfect conductor. The only way left for a current to exist is in the form of a surface current, $\mathbf{J}_s$. In this case, equation 9.91 becomes

$$E_{t1} = E_{t2} = 0 \quad \text{or} \quad \boxed{\mathbf{a}_n \times \mathbf{E}_1 = 0} \tag{9.95}$$

while equation 9.93 becomes

$$D_{n1} = \rho_s \quad \text{or} \quad \boxed{\mathbf{a}_n \cdot \mathbf{D}_1 = \rho_s} \tag{9.96}$$

and equation 9.94 becomes

$$B_{n1} = 0 \quad \text{or} \quad \boxed{\mathbf{a}_n \cdot \mathbf{B}_1 = 0} \tag{9.97}$$

In light of the existence of surface current (zero thickness layer), equation 9.46 is reapplied to Figure 9.8 giving

$$H_{t1} = J_s \quad \text{or} \quad \boxed{\mathbf{a}_n \times \mathbf{H}_1 = \mathbf{J}_s} \tag{9.98}$$

where the current flows perpendicular to H_{t1}. Examples of the use of these boundary conditions will occur frequently in the chapters to follow.

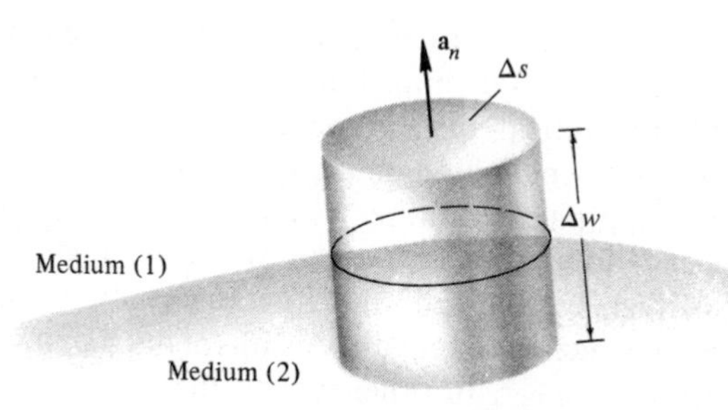

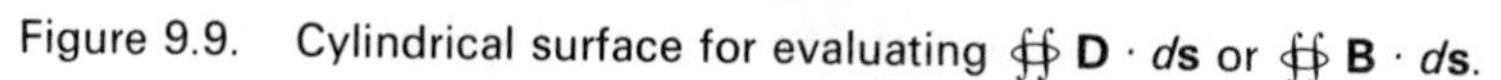

Figure 9.9. Cylindrical surface for evaluating $\oiint \mathbf{D} \cdot d\mathbf{s}$ or $\oiint \mathbf{B} \cdot d\mathbf{s}$.

9.8 CIRCUIT THEORY FROM FIELD THEORY

It is informative to show that circuit theory is a very special case of field theory. Put another way, the equations of circuit theory are Maxwell's equations recast into more applicable forms. This does not imply that one should start with Maxwell's equations to solve a circuit problem. What we intend to show here is where circuit equations come from and, in particular, what assumptions must be made for these equations to be valid.

EXAMPLE 6

Consider first the rather innocent looking geometry of Figure 9.10. A time dependent source is applied at the terminals in the form of a time dependent electric field. Some kind of current would be expected on and inside the conductor of finite conductivity. An internal electric field

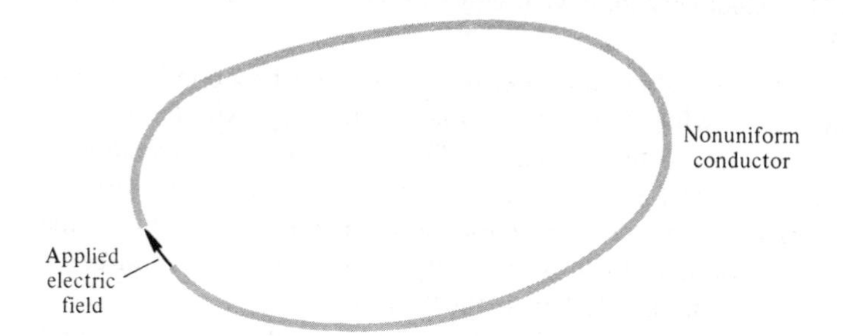

Figure 9.10. An electric field applied to a conducting body.

would be necessary to support this current. Some kind of nonuniform distributed self-inductance, both internal and external, would be associated with this geometry. Also, a nonuniform resistance (in the dc sense) would be expected. If the linear dimensions of the geometry are appreciable compared with $(f\sqrt{\mu\varepsilon})^{-1}$, *radiation* would occur introducing further complications. We would certainly hesitate to call this geometry a circuit. About the best we could do, and this only for sinusoidal time variation, would be to associate a resistance and reactance (inductive or capacitive?) or impedance, with the driven or "input" terminals. Any theoretical development past the starting equations would depend on having a nice simple geometry.

The preceding paragraph paints a rather bleak picture. Actually. many of the difficulties encountered there may be overcome if certain assumptions can be justified. First of all, and fundamentally, let us consider only those cases for which the linear dimensions of the geometry are small compared with $(f\sqrt{\mu\varepsilon})^{-1}$. In other words, the geometry is small physically, or the frequency is low, or both. This essentially eliminates radiation. Next, let us investigate the geometry of Figure 9.11 which

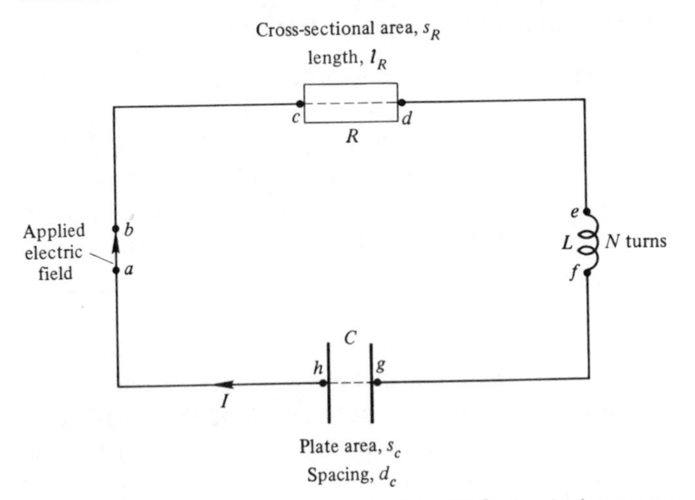

Figure 9.11. Geometry for developing *RLC* circuit theory equations from field theory.

certainly looks like a "circuit." Let the interconnecting conductors be filamentary everywhere, except for the small lossy cylinder of conductivity σ and the plates of areas s_c. The dimensions are indicated in Figure 9.11. This last requirement eliminates internal effects except for the lossy cylinder and the region between the plates. Let a time dependent electric field from some completely independent source be applied at the terminals *a-b*. This is the "input."

Let us evaluate the circulation of the electric field around the closed path, *l*, defined by the filamentary conductors and the dashed lines. Faraday's law, or Maxwell's equation, in integral form, is

$$\oint_l \mathbf{E} \cdot d\mathbf{l} = -\frac{d}{dt} \iint_s \mathbf{B} \cdot d\mathbf{s} \tag{9.4}$$

Let us consider the left side of the above equation and make additional assumptions as they seem necessary. The terminals *a-b* are very close so that the only effect here is that produced by the applied field, and

$$\int_a^b \mathbf{E} \cdot d\mathbf{l} = -V_{ba}$$

according to the definition, equation 9.80. V_{ba} is called the applied voltage (between points b and a) in circuit theory.

Let us next assume that the filamentary conductor is perfectly conducting or that the conductivity is large enough so that $\mathbf{E}$ is negligible on the conducting path. In this case,

$$\int_b^c \mathbf{E} \cdot d\mathbf{l} = \int_d^e \mathbf{E} \cdot d\mathbf{l} = \int_e^f \mathbf{E} \cdot d\mathbf{l} = \int_f^g \mathbf{E} \cdot d\mathbf{l} = \int_h^a \mathbf{E} \cdot d\mathbf{l} = 0$$

Notice particularly that the line integral, e to f, around the helical path is zero. Now

$$\int_c^d \mathbf{E} \cdot d\mathbf{l} = \int_c^d \frac{\mathbf{J}}{\sigma} \cdot d\mathbf{l} = \frac{J}{\sigma} \int_c^d dl = \frac{I l_R}{\sigma s_R} = IR$$

where uniform σ and E are assumed in the lossy material. R is the dc resistance of the lossy cylinder. We have assumed no displacement currents in the lossy material. In the region between the plates, we assume no conduction current. The current that flows between the plates is entirely displacement current, but it must equal (as we have seen before) the conduction current in the lossy material. Then

$$\int_g^h \mathbf{E} \cdot d\mathbf{l} = \int_g^h \frac{\mathbf{D}}{\varepsilon} \cdot d\mathbf{l} = \frac{D}{\varepsilon} \int_g^h dl = \frac{\rho_s s_c}{\varepsilon s_c} d_c = \frac{Q}{C}$$

where C is the "static" capacitance (geometry dependent). The current I can be related to the charge Q, because this current, flowing to the left *away from terminal* h, must equal the time rate of *increase* of charge at h (outside the plate). Thus,

$$I = \frac{dQ}{dt} \quad \text{or} \quad Q = \int I\, dt + Q_0$$

Notice that the same current (dQ/dt) is flowing *into terminal* g (outside the plate). This last result should be compared to equation 9.27. Finally, we have

$$\int_g^h \mathbf{E} \cdot d\mathbf{l} = \frac{1}{C} \int I\, dt + \frac{Q_0}{C}$$

and

$$\oint_l \mathbf{E} \cdot d\mathbf{l} = -V_{ba} + IR + \frac{1}{C} \int I\, dt + \frac{Q_0}{C} = -\frac{d}{dt} \iint_s \mathbf{B} \cdot d\mathbf{s}$$

There remains the right side which is a surface integral. The surface s is very complicated since it is that surface bounded by the path l, but at least it is fixed in time. It is reasonable to neglect the flux linking the open surface everywhere *except* through the N turn coil, for the magnetic field threading through the coil is much larger than anywhere else. Then,

$$-\frac{d}{dt}\iint_s \mathbf{B}\cdot d\mathbf{s} = -\frac{d\psi_m}{dt} = -\frac{d}{dt}(LI) = -L\frac{dI}{dt}$$

since L is the external inductance, determined only by the fixed coil geometry. The final result is

$$-V_{ba} + IR + \frac{1}{C}\int I\,dt + \frac{Q_0}{C} = -L\frac{dI}{dt}$$

or

$$V_{ba} = IR + L\frac{dI}{dt} + \frac{1}{C}\int I\,dt + \frac{Q_0}{C} \tag{9.99}$$

a well-known result. It is obvious from the development that we have a lumped R, L, and C forming our circuit.

It is extemely important to remember the large number of assumptions made in this development. To summarize, they are

1. All linear dimensions are much less than $(f\sqrt{\mu\varepsilon})^{-1}$.
2. A filamentary closed path is employed.
3. Perfect conductors exist everywhere in the circuit except at the input gap, between capacitor plates, and between the *resistor* terminals.
4. Displacement current is confined to the *capacitor*.
5. Magnetic flux is confined to the *inductor*.
6. The geometry is fixed in time.

These restrictions can be relaxed somewhat only after considerable experience has been gained with the simplest cases.

9.9 A NUMERICAL METHOD FOR QUASISTATIC ELECTRIC FIELD PROBLEMS

Consider an isotropic nonmagnetic region. Maxwell's second equation, equation 9.50, in phasor form is

$$\nabla \times \mathbf{H} = \mathbf{J} + j\omega\mathbf{D} = \sigma\mathbf{E} + j\omega\varepsilon\mathbf{E} = (\sigma + j\omega\varepsilon)\mathbf{E}$$

or

$$\nabla \times \mathbf{H} = \mathfrak{y}E \tag{9.100}$$

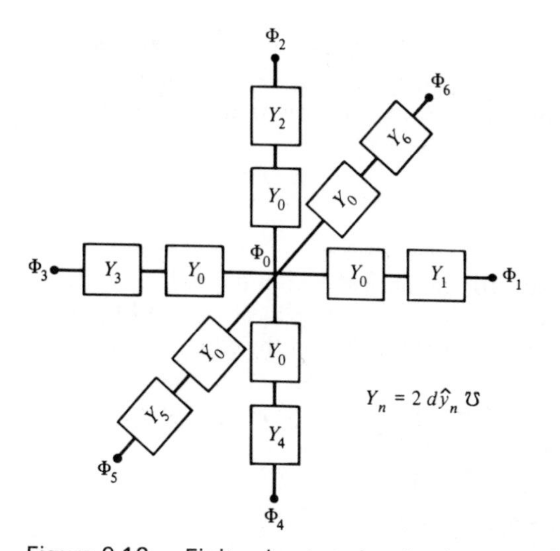

Figure 9.12. Finite element electric circuit model for $\oiint \hat{y}\nabla\Phi \cdot d\mathbf{s} = 0$.

where $\hat{y}$ is called the admittivity of the region. The divergence of equation 9.100 gives

$$\nabla \cdot (\hat{y}\mathbf{E}) = 0 \tag{9.101}$$

If magnetic effects are negligible (i.e., $\omega \approx 0$ or $j\omega\mathbf{A} \approx 0$), we have a quasistatic case, and equation 9.56 gives

$$\mathbf{E} = -\nabla\Phi \tag{9.102}$$

so,

$$\nabla \cdot (\hat{y}\nabla\Phi) = 0. \tag{9.103}$$

Expanding equation 9.103 using Appendix A, A.10, we obtain

$$\hat{y}\nabla^2\Phi + \nabla\Phi \cdot \nabla\hat{y} = 0 \tag{9.104}$$

If equation 9.104 is compared to equation 5.74, we see that they are dual if $\rho_v = 0$, and we can immediately write upon inspection of equation 5.80,

$$\Phi_0 = \frac{\displaystyle\sum_{n=1}^{6} \frac{\hat{y}_n}{\hat{y}_0 + \hat{y}_n}\Phi_n}{\displaystyle\sum_{n=1}^{6} \frac{\hat{y}_n}{\hat{y}_0 + \hat{y}_n}} \tag{9.105}$$

This is equivalent to a solution of $\oiint \hat{y}\nabla\Phi \cdot d\mathbf{s} = 0$. Inspection of Figures 5.8, 5.9, and 5.10 reveals that the finite element electric circuit model for equation 9.105 is as shown in Figure 9.12. From the circuit viewpoint we see nothing more than Kirchhoff's current law. Φ_n and $\hat{y}_n$ are the potential and admittivity, respectively, of the nth cell. The boundary conditions and methods of solution of the simultaneous equations resulting from equation 9.105 are the same as those discussed in Section 5.4. The principal difference here is the presence of a complex quantity, $\hat{y}$, and the absence of a "current source." The numerical method which has been briefly outlined here is useful in problems such as those arising in probe analysis with low-frequency excitation and inhomogeneous media.

EXAMPLE 7

A simple example to illustrate the numerical technique will now be given. Consider two large parallel plane electrodes as shown in Figure 9.13. As usual, we assume no fringing of the field at the edges. Note the

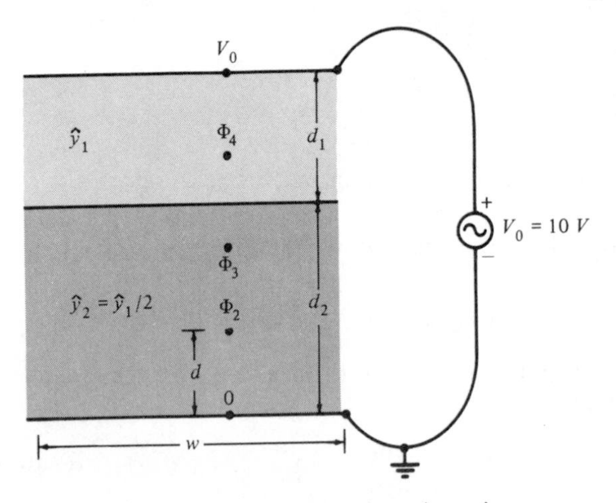

Figure 9.13. Two leaky capacitors in series.

similarity between this problem and that of Example 16, Chapter 8. The problem is one-dimensional, and using equation 9.105 (with only two terms per sum) for the unknowns, Φ_2, Φ_3, Φ_4, for points 2, 3, 4, respectively, gives

$$\Phi_2 = \tfrac{1}{2}\Phi_3 \qquad \Phi_2 = \tfrac{40}{13}\ \mathrm{V}$$

$$\Phi_3 = \tfrac{3}{7}\Phi_2 + \tfrac{4}{7}\Phi_4 \qquad \Phi_3 = \tfrac{80}{13}\ \mathrm{V}$$

$$\Phi_4 = \tfrac{2}{5}\Phi_3 + 6 \qquad \Phi_4 = \tfrac{110}{13}\ \mathrm{V}$$

Using Figure 9.12 for reference, the current from cell 5 into cell 4 is

$$\frac{\Phi_5 - \Phi_4}{Z_5 + Z_4} = \frac{\Phi_5 - \Phi_4}{1/2\ d\hat{y}_5 + 1/2\ d\hat{y}_4} = 2d\,\frac{10 - 110/13}{1/\hat{y}_1 + 1/\hat{y}_1} = d\hat{y}_1\,\frac{20}{13}$$

The total current for a width w and depth l is

$$I = d\hat{y}_1\,\frac{20\,wl}{13\,d^2} = \frac{20}{13}\,\hat{y}_1\,\frac{wl}{d}$$

so the admittance is ($V_0 = 10$ V)

$$Y = \frac{I}{V_0} = 0.154\hat{y}_1 \frac{wl}{d}$$

From the circuit viewpoint Figure 9.13 represents two "leaky" capacitors in series. Using the results from Chapter 6,

$$Y_1 = G_1 + j\omega C_1 \qquad\qquad Y_2 = G_2 + j\omega C_2$$

$$Y_1 = \sigma_1 \frac{wl}{1.5d} + j\omega\varepsilon_1 \frac{wl}{1.5d} \qquad\qquad Y_2 = \sigma_2 \frac{wl}{2.5d} + j\omega\varepsilon_2 \frac{wl}{2.5d}$$

$$Y_1 = \frac{wl}{1.5d}\hat{y}_1 \qquad\qquad Y_2 = \frac{wl}{5d}\hat{y}_1$$

and

$$Y = \frac{Y_1 Y_2}{Y_1 + Y_2} = 0.154 y_1 \frac{wl}{d}$$

as above. Notice that because of the one-dimensional nature, any pair of adjacent cells can be used to calculate the current.

EXAMPLE 8

A two-dimensional quasistatic boundary value problem is shown in Figure 9.14. Notice that the conducting sheet lying on the interface

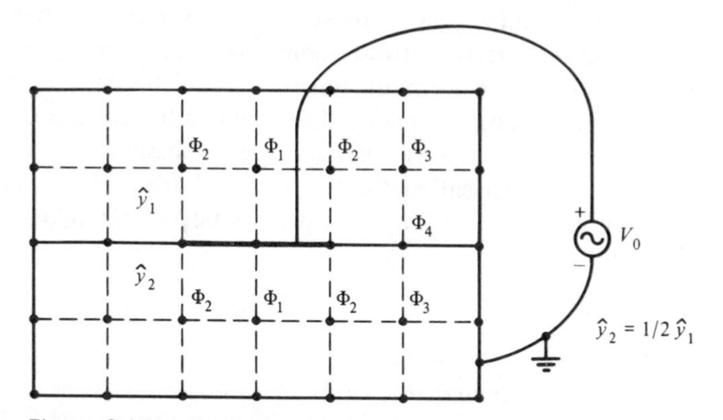

Figure 9.14. Two-dimensional quasistatic boundary value problem.

between the two media is at a potential V_0 with respect to the grounded outer conductor. Thus, we must place field points on the interface between the two media. This can be done by the second method described in Section 5.4. The appropriate equation is equation 5.87 with

$\langle \varepsilon_{0n} \rangle$ replaced by $\langle \hat{y}_{0n} \rangle$ (with no impressed current),

$$\Phi_0 = \frac{\sum_{n=1}^{4} \Phi_n \langle \hat{y}_{0n} \rangle}{\sum_{n=1}^{4} \langle \hat{y}_{0n} \rangle} \tag{9.106}$$

where $\langle \hat{y}_{0n} \rangle$ is the average admittivity of the two quadrants sharing the line 0-n. Because of the symmetry, we need only calculate Φ_1, Φ_2, Φ_3, and Φ_4 (even though $\hat{y}_1 \neq \hat{y}_2$). That is, the boundaries are symmetrical, Laplace's equation applies to the interior of both media, and the potential must be unique along the interface between the media. By the uniqueness theorem, therefore, the two solutions must be symmetrical. The equations, given by equation 9.106, and their solutions are

$$\begin{aligned} \Phi_1 &= \tfrac{1}{2}\Phi_2 + \tfrac{1}{4}V_0 & & \Phi_1 = 0.4555V_0 \\ \Phi_2 &= \tfrac{1}{4}\Phi_1 + \tfrac{1}{4}\Phi_3 + \tfrac{1}{4}V_0 & \Longrightarrow \quad & \Phi_2 = 0.4111V_0 \\ \Phi_3 &= \tfrac{1}{4}\Phi_2 + \tfrac{1}{4}\Phi_4 & & \Phi_3 = 0.1888V_0 \\ \Phi_4 &= \tfrac{1}{2}\Phi_3 + \tfrac{1}{4}V_0 & & \Phi_4 = 0.3444V_0 \end{aligned}$$

Notice that **E** will also be symmetrical in the two regions, but **D** will not. The admittance can be calculated as in Example 7.

9.10 CONCLUSIONS

The material presented in this chapter is extremely important. It is the essence of what we call electromagnetic field theory. Maxwell's equations have been developed starting with Faraday's law. These equations (with perhaps auxiliary potential functions derived from them) with the appropriate boundary conditions enable us to formally solve the problems of electromagnetics in the general dynamic case. They reduce identically to all electrostatic and magnetostatic results that were derived in earlier chapters.

A simple quasistatic numerical method was found from the static results of Chapter 5. It is useful in low-frequency probe analysis among other things.

REFERENCES

Bradshaw, M. D. and Byatt, W. J. *Introductory Engineering Field Theory*. Englewood Cliffs, N.J.: Prentice-Hall, 1967. It is interesting for the reader who has been with us through the first nine chapters to examine a textbook that begins with the dynamic equations.

Johnk, C. T. A. *Engineering Electromagnetic Fields and Waves*. New York: Wiley, 1975. The text begins with the dynamic equations.

Maxwell, J. C. *A Treatise on Electricity and Magnetism*, 3rd ed. New York: Oxford University Press, 1964, or New York: Dover, 1954.

Owen, G. E. *Introduction to Electromagnetic Theory*. Boston: Allyn and Bacon, 1963. Faraday's law is derived in a moving frame.

Rao, N. N. *Elements of Engineering Electromagnetics*. Englewood Cliffs, N.J.: Prentice-Hall, 1977. The text begins with the dynamic equations.

1. Evaluate the circulation, $\oint \mathbf{E} \cdot d\mathbf{l}$, around the closed path, (0, 0, 0) to (1, 0, 0) to (1, 1, 0) to (0, 1, 0) to (0, 0, 0), if $\mathbf{E} = 2x\mathbf{a}_x + x\mathbf{a}_y$. Find $\mathbf{B}$.
2. A resistor R replaces the voltmeter of Figure 9.3. Show that the power dissipated in the resistor equals the power required to move the end of the loop.
3. If $\mathbf{B} = B_0 \sin \omega t\, \mathbf{a}_x$ in Figure 9.4, find the emf.
4. A thin conducting strip is located in the $z = 0$ plane, $0 \le x \le l$, $-\infty < y < \infty$. The strip is moving with a velocity, $\mathbf{u} = u_0 \mathbf{a}_y$. A steady magnetic field, $\mathbf{B} = B_0 \mathbf{a}_z$, is applied. Compare this situation to that of the straight wire moving at right angles to a steady magnetic field. Calculate the emf induced in the strip. This emf may be picked off by a *fixed* voltmeter connected by brushes across the width of the strip. See Figure 9.15.

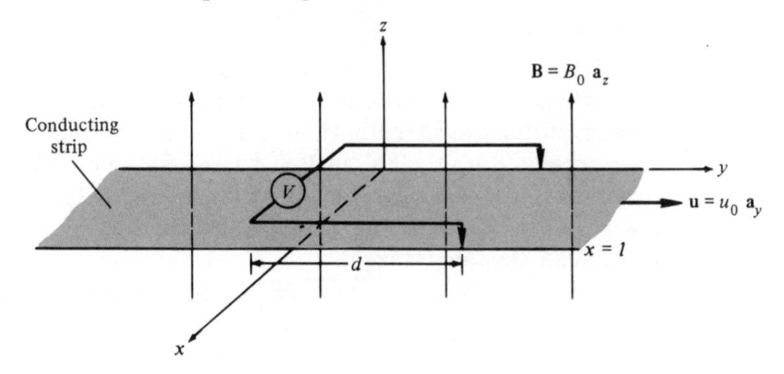

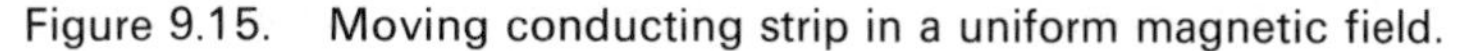
Figure 9.15. Moving conducting strip in a uniform magnetic field.

5. Repeat Problem 4, if $\mathbf{B} = B_m \cos \omega t\, \mathbf{a}_z$ and the plane containing the voltmeter leads is perpendicular to the strip.
6. Repeat Problem 4 if $\mathbf{B} = B_m \cos \omega t\, \mathbf{a}_z$ and the plane containing the voltmeter leads is parallel to the strip. The voltmeter leads define three sides of a rectangle (l in the x direction and d in the y direction). Is equation 9.4 applicable to this problem?
7. The Faraday disk generator is a practical version of Problem 4. It consists of a thin conducting disk rotating about its axis at an angular velocity, ω, while a magnetic field is applied perpendicular to the disk. The emf is picked off by brushes; one on the shaft, and one on the periphery of the disk. If the radius of the disk is a, the

voltmeter leads define three sides of a rectangle ($b \times a$) parallel to the plane of the disk. If $\mathbf{B} = B_0 \mathbf{a}_z$ and $\boldsymbol{\omega} = -\omega_0 \mathbf{a}_z$, find the emf. See Figure 9.16.

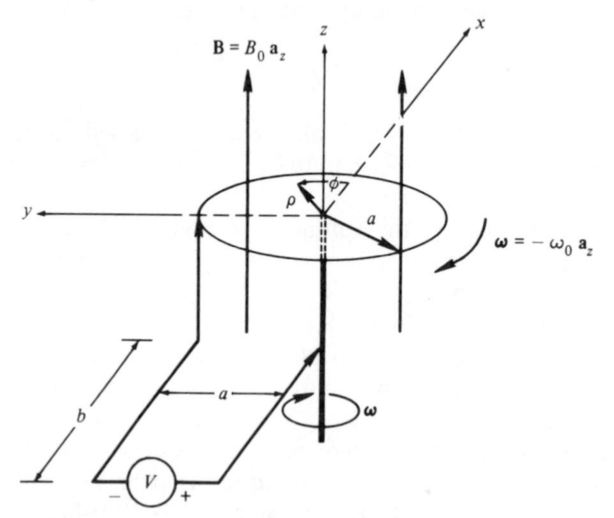

Figure 9.16. The Faraday disk generator.

8. Repeat Problem 7 if $\mathbf{B} = B_m \sin \omega_b t \mathbf{a}_z$.
9. A conducting loop of one turn and 10^{-1} m^2 area is exposed in air to a 10-MHz field. If 1 V rms is induced in the loop, find the rms value of the magnetic field strength, **H**, normal to the plane of the loop.
10. If $\mathbf{J} = x^2\mathbf{a}_x + yz\mathbf{a}_y + xyz\mathbf{a}_z$ in free space, find the total current flowing out of the unit cube situated in the first octant with one corner at (0, 0, 0). Use $\oiint \mathbf{J} \cdot d\mathbf{s}$. Next, apply the divergence theorem and evaluate the same current.
11. If $\partial/\partial\phi = 0$ and $B_z = B_0 \rho^2 t \cos 2\pi z/L$, $B_\phi = 0$, find E_ϕ and B_ρ. E_ϕ and B_ρ are finite for $\rho = 0$.
12. If $z > 0$ is free space, $z < 0$ is a perfect conductor, $\mathbf{J}_s = 50\mathbf{a}_x + 100\mathbf{a}_y$ A/m and $\rho_s = 10^{-7}$ C/m^2 on the interface at some instant of time, t, find **D**, **E**, **B**, and **H** at $(0, 0, 0^+, t)$.
13. Starting with one of Maxwell's equations in point form, derive the continuity equation.
14. If $E_x = E_0 u(t \pm z\sqrt{\mu_0 \varepsilon_0})$, $E_y = E_z = 0$, and

$$H_y = \mp E_0 \sqrt{\varepsilon_0/\mu_0}\, u(t \pm z\sqrt{\mu_0 \varepsilon_0}),\ H_x = H_z = 0$$

in free space (where u is the unit step function), show that Maxwell's equations are satisfied. Can a vector potential, **A**, be found from which this field is derivable?
15. A lossy parallel plate capacitor has conductivity σ and permittivity ε in the dielectric. Show that the ratio of conduction current density

to displacement current density is $\sigma/\omega\varepsilon$ (magnitudes only) in the time-harmonic case.

16. If a wavelength, λ, is $(f\sqrt{\mu_0\varepsilon_0})^{-1}$ in free space, and we arbitrarily decide that less than 0.1λ means "smaller" than a wavelength, how large is "appreciable large" at the usual power line frequency, 60 Hz? Comment.
17. Assume some *excess* charge is somehow placed in a homogeneous conductor. Using Ohm's law and the equation of continuity of current show that $\rho_v = \rho_{v0}e^{-\sigma t/\varepsilon}$. The time constant, ε/σ is called the *relaxation time*. Calculate this time constant for copper and interpret the result.
18. Using Maxwell's equations in a region where μ and ε are scalar constants:

    (a) Show that

$$\nabla^2\mathbf{E} + k^2\mathbf{E} = j\omega\mu\left[\mathbf{J} + \frac{1}{k^2}\nabla(\nabla\cdot\mathbf{J})\right]$$

    and

$$\nabla^2\mathbf{H} + k^2\mathbf{H} = -\nabla\times\mathbf{J}$$

    (b) Find Helmholtz integral solutions for **E** and **H**.

19. Show that the last equation of Problem 18 is satisfied by

$$H_y = [2u(z) - 1]e^{-jk[2u(z)-1]z} \qquad \text{if} \qquad J_x = -2\,\delta(z)$$

    (a current sheet over the $z = 0$ plane).
20. Assuming that the filamentary current loop of radius a lying in the $z = 0$ plane with its center at the origin carries the current density $\mathbf{J} = I\delta(\rho - a)\,\delta(z)\cos\phi\cos\omega t\mathbf{a}_\phi$, find **A** at $(0, 0, z)$.
21. Assuming that $\mathbf{A} = \cos\omega t\cos kz\,\mathbf{a}_x$ and $\sigma = 0$, (a) find **H**, (b) find **E**, and (c) find Φ.
22. Refer to Problem 24(b), Chapter 5, and Figure 5.25. Write the six simultaneous equations if the water in the trough is salty and an ac voltage is used.
23. The transformer of Figure 8.25 is shown schematically in Figure 9.17 using the dot convention described in Problem 24 of Chapter 8.

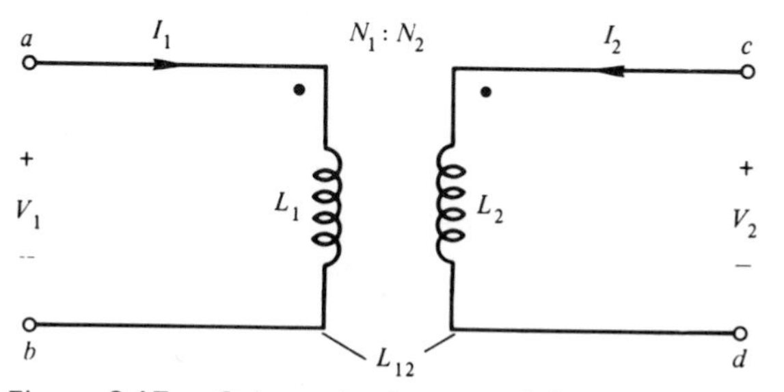

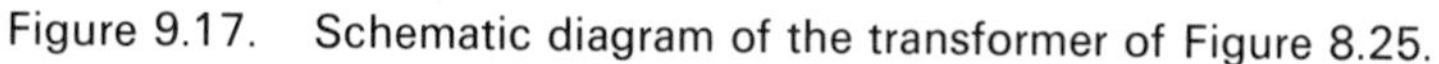
Figure 9.17. Schematic diagram of the transformer of Figure 8.25.

Using the definitions of inductance and voltage with Faraday's law:

(a) Derive the circuit equations for the transformer in terms of L_1, L_2, and L_{12}. Assume the transformer is *linear*. Use phasors.
(b) Assume the transformer has negligible leakage flux ($k_{mc} = 1$) so that it is *ideal*, and show that $I_1/I_2 = V_2/V_1 = -N_2/N_1$.

Chapter 10
Uniform Plane Wave Propagation

The uniform plane wave is introduced in this chapter because, first of all, it represents a very simple example of the use of Maxwell's equations in solving an electromagnetic field problem. A good understanding of the uniform plane wave is also necessary if one is to successfully master the more complex types of electromagnetic phenomena. Furthermore, it is very easy to move from a study of uniform plane waves into a study of transmission lines by way of the parallel plane system. Proceeding in this fashion gives one a better understanding of the relation between field quantities and circuit quantities as applied to the transmission line. The alternate approach, as far as the transmission line is concerned, is to assume an equivalent circuit and derive the voltage and current relationships. This seems rather incomplete and not related directly to previous work.

The uniform, undamped plane wave in unbounded space is considered first in this chapter. Next, the damped uniform plane wave in unbounded (but lossy) space is considered. We then have the necessary background to consider the case of reflection (first normal, then oblique incidence) from a plane interface. The material on either side of the interface may be quite general, but the only interface considered here is a plane one where the boundary conditions are easy to observe and apply. The treatment of plane wave reflection (scattering) from other geometries (an infinite cylinder, for example) is beyond the scope of this material and will not be pursued. With a good basic understanding of uniform plane waves, many avenues of investigation are open to us, and we will in the next chapter explore one; the transmission line.

The scalar and vector inhomogeneous *wave* equations for Φ and $\mathbf{A}$, respectively, were derived in the preceding chapter. Formal solutions (the Helmholtz integrals) to these equations were presented, but no explicit solutions showing a wave character were pursued. We are now ready to proceed toward the simplest practical cases of wave phenomena. Sinusoidal excitation in time and steady state conditions will usually be assumed so that the advantageous use of phasors is permissible. Phasor notation will be used throughout unless explicitly stated to the contrary.

A plane wave may be defined as a wave for which the electric and magnetic field vectors lie in a plane. This plane is perpendicular to the direction of wave motion or propagation. In the special case of time harmonic excitation (phasors), we may also say that contours of *constant phase* are these same planes. In a *cylindrical wave*, contours of constant phase are *cylinders*, and for a *spherical wave* contours of constant phase are *spheres*. Plane waves are difficult to generate in practice, but may be well approximated by cylindrical waves or spherical waves at large distances from their respective sources. The field at large distances from a radio antenna is essentially a plane wave over a limited region. These electromagnetic waves are examples of *transverse waves*, as opposed to an acoustic wave, which is an example of a longitudinal wave.

In the special case of a *uniform* plane wave, the electric and magnetic field vectors are *uniform* in a plane that is perpendicular to the direction of propagation. In the case of phasor quantities, the electric and magnetic fields are uniform in a plane of constant phase.

10.1 THE UNDAMPED UNIFORM PLANE WAVE

It would seem natural to pursue a solution for the electromagnetic field of an undamped uniform plane wave by way of the auxiliary potential functions. After all, this was the reason for their existence. In a sense, a general approach would incorporate the potential functions, but in the simplest cases, such as the one presently under consideration, it is more informative to employ Maxwell's equations directly. For present purposes equations 9.49 through 9.52 may be written in phasor form as

$$\nabla \times \mathbf{E} = -j\omega\mu\mathbf{H} \tag{10.1}$$

$$\nabla \times \mathbf{H} = (\sigma + j\omega\varepsilon)\mathbf{E} \tag{10.2}$$

$$\nabla \cdot \mathbf{D} = \rho_v \tag{10.3}$$

and

$$\nabla \cdot \mathbf{B} = 0 \tag{10.4}$$

where

ρ_v = isolated charge density (C/m^3)
μ = permeability (assumed to be a scalar constant) (H/m)
ε = permittivity (assumed to be a scalar constant) (F/m)
σ = conductivity (assumed to be a scalar constant) (V/m)

We are presently interested in the simplest possible cases and have over-simplified the situation somewhat insofar as the material parameters μ, ε, and σ are concerned. The assumptions we have made regarding these parameters can be relaxed after we have gained some knowledge of electromagnetic waves.

For present purposes, we assume that there are no isolated charges so that $\rho_v = 0$. We further simplify to the case for which the entire region of interest is conduction free, or $\sigma = 0$. Thus, equations 10.1 through 10.4 become

$$\nabla \times \mathbf{E} = -j\omega\mu\mathbf{H} \tag{10.5}$$

$$\nabla \times \mathbf{H} = j\omega\varepsilon\mathbf{E} \tag{10.6}$$

$$\nabla \cdot \mathbf{E} \equiv 0 \tag{10.7}$$

and

$$\nabla \cdot \mathbf{H} \equiv 0 \tag{10.8}$$

The *wave equation* for **E** is generated by taking the curl of both sides of equation 10.5,

$$\nabla \times (\nabla \times \mathbf{E}) = -j\omega\mu(\nabla \times \mathbf{H})$$

and substituting equation 10.6 for $\nabla \times \mathbf{H}$. Thus,

$$\nabla \times (\nabla \times \mathbf{E}) = k^2\mathbf{E} \tag{10.9}$$

where

$$\boxed{k = \omega\sqrt{\mu\varepsilon}} \quad \text{(rad/m)} \tag{10.10}$$

and k is called the *wave number* or *phase constant*. The left side of equation 10.9 may be expanded using a vector identity we have employed several times in previous chapters. This expansion, $\nabla \times (\nabla \times \mathbf{E}) \equiv \nabla(\nabla \cdot E) - \nabla^2\mathbf{E}$, gives

$$\nabla(\nabla \cdot \mathbf{E}) - \nabla^2\mathbf{E} - k^2\mathbf{E} = 0$$

Inspection of equation 10.7 reveals that the first term of the last equation is zero for *all* x, y, z, and ω. Therefore, the partial differential equation to be satisfied by **E** is

$$\nabla^2\mathbf{E} + k^2\mathbf{E} = 0 \tag{10.11}$$

A comparison of the preceding equation with equation 9.67 reveals that we have a *homogeneous vector wave equation* for **E**. Under the same assumed conditions the vector potential, **A**, would have to satisfy a partial differential equation *identical* with equation 10.11. That is,

$$\nabla^2 \mathbf{A} + k^2 \mathbf{A} = 0$$

If we follow the same procedure again, eliminating **E** rather than **H**, we obtain

$$\nabla^2 \mathbf{H} + k^2 \mathbf{H} = 0 \tag{10.12}$$

Thus, in an unbounded perfect dielectric (lossless) region (μ and ε, scalar constants) **A**, **E**, and **H** all satisfy the homogeneous vector wave equation. All of the results in this paragraph are completely general for the assumed conditions on the material parameters and are not specialized for plane waves.

We would next like to obtain a simple solution to equation 10.11 in the form of a uniform plane wave. To expedite this formulation, we *arbitrarily* choose $E_y = E_z \equiv 0$. In this case equation 10.11 becomes

$$\nabla^2 E_x + k^2 E_x = 0$$

or (see Appendix A), in rectangular coordinates,

$$\frac{\partial^2 E_x}{\partial x^2} + \frac{\partial^2 E_x}{\partial y^2} + \frac{\partial^2 E_x}{\partial z^2} + k^2 E_x = 0$$

If our plane wave is to be *uniform* in planes parallel to the $z = 0$ plane, for example, E_x cannot be dependent on x or y. In this special case,

$$\frac{\partial^2 E_x}{\partial z^2} + k^2 E_x = 0$$

or

$$\frac{d^2 E_x}{dz^2} + k^2 E_x = 0$$

The last equation is the ordinary differential equation of simple harmonic motion whose solution is well known and easily verified. One form of this solution is

$$E_x(z, \omega) = Ae^{-jkz} + Be^{+jkz} \tag{10.13}$$

where A and B are independent of z, but not necessarily independent of ω. The magnetic field that accompanies this solution for the electric field must have a similar form because both obey the same basic differential equation. It is now simpler to use equation 10.5 to obtain **H**, rather than solving equation 10.12.

$$\mathbf{H} = -\frac{1}{j\omega\mu}(\nabla \times \mathbf{E}) = -\frac{1}{j\omega\mu}\begin{vmatrix} \mathbf{a}_x & \mathbf{a}_y & \mathbf{a}_z \\ \dfrac{\partial}{\partial x} & \dfrac{\partial}{\partial y} & \dfrac{\partial}{\partial z} \\ E_x & 0 & 0 \end{vmatrix} = -\frac{1}{j\omega\mu}\left(\mathbf{a}_y \frac{\partial E_x}{\partial z} - \mathbf{a}_z \frac{\partial E_x}{\partial y}\right)$$

or

$$\mathbf{H} = -\frac{\mathbf{a}_y}{j\omega\mu}\frac{\partial E_x}{\partial z} \tag{10.14}$$

the other term in the expansion being zero because E_x is independent of y. Equation 10.14 may be written as a scalar equation since only one component is involved.

$$H_y = -\frac{1}{j\omega\mu}\frac{\partial E_x}{\partial z} = \frac{k}{\omega\mu}Ae^{-jkz} - \frac{k}{\omega\mu}Be^{+jkz}$$

Now, $k/\omega\mu = \omega\sqrt{\mu\varepsilon}/\omega\mu = \sqrt{\varepsilon/\mu}$, so

$$H_y(z, \omega) = \frac{A}{\sqrt{\mu/\varepsilon}}e^{-jkz} - \frac{B}{\sqrt{\mu/\varepsilon}}e^{+jkz} \tag{10.15}$$

A second order differential equation has two independent solutions, and we have shown both of these for **E** and **H**. If we had decided to work with the independent solutions *separately*, we would have certainly obtained the pairs of solutions,

$$\begin{aligned} E_x(z, \omega) &= Ae^{-jkz} \\ H_y(z, \omega) &= \frac{A}{\sqrt{\mu/\varepsilon}}e^{-jkz} \end{aligned} \tag{10.16}$$

and

$$\begin{aligned} E_x(z, \omega) &= Be^{+jkz} \\ H_y(z, \omega) &= -\frac{B}{\sqrt{\mu/\varepsilon}}e^{+jkz} \end{aligned} \tag{10.17}$$

The first *pair* represents an electromagnetic *traveling wave* propagating in the *positive* z direction, while the second pair represents an electromagnetic traveling wave propagating in the negative z direction. Notice particularly that both of these solutions meet the defined conditions for a uniform plane wave.

The best way to show that these solutions are traveling waves, and to show how a traveling wave behaves, is to consider a time domain solution explicitly. In particular, let us find general time forms corresponding to equations 10.16 and 10.17. The time domain forms of equations 10.5 and 10.6 are

$$\nabla \times \mathbf{E} = -\mu\frac{\partial \mathbf{H}}{\partial t} \tag{10.18}$$

and

$$\nabla \times \mathbf{H} = \varepsilon\frac{\partial \mathbf{E}}{\partial t} \tag{10.19}$$

Following the same procedure as that used in equations 10.5 through 10.17, and making the same assumptions, leads[1] to

$$\frac{\partial^2 E_x}{\partial z^2} - \mu\varepsilon \frac{\partial^2 E_x}{\partial t^2} = 0 \tag{10.20}$$

and

$$\frac{\partial^2 H_y}{\partial z^2} - \mu\varepsilon \frac{\partial^2 H_y}{\partial t^2} = 0 \tag{10.21}$$

which are also wave equations. It is left as an exercise to fill in the steps between equations 10.19 and 10.20. It is easy to verify that general solutions to equations 10.20 and 10.21 are (see Problem 4)

$$\boxed{E_x(z, t) = f(t - z\sqrt{\mu\varepsilon}) + g(t + z\sqrt{\mu\varepsilon})} \tag{10.22}$$

and

$$\boxed{H_y(z, t) = \frac{1}{\sqrt{\mu/\varepsilon}} f(t - z\sqrt{\mu\varepsilon}) - \frac{1}{\sqrt{\mu/\varepsilon}} g(t + z\sqrt{\mu\varepsilon})} \tag{10.23}$$

where f and g are (formally, at least) *any* well behaved functions.

Consider the Dirac delta function. According to equation 10.22, *one* possible solution for $E_x(z, t)$ is

$$E_x(z, t) = C\delta(t - z\sqrt{\mu\varepsilon}) \tag{10.24}$$

Suppose we could take a series of "snapshots" of this function at

$$t = -t_1, \qquad t = 0, \qquad t = t_1, \qquad t = t_2, \quad \text{etc.}$$

As can be seen in Figure 10.1(a) the impulse is at $z = -t_1/\sqrt{\mu\varepsilon}$ when $t = -t_1$. The impulse is at $z = 0$ when $t = 0$. The impulse is at $z = t_1/\sqrt{\mu\varepsilon}$ when $t = t_1$, and the impulse is at $z = t_2/\sqrt{\mu\varepsilon}$ when $t = t_2$. This wave is obviously traveling. It is traveling in the positive z direction with a velocity given by

$$\boxed{u = \frac{1}{\sqrt{\mu\varepsilon}}} \quad \text{(m/s)} \tag{10.25}$$

The traveling wave

$$E_x(z, t) = D\,\delta(t + z\sqrt{\mu\varepsilon}) \tag{10.26}$$

is moving in the negative z direction with the same velocity. This can be verified by again taking snapshots.

[1] The inverse Fourier transform of the two equations above equation 10.13 gives equations 10.20 and 10.21 almost immediately.

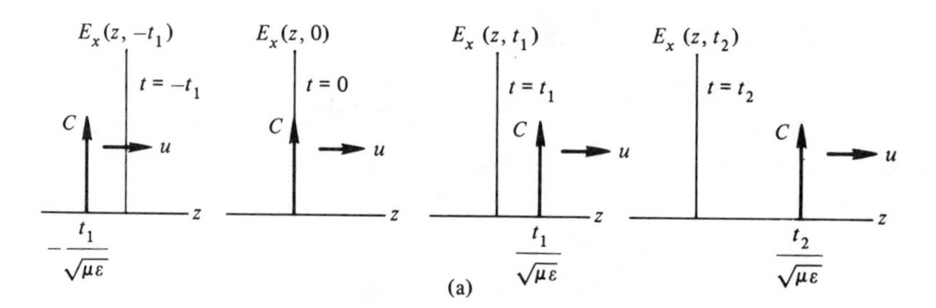

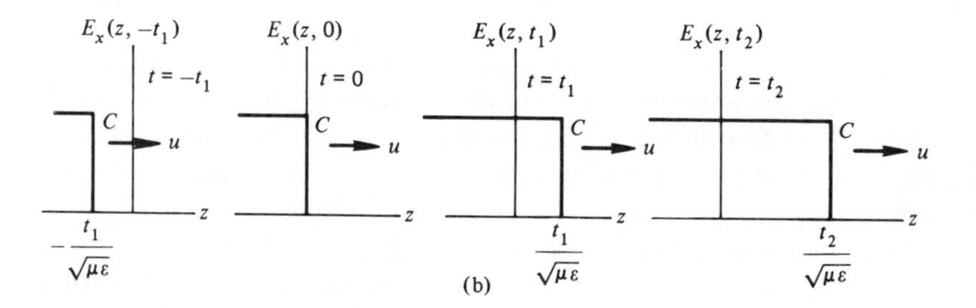

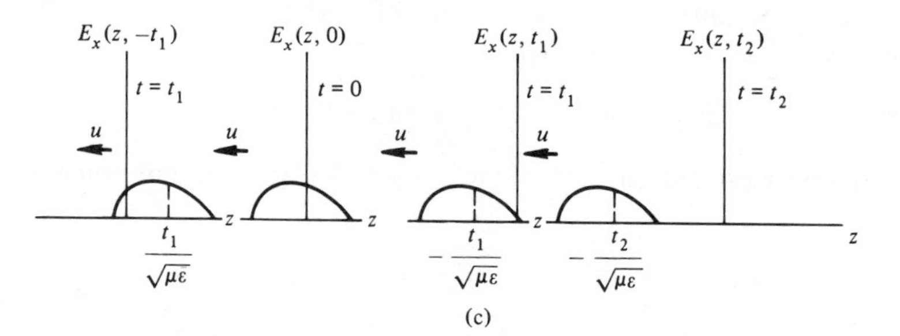

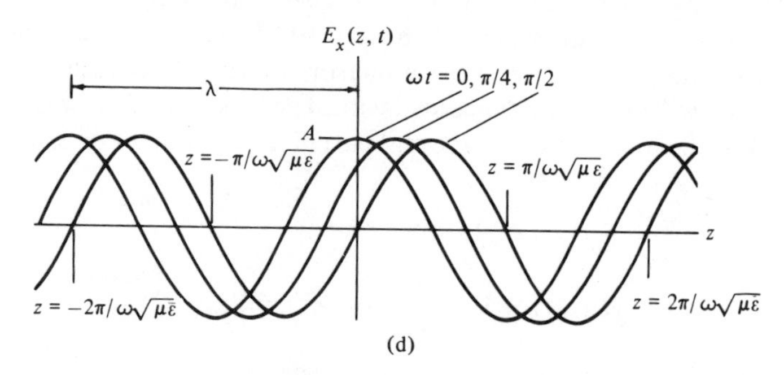

Figure 10.1. Traveling waves. (a) $E_x(z, t) = C\delta(t - z\sqrt{\mu\varepsilon})$, impulse traveling in the $+z$ direction. (b) $E_x(z, t) = Cu(t - z\sqrt{\mu_0\varepsilon_0})$, step traveling in the $+z$ direction in a vacuum. (c) $E_x(z, t) = f(t + z\sqrt{\mu_0\varepsilon_0})$, arbitrary pulse traveling in the $-z$ direction. (d) $E_x(z, t) = A\cos\omega(t - z\sqrt{\mu\varepsilon})$, cosinusoidal wave traveling in the $+z$ direction.

Consider the unit step function.

$$E_x(z, t) = Cu(t - z\sqrt{\mu_0 \varepsilon_0})$$

It is pictured in Figure 10.1(b) at various times. It too is a $+z$ traveling wave with velocity

$$u = \frac{1}{\sqrt{\mu_0 \varepsilon_0}} \approx \left(\frac{1}{4\pi \times 10^{-7} \times 1/36\pi \times 10^{-9}}\right)^{1/2}$$

$$u = c \approx 3 \times 10^8 \quad \text{(m/s)} \tag{10.27}$$

Thus, the plane wave velocity is the same as the velocity of light for the medium.

The arbitrary pulse $f(t + z\sqrt{\mu\varepsilon})$ propagating in the $-z$ direction is shown in Figure 10.1(c). Notice that pulse shape is preserved. This behavior is a direct result of the fact that we have no losses in the medium, and pulse shape is usually not preserved when the medium is lossy.

The general solution for E_x (equation 10.22) is a $+z$ traveling wave plus a $-z$ traveling wave. The same thing is true for H_y (equation 10.23). Actually, we could have arrived at these time domain forms from the phasor forms very easily. Take, for example, equation 10.13 (rewritten here to show completely the most general frequency dependence),

$$E_x(z, \omega) = A(\omega)e^{-j\omega\sqrt{\mu\varepsilon}z} + B(\omega)e^{+j\omega\sqrt{\mu\varepsilon}z}$$

The inverse Fourier transform of this function gives the general time domain form

$$E_x(z, t) = A(t - z\sqrt{\mu\varepsilon}) + B(t + z\sqrt{\mu\varepsilon})$$

by means of the time shift relation for Fourier transforms. (See Appendix E.) This is the same as equation 10.22 if $f(t) = A(t)$ and $g(t) = B(t)$!

Returning now to equations 10.16 and 10.17, we should determine precisely what time domain forms these phasors represent because it is the phasor forms we will usually be concerned with. The procedure followed is the usual one. Multiply by $e^{j\omega t}$ and take the real part of the resultant. For equation 10.16 this gives

$$E_x(z, t) = \text{Re}\ \{Ae^{-jkz}e^{j\omega t}\}$$

$$H_y(z, t) = \text{Re}\left\{\frac{A}{\sqrt{\mu/\varepsilon}}\, e^{-jkz}e^{j\omega t}\right\}$$

or, if A is real,

$$E_x(z, t) = A\cos(\omega t - kz)$$

$$H_y(z, t) = \frac{A}{\sqrt{\mu/\varepsilon}}\cos(\omega t - kz) \tag{10.28}$$

or

$$E_x(z, t) = A \cos \omega(t - z\sqrt{\mu\varepsilon})$$
$$H_y(z, t) = \frac{A}{\sqrt{\mu/\varepsilon}} \cos \omega(t - z\sqrt{\mu\varepsilon}) \tag{10.29}$$

Notice that this result is just a special case of equations 10.22 and 10.23! Figure 10.1(d) shows $E_x(z, t)$, or $H_y(z, t)$ except for the multiplying factor $\sqrt{\varepsilon/\mu}$, plotted against z for several successive values of ωt. An observer located at $z = 0$ with an instrument to measure $E_x(0, t)$ would record $A \cos \omega t$. If this same observer decided to take a trip down the positive z axis, such that his or her instrument recorded a constant value of E_x, then the observer must move at a velocity labeled u_p in Figure 10.1(d). Thus, the electric and magnetic field, at right angles to each other, do appear to move, or travel, in a direction at right angles to both of them. In other words, the direction of **E**, **H**, and the direction of propagation form a right-handed orthogonal set in that order. *Notice from equation* 10.28 *that* **E** *and* **H** *are in time phase*, and that $|\mathbf{E}/\mathbf{H}| = \sqrt{\mu/\varepsilon}$.

For equation 10.17, the time domain form is

$$E_x(z, t) = B \cos \omega(t + \sqrt{\mu\varepsilon} z)$$
$$H_y(z, t) = -\frac{B}{\sqrt{\mu/\varepsilon}} \cos \omega(t + \sqrt{\mu\varepsilon} z) \tag{10.30}$$

This electromagnetic wave is also a traveling wave, but it propagates in the negative z direction. Again, **E** and **H** are in time phase, and together with the direction of propagation form a right-handed orthogonal set.

10.2 WAVELENGTH

The distance between two successive points having a phase difference of 2π rad, as shown in Figure 10.1(d), is a wavelength, λ, in meters. Thus,

$$(\omega t - \omega\sqrt{\mu\varepsilon} z_1) - (\omega t - \omega\sqrt{\mu\varepsilon} z_2) = 2\pi$$

or

$$\omega\sqrt{\mu\varepsilon}(z_2 - z_1) = 2\pi = \omega\sqrt{\mu\varepsilon}\lambda$$

Therefore,

$$\boxed{\lambda = \frac{2\pi}{\omega\sqrt{\mu\varepsilon}} = \frac{1}{f\sqrt{\mu\varepsilon}} = \frac{2\pi}{k} \quad \text{(m)}} \tag{10.31}$$

Notice that λ depends on the material parameters μ and ε, as well as the frequency f.

10.3 PHASE VELOCITY

The velocity of a point of constant phase in the z direction is the phase velocity, u_p, in the z direction. Setting the phase equal to a constant and differentiating with time will give the phase velocity. Thus,

$$\omega t - \omega\sqrt{\mu\varepsilon}z = \text{constant}$$

and

$$\frac{d}{dt}(\omega t - \omega\sqrt{\mu\varepsilon z}) = 0$$

or

$$\omega - \omega\sqrt{\mu\varepsilon}\frac{dz}{dt} = 0, \qquad \frac{dz}{dt} = u_p$$

Then,

$$\boxed{u_p = \frac{1}{\sqrt{\mu\varepsilon}} = \frac{\omega}{k} = f\lambda} \quad \text{(m/s)} \tag{10.32}$$

Notice that u_p is also the velocity of a point of constant **E** field (or **H** field) amplitude in Figure 10.1(d).

In a vacuum $\varepsilon = \varepsilon_0 = 8.854 \times 10^{-12} \approx 10^{-9}/36\pi, \mu = \mu_0 \equiv 4\pi \times 10^{-7}$, and $u_p = 2.998 \times 10^8$ m/s $\approx 3 \times 10^8$ m/s, which has previously been recognized as the speed of light in the medium.

10.4 POWER DENSITY

It is convenient at this point to redefine for future use the peak values of the two independent solutions to the wave equation. Let $A \equiv E_1'$ and $B \equiv E_1$. Then, equations 10.16 and 10.17 become

$$E_x = E_1' e^{-jkz}$$

$$H_y = \frac{E_1'}{\sqrt{\mu/\varepsilon}} e^{-jkz}$$

and

$$E_x = E_1 e^{+jkz}$$

$$H_y = -\frac{E_1}{\sqrt{\mu/\varepsilon}} e^{+jkz}$$

for the waves traveling in the positive and negative z direction, respectively.

The time average Poynting vector is given by equation 9.90 as

$$\langle \mathbf{S} \rangle = \tfrac{1}{2}\,\text{Re}\,\{\mathbf{E} \times \mathbf{H}^*\} \ \ (\text{W/m}^2) \tag{9.90}$$

For the wave traveling in the positive z direction, equation 9.90 gives

$$\langle \mathbf{S} \rangle = \mathbf{a}_z \frac{(E_1')^2}{2\sqrt{\mu/\varepsilon}}$$

or simply

$$\langle S_z \rangle = \frac{(E_1')^2}{2\sqrt{\mu/\varepsilon}} \tag{10.33}$$

The interpretation of this result is quite simple. The power density flow is in the positive z direction (with the wave velocity). For the wave traveling in the negative z direction, we have

$$\langle S_z \rangle = -\frac{(E_1)^2}{2\sqrt{\mu/\varepsilon}} \tag{10.34}$$

indicating a power density flow in the negative z direction.

For *any* uniform plane wave the direction of power density flow may be obtained by the "right-hand rule" as indicated in equation 9.90. Rotate the fingers of the right hand from **E** to **H** and the thumb points in the direction of $\langle \mathbf{S} \rangle$ (which is also the direction of wave propagation).

10.5 VELOCITY OF ENERGY FLOW

The time average power flow in watts, $\langle P_f \rangle$, is simply the integral of the normal component of $\langle \mathbf{S} \rangle$ over whatever surface one wishes to investigate. It is the net power flowing through that surface. The time average total energy stored per unit length, $\langle W \rangle$, can be determined as the volume integral (for a length of unity in the propagation direction) of the time average energy densities in the electric and magnetic fields. These densities are discussed in Chapter 9, Section 9.6. From the preceding definitions,

$$\langle P_f \rangle = \iint_s \frac{1}{2} \operatorname{Re} \{\mathbf{E} \times \mathbf{H}^*\} \cdot d\mathbf{s} \quad \text{(W)} \tag{10.35}$$

and

$$\langle W \rangle = \iiint \frac{1}{2} \operatorname{Re} \left\{ \frac{\mathbf{D}}{2} \cdot \mathbf{E}^* + \frac{\mathbf{B}}{2} \cdot \mathbf{H}^* \right\} dx\, dy\, dz \quad \text{(J/m)} \tag{10.36}$$

It is implied in equation 10.36 that the limits of integration are to have a difference of unity for whichever dimension represents the direction of propagation. This ensures a per unit length basis so that $\langle W \rangle$ will have the units J/m.

We can now quite easily *define* the velocity of energy flow as

$$\boxed{u_e \equiv \frac{\langle P_f \rangle}{\langle W \rangle}} \quad \text{(W/J/m or m/s)} \tag{10.37}$$

For the wave traveling in the positive z direction, we may take a cube 1 m on a side and apply equation 10.37. This gives

$$u_e = \frac{\int_0^1\int_0^1 \frac{1}{2}\,\mathrm{Re}\,\{\mathbf{E}\times\mathbf{H}^*\}\cdot\mathbf{a}_z\,dx\,dy}{\int_0^1\int_0^1\int_0^1 \frac{1}{2}\,\mathrm{Re}\left\{\frac{\mathbf{D}}{2}\cdot\mathbf{E}^* + \frac{\mathbf{B}}{2}\cdot\mathbf{H}^*\right\}dx\,dy\,dz}$$

or

$$u_e = \frac{(E_1')^2/2(\sqrt{\mu/\varepsilon})}{(\varepsilon/4)(E_1')^2 + (\mu/4)(\varepsilon/\mu)(E_1')^2}$$

or

$$\boxed{u_e = \frac{1}{\sqrt{\mu\varepsilon}}} \qquad (10.38)$$

Notice that, for the case being considered, $u_e = u_p$. It is *not true in general* that the velocity of energy flow is the same as the phase velocity. This fact will be demonstrated in later sections dealing with the hollow waveguide. The concept of velocity of energy flow is usually reserved for lossless systems.

10.6 WAVE IMPEDANCE

Inspection of equations 10.16 and 10.17 reveals that E_x and H_y are very simply related for the special case that has been treated. In fact,

$$\frac{E_x}{H_y} = \sqrt{\frac{\mu}{\varepsilon}} \quad (\Omega)$$

for the wave traveling in the positive z direction, while

$$\frac{E_x}{H_y} = -\sqrt{\frac{\mu}{\varepsilon}} \quad (\Omega)$$

for the wave traveling in the negative z direction. Moreover, we will show at an appropriate time that

$$\mathbf{E} = \sqrt{\frac{\mu}{\varepsilon}}\,\mathbf{H}\times\mathbf{a}_n \qquad (10.39)$$

for a uniform plane wave traveling in the direction of the general unit vector $\mathbf{a}_n$ in a perfect dielectric region. The parameters, μ and ε, have been assumed to

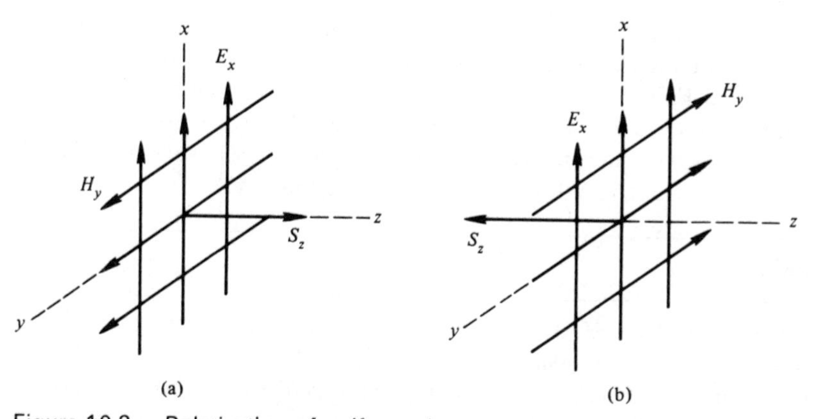

Figure 10.2. Polarization of uniform plane waves with t fixed showing **E** and **H** at $z = 0$ only for clarity. (a) $+z$ traveling wave. (b) $-z$ traveling wave.

be scalar constants to provide a perfect, or *lossless*, dielectric. These parameters are *intrinsic* to the medium through which the wave is propagating. Hence,

$$\boxed{\eta \equiv \sqrt{\frac{\mu}{\varepsilon}}} \quad (\Omega) \tag{10.40}$$

is called the *intrinsic impedance* of the medium. Here, the intrinsic impedance is purely real, but in general (with any loss mechanism in the material) it may be complex, as we shall shortly see. Notice that in free space

$$\eta_0 = \sqrt{\frac{\mu_0}{\varepsilon_0}} \approx 120\pi \approx 377 \quad (\Omega)$$

Another concept that will prove useful to us, particularly for more general waves, is that of *wave impedance*. The *transverse wave impedance* is defined in terms of a particular direction, usually the direction of power density flow. It is given by a ratio of $|\mathbf{E}|$ to $|\mathbf{H}|$ in a manner such that this ratio is greater than 0. **E** and **H** are *transverse* to the particular direction of power density flow. For the wave traveling in the $+z$ direction, we have

$$\eta_z^+ = \frac{E_x}{H_y} = \sqrt{\frac{\mu}{\varepsilon}} = \eta$$

while for the wave traveling in the $-z$ direction we have

$$\eta_z^- = -\frac{E_x}{H_y} = \sqrt{\frac{\mu}{\varepsilon}} = \eta$$

Results as simple as these will not always occur. Another view of the two waves we have been describing is given in Figure 10.2.

The uniform plane wave is an example of a *Transverse ElectroMagnetic mode* or simply a *TEM* mode. Our particular example has been a *TEM to z* mode. This designation implies that both the electric and magnetic fields lie in planes normal to the direction of propagation. Or, in other words, a TEM to z mode has $E_z = H_z = 0$. Of course, we could just as well have described a TEM to x mode, a TEM to y mode, or a mode TEM to any direction. The particular choice we have made in the preceding sections, a wave traveling in the positive or negative z direction, was purely arbitrary.

EXAMPLE 1

Suppose, for example, we have a uniform plane wave propagating in the $+z$ direction (in a lossless medium), and we know that the peak value of the electric field is 10 V/m and is aligned or *polarized* in the positive y direction. Then,

$$E_y = 10e^{+jkz}$$

The right-hand rule gives us the direction of the magnetic field. Thus,

$$H_x = -\frac{10}{\eta}e^{+jkz}$$

If the medium is a dielectric for which $\varepsilon_R = 4$ and $\mu = \mu_0$ and $f =$ 300 MHz, then from equation 10.31

$$\lambda = \frac{1}{f\sqrt{\mu\varepsilon}} = 0.5 \quad \text{(m)}$$

while from equation 10.40,

$$\eta = \sqrt{\frac{\mu}{\varepsilon}} = 60\pi \quad (\Omega)$$

Also,

$$u_p = 1.5 \times 10^8 \quad \text{(m/s)}$$

and

$$u_e = 1.5 \times 10^8 \quad \text{(m/s)}$$

from equations 10.32 and 10.38, respectively. From equation 9.90 the time average power density is

$$\langle S_z \rangle = \frac{100}{120\pi} = 0.265 \quad (\text{W/m}^2)$$

The wave impedances are

$$\eta_z^+ = -\frac{E_y}{H_x} = \sqrt{\frac{\mu}{\varepsilon}} = \eta \approx 60\pi \quad (\Omega)$$

and

$$\eta_z^- = \frac{E_y}{H_x} = \sqrt{\frac{\mu}{\varepsilon}} = \eta \approx 60\pi \quad (\Omega) \qquad \text{(for a} - z \text{ traveling wave)}$$

The equations for the electromagnetic field of a uniform plane wave propagating in *any* direction will be given when oblique incidence on a plane interface is considered in Section 10.16.

10.7 UNIFORM DAMPED PLANE WAVES

We next consider the effects of loss in the medium. This loss may be due to any of several different mechanisms, but the general effect may be observed in a simple manner if we simply allow σ to be nonzero. Loss is considered in more detail in Chapter 13. The medium is still linear, homogeneous, and isotropic, and ρ_v is still zero.

Equations 10.1 through 10.4 become

$$\nabla \times \mathbf{E} = -j\omega\mu\mathbf{H}$$

$$\nabla \times \mathbf{H} = (\sigma + j\omega\varepsilon)\mathbf{E}$$

$$\nabla \cdot \mathbf{D} = 0$$

$$\nabla \cdot \mathbf{B} = 0$$

Following exactly the same procedure as in Section 10.1,

$$\nabla \times (\nabla \times \mathbf{E}) = -j\omega\mu(\nabla \times \mathbf{H})$$

or

$$\nabla \times (\nabla \times \mathbf{E}) = -j\omega\mu(\sigma + j\omega\varepsilon)\mathbf{E}$$

Employing the vector identity on the left side,

$$\nabla(\nabla \cdot \mathbf{E}) - \nabla^2\mathbf{E} = -\gamma^2\mathbf{E}$$

where

$$\gamma^2 \equiv +j\omega\mu(\sigma + j\omega\varepsilon) \tag{10.41}$$

Now the divergence of $\mathbf{E}$ is still zero, so we obtain a complex wave equation,

$$\nabla^2\mathbf{E} - \gamma^2\mathbf{E} = 0$$

If we assume (again) that $E_y = E_z \equiv 0$, and there are no variations with x or y (for a uniform plane wave), then we obtain

$$\frac{d^2E_x}{dz^2} - \gamma^2 E_x = 0$$

The solution to this equation is well known, as is the corresponding equation for $\mathbf{H}$. The reader can easily verify that these solutions are

$$E_x(z, \omega) = Ae^{-\gamma z} + Be^{+\gamma z} \tag{10.42}$$

and

$$H_y(z, \omega) = \frac{Ae^{-\gamma z}}{j\omega\mu/\gamma} - \frac{B}{j\omega\mu/\gamma}e^{+\gamma z} \tag{10.43}$$

We define (see equation 10.41)

$$\gamma \equiv \alpha + j\beta = \sqrt{j\omega\mu(\sigma + j\omega\varepsilon)} = j\omega\sqrt{\mu\varepsilon}\sqrt{1 - \frac{j\sigma}{\omega\varepsilon}} \tag{10.44}$$

or

$$\alpha = \text{Re}\,(\gamma) = \text{Re}\left\{\sqrt{-\omega^2\mu\varepsilon\left(1 - j\frac{\sigma}{\omega\varepsilon}\right)}\right\} \quad (\text{Np/m}) \tag{10.45}$$

and

$$\beta = \text{Im}\,(\gamma) = \text{Im}\left\{\sqrt{-\omega^2\mu\varepsilon\left(1 - j\frac{\sigma}{\omega\varepsilon}\right)}\right\} \quad (\text{rad/m}) \tag{10.46}$$

α is called the *attenuation constant*, β the *phase constant*, and γ the *complex propagation constant*. One neper equals 8.69 dB and is actually dimensionless, as is the radian. See Problem 25. After some rather tedious complex algebra, equations 10.45 and 10.46 reduce to (see Problem 9)

$$\alpha = \omega\sqrt{\mu\varepsilon}\left[\frac{1}{2}\left(\sqrt{1 + \left(\frac{\sigma}{\omega\varepsilon}\right)^2} - 1\right)\right]^{1/2} \geq 0 \tag{10.47}$$

and

$$\beta = \omega\sqrt{\mu\varepsilon}\left[\frac{1}{2}\left(\sqrt{1 + \left(\frac{\sigma}{\omega\varepsilon}\right)^2} + 1\right)\right]^{1/2} > 0 \tag{10.48}$$

Notice that equation 10.44 may be the best form for numerical calculation.

Equations 10.42 and 10.43 may now be written as

$$E_x(z, \omega) = Ae^{-\alpha z}e^{-j\beta z} + Be^{+\alpha z}e^{+j\beta z} \tag{10.49}$$

and

$$H_y(z, \omega) = \frac{A}{\eta}e^{-\alpha z}e^{-j\beta z} - \frac{B}{\eta}e^{+\alpha z}e^{+j\beta z} \tag{10.50}$$

where

$$\eta = \frac{j\omega\mu}{\gamma} = \frac{\omega\mu}{\alpha^2 + \beta^2}(\beta + j\alpha) \quad (\Omega) \tag{10.51}$$

η is complex as long as σ is not zero. Interpretation of these equations is easy. The waves are traveling waves, but they are *damped* exponentially as they progress (*in either direction*). This damping is, of course, due to the finite conductivity of the medium which is the *only* loss mechanism (heating) present. The loss shows up in another way also, since **E** and **H** are no longer in phase for either the $+z$ traveling wave or the $-z$ traveling wave. That is, η is now complex, as predicted in Section 10.6.

The damped electromagnetic wave that is propagating in the positive z direction is

$$E_x(z, \omega) = Ae^{-\alpha z}e^{-j\beta z}$$
$$H_y(z, \omega) = \frac{A}{\eta} e^{-\alpha z}e^{-j\beta z} \tag{10.52}$$

whose time domain form is

$$E_x(z, t) = Ae^{-\alpha z} \cos(\omega t - \beta z)$$
$$H_y(z, t) = Ae^{-\alpha z} \operatorname{Re}\left\{\frac{1}{\eta} e^{j(\omega t - \beta z)}\right\} = Ae^{-\alpha z} \frac{1}{|\eta|} \cos\left(\omega t - \beta z - \tan^{-1}\frac{\alpha}{\beta}\right) \tag{10.53}$$

The phase velocity can be found by setting the phase of $E_x(z, t)$ equal to a constant and differentiating (as before). This results in

$$\boxed{u_p = \frac{\omega}{\beta}} \tag{10.54}$$

a *general result*, which reduces to equation 10.32 for the lossless case. The wavelength can be found (as before) to be

$$\boxed{\lambda = \frac{2\pi}{\beta}} \tag{10.55}$$

a general result, which reduces to equation 10.31 for the lossless case.

EXAMPLE 2

Consider the possibility of propagating a plane wave through seawater ($\sigma = 4$, $\varepsilon_R = 81$). We want to find α, β, u_p, and λ at 1 MHz. Direct calculation gives $\sigma/\omega\varepsilon = 889$ while equation 10.47 gives

$$\alpha = 3.972 \quad (\text{Np/m}), \qquad 34.5 \quad (\text{dB/m})$$

From equation 10.48

$$\beta = 3.976 \quad (\text{rad/m})$$

From equation 10.54

$$u_p = 1.58 \times 10^6 \quad (\text{m/sec})$$

From equation 10.55

$$\lambda = 1.58 \quad (\text{m})$$

At 100 KHz the same calculations give

$$\alpha = 1.26 \quad (\text{m}^{-1}), \qquad \beta = 1.26 \quad (\text{m}^{-1}), \qquad u_p = 0.501 \times 10^6 \quad (\text{m/sec}),$$
$$\lambda = 5.01 \quad (\text{m})$$

This example should make the reader wonder about some things. Which frequency would a submariner choose? Should there not be simpler formulas to use? The last question is answered in the next two sections. Another question that may not have occurred to all of us, is what is the effect of the different phase velocities for the two frequencies in the example? An obvious answer is that the two plane waves encounter very different time delays. The medium is said to be *dispersive* because of the loss.

Suppose a wave is propagating in a lossy or dispersive medium such as that in the preceding example. Further, suppose that this wave is the result of a modulated signal or perhaps it is a repetitive pulse. Using the methods of Fourier analysis we can decompose the signal into its sinusoidal components and each component (a phasor) produces a wave. Together, all the waves produce the composite wave, which is propagating. Notice particularly that the various parts do not propagate with the same velocity, and the wave will be dispersed. The result of this after propagation over some distance is that if the modulating signal is extracted or demodulated, it will no longer be the original modulating signal. It is distorted. In the case of a pulse, its shape will be changed. This distortion is due to the dispersion, but, more particularly, it is a direct result of the fact that the phase velocity is frequency dependent, or, since $u_p = \omega/\beta$, it is a result of the fact that β, as in equation 10.48, is not a linear function of frequency.

These ideas can often be expressed more concisely if a quantity called the group velocity, u_g, is introduced. In order to obtain an equation for the group velocity, consider a wave in a lossless medium traveling in the $+z$ direction and polarized in the x direction and consisting of two parts with equal amplitudes. Let the two frequencies be different (as in the discussion above), and let β be some general function of ω (as in the discussion above). One frequency is $\omega_0 - \Delta\omega$, and the other is $\omega_0 + \Delta\omega$, while the two phase constants are $\beta_0 - \Delta\beta$ and $\beta_0 + \Delta\beta$, respectively. Thus, we have

$$\begin{aligned} E_x = {} & E_0 \cos\left[(\omega_0 - \Delta\omega)t - (\beta_0 - \Delta\beta)z\right] \\ & + E_0 \cos\left[(\omega_0 + \Delta\omega)t - (\beta_0 + \Delta\beta)z\right] \end{aligned}$$

Using some trigonometry and rearranging,

$$E_x = 2E_0 \cos\left[\Delta\omega\left(t - \frac{\Delta\beta}{\Delta\omega} z\right)\right] \cos\left[\Delta\omega\left(t - \frac{\beta_0}{\omega_0} z\right)\right]$$

In this form, the wave is the product of two traveling waves, the second cosine term being a wave traveling with velocity ω_0/β_0, while the first cosine term is a wave traveling with velocity $(\Delta\omega)/(\Delta\beta)$. Thinking of the first term as being like a modulation envelope, the "envelope" of the overall wave moves with velocity $(\Delta\omega)/(\Delta\beta)$ and the "carrier" moves with velocity ω_0/β_0. The envelope velocity is defined as the group velocity for *small* excursions. That is,

$$u_g \equiv \lim_{\Delta\omega \to 0} \frac{\Delta\omega}{\Delta\beta} = \frac{d\omega}{d\beta} = \left(\frac{d\beta}{d\omega}\right)^{-1} \quad \text{(group velocity)}$$

The last form is usually most useful, as equation 10.48 shows.

Thus, if the dispersion in frequency is small, a group of waves will have an envelope which travels at the group velocity. The "signal" travels at the group velocity, not the phase velocity. If the dispersion in frequency is large, the concept is rather meaningless.[2] The *group time delay* over a distance l is

$$\tau_D = \frac{l}{u_g} = l\frac{d\beta}{d\omega} \quad \text{(group time delay)}$$

A somewhat analogous situation to that above occurs when a caterpillar is propagating itself along. The caterpillar is moving at group velocity, while the waves which ripple along its back from tail to head move at the phase velocity.[3]

10.8 GOOD DIELECTRICS

It is customary to consider two special cases explicitly. The dividing line between these cases occurs at the *frequency* at which conduction current density equals displacement current density in magnitude, or $\sigma = \omega\varepsilon$ (see equation 10.2). A "good" dielectric is a material for which displacement currents dominate, or $\omega\varepsilon \gg \sigma$. Notice that this means that even copper, normally considered a good conductor, may be a good dielectric if the frequency is high enough! This is one way to explain x-ray penetration of metals. If $\omega\varepsilon \gg \sigma$, but σ is not zero, and we would like to include its first order effects, then binomial expansions of equations 10.47, 10.48, and 10.51 are in order. That is, we use the fact that $(1 + x)^n \approx 1 + nx + [n(n-1)/2]x^2$, $x \ll 1$. From equation 10.47 we find that

$$\alpha \approx \omega\sqrt{\mu\varepsilon}\left[\frac{1}{2}\left(\frac{\sigma}{\omega\varepsilon}\right)\right] \tag{10.56}$$

while from equation 10.48 we have

$$\beta \approx \omega\sqrt{\mu\varepsilon}\left[1 + \frac{1}{8}\left(\frac{\sigma}{\omega\varepsilon}\right)^2\right] \tag{10.57}$$

Equation 10.51 gives

$$\eta \approx \sqrt{\frac{\mu}{\varepsilon}}\left[1 - \frac{3}{8}\left(\frac{\sigma}{\omega\varepsilon}\right)^2\right] + j\sqrt{\frac{\mu}{\varepsilon}}\left[\frac{1}{2}\left(\frac{\sigma}{\omega\varepsilon}\right)\right] \tag{10.58}$$

[2] This will be discussed again in Section 10.18.

[3] See Skilling, in the references at the end of Chapter 3.

Notice that $\sigma/\omega\varepsilon$ (called the *loss tangent*) has been kept intact because we probably will calculate it first to verify that it is, in fact, much less than one. In Chapter 9 we called $\hat{y} = \sigma + j\omega\varepsilon$ the admittivity. The phase velocity and wavelength can be determined from equations 10.54 and 10.55, respectively.

10.9 GOOD CONDUCTORS

For "good" conductors, conduction currents dominate, and so $\sigma \gg \omega\varepsilon$. In this case, equations 10.47 and 10.48 give

$$\alpha = \beta = \sqrt{\frac{\omega\mu\sigma}{2}} \tag{10.59}$$

while equation 10.54 gives

$$u_p = \sqrt{\frac{2\omega}{\mu\sigma}}, \qquad (=\tfrac{1}{2}u_g) \tag{10.60}$$

and equation 10.51 gives

$$\eta = \sqrt{\frac{\omega\mu}{\sigma}}\, e^{j\pi/4} = \sqrt{\frac{\omega\mu}{2\sigma}} + j\sqrt{\frac{\omega\mu}{2\sigma}} \tag{10.61}$$

It should be pointed out that the attenuation at reasonable frequencies is very high. For example, a plane wave propagating at 100 MHz in copper ($\sigma = 5.8 \times 10^7$), ($\mu = \mu_0 = 4\pi \times 10^{-7}$) is attenuated at the rate (α) of 15.2×10^4 Np/m or about 13.2×10^5 dB/m! At 10 GHz (x-band) these numbers are only increased by a factor of 10.

10.10 REFLECTION OF PLANE WAVES

It is helpful in viewing certain phenomena in transmission lines and other guiding systems to have a feeling for conditions that occur when a plane wave strikes a plane interface separating two different media. The simplest case, that of normal incidence, is sufficient to not only utilize previously established boundary conditions, but to introduce new parameters that will be very useful. For the time being, at least, we will only consider nonmagnetic media ($\mu = \mu_0$), permittivities that are simple scalar constants, and conductivities that are scalar constants. These assumptions still allow sufficient freedom to investigate the most important cases. Oblique incidence is considered in Section 10.16 of this chapter.

10.11 DIELECTRIC-CONDUCTOR INTERFACE

Consider a situation as shown in Figure 10.3. It is ultimately more convenient, but certainly not necessary, to choose the positive (or increasing) z direction as shown. For $z > 0$ the medium is a loss-free dielectric having parameters μ_0,

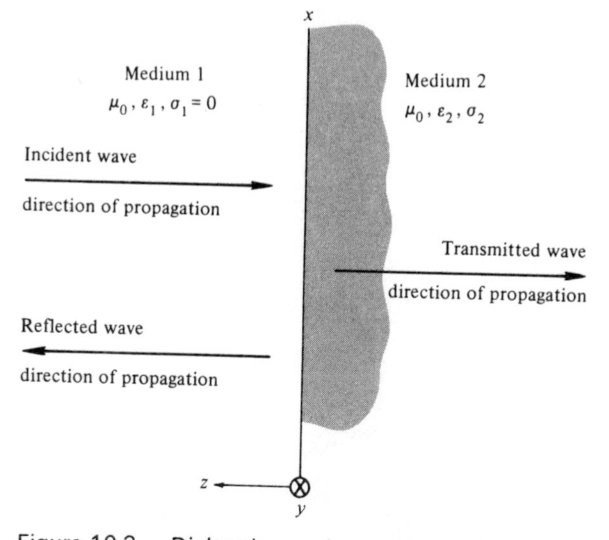

Figure 10.3. Dielectric-conductor plane interface.

ε_1, and $\sigma_1 = 0$, while for $z < 0$, the medium is a lossy dielectric having parameters μ_0, ε_2, and σ_2. Suitable representations for a uniform plane wave for $z > 0$ are certainly given by equations 10.13 and 10.15 with $A \equiv E_1'$ and $B \equiv E_1$.

$$E_x = E_1 e^{+jk_1 z} + E_1' e^{-jk_1 z}, \qquad z > 0 \tag{10.62}$$

and

$$H_y = -\frac{E_1}{\eta_1} e^{+jk_1 z} + \frac{E_1'}{\eta_1} e^{-jk_1 z}, \qquad z > 0 \tag{10.63}$$

We have implicitly assumed that $E_y = E_z \equiv 0$, as before. Notice that subscripts refer to medium 1. The interpretation of equations 10.62 and 10.63 is as follows. The first terms of E_x and H_y represent an undamped electromagnetic wave traveling in the negative z direction which is incident on the interface. This wave is called the *incident* wave. The second terms are an undamped electromagnetic wave traveling in the positive z direction, representing the *reflected wave*. Since at least part of the incident wave is apparently reflected, we suspect that part is also transmitted into medium 2. A suitable representation for the field in medium 2 must be a damped electromagnetic wave traveling in the negative z direction. The *transmitted wave* must have the same polarization as the incident and reflected waves in order to preserve continuity of tangential components at the interface, as required by equations 9.91 and 9.92. That is, if $\mathbf{E}$ (for $z < 0$) has a component other than E_x at $z = 0$, it cannot be canceled by the field from the other side at $z = 0$. It must be of the form $e^{+\gamma_2 z}$ (e.g., not $e^{+\gamma_2 y}$) in order to reduce to a constant at $z = 0$ and allow the cancellation to occur. No new frequencies are created by linear media, so the transmitted wave will be at the same frequency (in γ_2) as the incident and reflected waves. Thus,

$$E_x = E_2 e^{+\alpha_2 z} e^{+j\beta_2 z}, \qquad z < 0 \tag{10.64}$$

and

$$H_y = \frac{-E_2}{\eta_2} e^{+\alpha_2 z} e^{+j\beta_2 z}, \qquad z < 0 \tag{10.65}$$

where α_2, β_2, and η_2 are determined from appropriate forms depending on $\sigma_2/\omega\varepsilon_2$.

Setting $z = 0$ and equating tangential components gives

$$E_x\big|_{z=0^+} = E_x\big|_{z=0^-}$$

and

$$H_y\big|_{z=0^+} = H_y\big|_{z=0^-}$$

or

$$E_1 + E_1' = E_2 \tag{10.66}$$

and

$$-\frac{E_1}{\eta_1} + \frac{E_1'}{\eta_1} = -\frac{E_2}{\eta_2} \tag{10.67}$$

In equations 10.66 and 10.67 there are apparently three unknowns, E_1, E_1', and E_2. E_1 is not actually an unknown because the peak value of the incident wave is usually a given quantity. In other words, the incident wave must have come from some source to the left ($z \gg 0$), and the incident wave assumes the role of forcing function. In any case, the logical procedure is to solve for E_1' and E_2 in terms of E_1. This results in

$$E_1' = E_1 \frac{\eta_2 - \eta_1}{\eta_2 + \eta_1} \tag{10.68}$$

and

$$E_2 = E_1 \frac{2\eta_2}{\eta_2 + \eta_1} \tag{10.69}$$

The *coefficient of reflection* is defined as the ratio of the reflected to incident electric fields at $z = 0$, or

$$\boxed{\Gamma \equiv \frac{E_1'}{E_1} = \frac{\eta_2 - \eta_1}{\eta_2 + \eta_1} = |\Gamma| e^{j\phi_\kappa}} \tag{10.70}$$

and the *coefficient of transmission* is defined as the ratio of the transmitted to incident electric field at $z = 0$, or

$$\boxed{T \equiv \frac{E_2}{E_1} = \frac{2\eta_2}{\eta_2 + \eta_1}} \tag{10.71}$$

The field equations may be written in a more convenient form with these definitions. Then,

$$E_x = E_1 e^{jk_1 z}(1 + |\Gamma| e^{j(\phi_K - 2k_1 z)}), \qquad z > 0 \tag{10.72}$$

$$H_y = -\frac{E_1}{\eta_1} e^{jk_1 z}(1 - |\Gamma| e^{j(\phi_K - 2k_1 z)}), \qquad z > 0 \tag{10.73}$$

and

$$E_x = TE_1 e^{\alpha_2 z} e^{j\beta_2 z}, \qquad z < 0 \tag{10.74}$$

$$H_y = -\frac{TE_1}{\eta_2} e^{\alpha_2 z} e^{j\beta_2 z}, \qquad z < 0 \tag{10.75}$$

EXAMPLE 3

Suppose a 100-MHz plane wave is normally incident on a plane copper interface from air. If the incident **E** field has an rms value of 1 V/m, how large are the reflected and transmitted waves?

From equation 10.61 ($\sigma_{cu} = \sigma_2 = 5.8 \times 10^7$, $\varepsilon_{cu} = \varepsilon_2 = \varepsilon_0 = (1/36\pi) \times 10^{-9}$)

$$\eta_{cu} = \eta_2 = 2.61 \times 10^{-3} + j2.61 \times 10^{-3} \quad (\Omega)$$

From equation 10.71

$$\Gamma = \frac{2.61 \times 10^{-3} + j2.61 \times 10^{-3} - 377}{2.61 \times 10^{-3} + j2.61 \times 10^{-3} + 377} \approx -1$$

and from equation 10.71

$$T = \frac{5.22 \times 10^{-3} + j5.22 \times 10^{-3}}{2.61 + 10^{-3} + j2.61 \times 10^{-3} + 377} \approx 19.6 \times 10^{-6} \underline{|45^\circ}$$

So, the incident field is essentially totally reflected!

Inspection of the terms in parentheses in equations 10.72 or 10.73 reveals that $|E_x|$(or $|H_y|$) varies between maxima and minima ($z > 0$) in what is called a *standing wave pattern.* In this regard, it is convenient to define the *standing wave ratio* as

$$\text{SWR} \equiv \frac{|E_x|_{\max}}{|E_x|_{\min}} = \frac{|1 + |\Gamma| e^{j(\phi_K - 2k_1 z)}|_{\max}}{|1 + |\Gamma| e^{j(\phi_K - 2k_1 z)}|_{\min}} \tag{10.76}$$

$|E_x|_{max}$ obviously occurs when $\phi_k - 2k_1 z = -2n\pi, n = 0, 1, 2, \ldots$, therefore, $|E_x|_{max}$ occurs at $z_{max} = (\phi_k + 2n\pi)/2k_1$, $z_{max} > 0$. $|E_x|_{min}$ occurs when $\phi_k - 2k_1 z = -(2n - 1)\pi$, $n = 0, 1, 2, \ldots$, so $|E|_{min}$ occurs at $z_{min} = [\phi_k + (2n - 1)\pi]/2k_1$, $z_{min} > 0$. We must agree to use only positive values of ϕ_k, and, in any case, must use only positive values of z_{max} or z_{min}. It is easy to show (with equation 10.76) that

$$\boxed{\text{SWR} = \frac{1 + |\Gamma|}{1 - |\Gamma|}} \tag{10.77}$$

The distance between successive peaks ($|E|_{max}$) or valleys ($|E|_{min}$) corresponds to a change in phase of 2π rad of the numerator or denominator, respectively, of equation 10.76. Then,

$$2k_1(z_2 - z_1) = 2\pi$$

or

$$z_2 - z_1 = \frac{\pi}{k_1} = \frac{\lambda_1}{2} \tag{10.78}$$

where we have used equation 10.31. Thus, the spacing between adjacent maxima (or minima) is one-half wavelength, and either a maximum or minimum will always lie less than, or at most equal to, one-quarter wavelength from the interface at $z = 0$. It is helpful to examine Figure 10.4 to observe the quantities we are presently concerned with.

It is worthwhile to digress and write an equation for the reflected wave in terms of equations 10.62 and 10.70. It is

$$E_x^{\text{ref}} = E_1' e^{-jk_1 z} = \Gamma E_1 e^{-jk_1 z}$$

or, being more explicit about ω,

$$E_x^{\text{ref}}(z, \omega) = \Gamma(\omega) E_1(\omega) e^{-j\omega\sqrt{\mu_0 \varepsilon_1} z}$$

To find the general time domain form of the reflected wave, in case we have a nonsinusoidal incident wave, we must find the inverse Fourier transform of the preceding equation. It is (using the time-shift relation again)

$$E_x^{\text{ref}}(z, t) = \mathscr{F}^{-1}\{\Gamma(\omega) E_1(\omega)\}_{t \to t - \sqrt{\mu_0 \varepsilon_1} z}$$

But, the inverse transform of a product can be found by convolution (see Appendix E) so,

$$E_x^{\text{ref}}(z, t) = \int_{-\infty}^{\infty} \Gamma(\tau) E_1(t - \tau)\, d\tau \Big|_{t = t - \sqrt{\mu_0 \varepsilon_1} z} \tag{10.79}$$

where

$$\Gamma(t) = \frac{1}{2\pi} \int_{-\infty}^{\infty} \Gamma(\omega) e^{jt\omega}\, d\omega \tag{10.80}$$

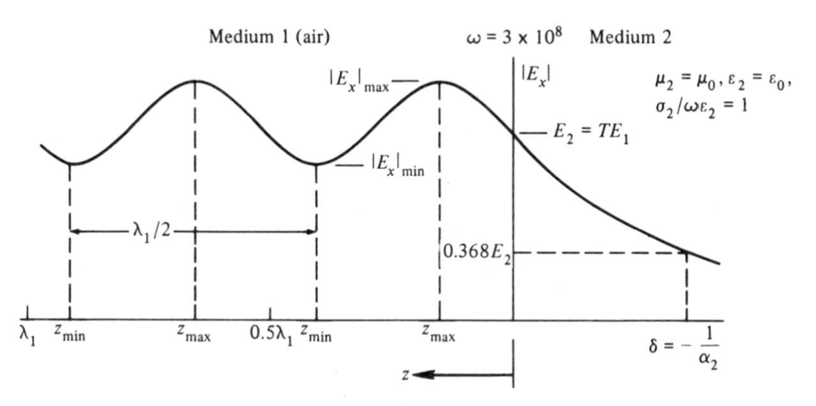

Figure 10.4. Reflection and transmission at a dielectric-conductor interface, $|E_x|$ versus z. Medium 1, air: Medium 2, $\sigma_2/\omega\varepsilon_2 = 1$.

and

$$\Gamma(\omega) = \frac{n_2(\omega) - n_1(\omega)}{n_2(\omega) + n_1(\omega)} \tag{10.81}$$

A general procedure, then, is to find the unit impulse response of the coefficient of reflection, equation 10.80, and then use this result in equation 10.79 [having been given $E_1(t)$]. The transmitted wave can be treated in exactly the same manner, but it may be a much more complicated problem to solve if loss is present, as in the current example. That is, both α and β may be frequency dependent and the resulting integrals for finding the inverse transforms may be very cumbersome. The presence of loss upsets our relatively simple picture of traveling waves. This will be inspected in the next section.

10.12 SKIN DEPTH AND SURFACE IMPEDANCE

It is customary to define the "skin depth," or depth of penetration, δ (not to be confused with the impulse function), as that distance for which the transmitted wave has been attenuated to $1/e = e^{-1}$ of its entering value at $z = 0$. From equation 10.74,

$$e^{+\alpha_2\delta} = e^{-1}$$

or

$$\delta = -\frac{1}{\alpha_2} \quad \text{(m)} \tag{10.82}$$

a negative number (since $z < 0$) for the transmitted wave we are dealing with here. A plot of $|E_x|$ versus z for an arbitrary choice of σ and ε is shown in Figure 10.4.

The concept of surface impedance is now introduced. We have previously called η_2 the intrinsic impedance of medium 2, but it may also be interpreted as a *surface impedance*. This concept will be explained in what follows. Using equation 10.51, we have

$$\eta_2 = \frac{j\omega\mu_2}{\gamma_2} = \frac{\omega\mu_2}{|\gamma_2|^2}(\beta_2 + j\alpha_2) \equiv R_s + jX_s \tag{10.83}$$

where R_s and X_s are the surface resistance and reactance, respectively. The time average power density *in the transmitted wave is* $-\langle S_z\rangle = -\frac{1}{2}\,\mathrm{Re}\,\{E_x H_y^*\}$, and its entering value at $z = 0$ (using equations 10.64, 10.65, and 10.83)

$$\begin{aligned} -\langle S_z\rangle\big|_{z=0} &= -\tfrac{1}{2}\,\mathrm{Re}\,\{E_x H_y^*\}\big|_{z=0} \\ &= +\frac{1}{2}E_2^2\,\mathrm{Re}\left\{\frac{1}{\eta_2^*}\right\} = \frac{1}{2}E_2^2\frac{R_s}{|\eta_2|^2} \\ &= \frac{1}{2}E_2^2\frac{\beta_2}{\omega\mu_2} = \frac{1}{2}|H_y|_{z=0}^2 R_s \end{aligned} \tag{10.84}$$

must ultimately appear as heat in lossy region 2. Notice the similarity between the last form and results from circuit theory. The reason for the terminology, *surface impedance*, should now be obvious. The next paragraph will clarify these ideas even more.

As shown in Figure 10.5, there is a current density in the x direction given by the point form of Ohm's law,

$$J_x = \sigma_2 E_x = \sigma_2 E_2 e^{\alpha_2 z} e^{j\beta_2 z} \tag{10.85}$$

The current itself in the differential strip shown is $J_x\,dz(1)$, so

$$dI_x = \sigma_2 E_2 e^{\alpha_2 z} e^{j\beta_2 z}\,dz(1)$$

The time average power dissipated in this strip is the effective current squared times the *effective* resistance, so

$$\langle dP\rangle = \left|dI_x\right|_{\text{eff}}^2 R_{\text{eff}}$$

but

$$R_{\text{eff}} = \frac{\text{length}}{\sigma_2(\text{area})} = \frac{1}{\sigma_2\,dz(1)} = \frac{1}{\sigma_2\,dz}$$

and

$$|dI_x|_{\text{eff}} = \frac{\sigma_2 E_2 e^{+\alpha_2 z}}{\sqrt{2}}\,dz$$

Therefore,

$$\langle dP\rangle = \frac{\sigma_2 E_2^2 e^{2\alpha_2 z}}{2}\,dz$$

or

$$\langle P\rangle = \frac{1}{2}\sigma_2 E_2^2 \int_{-\infty}^{0} e^{2\alpha_2 z}\,dz = \frac{\sigma_2 E_2^2}{4\alpha_2} \tag{10.86}$$

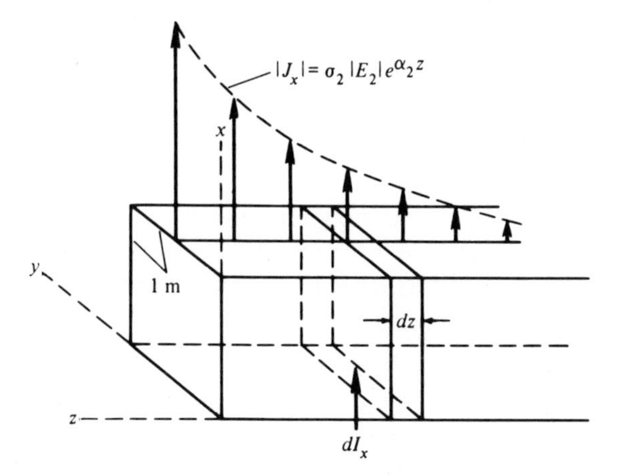

Figure 10.5. Nonuniform current density flowing in conducting region 2.

This is the average power per unit area dissipated in the medium, $z < 0$, and must equal the average power density entering at $z = 0$ which is given by any of the forms in equation 10.84. Thus, we are required to show that

$$\frac{\sigma_2 E_2^2}{4\alpha_2} = \frac{\beta_2 E_2^2}{2\omega\mu_2}$$

or (without subscripts)

$$\omega\mu\sigma = 2\beta\alpha$$

and this is easy to do with equations 10.47 and 10.48. Thus, the average power density is the same, regardless of which way we calculate it.

Next, suppose that instead of the actual exponentially distributed current we have a *uniform* distribution of current to a depth δ (only). The effective resistance in this case is

$$R_{\text{eff}} = \frac{1}{\sigma_2 \delta \cdot 1} = \frac{\alpha_2}{\sigma_2}$$

and the *uniform* current is calculated as

$$I_{\text{eff}} = \frac{1}{\sqrt{2}} \int_{-\infty}^{0} dI_x = \frac{1}{\sqrt{2}} \int_{-\infty}^{0} \sigma_2 E_2 e^{(\alpha_2 + j\beta_2)z} \, dz \cdot 1$$

or

$$I_{\text{eff}} = \frac{\sigma_2 E_2}{\sqrt{2}} \frac{1}{\gamma_2}$$

Then,

$$\langle P \rangle = |I_{\text{eff}}|^2 R_{\text{eff}} = \frac{\sigma_2 \alpha_2 E_2^2}{2|\gamma_2|^2} \tag{10.87}$$

The concept of skin depth is only useful for good conductors, in which case $|\gamma_2|^2 = 2\alpha_2^2$ by equation 10.59, and

$$\langle P \rangle = \frac{\sigma_2 E_2^2}{4\alpha_2}$$

which is the same as equation 10.86. We can now interpret δ as the depth required for a *uniform current density* to give the same power density as the actual distribution. Using equations 10.59, 10.61, 10.82, and 10.83 we have

$$\boxed{R_s = R_{\text{eff}} = \frac{1}{\sigma\delta}} \tag{10.88}$$

where

$$\boxed{\delta = \sqrt{\frac{2}{\omega\mu\sigma}} = \sqrt{\frac{1}{\pi f \mu\sigma}}} \quad \text{(m)} \tag{10.89}$$

Notice that equation 10.89 shows that for good conductors (using equation 10.59) the *depth of penetration of the field* is equal to $1/\alpha_2$. These results have been derived for plane conductors, but we will continue to use them for curved surfaces so long as δ, as calculated by equation 10.89, is much less than the radius of curvature. The surface impedance for a good conductor is given by equations 10.61 and 10.83.

$$\boxed{Z_s = R_s + jX_s = \sqrt{\frac{\pi f \mu}{\sigma}} + j\sqrt{\frac{\pi f \mu}{\sigma}} = \frac{1+j}{\sigma\delta}} \quad (\Omega) \tag{10.90}$$

We now have more than one method for calculating loss due to conductivity. The basic formula is Joule's law which appears in Poynting's theorem, equation 9.88. It is

$$\langle P_d \rangle = \frac{1}{2} \text{Re} \left\{ \iiint_{\text{vol}} \mathbf{J} \cdot \mathbf{E}^* \, dv \right\}$$

and by Ohm's law,

$$\langle P_d \rangle = \frac{\sigma}{2} \iiint_{\text{vol}} |\mathbf{E}|^2 \, dv = \frac{1}{2\sigma} \iiint_{\text{vol}} |\mathbf{J}|^2 \, dv \quad \text{(W)} \tag{10.91}$$

Equation 10.91 is general. It can be used to calculate power loss in a good dielectric or a good conductor. We *must* use this equation to calculate the loss in a good dielectric. If we are working with a good conductor, equation 10.91 still applies, and the difference between the upper and lower limit for one of the variables will always be δ, the skin depth. On the other hand, for the good conductor case, we can also use

$$\langle P_d \rangle = \frac{R_s}{2} \iint_s \left| \mathbf{J}_s \right|_s^2 ds = \frac{R_s}{2} \iint_s \left| \mathbf{H}_{\tan} \right|_s^2 ds \quad \text{(W)} \tag{10.92}$$

The equivalence of equations 10.91 and 10.92 for the good conductor case was established in the preceding paragraphs. We will use both in lossy guiding systems.

As an example of the practical importance of the skin effect, consider the outdoor television receiving antenna which is usually constructed of hollow tubular aluminum conductors. This construction is mechanically better able to withstand wind loading, and there is no need electrically to have solid conductors anyway because of the skin effect. There is essentially no current in the interior of the conductors.

10.13 DIELECTRIC-DIELECTRIC INTERFACE

A dielectric-dielectric interface is shown in Figure 10.6. This situation is easy to handle because we already have the necessary tools.

From equation 10.70,

$$\Gamma = \frac{\eta_2 - \eta_1}{\eta_2 + \eta_1} = \frac{1 - \eta_1/\eta_2}{1 + \eta_1/\eta_2}, \qquad z > 0$$

but

$$\frac{\eta_1}{\eta_2} = \frac{\sqrt{\mu_0/\varepsilon_1}}{\sqrt{\mu_0/\varepsilon_2}} = \sqrt{\frac{\varepsilon_2}{\varepsilon_1}}$$

so that

$$\Gamma = \frac{1 - \sqrt{\varepsilon_2/\varepsilon_1}}{1 + \sqrt{\varepsilon_2/\varepsilon_1}}, \qquad z > 0$$

If $\varepsilon_2 > \varepsilon_1$, then the region $z < 0$ is said to be more "dense" than that for $z > 0$, and

$$|\Gamma| = \frac{\sqrt{\varepsilon_2/\varepsilon_1} - 1}{\sqrt{\varepsilon_2/\varepsilon_1} + 1}, \qquad \varepsilon_2 > \varepsilon_1$$

The standing wave ratio, given by equation 10.77, becomes

$$\text{SWR} = \sqrt{\frac{\varepsilon_2}{\varepsilon_1}}, \qquad \varepsilon_2 > \varepsilon_1 \tag{10.93}$$

If $\varepsilon_1 > \varepsilon_2$, then

$$|\Gamma| = \frac{1 - \sqrt{\varepsilon_2/\varepsilon_1}}{1 + \sqrt{\varepsilon_2/\varepsilon_1}}$$

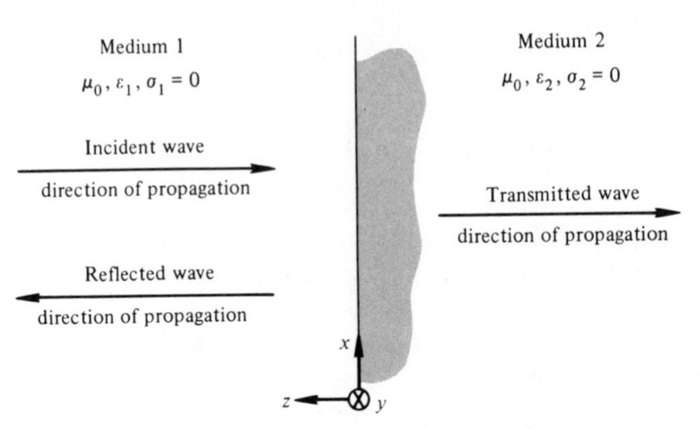

Figure 10.6. Dielectric-dielectric plane interface.

and

$$\text{SWR} = \sqrt{\frac{\varepsilon_1}{\varepsilon_2}}, \qquad \varepsilon_1 > \varepsilon_2 \tag{10.94}$$

At the interface, $z = 0$, the electric field is

$$E_2 = TE_1 = \frac{2E_1}{1 + \sqrt{\varepsilon_2/\varepsilon_1}} \tag{10.95}$$

EXAMPLE 4

As an example suppose that $\varepsilon_2 = 4\varepsilon_1$. Then,

$$\Gamma = \frac{1-2}{1+2} = -\frac{1}{3} = \frac{1}{3}e^{j\pi}$$

$$T = \frac{2}{3}$$

and

$$\text{SWR} = 2$$

From equations 10.72 through 10.75,

$$E_x = E_1 e^{jk_1 z}\left(1 + \frac{1}{3}e^{j(\pi - 2k_1 z)}\right), \qquad z > 0$$

$$H_y = -\frac{E_1}{\eta_1} e^{jk_1 z}\left(1 - \frac{1}{3}e^{j(\pi - 2k_1 z)}\right), \qquad z > 0$$

$$E_x = \frac{2E_1}{3} e^{jk_2 z}, \qquad z < 0$$

$$H_y = -\frac{2E_1}{3\eta_2} e^{jk_2 z}, \qquad z < 0$$

and

$$k_2 = \omega\sqrt{\mu_0 \varepsilon_2} = 2k_1 = \frac{4\pi}{\lambda_1} = \frac{2\pi}{\lambda_2}$$

and

$$\eta_2 = \sqrt{\frac{\mu_0}{\varepsilon_2}} = \frac{\eta_1}{2}$$

A plot of $|E_x|$ and $|H_y|$ is shown in Figure 10.7.

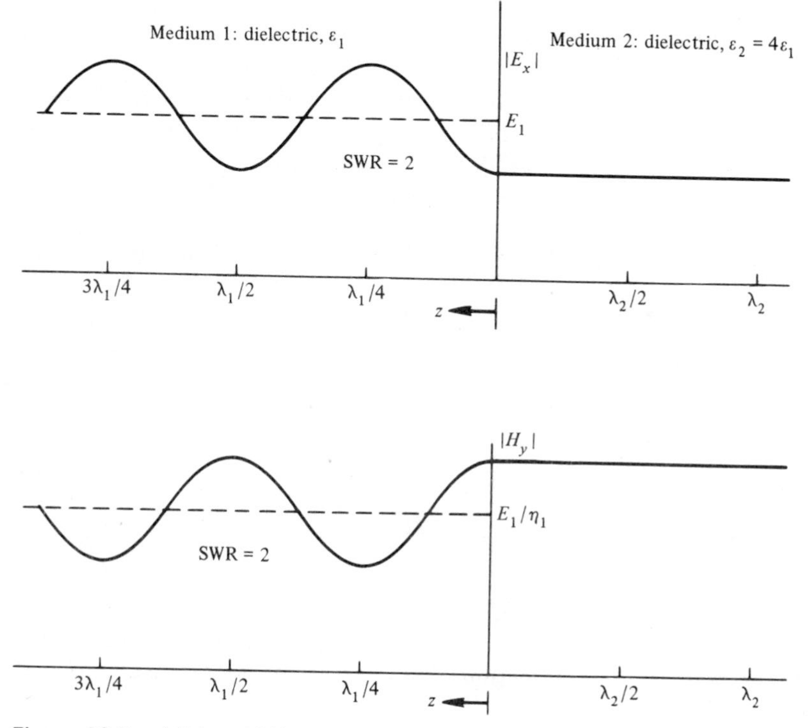

Figure 10.7. $|E_x|$ and $|H_y|$ as functions of z for a dielectric-dielectric interface, $\varepsilon_2 = 4\varepsilon_1$, SWR = 2.

It is interesting and informative to calculate the time average power densities for this example.

$$\langle \mathbf{S} \rangle = \frac{1}{2} \operatorname{Re} \{\mathbf{E} \times \mathbf{H}^*\}$$

or

$$\langle S_z \rangle = -\frac{1}{2} \operatorname{Re} \left\{ \frac{E_1^2}{\eta_1} \left(1 + \frac{1}{3} e^{+j(\pi - 2k_1 z)}\right)\left(1 - \frac{1}{3} e^{-j(\pi - 2k_1 z)}\right)\right\}, \qquad z > 0$$

or

$$\langle S_z \rangle = -\frac{1}{2}\frac{E_1^2}{\eta_1}\left(1 - \frac{1}{9}\right) = -\frac{4}{9}\frac{E_1^2}{\eta_1}, \qquad z > 0$$

This is a *net* power flow in the $-z$ direction. Now,

$$\langle S_z \rangle = -\frac{1}{2}\,\text{Re}\left\{\frac{4}{9}\frac{E_1^2}{\eta_2}\right\} = -\frac{4}{18}\frac{E_1^2}{\eta_2}, \qquad z < 0$$

but $\eta_2 = \eta_1/2$, so

$$\langle S_z \rangle = -\frac{4}{9}\frac{E_1^2}{\eta_1}, \qquad z < 0$$

Thus the net power flow in the $-z$ direction is the same for $z < 0$ and $z > 0$. The *incident* time average power density *on* the interface is $\langle S_z \rangle_{\text{inc}} = -E_1^2/2\eta_1$, while the *reflected* power density *from* the interface is $\langle S_z \rangle_{\text{ref}} = E_1^2/18\eta_1$. The net, $(\langle S_z \rangle_{\text{inc}} - \langle S_z \rangle_{\text{ref}})$, or transmitted power density is $\langle S_z \rangle_{\text{tran}} = -(4/9)(E_1^2/\eta_1)$ as before.

10.14 DIELECTRIC-PERFECT CONDUCTOR INTERFACE

Referring to Figure 10.3 again, if we let $\sigma_2 \to \infty$ (perfect conductor) then $\eta_2 \to 0$ (see equations 10.44 and 10.51). In this case, $\Gamma = -1$, $T = 0$, and SWR $\to \infty$. This is an example of total reflection, and E_x and H_y become *pure standing waves*, giving no net transmitted power density. The field $(z > 0)$ is (see equations 10.72 and 10.73)

$$|E_x| = 2E_1|\sin k_1 z| \tag{10.96}$$

$$|H_y| = \frac{2E_1}{\eta_1}|\cos k_1 z| \tag{10.97}$$

and is plotted in Figure 10.8(a).

It is worthwhile at this point to examine the behavior of a pure standing wave (SWR $\to \infty$) as a function of time. Consider E_x. Using equation 10.72 with $\Gamma = -1 = e^{j\pi}$,

$$E_x = Ee^{jk_1 z}(1 - e^{-j2k_1 z})$$

or

$$E_x = E_1(e^{jk_1 z} - e^{-jk_1 z})$$

We obtain the time domain form by multiplying by $e^{j\omega t}$ and then taking the real part. Thus,

$$\begin{aligned} E_x(z, t) &= E_1\,\text{Re}\,\{e^{j(\omega t + k_1 z)} - e^{j(\omega t - k_1 z)}\} \\ &= E_1[\cos(\omega t + k_1 z) - \cos(\omega t - k_1 z)] \end{aligned}$$

or

$$E_x(z, t) = -2E_1 \sin \omega t \sin k_1 z$$

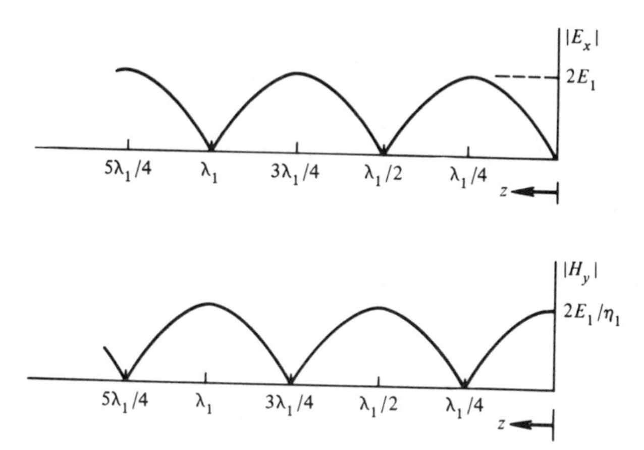

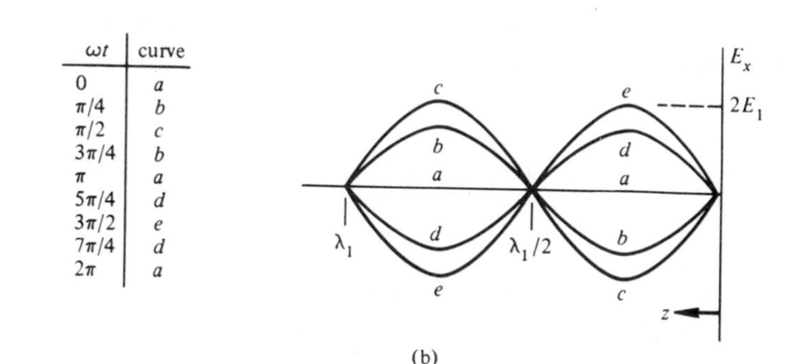

ωt	curve
0	a
$\pi/4$	b
$\pi/2$	c
$3\pi/4$	b
π	a
$5\pi/4$	d
$3\pi/2$	e
$7\pi/4$	d
2π	a

Figure 10.8. Pure standing waves, SWR $\to \infty$. (a) $|E_x|$ and $|H_y|$ versus z, dielectric-perfect conductor interface. (b) E_x versus z for various values of t, $E_x(z, t) = -2E_1 \sin \omega t \sin k_1 z$.

An observer with an electric field strength meter would measure nothing at $z = 0$, while at $z = \lambda/8$, he or she would measure $-\sqrt{2}E_1 \sin \omega t$, and at $z = \lambda/4$, he or she would measure $-2E_1 \sin \omega t$. Figure 10.8(b) shows this particular pure standing wave. Compare this standing wave to the traveling wave of Figure 10.1(d). Does this standing wave behavior resemble that of a vibrating string?

In the case of a perfect conductor, the skin depth is zero, and the current flow is in the form of a surface current. Multiplying this current times the *width* across which it flows gives the *total* current. Formally, the surface current density is written as

$$\mathbf{J}_s = \mathbf{a}_n \times \mathbf{H}\big|_{\text{on } s} \quad \text{(A/m)} \tag{10.98}$$

where $\mathbf{a}_n$ is a unit vector normal to the perfect conductor surface s and pointing *into* the region of interest. Thus, as we saw in the previous chapter, the magnetic field is *discontinuous* at the surface of a perfect conductor. For Figure 10.8(a), we have

$$\mathbf{J}_s = \mathbf{a}_z \times \mathbf{a}_y\left(-\frac{2E_1}{\eta_1}\right) = \mathbf{a}_x \frac{2E_1}{\eta_1}$$

or

$$J_{sx} = \frac{2E_1}{\eta_1} \tag{10.99}$$

Compare the results of this section to Example 3 in Section 10.11.

10.15 CONDUCTOR-PERFECT CONDUCTOR INTERFACE

There will again be no field for $z < 0$ in this example because total reflection again occurs. For $z > 0$, the situation is now more complicated. Since $\eta_2 = 0$, it follows that $\Gamma = -1$ and $T = 0$. The field quantities for $z > 0$ become

$$E_x = E_1 e^{+\alpha_1 z} e^{+j\beta_1 z} + E_1' e^{-\alpha_1 z} e^{-j\beta_1 z} \tag{10.100}$$

and

$$H_y = -\frac{E_1}{\eta_1} e^{+\alpha_1 z} e^{+j\beta_1 z} + \frac{E_1'}{\eta_1} e^{-\alpha_1 z} e^{-j\beta_1 z} \tag{10.101}$$

or

$$E_x = E_1 e^{(\alpha_1 + j\beta_1)z}(1 - e^{-2(\alpha_1 + j\beta_1)z}) \tag{10.102}$$

and

$$H_y = -\frac{E_1}{\eta_1} e^{(\alpha_1 + j\beta_1)z}(1 + e^{-2(\alpha_1 + j\beta_1)z}) \tag{10.103}$$

Figure 10.9. Conductor-perfect conductor interface.

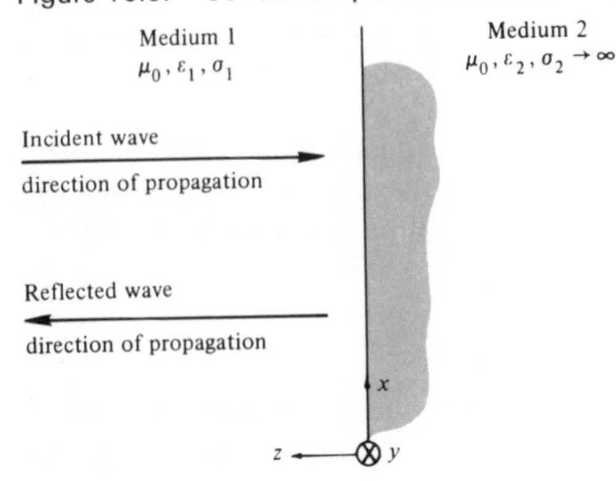

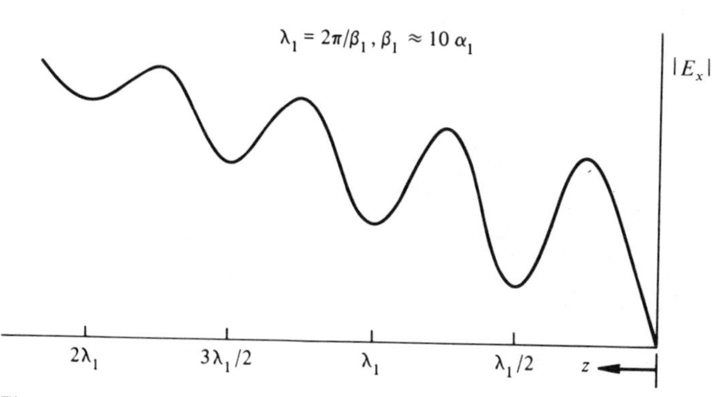

Figure 10.10 $|E_x|$ versus z, conductor-perfect conductor interface, $\beta_1 \approx 10\alpha_1$.

The *magnitudes* of the two quantities are

$$|E_x| = E_1 e^{\alpha_1 z}|(1 - e^{-2(\alpha_1 + j\beta_1)z})| \quad (10.104)$$

and

$$|H_y| = \frac{E_1}{|\eta_1|} e^{\alpha_1 z}|(1 + e^{-2(\alpha_1 + j\beta_1)z})| \quad (10.105)$$

$|E_x|$ and $|H_y|$ are no longer periodic with z ($z > 0$) and the SWR concept becomes rather meaningless. A plot of $|E_x|$ against z is shown in Figure 10.10 for $\beta_1 \approx 10\alpha_1$. Notice that the effect of the reflected wave is rapidly disappearing as the source of the incident wave is approached (large z).

10.16 OBLIQUE INCIDENCE

It is not too difficult to show from the equation of a plane that a general expression for a uniform plane wave is (phasor form)

$$\boxed{\mathbf{E} = \mathbf{E}_0 e^{-\gamma \mathbf{a}_n \cdot \mathbf{r}}} \quad (10.106)$$

for a wave propagating in the $\mathbf{a}_n$ direction. In equation 10.106, $\mathbf{a}_n$ is a unit vector normal to an equiphase plane and making angles θ_x, θ_y, and $\theta_z = \theta$ with the x, y, and z axes, respectively. $\mathbf{r}$ is a radial vector to any point on the equiphase plane. Then,

$$\mathbf{a}_n = \mathbf{a}_x \cos\theta_x + \mathbf{a}_y \cos\theta_y + \mathbf{a}_z \cos\theta_z \quad (10.107)$$

$$\mathbf{r} = \mathbf{a}_x x + \mathbf{a}_y y + \mathbf{a}_z z \quad (10.108)$$

and

$$\mathbf{a}_n \cdot \mathbf{r} = x\cos\theta_x + y\cos\theta_y + z\cos\theta_z \quad (10.109)$$

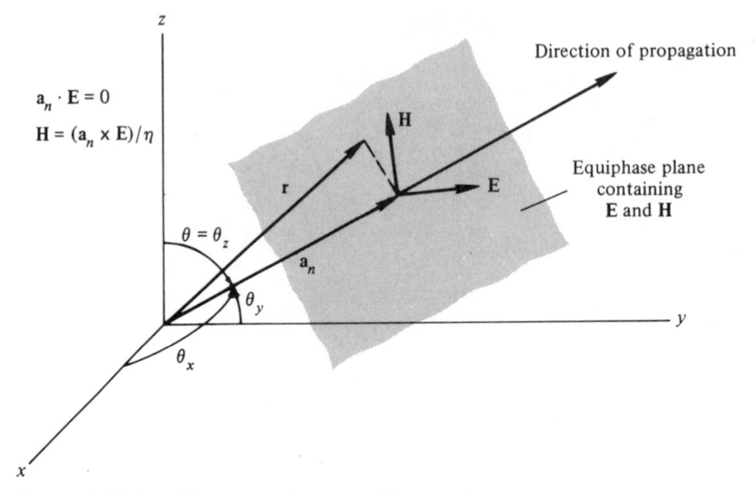

Figure 10.11. Geometry for an arbitrary plane wave.

The vector $\mathbf{E}_0$ may be complex, but in any case both it and $\mathbf{H}$ lie in the equiphase plane, so that

$$\mathbf{a}_n \cdot \mathbf{E} = 0 \quad \text{and} \quad \frac{\mathbf{a}_n \times \mathbf{E}}{\eta} = \mathbf{H} \tag{10.110}$$

This is shown in Figure 10.11.

Let us now consider the case of oblique incidence on a plane interface. We will explicitly consider the cases of $\mathbf{E}$ *parallel* to the interface and $\mathbf{H}$ *parallel* to the interface, arbitrary polarization being a superposition of these two cases. We will follow the same procedure as that used for normal incidence in Section 10.11. Consider Figure 10.12.

(1) For $\mathbf{E}$ parallel to the $z = 0$ plane, $\mathbf{E}$ will be in the $\pm x$ direction. We choose the positive direction. Using equations 10.107 through 10.110, we have for the incident wave

$$\theta_x = \frac{\pi}{2}, \qquad \theta_y = \frac{\pi}{2} - \theta_i, \qquad \theta_z = \pi - \theta_i$$

$$E_x = E_1 e^{-jk_1(y \sin\theta_i - z\cos\theta_i)} \tag{10.111}$$

and

$$\mathbf{H} = (-\mathbf{a}_y \cos\theta_i - \mathbf{a}_z \sin\theta_i)\frac{E_1}{\eta_1} e^{-jk_1(y\sin\theta_i - z\cos\theta_i)} \tag{10.112}$$

if medium (1) is lossless ($\gamma = jk$). The *incident wave impedance in the* $-z$ direction is (see Section 10.6).

$$\eta_z^- \equiv -\frac{E_x}{H_y} \equiv \frac{\eta_1}{\cos\theta_i} \tag{10.113}$$

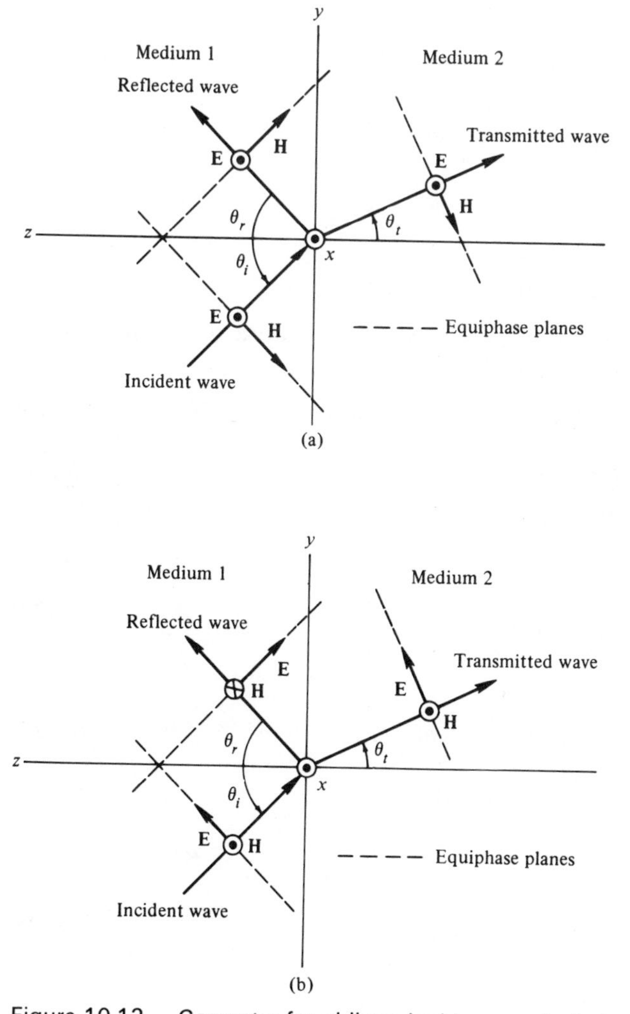

Figure 10.12. Geometry for oblique incidence calculations. (a) **E** parallel to the $z = 0$ plane. (b) **H** parallel to the $z = 0$ plane.

For the reflected wave

$$E_x = E_1' e^{-jk_1(y \sin \theta_r + z \cos \theta_r)} \tag{10.114}$$

$$\mathbf{H} = (\mathbf{a}_y \cos \theta_r - \mathbf{a}_z \sin \theta_r) \frac{E_1'}{\eta_1} e^{-jk_1(y \sin \theta_r + z \cos \theta_r)} \tag{10.115}$$

and

$$\eta_z^+ \equiv \frac{E_x}{H_y} = \frac{\eta_1}{\cos \theta_r} \tag{10.116}$$

For the transmitted wave

$$E_x = E_2 e^{-jk_2(y \sin\theta_t - z\cos\theta_t)} \tag{10.117}$$

$$\mathbf{H} = (-\mathbf{a}_y \cos\theta_t - \mathbf{a}_z \sin\theta_t)\frac{E_2}{\eta_2} e^{-jk_2(y\sin\theta_t - z\cos\theta_t)} \tag{10.118}$$

and

$$\eta_z^- \equiv -\frac{E_x}{H_y} = \frac{\eta_2}{\cos\theta_t} \tag{10.119}$$

(2) For **H** parallel to the interface, we again choose, or assume, positive tangential **E** components. Then, for the incident wave,

$$\mathbf{E} = (\mathbf{a}_y \cos\theta_i + \mathbf{a}_z \sin\theta_i)E_1 e^{-jk_1(y\sin\theta_i - z\cos\theta_i)} \tag{10.120}$$

$$H_x = \frac{E_1}{\eta_1} e^{-jk_1(y\sin\theta_i - z\cos\theta_i)} \tag{10.121}$$

and

$$\eta_z^- \equiv \frac{E_y}{H_x} = \eta_1 \cos\theta_i \tag{10.122}$$

For the reflected wave,

$$\mathbf{E} = (\mathbf{a}_y \cos\theta_r - \mathbf{a}_z \sin\theta_r)E_1' e^{-jk_1(y\sin\theta_r + z\cos\theta_r)} \tag{10.123}$$

$$H_x = -\frac{E_1'}{\eta_1} e^{-jk_1(y\sin\theta_r + z\cos\theta_r)} \tag{10.124}$$

and

$$\eta_z^+ \equiv -\frac{E_y}{H_x} = \eta_1 \cos\theta_r \tag{10.125}$$

For the transmitted wave,

$$\mathbf{E} = (\mathbf{a}_y \cos\theta_t + \mathbf{a}_z \sin\theta_t)E_2 e^{-jk_2(y\sin\theta_t - z\cos\theta_t)} \tag{10.126}$$

$$H_x = \frac{E_2}{\eta_2} e^{-jk_2(y\sin\theta_t - z\cos\theta_t)} \tag{10.127}$$

and

$$\eta_z^- \equiv \frac{E_y}{H_x} = \eta_2 \cos\theta_t \tag{10.128}$$

Notice that the directions of **H** and **S** with respect to **E** are determined by $\mathbf{S} = \mathbf{E} \times \mathbf{H}$, where **S** is the power density vector, pointing in the direction of propagation, and *positive* tangential components of **E** were *assumed* in all cases. The transverse wave impedances were defined in Section 10.6.

Boundary conditions at the interface simply require that the tangential components of **E** and **H** are continuous at $z = 0$. Thus, for (1) **E** parallel to the interface,

$$E_1 e^{-jk_1 y \sin\theta_i} + E_1' e^{-jk_1 y \sin\theta_r} = E_2 e^{-jk_2 y \sin\theta_t} \tag{10.129}$$

$$-\cos\theta_i \frac{E_1}{\eta_1} e^{-jk_1 y \sin\theta_i} + \cos\theta_r \frac{E_1'}{\eta_1} e^{-jk_1 y \sin\theta_r} = -\cos\theta_t \frac{E_2}{\eta_2} e^{-jk_2 y \sin\theta_t} \tag{10.130}$$

and for (2) **H** parallel to the interface,

$$\cos\theta_i E_1 e^{-jk_1 y \sin\theta_i} + \cos\theta_r E_1' e^{-jk_1 y \sin\theta_r} = \cos\theta_t E_2 e^{-jk_2 y \sin\theta_t} \tag{10.131}$$

$$\frac{E_1}{\eta_1} e^{-jk_1 y \sin\theta_i} - \frac{E_1'}{\eta_1} e^{-jk_1 y \sin\theta_r} = \frac{E_2}{\eta_2} e^{-jk_2 y \sin\theta_t} \tag{10.132}$$

The phenomena we are observing at the interface must be independent of y, and this can only occur if all the phase angles of the phasor quantities are the same in the last four equations. Therefore,

$$k_1 \sin\theta_i = k_1 \sin\theta_r = k_2 \sin\theta_t \tag{10.133}$$

From the first of equations 10.133

$$\boxed{\theta_i = \theta_r} \tag{10.134}$$

or the angle of incidence is equal to the angle of reflection. From the second of equations 10.133,

$$\boxed{\frac{\sin\theta_t}{\sin\theta_r} = \frac{\sin\theta_t}{\sin\theta_i} = \frac{k_1}{k_2} = \frac{u_2}{u_1} = \sqrt{\frac{\mu_1\varepsilon_1}{\mu_2\varepsilon_2}}} \tag{10.135}$$

This result is known as *Snell's law*. In the most common case, $\mu_1 = \mu_2 = \mu_0$, for which

$$\sin\theta_t = \sin\theta_i \sqrt{\frac{\varepsilon_1}{\varepsilon_2}} \tag{10.136}$$

We now have for (1) **E** parallel to the interface,

$$E_1 + E_1' = E_2 \tag{10.137}$$

$$\frac{\cos\theta_i}{\eta_1}(E_1 - E_1') = \cos\theta_t \frac{E_2}{\eta_2} \tag{10.138}$$

and for (2) **H** parallel to the interface,

$$\cos\theta_i (E_1 + E_1') = \cos\theta_t E_2 \tag{10.139}$$

$$\frac{E_1}{\eta_1} - \frac{E_1'}{\eta_1} = \frac{E_2}{\eta_2} \tag{10.140}$$

The coefficients of reflection (E_1'/E_1) are obtained by solving the preceding pairs of equations simultaneously. They are (from equations 10.137 and 10.138)

$$\Gamma_E = \frac{\eta_2 \cos \theta_i - \eta_1 \cos \theta_t}{\eta_2 \cos \theta_i + \eta_1 \cos \theta_t} \quad (\mathbf{E} \text{ parallel}) \tag{10.141}$$

and (from equations 10.139 and 10.140)

$$\Gamma_H = \frac{\eta_2 \cos \theta_t - \eta_1 \cos \theta_i}{\eta_2 \cos \theta_t + \eta_1 \cos \theta_i} \quad (\mathbf{H} \text{ parallel}) \tag{10.142}$$

for **E** parallel to the interface and **H** parallel to the interface, respectively. Notice that these equations are in the same form as equation 10.70 (normal incidence), to which equations 10.141 and 10.142 reduce when $\theta_i = \theta_t = 0$

Total transmission occurs when $\Gamma = 0$, or

$$\sin \theta_i = \sqrt{\frac{\varepsilon_2/\varepsilon_1 - \mu_2/\mu_1}{\mu_1/\mu_2 - \mu_2/\mu_1}} \quad (\mathbf{E} \text{ parallel}) \tag{10.143}$$

and

$$\sin \theta_i = \sqrt{\frac{\varepsilon_2/\varepsilon_1 - \mu_2/\mu_1}{\varepsilon_2/\varepsilon_1 - \varepsilon_1/\varepsilon_2}} \quad (\mathbf{H} \text{ parallel}) \tag{10.144}$$

Equations 10.143 and 10.144 may not have solutions in all cases, but one special case is important. For **H** parallel to the interface and $\mu_1 = \mu_2 = \mu_0$, equation 10.144 gives

$$\boxed{\theta_i = \sin^{-1} \sqrt{\frac{\varepsilon_2}{\varepsilon_1 + \varepsilon_2}} = \tan^{-1} \sqrt{\frac{\varepsilon_2}{\varepsilon_1}}} \tag{10.145}$$

This angle is the *polarizing angle* or *Brewster* angle because an arbitrarily polarized wave incident at this angle will be reflected with **E** polarized parallel to the interface. That is, the other part of **E** is totally transmitted.

Total reflection occurs for $|\Gamma| = 1$. Inspection of equations 10.141 and 10.142 reveals that this cannot occur for real values of θ_i and θ_t. Suppose that θ_t is imaginary, or $\sin \theta_t > 1$. For convenience let

$$k_2 \sin \theta_t \equiv \beta$$

and

$$k_2 \cos \theta_t = k_2 \sqrt{1 - \sin^2 \theta_t} \equiv \pm j\alpha$$

Choosing the minus sign to give a damped or decreasing wave and using these definitions in equations 10.117 and 10.118 for the transmitted wave (**E** parallel), we have

$$E_x = E_2 e^{+\alpha z} e^{-j\beta y} \tag{10.146}$$

$$\mathbf{H} = \left(\mathbf{a}_y \frac{j\alpha}{k_2} - \mathbf{a}_z \frac{\beta}{k_2} \right) \frac{E_2}{\eta_2} e^{+\alpha z} e^{-j\beta y} \tag{10.147}$$

The wave impedance *into medium* 2 is

$$\eta_z^- \equiv -\frac{E_x}{H_y} = j\frac{\eta_2 k_2}{\alpha} \tag{10.148}$$

which is imaginary (η_2, k_2 real) indicating *no power flow* into medium 2!

It is a straightforward matter to show that equations 10.146 and 10.147 simultaneously satisfy Maxwell's equations. These equations represent waves propagating in the $+y$ direction and are *evanescent* (exponential damping only) as far as the z direction is concerned. Now, $\sin\theta_t$ becomes greater than one (for *both* types of polarization) when $\sin\theta_i$ becomes greater than $\sqrt{\mu_2\varepsilon_2/\mu_1\varepsilon_1}$, as equation 10.135 shows. *The critical angle* then is

$$\theta_i = \sin^{-1}\sqrt{\frac{\mu_2\varepsilon_2}{\mu_1\varepsilon_1}} \qquad (\sin\theta_i < 1) \tag{10.149}$$

and a wave incident on a plane interface at an angle equal to or greater than this angle will be totally reflected. When $\sin\theta_t$ is greater than one, equation 10.141, for example, becomes

$$\Gamma_E = \frac{\eta_2\sqrt{1-(\mu_2\varepsilon_2/\mu_1\varepsilon_1)^2} + j(\eta_1\alpha/k_2)}{\eta_2\sqrt{1-(\mu_2\varepsilon_2/\mu_1\varepsilon_1)^2} - j(\eta_1\alpha/k_2)} = \frac{a+jb}{a-jb}$$

so $|\Gamma_E|$ is indeed unity when $\sqrt{(\mu_2\varepsilon_2)/(\mu_1\varepsilon_1)} < 1$. All of the results of this section can be extended to the more general case of dissipative media if the angles and impedance are allowed to become complex.

We next consider a somewhat oversimplified example which will be very useful to us when rectangular waveguides are studied. Suppose medium (2) is perfectly conducting so that $\eta_2 = 0$. Then, for the case of $\mathbf{E}$ parallel to the interface we have

$$\eta_2 = 0$$

$$\Gamma_E = -1$$

$$\mathbf{E}, \mathbf{H} = 0, \qquad z < 0 \text{ (medium 2)}$$

$$E_x = E_1 e^{-jk_1 y\sin\theta_i}[2j\sin(k_1 z\cos\theta_i)], \qquad z > 0 \tag{10.150}$$

and

$$\mathbf{H} = -\frac{E_1}{\eta_1}e^{-jk_1 y\sin\theta_i}[2\mathbf{a}_y\cos\theta_i\cos(k_1 z\cos\theta_i) + 2j\mathbf{a}_z\sin\theta_i\sin(k_1 z\cos\theta_1)], \qquad z > 0 \tag{10.151}$$

The time average power densities are

$$\langle S_z\rangle = \frac{1}{2}\operatorname{Re}\{E_x H_y^*\} = 0 \tag{10.152}$$

and

$$\langle S_y\rangle = -\frac{1}{2}\operatorname{Re}\{E_x H_z^*\} = \frac{2E_1^2}{\eta_1}\sin\theta_i\sin^2(k_1 z\cos\theta_i) \tag{10.153}$$

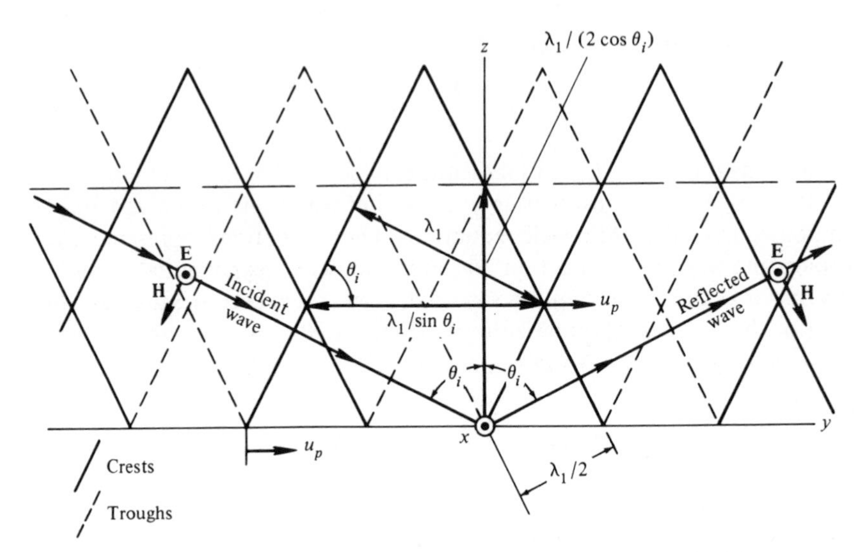

Figure 10.13. Oblique incidence, plane wave in a lossless dielectric onto a perfect conductor.

Inspection of equations 10.150 and 10.151 reveals, first of all, through the term [$\sin(k_1 z \cos\theta_i)$] that a *standing* wave pattern exists in the z direction with nulls where $k_1 z \cos\theta_i = p\pi$, $p = 0, 1, 2, \ldots$, or $z = p/(2f\sqrt{\mu_1\varepsilon_1}\cos\theta_i)$. One of these loci is indicated in Figure 10.13 by the horizontal (long) dashed line. Secondly, the term $e^{-jk_1 y \sin\theta_i}$ reveals that a *traveling wave* exists in the y direction with a phase velocity given by setting the phase constant, $k_1 \sin\theta_i$, equal to ω/u_p. That is, $u_p = \omega/(k_1 \sin\theta_i)$, or

$$u_p = \frac{1}{\sqrt{\mu_1\varepsilon_1}\sin\theta_i} \geq \frac{1}{\sqrt{\mu_1\varepsilon_1}} \tag{10.154}$$

The most striking feature of equation 10.154 is the fact that the phase velocity is greater ($\theta_i < \pi/2$) that the speed of light, or "intrinsic" velocity in the same medium, and may even approach infinity. This feature does not violate anything of a fundamental nature. Figure 10.13 reveals that the phase velocity in the y direction *must* be greater than the intrinsic velocity. The field we are considering here is the superposition of two simple plane waves, each having a phase velocity of $1/\sqrt{\mu_1\varepsilon_1}$. The *intersection* of their equiphase planes at a crest (for example) moves a distance $\lambda_1/\sin\theta_i$, while over the same time interval the same point of constant phase in the simple plane wave moves only a distance λ_1. Thus, the phase velocity in the composite wave is given by the plane wave phase velocity divided by $\sin\theta_i$. The intersection between one of the equiphase planes and the $z = 0$ plane also moves at this velocity. Many of us have observed an ocean wave striking a straight shoreline at an oblique angle such that the point of intersection between the crest of a wave and the shoreline moves with a greater velocity than the crest itself.

The concept of velocity of energy flow is ordinarily reserved for guiding systems, but if it were calculated according to equation 10.37 for the system under present discussion, we would have

$$u_e = u_g = \frac{\sin\theta_i}{\sqrt{\mu_1\varepsilon_1}} \le \frac{1}{\sqrt{\mu_1\varepsilon_1}}$$

which is comforting.

Notice that since the field is zero where $z = p/(2f\sqrt{\mu_1\varepsilon_1}\cos\theta_i)$, a perfectly conducting plane could be placed there *without changing* the field between the planes. Thus, we could construct a two-conductor waveguide with little more mathematical effort. These concepts will be examined in more detail in Chapter 14, but it can be pointed out at this time that when $\theta_i = 0$, $u_p \to \infty$, $u_e = 0$, and there is *no net propagation in the y direction.* Insofar as waveguide concepts are concerned, we then say that the waveguide is *cutoff* or nonpropagating in the y direction. It is also true that $u_g = u_e$.

10.17 POLARIZATION

In several instances we have referred to *polarization.* In fact, we have considered several cases of *linear* polarization, where the lines of electric field $\mathbf{E}$ have been parallel to some fixed axis. Figure 10.2 is a good example of this behavior. In order to examine other types of polarization, consider the amplitudes A and B (which may be complex) and the field

$$\mathbf{E} = (\mathbf{a}_x A + \mathbf{a}_y B)e^{-jkz} \tag{10.155}$$

$$\mathbf{H} = (\mathbf{a}_z \times \mathbf{E})/\eta \tag{10.156}$$

(1) If A and B have the same phase angle, the wave is *linearly polarized* and $\mathbf{E}$ will always lie in the plane containing the z axis, but inclined at an angle whose tangent is B/A from the $y = 0$ plane. This is seen in Figure 10.14. If $B = 0$, the wave is obviously polarized in the x direction, while if $A = 0$, the wave is obviously polarized in the y direction.

(2) If A and B are complex and have different phase angles, then $\mathbf{E}$ will no longer remain in one plane. Suppose $A = |A|e^{ja}$ and $B = |B|e^{jb}$. Then,

$$\mathbf{E} = \mathbf{a}_x|A|e^{j(a-kz)} + \mathbf{a}_y|B|e^{j(b-kz)} \tag{10.157}$$

and

$$\mathbf{H} = -\mathbf{a}_x\frac{|B|}{\eta}e^{j(b-kz)} + \mathbf{a}_y\frac{|A|}{\eta}e^{j(a-kz)} \tag{10.158}$$

In the time domain,

$$E_x(z,t) = |A|\cos(\omega t + a - kz) \tag{10.159}$$

and

$$E_y(z,t) = |B|\cos(\omega t + b - kz) \tag{10.160}$$

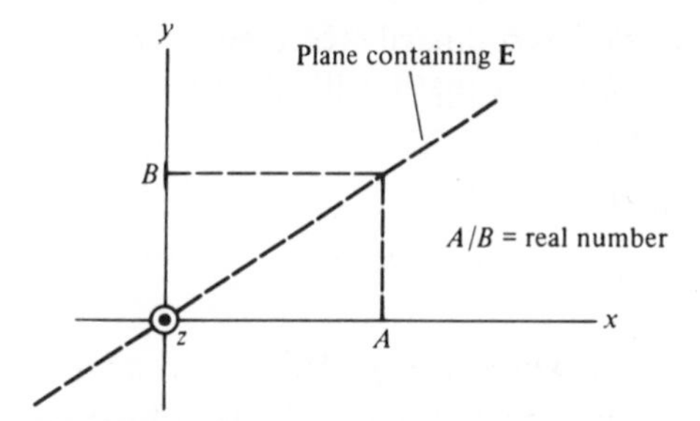

Figure 10.14. Linear polarization.

The locus of the endpoint of the vector $\mathbf{E}(z, t)$ will trace out an ellipse once each cycle, giving *elliptical polarization.* For example, suppose, $|A| = 2|B| = 1$, $a = 0$ and $b = \pi/2$. Then, we have left-handed elliptic polarization where

$$E_x(z, t) = \cos(\omega t - kz)$$

and

$$E_y(z, t) = -\tfrac{1}{2}\sin(\omega t - kz)$$

A plot of $\mathbf{E}(z, t)$ in the $z = 0$ plane is given in Figure 10.15.

(3) If A and B are equal in amplitude and differ in phase angle by $\pi/2$, the ellipse becomes a circle, and we have the special case of circular polarization. If the thumb of the right hand points in the direction of propagation while the fingers point in the direction of rotation of $\mathbf{E}$, the polarization is said to be *right-handed circular*. For example, if $A = jB$, or $|A| = |B| = 1$, $a = 0$, $b = -\pi/2$, we have from equations 10.159 and 10.160

and

$$E_y(z, t) = \sin(\omega t - kz)$$

Figure 10.15. Left-handed elliptic polarization.

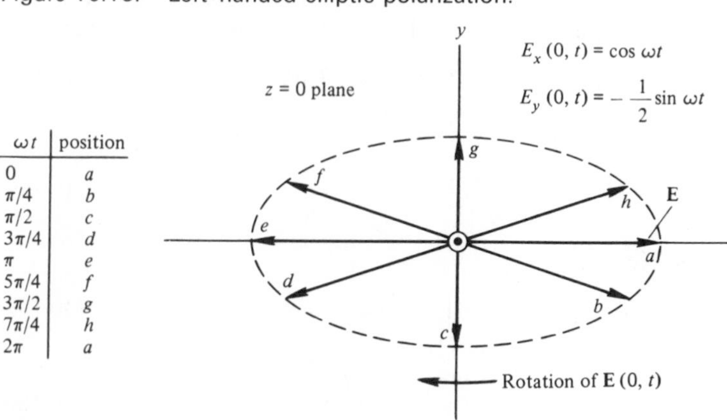

ωt	position
0	a
$\pi/4$	b
$\pi/2$	c
$3\pi/4$	d
π	e
$5\pi/4$	f
$3\pi/2$	g
$7\pi/4$	h
2π	a

It is easy to see that this is an example of right-handed circular polarization. If t were held constant and a plot of **E** versus z made, the endpoint of **E** would trace out a helix. A circularly polarized wave also has the interesting properties that both the power density and energy density are independent of time and space!

10.18 DISPERSION

It was mentioned in Section 10.7, where group velocity was defined, that the frequency spread of the group must be small, or else the concept is meaningless. This needs further comment. A *normally* dispersive medium or system is one for which $du_p^{-1}/d\omega > 0$ and $u_g < u_p$, and an *anomalously* dispersive medium or system is one for which $du_p^{-1}/d\omega < 0$ and $u_g > u_p$. Systems that are anomalously dispersive can have $u_g > 1/\sqrt{\mu\varepsilon}$, which is clearly impossible. This occurs because of the way in which u_g was defined, but simply means that the derivation of the equation for u_g is not valid because the frequency spread is not small. Thus, the identity of group velocity is not valid for systems with anomalous dispersion. The uniform plane wave propagating in a lossy dielectric,[4] or TEM waves on a lossy transmission line can have $u_g > 1/\sqrt{\mu\varepsilon}$ or $u_g > 1/\sqrt{LC}$, respectively. This is best seen in the $\omega - \beta$ diagram of Figure 10.16. Notice that for oblique incidence on a perfectly conducting

Figure 10.16. The ω-β diagram for various systems. (a) Plane wave in a lossless dielectric, or a lossless transmission line, $u_p = u_e = u_g = 1/\sqrt{\mu\varepsilon}$, $\beta = \omega\sqrt{\mu\varepsilon}$. (b) Plane wave in a lossy dielectric, or a lossy transmission line, $u_g = d\omega/d\beta$. (c) Plane wave with oblique incidence on a perfectly conducting plane, or a lossless waveguide, or a plane wave in an ionized region, $u_g = (1/\sqrt{\mu\varepsilon})\sqrt{1 - (\omega_c/\omega^2)}$.

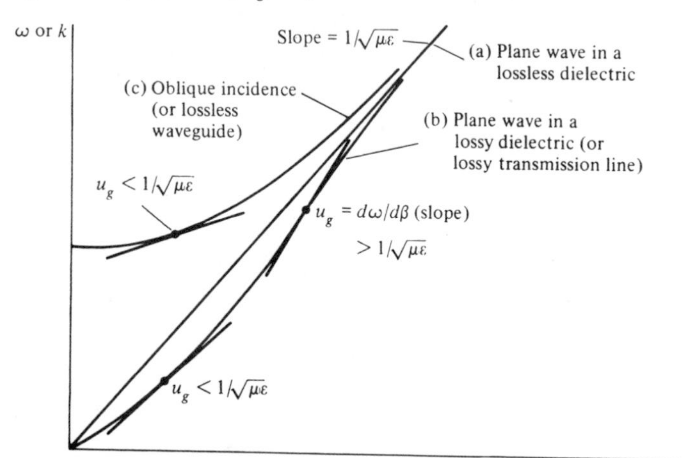

[4] See Problem 22.

plane (Section 10.16), $u_g \leq 1/\sqrt{\mu\varepsilon}$. The same is true of the lossless waveguide to be investigated later.

Another way to reach the conclusions of the preceding paragraph occurs when we realize that a signal (or group) can never travel at a velocity greater than that of light, for if it did, some observers would detect some effects *before* their causes, thus violating the *principle of causality*. All we need in order to test causality for the systems we are discussing is their unit impulse response. The results of Problem 30 (given) show that a lossy dielectric region is a causal system for uniform plane waves. Problem 31 shows that the same is true for an ionized region. Problem 13, Chapter 13 gives the same result for lossy transmission lines, while Problem 13, Chapter 14 applies to the lossless waveguide.

10.19 CONCLUDING REMARKS

It was the main objective of this chapter to develop a good understanding of uniform plane waves and normal incidence of plane waves on plane interfaces. This was our primary goal because of the direct relationship between these phenomena and the phenomena occurring in guiding systems. More general problems such as oblique incidence with arbitrary **E** field alignment or polarization, or problems including more general media (such as magnetic materials or dielectrics with complex permittivities), are not included here. It should be clearly understood that our choice of the $\pm z$ axis as the direction of propagation is completely arbitrary. Oblique incidence was briefly considered because it applies to optics and to the field behavior in rectangular waveguides. We could erect a parallel perfectly conducting plane in Figure 10.13 which would reflect the original plane wave back toward the $z = 0$ plane where it is reflected back and forth between the two planes while progressing in the y direction. In other words, we are now predicting the possibility of an elementary two-conductor waveguide system. This point will be mentioned again when we explicitly investigate waveguide systems in Chapter 14.

REFERENCES

Harrington, R. F. *Time Harmonic Electromagnetic Fields.* New York: McGraw-Hill, 1961. An advanced textbook. The technique of obtaining the electromagnetic field from both an electric and magnetic vector potential is extensively used.

Jordan, E. C. and Balmain, K. G. *Electromagnetic Waves and Radiating Systems*, 2nd ed. Englewood Cliffs, N.J.: Prentice-Hall, 1968. A revised edition of a popular graduate level textbook.

Pierce, J. R. *Almost All About Waves.* Cambridge, Mass.: The MIT Press, 1974. A very readable little book more general in nature than the material being presented herein.

Ramo. (See references for Chapter 4).

PROBLEMS

1. Verify equation 10.13.
2. Verify equation 10.22.
3. Verify equation 10.23.
4. Verify equation 10.24.
5. Verify equations 10.42 and 10.43.
6. If $\mathbf{H} = 10\mathbf{a}_z e^{-j10y}$, find $\mathbf{E}$ (plane wave).
7. (a) If $\mathbf{E} = 10\mathbf{a}_y \sin(\omega t - x)$ is propagating in free space, what is $\mathbf{H}$?
   (b) Repeat if $\varepsilon_R = 4$.
8. For Problem 7a, find (a) ω, (b) λ, (c) $\langle \mathbf{S} \rangle$, (d) u_p, (e) u_e
9. Derive equations 10.47 and 10.48.
10. According to equation 10.90, surface resistance is $R_s = \sqrt{\omega\mu/2\sigma}$ (Ω) for a plane surface. Show that

$$\begin{aligned} R_s &= 2.52 \times 10^{-7}\sqrt{f} \quad \text{(silver)} \\ &= 2.61 \times 10^{-7}\sqrt{f} \quad \text{(copper)} \\ &= 3.12 \times 10^{-7}\sqrt{f} \quad \text{(gold)} \\ &= 3.26 \times 10^{-7}\sqrt{f} \quad \text{(aluminum)} \\ &= 5.01 \times 10^{-7}\sqrt{f} \quad \text{(brass)} \end{aligned}$$

    if $(\mu = \mu_0)$.
11. Find α and β for copper at (a) 10 MHz, (b) 10 GHz, (c) 10^{13} GHz.
12. A plane wave is incident normally from air onto a material for which $\eta = 100\underline{|30^\circ}$. Find (a) Γ, (b) T, (c) SWR.
13. The induced current density at the surface of a copper plate from a normally incident plane wave is 10^3 A/m^2 at $f = 10^4$ Hz. What is the current density 1 cm into the copper?
14. What percentage of the transmitted power density in a conductor is dissipated in heat over the distance δ (*normal incidence*)?
15. Fifty percent of the power density is reflected when a plane wave is normally incident on an air-dielectric interface. Plot $|\mathbf{E}|$ versus z as in Figure 10.7.
16. Find an electric and magnetic field (in phasor form) for a uniform plane wave propagating in the direction $\mathbf{a}_n = (\mathbf{a}_x + 2\mathbf{a}_y + 3\mathbf{a}_z)/\sqrt{14}$ in free space. Notice that the solution is not unique with respect to polarization.
17. (a) Calculate the polarizing angle for a plane interface between air and water. ($\varepsilon_R = 80$, water.)
    (b) Repeat part (a) if the wave enters from the water side.
18. (a) Calculate the critical angle for a plane interface between water and air.
    (b) Repeat part (a) for polystyrene and air.

19. In viewing an object located under water, does it appear closer, or more distant from the surface, than it actually is (*oblique incidence*)?
20. Show that a linearly polarized uniform plane wave can be expressed as the sum of a right-handed circularly polarized wave and a left-handed circularly polarized wave.
21. A standing wave is the sum of two traveling waves. Show that a traveling wave is the sum of two standing waves.

22. (a) For a good conductor show that

$$u_p \approx \sqrt{\frac{2\omega}{\mu\sigma}}, \qquad u_g \approx 2\sqrt{\frac{2\omega}{\mu\sigma}}$$

(b) For a good dielectric show that

$$u_p \approx \frac{1}{\sqrt{\mu\varepsilon}}\left[1 + \frac{1}{8}\left(\frac{\sigma}{\omega\varepsilon}\right)^2\right]^{-1}, \qquad u_g \approx \frac{1}{\sqrt{\mu\varepsilon}}\left[1 - \frac{1}{8}\left(\frac{\sigma}{\omega\varepsilon}\right)^2\right]^{-1}$$

Notice that in (b), and possibly in (a), $u_g > 1/\sqrt{\mu\varepsilon}$, so that over much of the frequency range a lossy dielectric is anomalously dispersive.

23. If copper and silver shields are to have the same effectiveness, compare their thicknesses.
24. Plot the resistance of copper wire of 0.5 cm diameter versus frequency (*surface resistance*). Comment on approximations made.
25. Prove that 1 Np = 8.69 dB.

26. (a) Find the time domain form for a linearly polarized plane wave traveling in any direction in a lossy dielectric. The time variation is sinusoidal.
    (b) Repeat (a), if the medium is lossless.
    (c) Repeat, if the medium is lossless and the time variation is e^{-vt}.
    In all cases, μ, ε, and σ are scalar constants.

27. (a) A uniform plane wave in the form of an impulse of strength 10 V/m is propagating in air in the negative z direction. It strikes a plane dielectric interface ($\varepsilon_R = 4$) at $z = 0$. Find the reflected and transmitted waves. Notice that in general Γ and T are frequency dependent.
    (b) Repeat part (a) if the incident wave is a half-cosine pulse (positive only) of base width 10 μs and peak amplitude 10 V/m.

28. Find $\Gamma(\omega)$, equation 10.81, if region (1) is air and region (2) has $\mu = \mu_0$, $\varepsilon = \varepsilon_0$, $\sigma \neq 0$. Now repeat Problem 17a for this interface.
29. Show that a current sheet over the $z = 0$ plane described by $\mathbf{J} = -(2E_0/\eta)\delta(z)\mathbf{a}_x$ A/m^2 produces the plane waves

$$E_x(z) = \begin{cases} E_0 e^{-jkz}, & z > 0 \\ E_0 e^{+jkz}, & z < 0 \end{cases}$$

or

$$E_x(z) = E_0 e^{-jkz[2u(z)-1]}$$

Hint: use the results of Problem 18, Chapter 9. That is, show that

$$\nabla^2 \mathbf{E} + k^2 \mathbf{E} = j\omega\mu\left[\mathbf{J} + \frac{1}{k^2}\nabla(\nabla \cdot \mathbf{J})\right]$$

or

$$\frac{d^2 E_x}{dz} + k^2 E_x = j\omega\mu J_x = -j2kE_0\,\delta(z)$$

30. An x-polarized uniform plane wave is propagating in the $+z$ direction in a medium characterized by σ, μ, ε (all assumed constant). That is,

$$E_x(z, \omega) = E_0(\omega)e^{-\gamma(\omega)z}$$

where $\gamma(\omega) = \sqrt{j\omega\mu(\sigma + j\omega\varepsilon)}$.

(a) Find the impulse response $[E_0(\omega) = 1]$ for E_x by using Table E.1 (Appendix E) and show that it is

$$E_x(z, t) = e^{-\sigma t/2\varepsilon}\,\delta(t - \sqrt{\mu\varepsilon}z)$$
$$+ \frac{1}{4}\left(\frac{\sigma}{\varepsilon}\right)^2 \sqrt{\mu\varepsilon}ze^{-\sigma t/2\varepsilon}\,\frac{I_1[(\sigma/2\varepsilon)\sqrt{t^2 - \mu\varepsilon z^2}]}{(\sigma/2\varepsilon)\sqrt{t^2 - \mu\varepsilon z^2}}\,u(t - \sqrt{\mu\varepsilon}z)$$

(b) Show that $[E_0(\omega) = 1)]$

$$H_y(z, t) = \frac{e^{-\sigma t/2\varepsilon}}{\sqrt{\mu/\varepsilon}}\,\delta(t - \sqrt{\mu\varepsilon}z)$$
$$+ \frac{1}{4}\left(\frac{\sigma}{\varepsilon}\right)^2 t\,\frac{e^{-\sigma t/2\varepsilon}}{\sqrt{\mu/\varepsilon}}\,\frac{I_1[(\sigma/2\varepsilon)\sqrt{t^2 - \mu\varepsilon z^2}]}{(\sigma/2\varepsilon)\sqrt{t^2 - \mu\varepsilon z^2}}\,u(t - \sqrt{\mu\varepsilon}z)$$
$$+ \frac{\sigma}{2\varepsilon}\frac{e^{-\sigma t/2\varepsilon}}{\sqrt{\mu/\varepsilon}}\,I_0\left(\frac{\sigma}{2\varepsilon}\sqrt{t^2 - \mu\varepsilon z^2}\right)\quad u(t - \sqrt{\mu\varepsilon}z)$$

(c) Interpret these results.

31. (a) Using $\mathbf{J} = \rho_v\mathbf{u}$ and $\mathbf{F} = q\mathbf{E} = m\mathbf{a}$ show that

$$\mathbf{J} = -j\frac{Nq^2}{\omega m}\mathbf{E}$$

in a region containing N electrons per cubic meter (charge q, mass m) if the excitation is sinusoidal in time and electronic collisions with molecules may be neglected.

(b) Replacing conduction current density with that in part (a), show that a plane wave solution for the region is

$$E_x(z, \omega) = E_0(\omega)e^{-\gamma(\omega)z}$$

$$H_y(z, \omega) = \frac{E_0(\omega)}{\eta(\omega)} e^{-\gamma(\omega)z}$$

where

$$\gamma(\omega) = \sqrt{\mu\varepsilon}\sqrt{\omega_p^2 - \omega^2}$$

$$\eta(\omega) = \sqrt{\mu/\varepsilon}\sqrt{\frac{1}{1 - (\omega_p/\omega)^2}}$$

$$\omega_p^2 \equiv \frac{Nq^2}{m\varepsilon}, \quad \text{plasma frequency}$$

(c) Using Table E.1, Appendix E, show that the impulse response $[E_0(\omega) = 1]$ for the electric field is

$$E_x(z, t) = \delta(t - \sqrt{\mu\varepsilon}z)$$

$$- \omega_p^2\sqrt{\mu\varepsilon}z \frac{J_1(\omega_p\sqrt{t^2 - \mu\varepsilon z^2})}{\omega_p\sqrt{t^2 - \mu\varepsilon z^2}} u(t - \sqrt{\mu\varepsilon}z)$$

(*plane wave in an ionized region*).

Chapter 11
Lossless Two-Wire Systems (Transmission Lines)

The characteristics of uniform plane waves were developed in some detail in the previous chapter. In this chapter we will use these results to develop a two-conductor guiding system, commonly called a transmission line. Transmission lines are important in those applications where it is necessary to convey energy from one point to another at frequencies from 0 Hz to 3 or 4 GHz. The treatment here is a basic one, but the emphasis is certainly aimed at "high-frequency" transmission lines. This chapter is concerned only with the idealized case of lossless uniform transmission lines. That is, the conductors are perfect ($\sigma \to \infty$) and the dielectric in which they are embedded is lossless ($\sigma = 0$). The effect of the generator at the sending end and the load at the receiving end will be considered. Many of the parameters we will encounter will be very similar to those that occur for plane waves with normal incidence on a plane interface.

11.1 THE PARALLEL PLANE GUIDING SYSTEM

Consider a uniform plane wave propagating in a lossless dielectric, free of conductors, as shown in Figure 11.1. As defined in Chapter 10, this is a TEM to z mode of propagation since **E** and **H** lie in a plane (which extends to

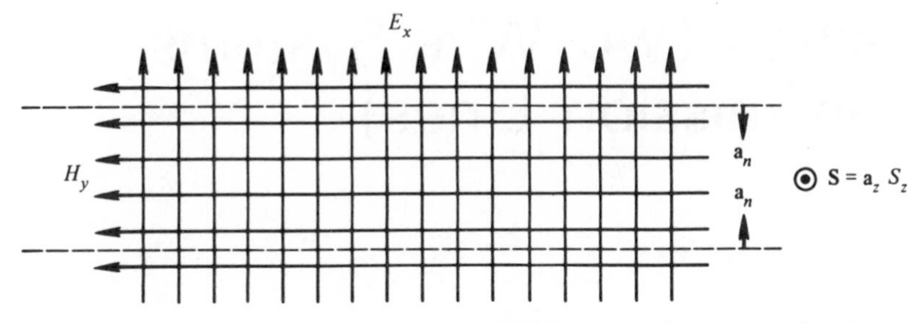

Figure 11.1. A uniform plane wave or TEM to z mode propagating in a lossless dielectric (time = constant).

infinity in two dimensions) perpendicular to the z direction of propagation or power density flow. Is it possible to place perfectly conducting planes of infinite extent along the dashed lines in Figure 11.1 without disturbing the field distribution? A little thought should convince us that the answer is yes. We conclude that a positive surface charge density exists on the lower conductor, while a (equal magnitude) negative surface charge density exists on the upper conductor. These charges fulfill the boundary condition (see equations 9.95 through 9.98),

$$\rho_s = \mathbf{a}_n \cdot \mathbf{D} \quad (\mathrm{C/m^2}) \tag{11.1}$$

or simply

$$|\rho_s| = |D_n| \tag{11.2}$$

where $\mathbf{D} = \varepsilon\mathbf{E}$ and the unit vector $\mathbf{a}_n$ is as shown in Figure 11.1. Also, these charges are changing in time (since $\mathbf{E}$ is changing in time) and therefore represent surface currents. The relation between the currents and charges is the *continuity of current* equation, or as it is often called, the *conservation of charge* equation, equation 9.30, which for surface currents and charges becomes

$$\nabla \cdot \mathbf{J}_s = -j\omega\rho_s \tag{11.3}$$

Simply stated, the surface currents obey the boundary condition

$$\mathbf{J}_s = \mathbf{a}_n \times \mathbf{H} \quad (\mathrm{A/m}) \tag{11.4}$$

which was introduced as equation 9.98. Thus, a surface current ($|J_s| = |H_{\tan}|$) will flow *out* of the page on the *inside* of the *lower* conductor and the "return" current will flow *into* the page on the *inside* of the *upper* conducting plane. The current flow is pictured in Figure 11.2.

As far as the *region between the conducting planes* is concerned, we must have a possible solution, because we have satisfied Maxwell's equations *in* the region and *boundary* conditions *on* the boundaries. The mode of propagation is still *TEM*. Other modes, called TE and TM modes, can exist for the system

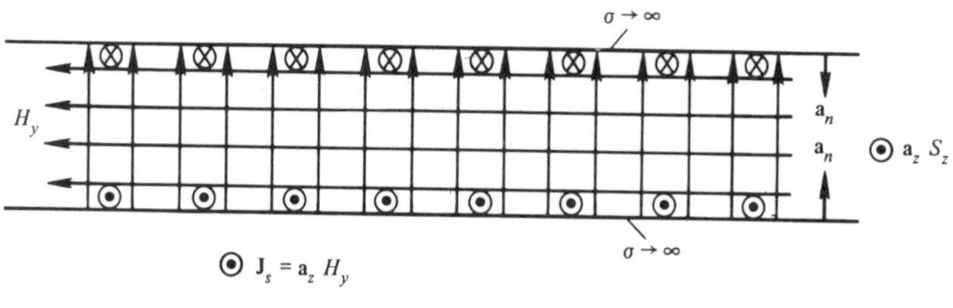

Figure 11.2. Surface current densities in an infinite parallel plane guiding system (TEM to z mode).

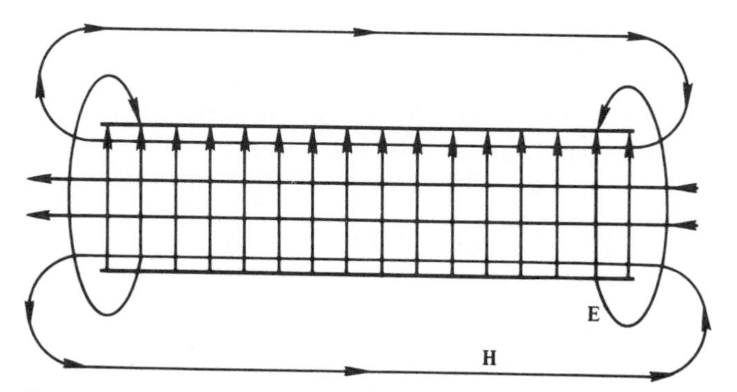

Figure 11.3. The field around a finite (width) parallel plane guiding system (TEM to z mode).

we have constructed. They will be investigated at an appropriate time, and are not considered at this point. See Figure 10.13.

Suppose next that we remove a *finite* width from this system and examine those changes that must occur. We expect that a TEM mode will *still* exist because we *still* essentially have a two-dimensional system. There must be some fringing or distortion of the fields near the edges of the finite planes as indicated in Figure 11.3. This *new* problem can be solved exactly, but at the present time that procedure would merely complicate the simple concept we are presenting. The final system which has evolved is a practical two-wire transmission line with *no loss.*

11.2 THE GENERAL LOSSLESS LINE

Let us now consider two uniform perfect conductors of infinite length located in a perfect dielectric such that the conductors form cylinders with axes parallel to the z axis. The cross-sectional geometry is shown in Figure 11.4. Further, assume $E_z = H_z = 0$, and z dependence of the form $e^{-\gamma z}$. We have already seen that a traveling wave can take this form. γ may be imaginary (lossless case), or complex (general case). It is important to recognize that an

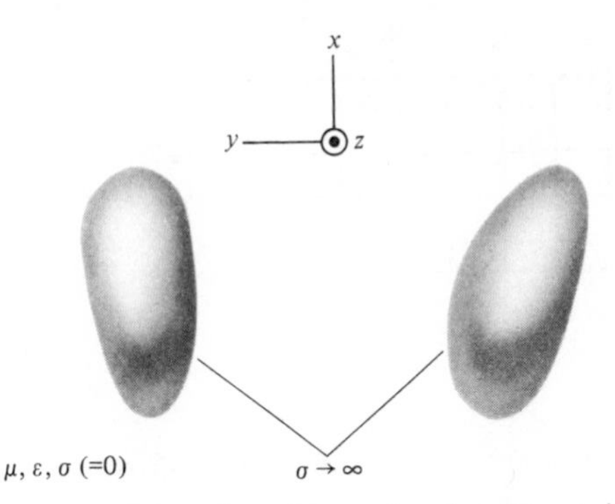

Figure 11.4. An arbitrary lossless two-wire uniform guiding system or transmission line in a lossless medium.

identical geometry existed when we considered "static" capacitance and inductance in Figure 8.14 and Figure 8.15. We next apply Maxwell's equations in rectangular coordinates. The fields are assumed to be of the form $e^{-\gamma z}$, and with $E_z = H_z = 0$, rewriting Maxwell's equations in components yields

$$\gamma E_y = -j\omega\mu H_x \tag{11.5}$$

$$\gamma E_x = j\omega\mu H_y \tag{11.6}$$

$$\frac{\partial E_y}{\partial x} - \frac{\partial E_x}{\partial y} = 0 \tag{11.7}$$

$$\gamma H_y = j\omega\varepsilon E_x \tag{11.8}$$

$$\gamma H_x = -j\omega\varepsilon E_y \tag{11.9}$$

and

$$\frac{\partial H_y}{\partial x} - \frac{\partial H_x}{\partial y} = 0 \tag{11.10}$$

It is easy to show that $\gamma = j\omega\sqrt{\mu\varepsilon} = jk$, k being the *intrinsic* phase constant. Thus the z dependence is e^{-jkz}, or, in other words, the fields are undamped traveling waves in the positive z direction. We take note of the fact that we could just as easily have obtained traveling waves in the negative z direction (e^{+jkz}), or, in general, waves traveling in *both* directions simultaneously. Note also that equations 11.5 through 11.10 may be described by

$$\boxed{\mathbf{E} = \eta\mathbf{H} \times \mathbf{a}_z} \tag{11.11}$$

and

$$\boxed{\mathbf{H} = \frac{1}{\eta}\mathbf{a}_z \times \mathbf{E}} \tag{11.12}$$

where $\eta = \sqrt{\mu/\varepsilon}$, the intrinsic impedance. It is then apparent that

$$\mathbf{E} \cdot \mathbf{H} = 0 \tag{11.13}$$

or **E** and **H** are everywhere *perpendicular*. Solving equations 11.5 through 11.10 simultaneously yields Laplace's equation in two dimensions for all four field components.

$$\frac{\partial^2 E_x}{\partial x^2} + \frac{\partial^2 E_x}{\partial y^2} = 0 \tag{11.14}$$

$$\frac{\partial^2 E_y}{\partial x^2} + \frac{\partial^2 E_y}{\partial y^2} = 0 \tag{11.15}$$

$$\frac{\partial^2 H_x}{\partial x^2} + \frac{\partial^2 H_x}{\partial y^2} = 0 \tag{11.16}$$

and

$$\frac{\partial^2 H_y}{\partial x^2} + \frac{\partial^2 H_y}{\partial y^2} = 0 \tag{11.17}$$

The field components must also satisfy the boundary conditions,

$$\begin{aligned} E_{\text{tan}} &= 0 \\ H_{\text{norm}} &= 0 \end{aligned} \tag{11.18}$$

on the perfect conductor surfaces.

Equations 11.14, 11.15, and 11.18 represent exactly the same boundary value problem for **E** that we have already solved for the *electrostatic* case. Equations 11.16, 11.17, and 11.18 represent exactly the same boundary value problem for **H** that we have already solved for the *magnetostatic* case. Figure 11.5 may now be compared with Figure 8.14 and Figure 8.15. Electric field lines in Figure 11.5 will be the same (for conductors with the same geometry, of course) as the electric field lines in Figure 8.15 *or* the lines of Φ_m = constant in Figure 8.14. Likewise, magnetic field lines in Figure 11.5 will be the same as the magnetic field lines in Figure 8.14 or the lines of Φ = constant in Figure 8.15. Therefore, we are permitted to use "static" capacitance and "static" external inductance, when the need for their use arises in this system, even though the fields are varying sinusoidally in time. This rather unusual result has occurred because the *mode* we are describing is TEM to z ($E_z = H_z \equiv 0$), whose z variation is simply $e^{-\gamma z}$.

A voltage, or emf, and current at any z for our infinite lossless two-wire line can be found by integrating along the contours shown in Figure 11.5. Notice that z is constant on these contours. Notice also that with z directed currents the fields are derivable from a vector potential **A** lying entirely in the z direction, and according to equation 9.82 and the integrating contours of Figure 11.5, voltage and potential difference will be the same. This is another property of TEM waves.

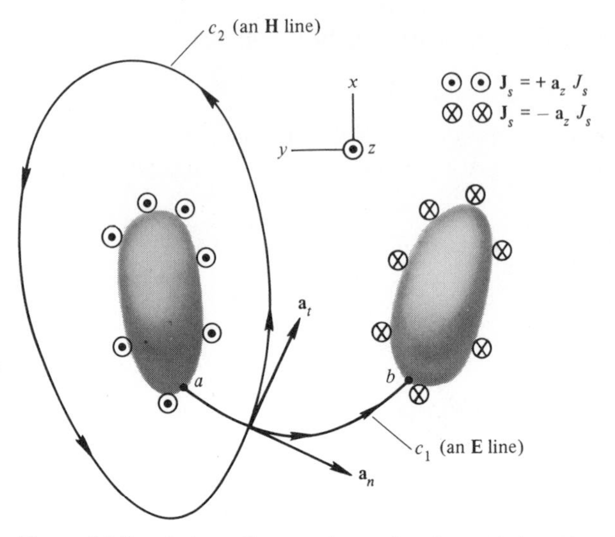

Figure 11.5. Integrating contours for determining V and I for the two-wire line with surface currents.

The appropriate defining equations are

$$V = \int_{a,\,c1}^{b} \mathbf{E} \cdot d\mathbf{l}_1 \quad \text{(V)}, \qquad (= V_{ab}) \tag{11.19}$$

the voltage between the conductor carrying the $+z$ directed current and the conductor carrying the $-z$ directed current, and

$$I = \oint_{c2} \mathbf{H} \cdot d\mathbf{l}_2 \tag{11.20}$$

from Ampere's law. Notice that displacement current does not enter equation 11.20 (Why not?). Since **E** and **H** represent traveling waves, it follows from equations 11.19 and 11.20 that V *and* I *also represent traveling waves*! That is, V and I must have the same z variation as **E** and **H**, respectively.

Substituting equation 11.12 into equation 11.20 gives

$$I = \frac{1}{\eta} \oint_{c2} \mathbf{a}_z \times \mathbf{E} \cdot d\mathbf{l}_2$$

but $\mathbf{E} = \mathbf{a}_n E_n$ at the intersection of c_1 and c_2, where $\mathbf{a}_n$ and $\mathbf{a}_t$ are unit vectors, normal and tangent to c_2, respectively, and are *not constant* vectors. Also, $\mathbf{H} = \mathbf{a}_t H_t$ at the intersection of c_1 and c_2. Equation 11.20 can thus be written ($d\mathbf{l}_2 = \mathbf{a}_t\, dl_2$)

$$I = \frac{1}{\eta} \oint_{c2} E_n\, dl_2 \tag{11.21}$$

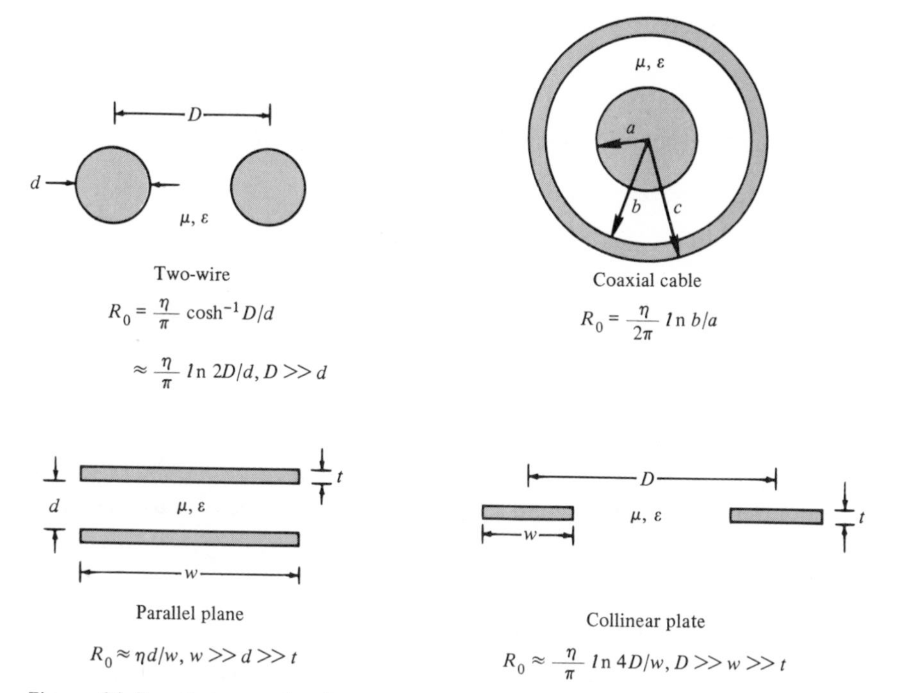

Figure 11.6. Common lossless two-conductor transmission lines and their characteristic resistances.

From electrostatics, the definition of capacitance is $C = Q/V$, where, for a unit length in the z direction, $Q = \varepsilon \oint_{c_2} E_n \, dl_2(1)$. Therefore, the capacitance per unit length is (see equation 8.43)

$$C = \frac{\varepsilon}{V} \oint E_n \, dl_2 \quad \text{(F/m)} \tag{11.22}$$

Comparing equations 11.21 and 11.22, it is easy to see that we may find the ratio of voltage to current as

$$\boxed{R_0 \equiv \frac{V}{I} = \eta \frac{\varepsilon}{C}} \quad (\Omega) \tag{11.23}$$

and call it the characteristic resistance of the lossless line. In the same way ($d\mathbf{l}_1 = \mathbf{a}_n \, dl_1$),

$$V = \eta \int_{a, c_1}^{b} \mathbf{H} \times \mathbf{a}_z \cdot d\mathbf{l}_1 = \int_{c_1} H_t \, dl_1 \tag{11.24}$$

and inductance is $L \equiv \psi_m/I$, where ψ_m, the flux, for a unit length in the z direction, is $\psi_m = \mu \int_{c_1} H_t \, dl_1(1)$. Therefore, the inductance per unit length is

(see equation 8.44)

$$L = \frac{\mu}{I} \int_{c_1} H_t \, dl_1 \quad \text{(H/m)} \tag{11.25}$$

Comparing equations 11.24 and 11.25, it is easy to show that

$$\boxed{R_0 = \frac{V}{I} = \eta \frac{L}{\mu}} \quad (\Omega) \tag{11.26}$$

Comparing equations 11.23 and 11.26, it follows, as we have already seen in equation 8.45, that

$$\boxed{LC = \mu\varepsilon} \quad (L = L_{\text{ext}}) \tag{11.27}$$

Then, if *either* L or C is known, the other may be found using equation 11.27. Remember that the values of L and C used here are *per unit length* values. Some common geometries, with their characteristic resistances, are shown in Figure 11.6.

11.3 THE LOSSLESS LINE EQUIVALENT CIRCUIT

We have shown that the voltage and current at any point on the infinite lossless line are traveling waves, as long as the z dependence is $e^{-\gamma z}$. The ratio of V to I is R_0, where

$$R_0 = \eta \frac{\varepsilon}{C} = \eta \frac{L}{\mu}$$

L and C are the static *external* inductance and static capacitance, respectively, per unit length. They are *distributed* parameters. Since the line is of infinite length, the ratio of V to I at any point z is also the *input impedance* at the point z. Apparently, then, V, I, and R_0 play roles similar to those of $\mathbf{E}$, $\mathbf{H}$, and η. A distributed system, which has been described above, is shown in Figure 11.7. We would like to show that this system represents the equivalent circuit of a lossless two-conductor line when supporting the TEM to z mode. It should be pointed out again that such an equivalent circuit is not unique to the geometry, for this geometry can support other modes. The TEM mode is, however, by far the most important mode. It is instructive to first calculate the input impedance to such a system assuming that it extends to infinity as indicated in Figure 11.8. If the system is broken at $B - B'$, the input impedance at $B - B'$ is *still* that being sought. Then,

$$Z_{\text{in}} = \frac{j\omega L}{2} \Delta z + \frac{(1/j\omega C \, \Delta z)(Z_{\text{in}} + j\omega L \, \Delta z/2)}{1/j\omega C \, \Delta z + Z_{\text{in}} + j\omega L \, \Delta z/2}$$

or

$$Z_{\text{in}}^2 = \frac{L}{C} - \frac{\omega^2 L^2}{4} (\Delta z)$$

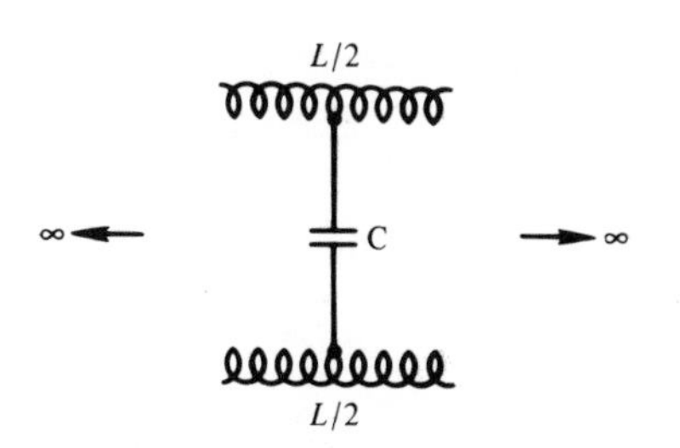

Figure 11.7. The distributed parameter representation of a lossless two-wire line. L = total external inductance per unit length. C = capacitance per unit length.

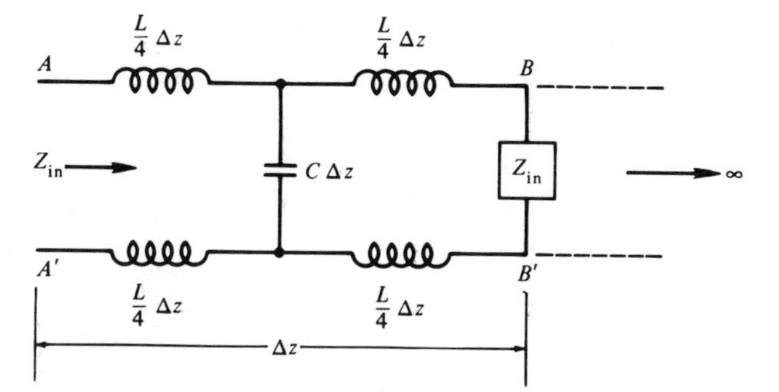

Figure 11.8. Equivalent circuit for determining the input impedance, Z_{in}, for the system of Figure 11.7.

Figure 11.9. Equivalent circuit for determining the voltage and current for a lossless two-wire line.

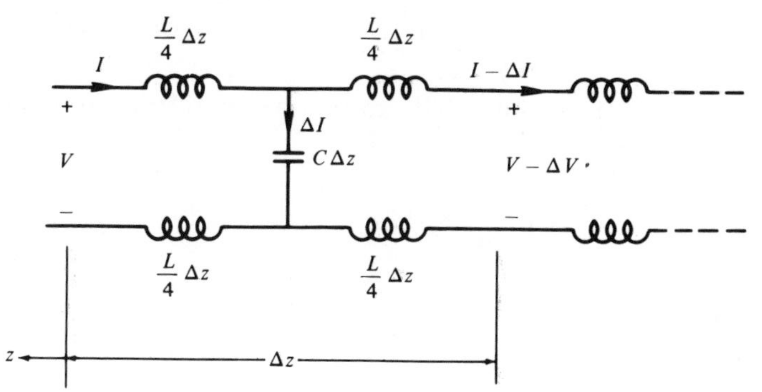

The line does not consist of lumped parameters, but is a uniformly distributed system. Hence, we must let Δz approach zero; and in the limit

$$Z_{in} = \sqrt{\frac{L}{C}} \quad (\Omega) \tag{11.28}$$

Now, from equations 11.23 and 11.26,

$$\boxed{R_0 = \sqrt{\frac{L}{C}}} \quad (\Omega) \tag{11.29}$$

and R_0 is indeed the input impedance to an infinite length of the distributed system, as was noted in the previous paragraph.

The circuit voltage and current differential equations may also be derived from the distributed system of Figure 11.8, redrawn and labeled in Figure 11.9. Considering a length Δz of the system and applying Kirchhoff's laws,

$$V - j\omega \frac{L}{2} \Delta z\, I - j\omega \frac{L}{2} \Delta z(I - \Delta I) - (V - \Delta V) = 0 \tag{11.30}$$

and

$$I - j\omega C\, \Delta z \left(V - \frac{\Delta V}{2}\right) - (I - \Delta I) = 0 \tag{11.31}$$

or

$$\frac{\Delta V}{\Delta z} = j\omega LI - j\omega \frac{L}{2} \Delta I \tag{11.32}$$

and

$$\frac{\Delta I}{\Delta z} = j\omega CV - j\omega C \frac{\Delta V}{2} \tag{11.33}$$

Now, in the limit as $\Delta z \to 0$ both ΔI and ΔV approach zero also. Recalling the definition of a derivative, we obtain in the limit,

$$\frac{dV}{dz} = j\omega LI \tag{11.34}$$

and

$$\frac{dI}{dz} = j\omega CV \tag{11.35}$$

Differentiating equation 11.34 with z, and substituting for dI/dz from equation 11.35 gives

$$\frac{d^2V}{dz^2} = -\omega^2 LCV$$

or

$$\frac{d^2V}{dz^2} + k^2 V = 0 \tag{11.36}$$

since $k = \omega\sqrt{LC} = \omega\sqrt{\mu\varepsilon}$, from equation 11.27. In the same way,

$$\frac{d^2 I}{dz^2} + k^2 I = 0 \tag{11.37}$$

Equations 11.36 and 11.37 are the *homogeneous wave equations* for V and I, respectively. We already know that V and I are traveling waves. We also know from Chapter 10 that solutions to equations 11.36 and 11.37 will be traveling waves with phase velocity $u_p = 1/\sqrt{\mu\varepsilon} = 1/\sqrt{LC}$. The distributed system of Figure 11.7 thus leads to an accurate representation of V and I, and therefore, does indeed represent the equivalent circuit of the lossless two-wire line when it is supporting the TEM mode. Consideration of Problem 3 at the end of this chapter also demonstrates the validity of the equivalent circuit concept.

11.4 GENERAL SOLUTIONS FOR V AND I—LOSSLESS LINE

The wave equations for the uniform plane wave propagating in a lossless medium as given in Chapter 10 were

$$\frac{d^2 E_x}{dz^2} + k^2 E_x = 0$$

and

$$\frac{d^2 H_y}{dz^2} + k^2 H_y = 0$$

The latter equation was not actually derived, but follows almost immediately from equation 10.12. The solutions to these equations are equations 10.13 and 10.15.

$$E_x = Ae^{-jkz} + Be^{+jkz} \tag{10.13}$$

and

$$H_y = \frac{Ae^{-jkz}}{\sqrt{\mu/\varepsilon}} - \frac{Be^{+jkz}}{\sqrt{\mu/\varepsilon}} \tag{10.15}$$

Comparing the last six equations leads to the conclusion that *general* solutions to equations 11.36 and 11.37 for V and I at any point z must be

$$V = C_1 e^{+jkz} + C_2 e^{-jkz} \tag{11.38}$$

and

$$I = C_3 e^{+jkz} + C_4 e^{-jkz} \tag{11.39}$$

Remember that

$$R_0 = \eta\frac{\varepsilon}{C} = \eta\frac{L}{\mu} = \sqrt{\frac{L}{C}}$$

and

$$k = \omega\sqrt{\mu\varepsilon} = \omega\sqrt{LC}$$

Again, the first terms of V and I represent undamped (no losses) traveling waves in the $-z$ direction, while the second terms represent undamped traveling waves in the $+z$ direction. Using equations 11.34, 11.38, and 11.29, we have

$$I = \frac{1}{j\omega L}\frac{dV}{dz} = \frac{1}{j\omega L}(j\omega\sqrt{LC}C_1 e^{j\omega\sqrt{LC}z} - j\omega\sqrt{LC}C_2 e^{-j\omega\sqrt{LC}z})$$

or

$$I = \frac{C_1}{R_0} e^{j\omega\sqrt{LC}z} - \frac{C_2}{R_0} e^{-j\omega\sqrt{LC}z} \tag{11.40}$$

Notice that for the $-z$ traveling wave (alone)

$$\frac{V^-}{I^-} = R_0 \tag{11.41}$$

while for the $+z$ traveling wave

$$\frac{V^+}{I^+} = -R_0 \tag{11.42}$$

The constant C_2 (independent of z) can be determined in terms of the constant C_1 (independent of z) if conditions at the load end ($z = 0$) are known. Knowledge of the load impedance, Z_l, is all that is necessary. Then,

$$\left.\frac{V}{I}\right|_{z=0} = Z_l = \frac{C_1 + C_2}{C_1 - C_2} R_0$$

where $z = 0$ is located at the load, and z increases toward the generator as in Figure 11.10. Solving for C_2/C_1 we obtain

$$\frac{C_2}{C_1} = \frac{Z_l - R_0}{Z_l + R_0} \tag{11.43}$$

This ratio is quite logically called the *voltage coefficient of reflection at the load*, Γ_l. That is,

$$\Gamma_l = \frac{Z_l - R_0}{Z_l + R_0} = |\Gamma_l| e^{j\phi_l} \tag{11.44}$$

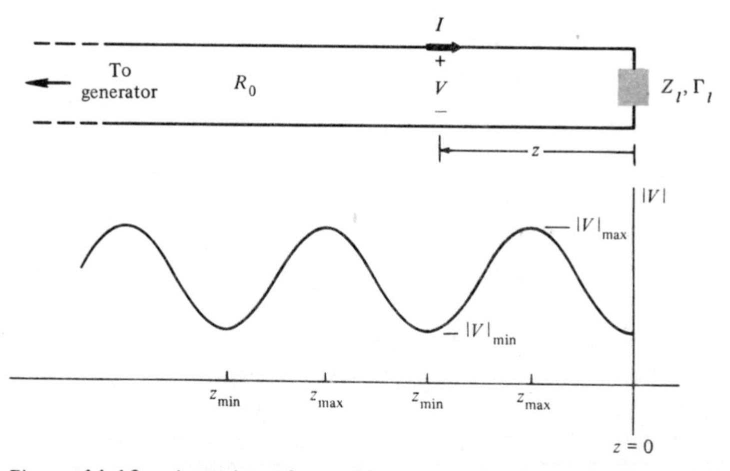

Figure 11.10. Location of an arbitrary load on a lossless line and the resultant voltage standing wave.

Compare this result to equation 10.70 for plane wave reflection. The phase constant is $k = \omega\sqrt{LC} = \omega\sqrt{\mu\varepsilon} = \omega/u_p$. Equations 11.38 and 11.39 may now be written

$$V(z, \omega) = C_1 e^{jkz} + \Gamma_l C_1 e^{-jkz} \tag{11.45}$$

and

$$I(z, \omega) = \frac{C_1}{R_0} e^{jkz} - \frac{\Gamma_l C_1}{R_0} e^{-jkz} \tag{11.46}$$

or

$$V(z, \omega) = C_1 e^{jkz}(1 + |\Gamma_l| e^{j(\phi_l - 2kz)}) \tag{11.47}$$

and

$$I(z, \omega) = \frac{C_1}{R_0} e^{jkz}(1 - |\Gamma_l| e^{j(\phi_l - 2kz)}) \tag{11.48}$$

11.5 VOLTAGE STANDING WAVE RATIO

The voltage standing wave ratio, VSWR, is defined as the ratio $|V|_{max}/|V|_{min}$, and from equation 11.47, it is

$$\boxed{\text{VSWR} = \frac{1 + |\Gamma_l|}{1 - |\Gamma_l|}} \tag{11.49}$$

$|V|_{max}$ and $|V|_{min}$ are identified in Figure 11.10. $|V|_{max}$ occurs where

$$\phi_l - 2kz = -2n\pi, \qquad n = 0, 1, 2, \ldots, \qquad (z > 0)$$

or

$$z_{max} = (\phi_l + 2n\pi)/2k \tag{11.50}$$

$|V|_{min}$ occurs where

$$\phi_l - 2kz = -(2n - 1)\pi, \qquad n = 0, 1, 2, \ldots, \qquad (z > 0)$$

or

$$z_{min} = [\phi_l + (2n - 1)\pi]/2k \tag{11.51}$$

We will agree to use only positive values for ϕ_l, and, in any case, must use positive values for z_{max} or z_{min}. The spacing between adjacent crests ($|V|_{max}$) or adjacent troughs ($|V|_{min}$) will be $\lambda/2$. Either a crest or a trough will occur at a distance less than (or at most, equal to) $\lambda/4$ from the load. See Figure 11.10.

Knowledge of the location of $|V|_{min}$ or $|V|_{max}$ determines ϕ_l by equations 11.51 or 11.50, while knowledge of the VSWR determines $|\Gamma_l|$ by equation 11.49. Then, Γ_l can be completely determined from both pieces of information. If Γ_l is known, and if R_0 is known, then Z_l can be found from equation 11.44. This is an important practical consideration, for in many cases the load impedance is not easy to identify. An antenna is a good example. It certainly represents a load on the end of the line, but the impedance of this kind of load is certainly not obvious by visual inspection. The similarity between the preceding equations (and the interpretation of them) and those of Chapter 10 for plane waves with normal incidence on an interface is obvious.

11.6 INPUT IMPEDANCE

The input impedance for an infinite length of line has been found to be R_0. In this section, the input impedance for a finite length line terminated in a load impedance Z_l will be considered. The input impedance can be measured with a bridge, and can be calculated from equations 11.45 and 11.46. It is important to recognize at this point that we are implicitly assuming that the wave reflected from the load end ($z = 0$) is completely absorbed at the source or generator end which is somewhere to the left of the point where we are calculating the input impedance. Put another way, we are assuming that the reflected wave is itself not reflected at the source to give another incident wave. The input impedance is

$$Z_{in} = \frac{V}{I} = R_0 \frac{e^{jkz} + \Gamma_l e^{-jkz}}{e^{jkz} - \Gamma_l e^{-jkz}}$$

or

$$Z_{in} = R_0 \frac{(Z_l + R_0)e^{jkz} + (Z_l - R_0)e^{-jkz}}{(Z_l + R_0)e^{jkz} - (Z_l - R_0)e^{-jkz}}$$

or

$$Z_{\text{in}} = R_0 \frac{Z_l \cos kz + jR_0 \sin kz}{R_0 \cos kz + jZ_l \sin kz} \tag{11.52}$$

If the line is terminated in a short circuit ($Z_l = 0$), equation 11.52 gives

$$Z_{\text{in}}^{\text{sc}} = jR_0 \tan kz \tag{11.53}$$

and if the line is terminated in an open circuit ($Z_l \to \infty$), equation 11.52 gives

$$Z_{\text{in}}^{\text{oc}} = -jR_0 \operatorname{ctn} kz \tag{11.54}$$

Combining equations 11.53 and 11.54 gives

$$R_0 = \sqrt{Z_{\text{in}}^{\text{sc}} Z_{\text{in}}^{\text{oc}}} \tag{11.55}$$

An example will be considered at this point.

EXAMPLE 1

The following measurements are made on a lossless line, Here, we are ignoring the fact that *all* real lines have some loss.

$$Z_{\text{in}}^{\text{sc}} = j50\ \Omega \qquad Z_{\text{min}} = 0, \frac{\lambda}{2}, \lambda, \ldots$$

$$Z_{\text{in}}^{\text{oc}} = -j50\ \Omega \qquad \text{VSWR} = 2$$

From equation 11.55,

$$R_0 = \sqrt{(j50)(-j50)} = 50 \quad (\Omega)$$

From equation 11.49,

$$2 = \frac{1 + |\Gamma_l|}{1 - |\Gamma_l|} \quad \text{or} \quad |\Gamma_l| = \frac{1}{3}$$

and from the equation preceding equation 11.51

$$\phi_l = 2kz_{\text{min}} - (2n - 1)\pi = \frac{4\pi}{\lambda}(0) - (2n - 1)\pi = \pi, -\pi, -3\pi, \ldots = \pi$$

It should be pointed out that the other values of z_{min} ($\lambda/2$, λ, etc.) give the same value of ϕ_l. Thus,

$$\Gamma_l = \frac{1}{3} e^{j\pi} = -\frac{1}{3}$$

Then, from equation 11.44,

$$-\frac{1}{3} = \frac{Z_l - R_0}{Z_l + R_0}$$

or

$$Z_l = 25 \quad (\Omega)$$

Some important practical questions arise in considering the preceding example.

1. How is the position of $|V|_{\min}$ (that is, $z_{\min}$) actually measured?
2. How is the voltage standing wave ratio measured?
3. Is it possible to develop a simpler method to perform the preceding calculations?

These questions will all be answered as we proceed.

11.7 EFFECT OF THE GENERATOR

It was mentioned in the preceding section that the effect of the generator must ultimately be considered. This point is further emphasized when we realize that equations 11.45 and 11.46 for the voltage and current, respectively, *still* contain the unknown constant C_1. C_1 must be determined by what happens at the generator. Consider the complete system in Figure 11.11. If $Z_l \neq R_0$, then $\Gamma_l \neq 0$, and there will be a reflection from the load end. This reflected wave will travel to the generator and will be reflected itself if $Z_g \neq R_0$. This second reflection travels back toward the load and is reflected, and so on. Since

$$\Gamma_l = \frac{Z_l - R_0}{Z_l + R_0}$$

it follows that

$$\Gamma_g = \frac{Z_g - R_0}{Z_g + R_0} \tag{11.56}$$

for the generator end reflection coefficient.

The first step in the process may be analyzed with the equivalent circuit of Figure 11.12, where it is obvious that

$$V_s' = \frac{V_g R_0}{Z_g + R_0} \tag{11.57}$$

and

$$I_s' = \frac{V_g}{Z_g + R_0} \tag{11.58}$$

Thus V_s' and I_s' are the amplitudes of the *first* incident (on the load) voltage and current waves, respectively, on the line. They are the same as the *only* incident waves on an infinitely long line (no reflection at the load). At z the first incident

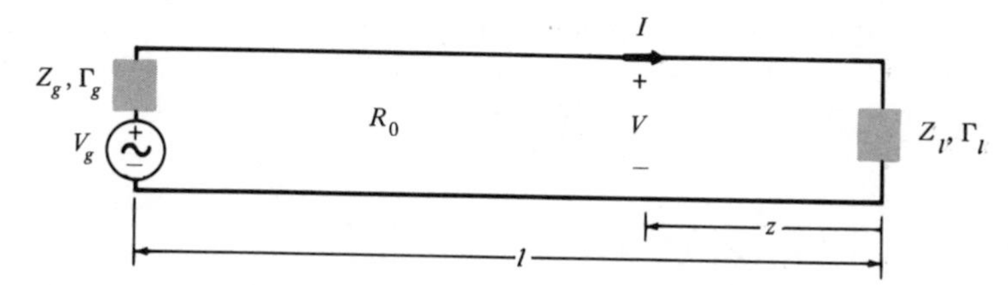

Figure 11.11. Complete two-wire transmission line system.

wave is $V_s' e^{-jk(l-z)}$. Traveling toward the load, the preceding wave is reflected, whereupon at z it is $\Gamma_l V_s' e^{-jk(l+z)}$. Continuing toward the generator the wave is also reflected, and at z it is $\Gamma_l \Gamma_g V_s' e^{-jk(3l-z)}$. This process continues, and tracing the wave gives the infinite series,

$$V(z, \omega) = V_s' e^{-jk(l-z)} + \Gamma_l V_s' e^{-jk(l+z)} + \Gamma_l \Gamma_g V_s' e^{-jk(3l-z)} + \Gamma_l \Gamma_g \Gamma_l V_s' e^{-jk(3l+z)} + \Gamma_l \Gamma_g \Gamma_l \Gamma_g V_s' e^{-jk(5l-z)} + \cdots \tag{11.59}$$

or

$$V(z, \omega) = V_s' e^{-jkl}(e^{jkz} + \Gamma_l e^{-jkz}) \times [1 + \Gamma_l \Gamma_g e^{-j2kl} + (\Gamma_l \Gamma_g e^{-j2kl})^2 + (\Gamma_l \Gamma_g e^{-j2kl})^3 + \cdots] \tag{11.60}$$

The second bracketed term in equation 11.60 is a geometric series, and the sum of such a series is

$$\text{sum} = \frac{a_0}{1 - a_{n+1}/a_n} \quad \text{(infinite geometric series)}$$

where a_n is the nth term in the series. Then,

$$V(z, \omega) = V_s' e^{-jkl}(e^{jkz} + \Gamma_l e^{-jkz}) \frac{1}{1 - \Gamma_l \Gamma_g e^{-j2kl}}$$

or

$$\boxed{V(z, \omega) = \frac{V_g R_0}{R_0 + Z_g} \frac{e^{-jkl}}{1 - \Gamma_l \Gamma_g e^{-j2kl}} (e^{jkz} + \Gamma_l e^{-jkz})} \tag{11.61}$$

Figure 11.12. Equivalent circuit for the first incident wave on a two-wire line.

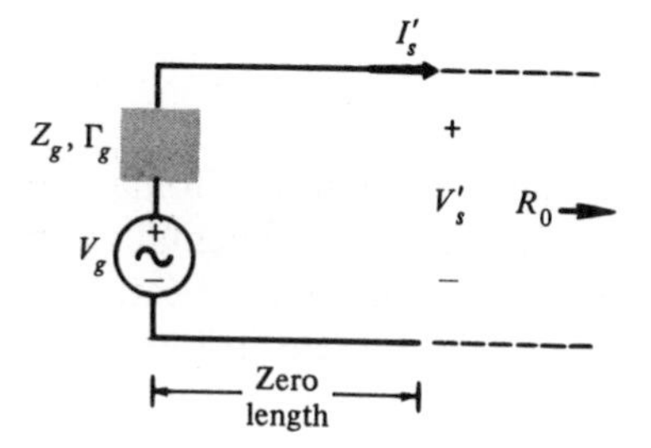

The current is easily found from equation 11.61 by simply noting that the coefficient of reflection for current is opposite to that for voltage. Thus, in equation 11.61, we replace V'_s by I'_s, Γ_l by $-\Gamma_l$ and Γ_g by $-\Gamma_g$. Thus,

$$I(z, \omega) = \frac{V_g}{R_0 + Z_g} \frac{e^{-jkl}}{1 - \Gamma_l \Gamma_g e^{-j2kl}} (e^{jkz} - \Gamma_l e^{-jkz}) \tag{11.62}$$

How does the input impedance as calculated with equations 11.61 and 11.62 compare with equation 11.52? Comparing equations 11.61 and 11.62 with 11.47 and 11.48, it is readily apparent that the constant C_1 is

$$C_1 = \frac{V_g R_0}{R_0 + Z_g} \frac{e^{-jkl}}{1 - \Gamma_l \Gamma_g e^{-j2kl}} \tag{11.63}$$

In order to eliminate multiple reflections, it is highly desirable to make the impedance looking into the generator, Z_g, equal to the line impedance, R_0. In this case, $\Gamma_g = 0$ so that

$$V(z, \omega) = \frac{V_g}{2} e^{-jkl}(e^{jkz} + \Gamma_l e^{-jkz}), \qquad Z_g = R_0 \tag{11.64}$$

and

$$I(z, \omega) = \frac{V_g}{2R_0} e^{-jkl}(e^{jkz} - \Gamma_l e^{-jkz}), \qquad Z_g = R_0 \tag{11.65}$$

11.8 SPECIAL LOAD IMPEDANCES

An examination of several special values of Z_l will shed additional light on the behavior of transmission lines. *Assume* $Z_g = R_0$.

1. *Open Circuit*

$$Z_l \to \infty$$

$$\Gamma_l = +1$$

$$\text{VSWR} \to \infty$$

$$Z_{\text{in}}^{\text{oc}} = -jR_0 \operatorname{ctn} kz \tag{11.66}$$

$$|V| = V_g |\cos kz|$$

$$|I| = \frac{V_g}{R_0} |\sin kz|$$

A plot of $|V|$, $|I|$, and $Z_{\text{in}}^{\text{oc}}$ against z is shown in Figure 11.13. It is customary to plot magnitudes for V and I since normal detectors in an experimental setup do not reproduce phase. The infinite standing wave ratios are indicative of the lack of power transfer to the open circuit. That is, there is just as much power flowing toward the open circuit as there is reflected back toward the generator where it is totally absorbed ($Z_g = R_0$). The input impedance indicates total reflection also, but in a slightly different manner. Z_{in}^{oc} is always a pure reactance, $-\infty \le X \le +\infty$. Depending on the distance to the load, z, the line can look like *any* capacitance or *any* inductance *at the terminals* where Z_{in}^{oc} is measured. At $z = 0, \lambda/2, \lambda, 3\lambda/2, 2\lambda, \ldots$, the line looks like a lossless parallel resonant circuit ($Z_{in}^{oc} \rightarrow \infty$), while for $z = \lambda/4, 3\lambda/4, 5\lambda/4, \ldots$, the line looks like a lossless series resonant circuit ($Z_{in}^{oc} = 0$).

2. *Short Circuit*

$$\begin{aligned} Z_l &= 0 \\ \Gamma_l &= -1 \\ \text{VSWR} &\rightarrow \infty \\ Z_{in}^{sc} &= +jR_0 \tan kz \\ |V| &= V_g |\sin kz| \\ |I| &= \frac{V_g}{R_0} |\cos kz| \end{aligned} \tag{11.67}$$

The equations, as well as Figure 11.14, indicate that the short-circuited line is complementary to the open-circuited line. There is no power transfer to the load, resulting in infinite standing wave ratios. Z_{in}^{sc} is always a pure reactance, so the line always looks like a capacitance or inductance at the terminals where Z_{in}^{sc} is measured. At $z = 0, \lambda/2, \lambda, 3\lambda/2, \ldots$ the line looks like a lossless series resonant circuit ($Z_{in}^{sc} = 0$), while for $z = \lambda/4, 3\lambda/4, 5\lambda/4, \ldots$ the line looks like a lossless parallel resonant circuit ($Z_{in}^{sc} \rightarrow \infty$).

3. *Mismatched Case*

$$\begin{aligned} Z_l &= 2R_0 \\ \Gamma_l &= +\frac{1}{3} \\ \text{VSWR} &= 2 \\ Z_{in} &= R_0 \frac{2 \cos kz + j \sin kz}{\cos kz + j2 \sin kz} \\ |V| &= \frac{V_g}{2} \left| 1 + \frac{1}{3} e^{-j2kz} \right| \\ |I| &= \frac{V_g}{2R_0} \left| 1 - \frac{1}{3} e^{-j2kz} \right| \end{aligned} \tag{11.68}$$

Figure 11.15 shows plots of R_{in}, X_{in}, $|V|$, and $|I|$ against z. Notice that these functions are repetitive with a period of $\lambda/2$. It is easy to show that this is *always* the case when there is a reflection from the load (no line loss).

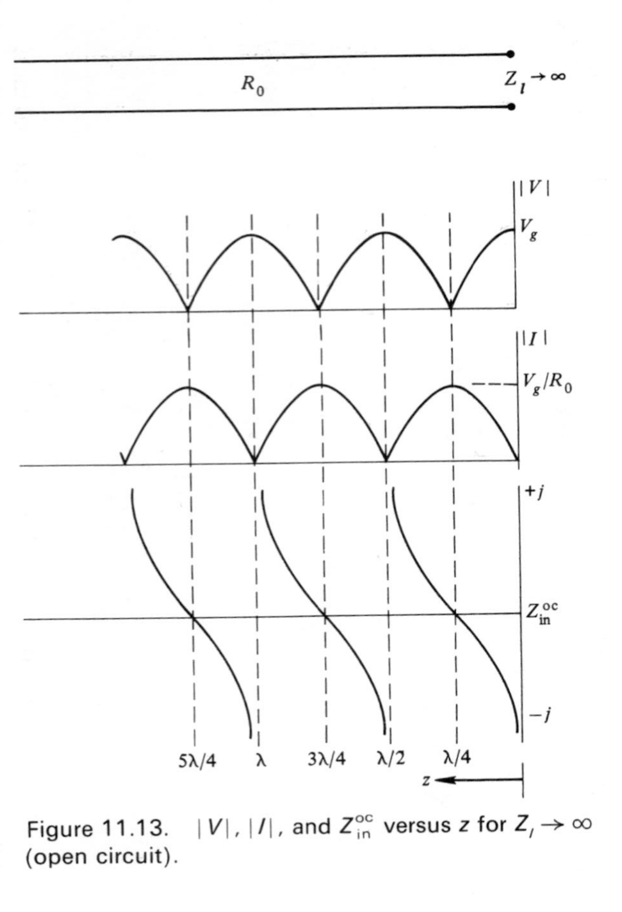

Figure 11.13. $|V|$, $|I|$, and Z_{in}^{oc} versus z for $Z_l \to \infty$ (open circuit).

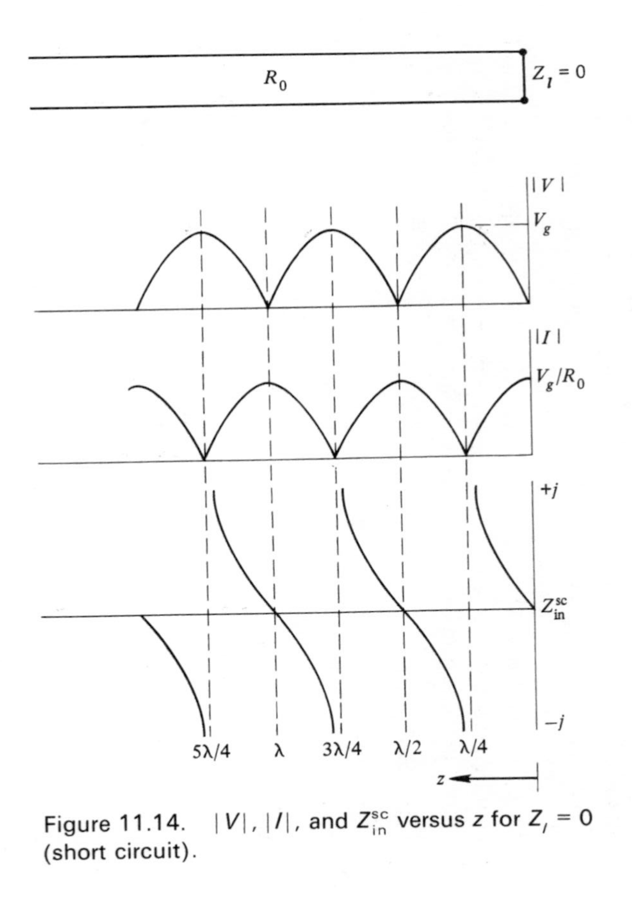

Figure 11.14. $|V|$, $|I|$, and Z_{in}^{sc} versus z for $Z_l = 0$ (short circuit).

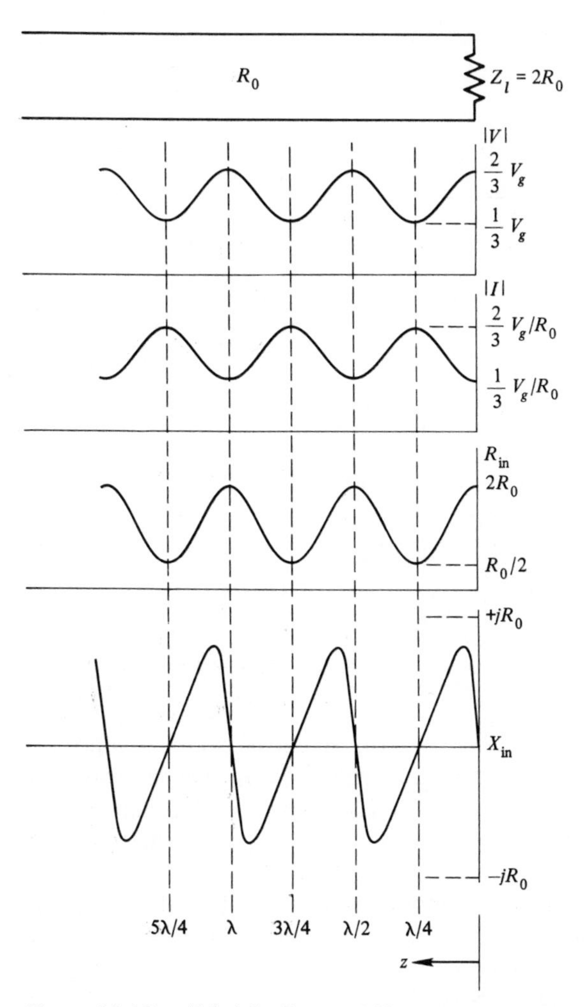

Figure 11.15. $|V|$, $|I|$, R_{in}, and X_{in} versus z for $Z_l = 2R_0$ ($Z_{in} = R_{in} + jX_{in}$).

4. *Matched Case*

$$\begin{aligned} Z_l &= R_0 \\ \Gamma_l &= 0 \\ \text{VSWR} &= 1 \\ Z_{in} &= R_0 \\ |V| &= \frac{V_g}{2} \\ |I| &= \frac{V_g}{2R_0} \end{aligned} \tag{11.69}$$

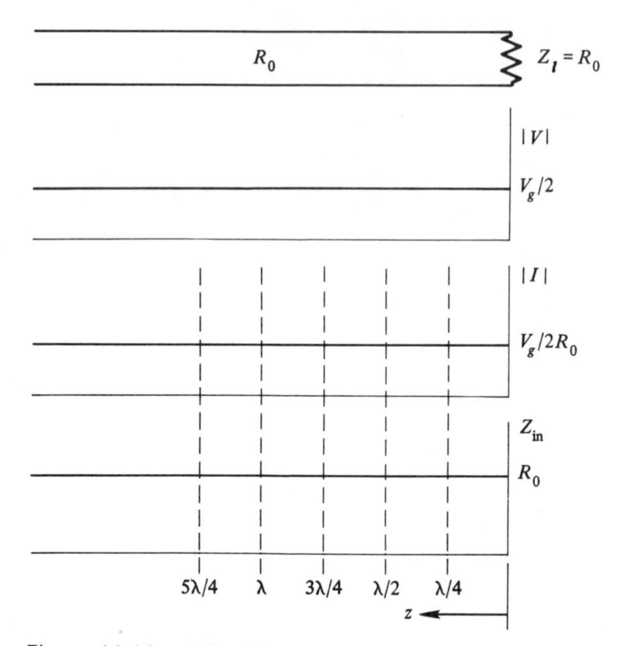

Figure 11.16. $|V|$, $|I|$, and Z_{in} versus z for $Z_l = R_0$ (matched case).

In this special and highly desirable case, the length of line is unimportant except in terms of phase shift or total time delay. To the incident wave, the load, $Z_l = R_0$, looks just like more transmission line with R_0 characteristic resistance. The line is said to be *terminated in a match*. Results are shown in Figure 11.16.

Since the load does look just like more transmission line, we may simplify the system to the equivalent circuit of *zero length* shown in Figure 11.17. In this equivalent circuit, maximum power transfer obviously occurs as long as $Z_g = R_0$. The time average load power is

$$\langle P_l \rangle = \frac{V_g^2}{8R_0} \quad \text{(W)} \tag{11.70}$$

Figure 11.17. Equivalent circuit (omitting the time delay) for the matched two-wire line.

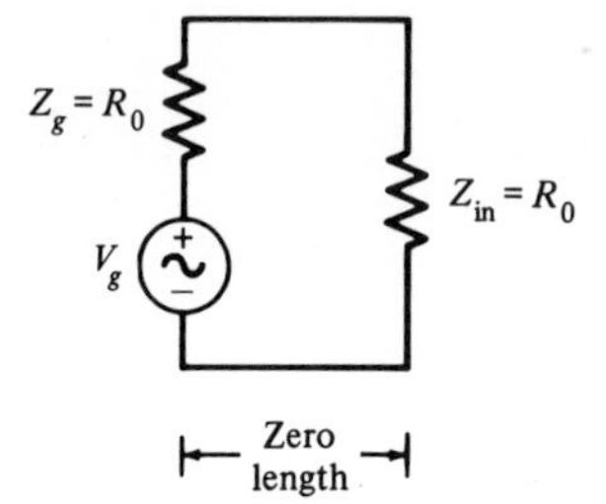

since V_g is a *peak value*. Notice that the original time delay is not retained in our equivalent circuit.

For the case $Z_g = R_0, Z_l \neq R_0$, it is not difficult to show that

$$\langle P_{\text{inc}} \rangle = \frac{V_g^2}{8R_0} \tag{11.71}$$

which is obviously the same as the power to a matched load (equation 11.70). It is also easy to show that

$$\langle P_{\text{ref}} \rangle = \frac{|\Gamma_l|^2 V_g^2}{8R_0} \tag{11.72}$$

The *net* load power is then

$$\langle P_l \rangle = \langle P_{\text{inc}} \rangle - \langle P_{\text{ref}} \rangle$$

or

$$\boxed{\langle P_l \rangle = \frac{V_g^2}{8R_0}(1 - |\Gamma_l|^2)} \tag{11.73}$$

Notice that equation 11.73 reduces to equation 11.70 if $\Gamma_l = 0$ (or $Z_l = R_0$).

11.9 EQUIVALENT LUMPED PARAMETERS

The so-called Q, or quality factor, of a lossless line is infinite by definition. Practical lines operated at reasonably high frequencies have small loss, so that, even though the Q is not infinite, it is very *high*. In this case, we mean high compared to a *lumped* parameter in the same application. As an example of a situation where this is important, consider the electronics engineer designing a UHF "tank" circuit for an oscillator or amplifier. She finds that any lumped inductance available represents a reactance which is too large. A possible solution is the use of a section of silver-plated transmission line whose length, z, is less than $\lambda/4$, representing the desired inductance. This example, along with some others is pictured in Figure 11.18.

In Figure 11.18(a) the length $z < \lambda/4$ is adjusted so that $X_L = X_C$, or (equation 11.53) $1/\omega C = R_0 \tan(2\pi z/\lambda)$. When this is done, parallel resonance occurs at the $x - x$ terminals. In Figure 11.18(b) the impedance at $x - x$ is the parallel combination of that from a section $d < \lambda/4$ with $Z_l = 0$, and that from a section $(\lambda/4 - d) < \lambda/4$ with $Z_l \to \infty$. Equations 11.53 and 11.54 are appropriate and give $Z_{\text{in}} \to \infty$ (parallel resonance) *for any value of* d. There are various ways in which this section may be used as a bandpass filter, where the center of the band will be $f = u_p/\lambda = 1/\lambda\sqrt{LC}$.

Using equation 11.52 with $z = \lambda/4$, it is easy to show that the input impedance at $x - x$ (looking to the right) in Figure 11.18(c) is R_{01}, and hence the system is matched, if $R_{02} = \sqrt{R_{01}R_l}$. Notice that this technique only

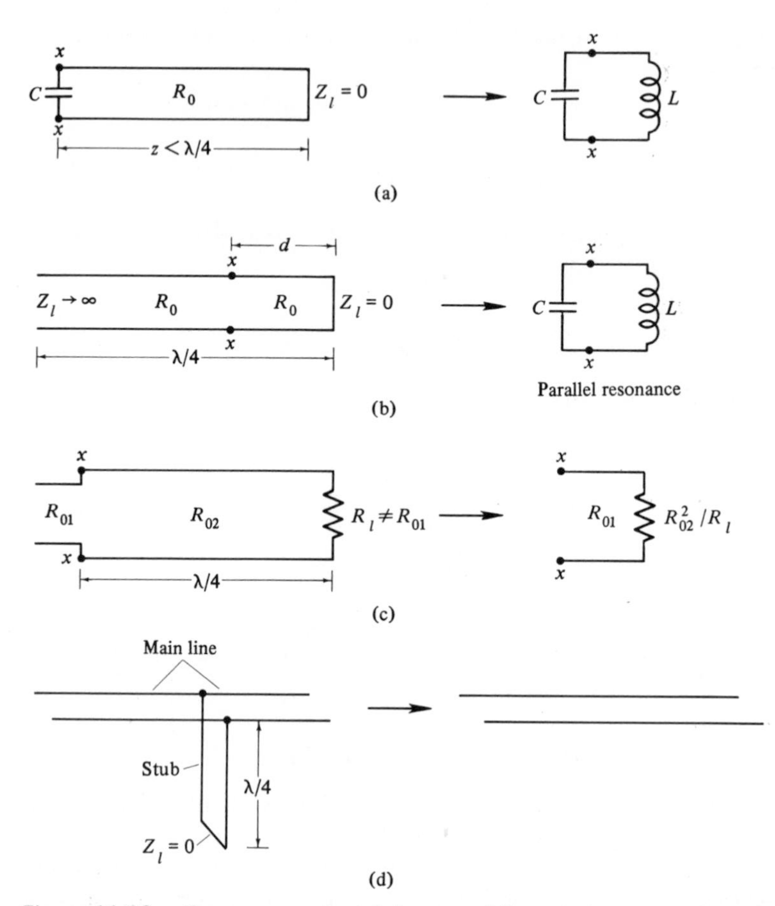

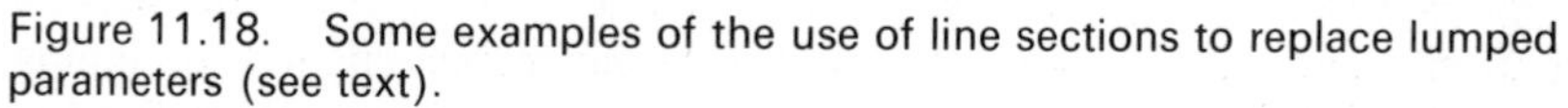

Figure 11.18. Some examples of the use of line sections to replace lumped parameters (see text).

applies to purely resistive loads, R_l. The impedance looking *into* the $\lambda/4$ section or "stub" of Figure 11.18(d) is infinite (equation 11.53), hence the stub has no effect on the main line. It should be emphasized that all of the systems in Figure (11.18) are *inherently* narrow-band.

11.10 PULSES ON THE LOSSLESS LINE

It is not difficult to show that the lossless line is also a distortionless line. The time delay, τ_D, encountered by a sinusoidal signal traveling a distance l on a lossless line is

$$\tau_D = \frac{l}{u_e} = \frac{l}{u_p} = \frac{l}{u_g} \tag{11.74}$$

The group velocity is $u_g = 1/(d\beta/d\omega) = 1/(dk/d\omega) = 1/\sqrt{\mu\varepsilon}$, so all the velocities are the same and independent of frequency. Thus,

$$\boxed{\tau_D = l\sqrt{\mu\varepsilon} = l\sqrt{LC}} \tag{11.75}$$

which is *frequency independent*. Another way to reach the same conclusion from a somewhat more general approach is to note that the phase constant for a lossless line is the *intrinsic* phase constant and is given by

$$k = \omega\sqrt{\mu\varepsilon} = \omega\sqrt{LC}$$

The phase shift over a distance l is $-kl = -\omega\sqrt{LC}l$ since it represents a phase *lag* in the direction of propagation. The time delay may now be defined in a general way as

$$\tau_D \equiv -\frac{d}{d\omega}(\text{phase shift}) = \text{group time delay}^1 \tag{11.76}$$

In the present case, equation 11.76 gives

$$\tau_D = -\frac{d}{d\omega}(-\omega\sqrt{LC}l) = l\sqrt{LC}$$

which is equation 11.75 again. We may now conclude that if the phase constant is a linear function of frequency, or equivalently, if the phase velocity is independent of frequency, then the time delay is independent of frequency and the line is distortionless.

Equation 11.60 can be written with the frequency dependence shown explicitly as

$$V(z, \omega) = \frac{V_g(\omega)R_0}{Z_g(\omega) + R_0} e^{-j\omega l\sqrt{LC}}[e^{j\omega z\sqrt{LC}} + \Gamma_l(\omega)e^{-j\omega z\sqrt{LC}}]$$
$$\cdot [1 + \Gamma_l(\omega)\Gamma_g(\omega)e^{-j2\omega l\sqrt{LC}} + \Gamma_l^2(\omega)\Gamma_g^2(\omega)e^{-j4\omega l\sqrt{LC}} + \cdots] \tag{11.77}$$

The current is

$$I(z, \omega) = \frac{V_g(\omega)}{Z_g(\omega) + R_0} e^{-j\omega l\sqrt{LC}}[e^{j\omega z\sqrt{LC}} - \Gamma_l(\omega)e^{-j\omega z\sqrt{LC}}]$$
$$\cdot [1 + \Gamma_l(\omega)\Gamma_g(\omega)e^{-j2\omega l\sqrt{LC}} + \Gamma_l^2(\omega)\Gamma_g^2(\omega)e^{-j4\omega l\sqrt{LC}} + \cdots] \tag{11.78}$$

These equations are rather formidable in appearance and, yet, these are probably the best forms to treat pulses on lines. The presence of the exponentials is expected because in the time domain these are merely time shifts. We can write a formal time domain solution for $v(z, t)$ and $i(z, t)$ by simply taking the inverse Fourier transform of equations 11.77 and 11.78, respectively,

[1] The group velocity and time delay were introduced following Example 2, Section 10.7.

and this will certainly be correct.

$$v(z, t) = \mathscr{F}^{-1}\{V(z, \omega)\} \tag{11.79}$$

and

$$i(z, t) = \mathscr{F}^{-1}\{I(z, \omega)\} \tag{11.80}$$

One fact is immediately obvious. The inversion will not be simple if $Z_g(\omega)$ [or $\Gamma_g(\omega)$] and $Z_l(\omega)$ [or $\Gamma_l(\omega)$] are complicated functions of ω.

EXAMPLE 2

We desire the voltage impulse response at the middle of a 600-m long lossless air line if $R_0 = Z_g = 50\ \Omega$ and $Z_l = 0$. We can say that the line is 2 μs long since $u_e = 300 \times 10^6$ m/s $= 300$ m/μs. Also, $\Gamma_g = 0$ and $\Gamma_l = -1$. This is a relatively simple problem, so we can either use the equations directly or use common sense. Equation 11.77 gives

$$V(300, \omega) = \frac{V_g(\omega)}{2} e^{-j\omega(2 \times 10^{-6})} [e^{j\omega(1 \times 10^{-6})} - e^{-j\omega(1 \times 10^{-6})}]$$

$$V(300, \omega) = \frac{V_g(\omega)}{2} e^{-j\omega(1 \times 10^{-6})} - \frac{V_g(\omega)}{2} e^{-j\omega(3 \times 10^{-6})}$$

Figure 11.19. Voltage (unit) impulse response at the middle of a lossless line.

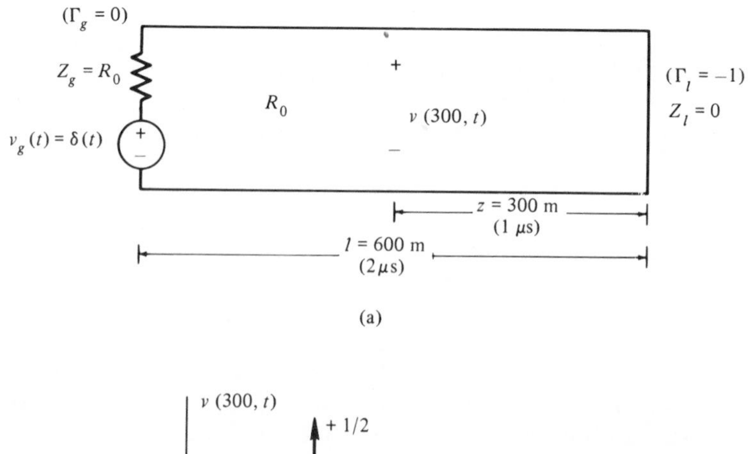

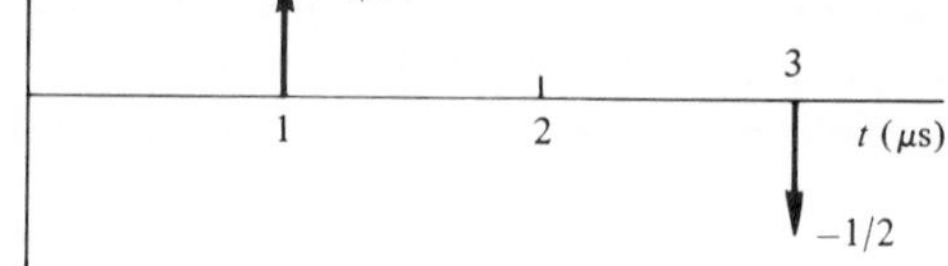

Now since we wanted the (unit) impulse response, $v_g(t) = \delta(t)$ and $V_g(\omega) = 1$. Thus, by equation 11.79.

$$v(300, t) = \mathscr{F}^{-1}\{V(300, \omega)\} = \mathscr{F}^{-1}\{\tfrac{1}{2}e^{-j\omega(1 \times 10^{-6})} - \tfrac{1}{2}e^{-j\omega(3 \times 10^{-6})}\}$$
$$v(300, t) = \tfrac{1}{2}\,\delta(t - 10^{-6}) - \tfrac{1}{2}\,\delta(t - 3 \times 10^{-6}) \quad \text{(V)}$$

Next, let us try common sense. We can do this by starting in the time domain and *staying* there. The input source is a 1-V (unit) impulse which, by simple voltage division, appears as a $\frac{1}{2}$-V impulse at the input to the line since $Z_g = 50\ \Omega$ and $Z_{\text{in}} = 50\ \Omega$ (and at this point the impulse does not know how long the line is). In 1 μs this $\frac{1}{2}$-V impulse appears in the middle ($z = 300$ m) of the line. It proceeds to the end of the line where it arrives after (a total of) 2 μs, and is reflected with sign change ($\Gamma_l = -1$) and appears again at the middle of the line after (a total of) 3 μs. After a total of 4 μs the impulse is completely absorbed in Z_g ($\Gamma_g = 0$). This straightforward approach has given us an answer that agrees completely with the equations. Quite often, a time-domain approach using a unit impulse is very helpful, and, as a matter of fact, if we know the unit impulse response, convolution will give us the response for any excitation!

EXAMPLE 3

Let us attempt another example, which is not so simple, but has practical applications such as representing a pulse generator for radar systems. The system is shown in Figure 11.20(a). We want to find $v_1(t)$, the voltage across R_1. Since the generator impedance is $Z_g = R_0$ for $-\infty < t < \infty$, the circuit of Figure 11.20(b) is applicable and equation 11.77 gives

$$V(l, \omega) = \frac{V_g(\omega)}{2}(1 + e^{-j2\omega l\sqrt{LC}})$$

Now the battery has charged the line to V_0 before the switch is thrown at $t = 0$ (this is done electronically), so

$$v_g(t) = V_0 u(-t) = V_0[1 - u(t)]$$

We can find

$$V_g(\omega) = V_0\left[\pi\,\delta(\omega) - \frac{1}{j\omega}\right] = \mathscr{F}\{v_g(t)\}$$

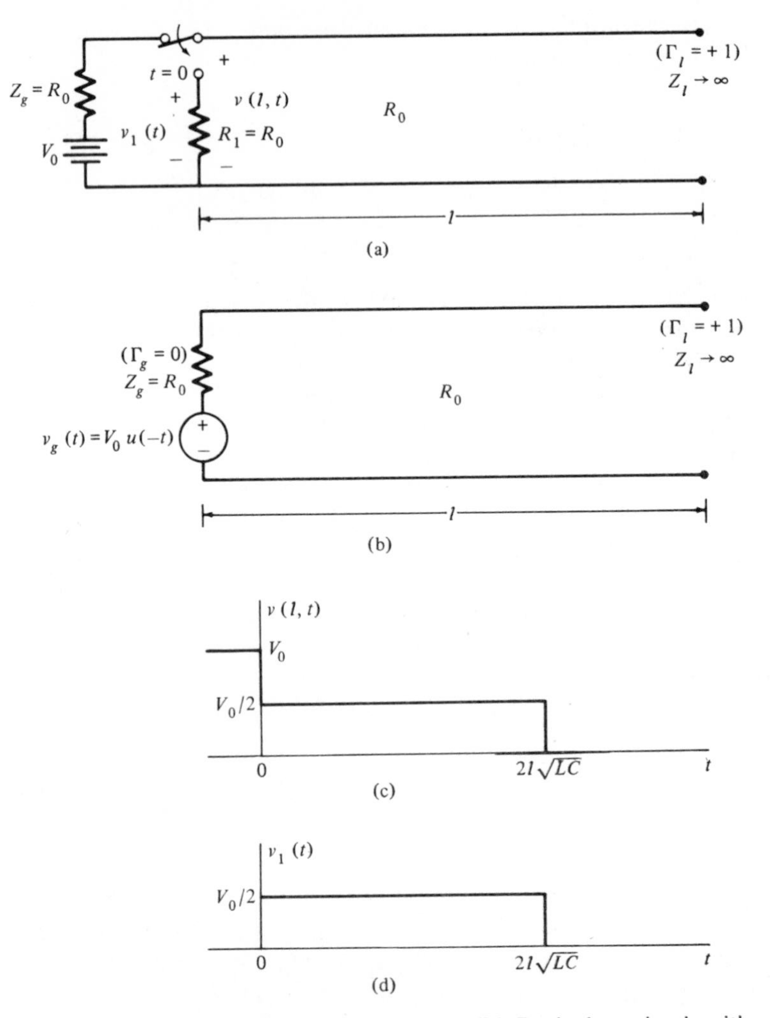

Figure 11.20. (a) Voltage pulse generator. (b) Equivalent circuit with no switch. (c) $v(l, t)$ versus t. (d) $v_1(t)$ versus t (output pulse).

by taking the Fourier transform of $v_g(t)$, but this is not necessary. The easiest way to find $v(l, t)$ is by way of convolution which gives

$$v(l, t) = \frac{1}{2} v_g(t) + \frac{1}{2} v_g(t - 2l\sqrt{LC})$$

$$v(l, t) = \frac{V_0}{2} - \frac{V_0}{2} u(t) + \frac{V_0}{2} - \frac{V_0}{2} u(t - 2l\sqrt{LC})$$

$$v(l, t) = V_0 - \frac{V_0}{2} [u(t) + u(t - 2l\sqrt{LC})]$$

The voltage across the resistor R_1, according to Figure 11.20(a), is the voltage pulse given by

$$v_1(t) = v(l, t)u(t) = \frac{V_0}{2}[u(t) - u(t - 2l\sqrt{LC})]$$

whose width $2l\sqrt{LC}$ is easily altered by changing l!

As a last example consider a case where neither end of the line is matched, but we can use a strictly time domain approach.

EXAMPLE 4

A popular coaxial cable, type *RG*-8A/U, is connected to a 150-Ω pulse generator. The open-circuit generator voltage is a 10-V peak, 1-μs wide pulse. It is assumed that the cable is lossless and that the period of the repetitive pulse is very long. For type RG-8A/U

$$R_0 \approx 50\ \Omega$$

$$u_p \approx \frac{1}{\sqrt{\mu_0 \varepsilon}} = \frac{1}{\sqrt{\mu_0 \varepsilon_R \varepsilon_0}} = \frac{3 \times 10^8}{\sqrt{\varepsilon_R}}$$

$$u_p = \frac{3 \times 10^8}{\sqrt{2.25}} = 2 \times 10^8 \text{ m/s} \quad \text{(polyethylene dielectric)}$$

We desire a plot of the voltage and current at the input to the line ($z = l$) as functions of *time* when $l = 400$ m and $Z_l = 0$.

The time delay from input to load is

$$\tau_D = \frac{l}{u_p} = \frac{400}{2 \times 10^8} = 2\ \mu\text{s}$$

The system is shown in Figure 11.21. A step-by-step procedure will be followed in the analysis.

1. A $10(50)/(150 + 50) = 2.5$-V pulse leaves the input terminals and reaches the load in 2 μs, where a −2.5-V pulse ($\Gamma_l = -1$) is reflected.
2. The −2.5-V pulse arrives at the input, after a *total* time delay of 4 μs, where a −1.25-V pulse $[\Gamma_g = (150 - 50)/(150 + 50) = \frac{1}{2}]$ is reflected. Thus, the input voltage *at this time* is −3.75 V.
3. At the end of 6 μs, a +1.25-V pulse is reflected from the load, and, at the end of 8 μs, arrives at the input, where a 0.625-V pulse is reflected. The input voltage at this time is 1.875 V.
4. This process repeats indefinitely.

For the plot of input current it is only necessary to notice that the coefficients of reflection have opposite signs for current compared to voltage. The current at the input terminals is initially $10/(150 + 50) =$ 50 mA. Results are shown in Figure 11.22.

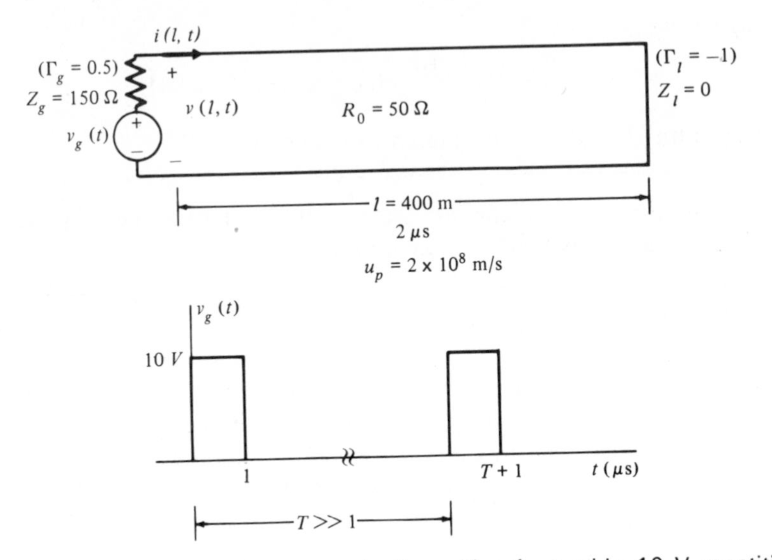

Figure 11.21. Lossless two-wire line with a 1-μs-wide, 10-V, repetitive pulse input.

Figure 11.22. Input voltage and current, 1-μs, 10-V generator pulse for the line of Figure 11.21.

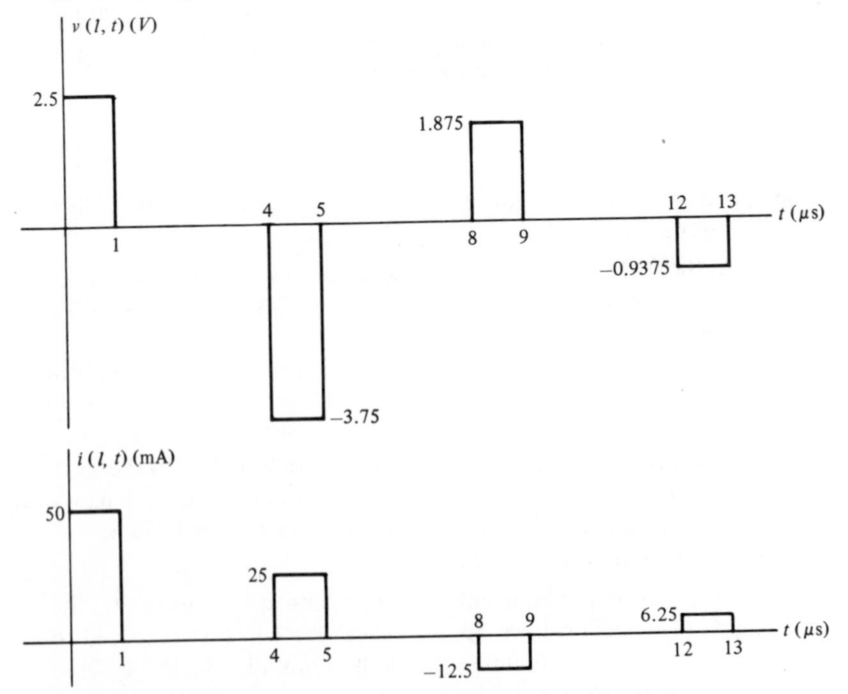

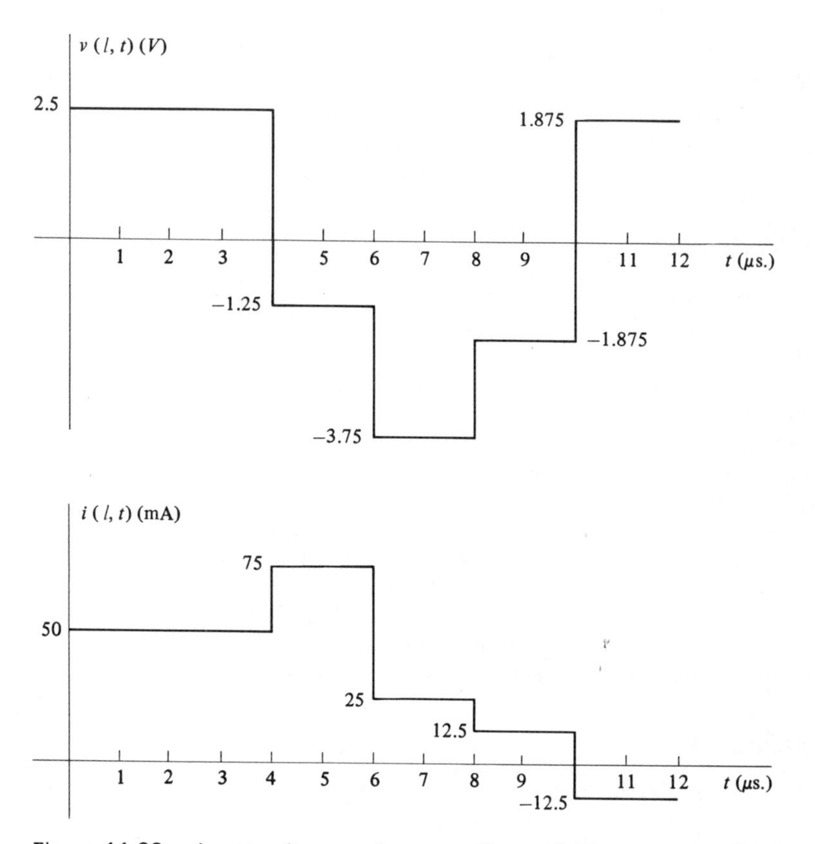

Figure 11.23. Input voltage and current, 6-μs, 10-V generator pulse for the line of Figure 11.21.

Another apparent difficulty arises if the pulse width is greater than twice the delay time, because overlapping will occur in the waveforms. This is not difficult to handle, however, because the procedure is exactly the same as before. We start by assuming a pulse width less than $2\tau_D$ and proceed as before. Each pulse on the waveform may then be expanded to its actual width and added to its adjacent pulse where the two overlap. As an example, the input voltage and current for the previous example are shown in Figure 11.23 for the case of a 6-μs-wide generated pulse.

11.11 CONCLUDING REMARKS

The general behavior of lossless transmission lines is similar to the behavior of plane waves in a lossless medium. Both can be described in terms of coefficient of reflection, standing wave ratio, and power (density) flow. Even though all transmission lines have *some* loss, it is usually small enough so that results given in this chapter are applicable, or at least useful. Several questions that were raised in this chapter will be answered in the chapters that follow.

The distributed capacitance and (external) inductance per unit length for a lossless two-wire system may be calculated using *static* techniques. This makes the work much easier, and in addition, values for some of the most common geometries have already been calculated in previous chapters. We predicted the usefulness of such a calculation at the time it was made.

REFERENCES

Brown, R. G., Sharpe, R. A., and Hughes, W. L. *Lines, Waves, and Antennas*, 2nd ed. New York: Ronald Press, 1973. A good reference for transmission lines and waveguides.

Chipman, R. A. *Transmission Lines*, Schaum Outline Series. New York: McGraw-Hill, 1968. Many solved problems are included.

Harrington. (See references for Chapter 10.)

International Telephone and Telegraph Co. Inc. *Reference Data Radio Engineers*, 5th ed. Indianapolis: Howard W. Sams and Co., 1968. A handbook containing much data on transmission lines of various shapes and dielectric materials.

Moore, R. K. *Traveling-Wave Engineering*. New York: McGraw-Hill, 1960. A simultaneous development of plane waves and transmission lines is given.

Ryder, J. D. *Networks, Lines and Fields*. Englewood Cliffs, N.J.: Prentice Hall, 1949. A still valuable book for transmission line and waveguide studies.

Skilling, H. H. *Electric Transmission Lines*. New York: McGraw-Hill, 1951.

PROBLEMS

1. The distributed capacitance per unit length for a certain lossless line is 60 pF/m in air. Find R_0 and L.
2. If $R_0 = 100\ \Omega$ and $Z_l = 50 - j50\ \Omega$ for a lossless line, find Γ_l and the VSWR. Find the input impedance at $z = 0.15\lambda$ from the load.
3. The following measurements are made on a lossless line: $Z_{\text{in}}^{\text{sc}} = j100$, $Z_{\text{in}}^{\text{oc}} = -j25$, $z_{\text{min}} = 0.1\lambda, 0.6\lambda, \ldots$, VSWR = 3. Find Z_l.
4. The following measurements are made on a lossless line: $Z_{\text{in}}^{\text{sc}}$, $Z_{\text{in}}^{\text{oc}}$, and Z_{in} (actual load in place). Starting with equation 11.52 show that
$$Z_l = Z_{\text{in}}^{\text{oc}} \frac{Z_{\text{in}}^{\text{sc}} - Z_{\text{in}}}{Z_{\text{in}} - Z_{\text{in}}^{\text{oc}}}$$
5. If $Z_{\text{in}}^{\text{sc}} = j100$, $Z_{\text{in}}^{\text{oc}} = -j25$ and $Z_{\text{in}} = 75\underline{|30^\circ}$, find Z_l.
6. Find the phasor voltage and current in the middle of a lossless 50-Ω line one wavelength long if $V_g = 1\underline{|0^\circ}$, $Z_g = 100\ \Omega$, and $Z_l = 0$.
7. Find the time domain voltage in the middle of a lossless 50-Ω air line 300 m long if $v_g(t) = \delta(t)$, $Z_g = 100\ \Omega$, and $Z_l = 0$. Find this voltage by starting from "scratch" and also by using equation 11.79.
8. The time average incident power on a certain lossless line is 100 W and the VSWR is 4. Find $\langle P_l \rangle$.
9. Find the length of a 300-Ω line required to resonate with an output capacitance of 5 pF from a push-pull amplifier. The frequency is 300 MHz and the dielectric is air. The line is short-circuited.

10. Show that the input impedance at the $x - x$ terminals of Figure 11.18(b) is infinite (parallel resonance) regardless of d.
11. Using the results of Problem 10 design a bandpass filter to pass 100 MHz and reject 120 MHz. The dielectric is air.
12. What characteristic resistance is required for a $\lambda/4$ transformer to match a 50-Ω lossless line to a 100-Ω load? Find the ratio of the magnitude of the load voltage to the magnitude of the voltage at the junction of the two lines.
13. (a) Plot $\langle P_l \rangle$ versus R_g/R_0 if $R_l = R_0$
    (b) Plot $\langle P_l \rangle$ versus R_l/R_0 if $R_g = R_0$. See Figure 11.24.

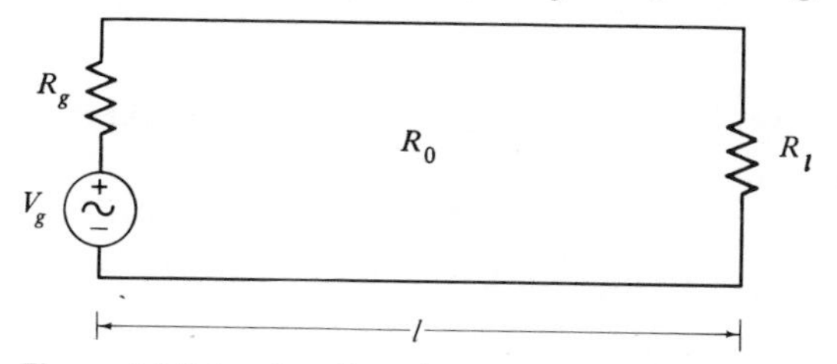

Figure 11.24. Lossless line.

14. (a) The load on a lossless line is $Z_l = R_l + j\omega L_l$. If the generator is matched, use equation 11.64 and find the voltage impulse response for any z.
    (b) Repeat if $Z_l = R_l + 1/j\omega C_l$.
    (c) Repeat if $Z_l = R_l + j\omega L_l + 1/j\omega C_l$.
15. Refer to Problem 24, Chapter 4, and show that for a lossless two-wire transmission line made up of conductors with diameters d_1 and d_2 and separated by D (center-to-center)

$$R_0 = \frac{\eta}{2\pi}\left(\sinh^{-1}\frac{2a}{d_1} \pm \sinh^{-1}\frac{2a}{d_2}\right) \quad (\Omega)$$

where

$$a = \frac{1}{2D}\left[D^4 - 2D^2(r_2^2 + r_1^2) + (r_2^2 - r_1^2)^2\right]^{1/2}$$

and the $-$ sign applies if the smaller wire is within the larger wire, but the $+$ sign applies otherwise.
16. Refer to Problem 24, Chapter 4, and find the characteristic resistance of an eccentric cable made of an inner conductor of radius r_1 within a hollow conductor of radius r_2 and a center-to-center spacing of D.
17. Derive a formula for $\langle P_l \rangle / \langle P_{inc} \rangle$ as a function of VSWR and plot this function.
18. Show that $|\Gamma_l| = 1$ for any purely reactive load on a lossless line.
19. If $Z_l = 50 + j50\ \Omega$ on a lossless 50-Ω line, what value of capacitance placed in series with the load will produce a matched load at 100 MHz?

20. Show that Z_{in} has a real part $R_{in} = R_0$ at a point $z = l_1$ from the end of a lossless line if

$$l_1 = \frac{\phi_l - \cos^{-1}|\Gamma_l|}{2k}$$

If a series reactance equal to $-X_{in}$ is added to the line at this point, what is the result?

21. Show that $Y_{in} = 1/Z_{in}$ has a real part $G_{in} = 1/R_0$ at a point $z = l_1$ from the end of a lossless line if

$$l_1 = \frac{\phi_l + \pi - \cos^{-1}|\Gamma_l|}{2k}$$

If a shunt susceptance equal to $-B_{in}$ is added to the line at this point, what is the result?

22. Why is it possible to use the equation $\mathbf{E} = -\nabla\Phi$ rather than $\mathbf{E} = -\nabla\Phi - \partial\mathbf{A}/\partial t$ to obtain $\mathbf{E}$ for lossless lines supporting TEM to z modes?

23. Refer to Example 7, Chapter 4, and show that

$$E_\rho(\rho, z, t) = \frac{v(z, t)}{\rho \ln b/a}$$

for a coaxial cable. $v(z, t)$ is the line *voltage*.

24. In Problem 23 the maximum value of E_ρ occurs at $\rho = a$. Show that a coaxial cable designed for maximum voltage breakdown (minimum E_ρ) has $R_0 = 60\ \Omega$ (*air dielectric*).

25. Refer to Example 8, Chapter 4, and show that

$$\Phi(x, y, z, t) = \frac{v(z, t)}{4\cosh^{-1} D/d} \ln \frac{(x + a)^2 + y^2}{(x - a)^2 + y^2}$$

for a two-wire line. $v(z, t)$ is the line *voltage* and $a = \sqrt{(D/2)^2 - (d/2)^2}$.

26. Calculate $\mathbf{E}$ for the two-wire line of Problem 25. See Problem 22, Chapter 4.

27. (a) Make a normalized plot of $|\langle \mathbf{S} \rangle|$ versus y with $x = 0$ for the two-wire line of Problem 25.
    (b) Make a normalized plot of $|\langle \mathbf{S} \rangle|$ versus x with $y = 0$ for the two-wire line of Problem 25.

28. Show that the energy dissipated in R_1 is equal to the initial energy stored in the line for the pulse forming line of Example 3.

29. (a) Derive the Thévenin equivalent circuit for the lossless transmission line of length l for any Z_g and Z_l using equations 11.61 and 11.62.
    (b) Show that the load current is

$$I_l = \frac{V_g}{(Z_g + Z_l)\cos kl + j(R_0 + Z_g Z_l/R_0)\sin kl}$$

Chapter 12
The Smith Chart

Several unanswered questions arose in Chapter 11. The first two questions were concerned with methods of measuring the VSWR and positions of $|V|_{\text{min}}$ on a transmission line, and the third question involved methods of making transmission line calculations more efficient. The first two questions are answered very quickly by considering a laboratory device called the slotted line. This is done in the first section of this chapter. Next, it is shown that a graphical technique using a transmission line calculator, or Smith chart, can be developed for making rapid transmission line calculations. These calculations are usually sufficiently accurate for engineering purposes. Some versions of today's pocket electronic calculator can be used very well for making transmission line calculations. It will be shown in later chapters that the Smith chart can be used for lossy transmission lines and for one-conductor waveguides.

12.1 THE SLOTTED LINE

The slotted line is a rigid section of shielded transmission line which is designed so that it can be probed to obtain a signal proportional to the line voltage at various points on the line. Thus, it can be used to measure the standing wave ratio and the location of $|V|_{min}$ and $|V|_{max}$, assuming, of course, that the length over which the slotted line can be probed is large enough to include both $|V|_{min}$ and $|V|_{max}$. Since this length is electrically $\lambda/4$ (that is, $|V|_{min}$ and $|V|_{max}$ occur $\lambda/4$ apart on a lossless line), a lowest frequency of operation is determined by the physical length over which the slotted line can be probed. As was demonstrated in the preceding chapter, knowledge of the standing wave ratio and the location of $|V|_{min}$ is sufficient to allow one to calculate the load impedance, or, as we will shortly see, use the Smith chart to graphically evaluate the load impedance. The operating wavelength (e.g., $\lambda/4$, or $\lambda/2$), and hence operating frequency can also be determined from the same measurements. The slotted line is inserted between the generator and the unknown load impedance, being located as close to the load as practical considerations allow. Its impedance should be the same as that of the rest of the line.

The success of *any* measurement scheme depends on the ability of the measurement equipment to be in position without appreciably disturbing the behavior of the phenomenon being examined. The slotted line meets these requirements for transmission line work. It usually consists of a section of rigid coaxial line with an air dielectric and a narrow longitudinal slot cut in the outer conductor. The dominant TEM mode field distribution is shown in Figure 12.1. Notice that this field distribution meets the requirements set forth in the first part of Chapter 11 for the general lossless line, The tangential electric field is zero at the conductor surface, the normal magnetic field is zero at the conductor surface, and **E** and **H** are everywhere perpendicular to each other. *Notice that* **E** *and* **H** *are distributed exactly as they were in the "static" cases considered earlier.*

The transverse magnetic field gives rise to (or is accompanied by) an *outward* (out of the paper, toward the reader) flowing longitudinal surface current on the *outside* of the *inner* conductor and an *inward* flowing longitudinal surface current on the *inside* of the *outer* conductor. Since we are assuming perfect conductors (lossless case), both surface currents are determined by $\mathbf{J}_s = \mathbf{a}_n \times \mathbf{H}|_s$. It is easy to show that the total current on each conductor is the same. Since the longitudinal slot is *parallel* to the direction of current flow, the slot does not appreciably disturb the flow of current or alter the fields.

A small electric field probe may be inserted in the slot without altering appreciably the field distribution so long as its depth of penetration is small. This probe is parallel to the electric field so that the signal induced on the probe *is proportional to the electric field*, which is in turn distributed axially exactly like the line voltage. The probe is mounted in a carriage which is free to move in the axial (z) direction. The generator is matched to the line ($Z_g = R_0$) by any one of several schemes, and the signal (usually 300 MHz or higher so that the line is not too long) is usually amplitude modulated at some convenient audio frequency (e.g., 1000 Hz). The probe is connected to a diode detector which demodulates the probe signal. In this way an audio voltage

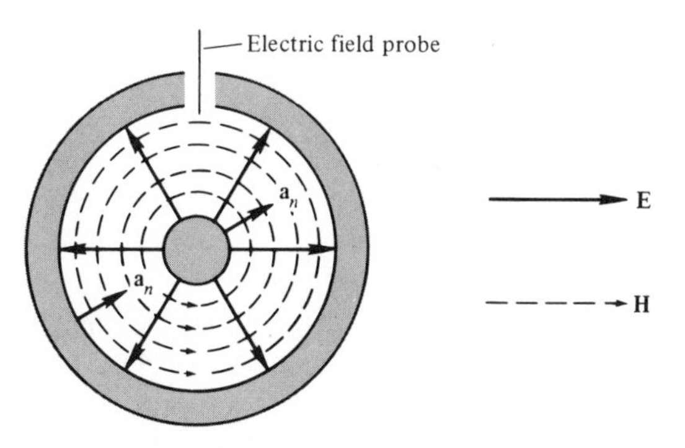

Figure 12.1. Coaxial slotted line with an electric field probe and electromagnetic field distribution.

whose amplitude is proportional to $|\mathbf{E}|$ is provided. Any detection scheme will suffice since the probe signal is proportional to the transmission line voltage. With a *calibrated* detection system, it is no problem to measure $|V|_{max}$ and $|V|_{min}$ (hence VSWR), or to measure the locations of $|V|_{max}$ and $|V|_{min}$. Since the axial (z) spacing between two adjacent locations of $|V|_{min}$ (or $|V|_{max}$) is $\lambda/2$, the slotted line can accurately measure the signal frequency. As a practical matter, more accuracy is obtained by making measurements of $|V|_{min}$ rather than $|V|_{max}$ (Why?).

A minor mechanical difficulty is that the probe can never be positioned *at* the load. A position $z = \lambda/2$ toward the generator serves just as well (if the slotted line losses are negligible) as the location of Z_l itself, since $|V|$, $|I|$, and Z_{in} repeat *every* $\lambda/2$ on the line. This artificial end of the line ($z = 0$) is easily determined by replacing the actual load with a short circuit and locating that voltage minimum occurring closest to the end of the line.

12.2 DERIVATION OF THE SMITH CHART[1]

Using the equations that determine transmission line behavior may become tedious, even in the lossless case. This is especially true when the same calculation is repeated many times. Fortunately, a transmission line calculator, known as the Smith chart after its inventor, is available. After only a small amount of experience, it becomes very easy to use.

The Smith chart results from a series of simple transformations. The starting point is the input impedance formula, given in Section 11.6,

$$Z_{\text{in}} = \frac{V}{I} = R_0 \frac{e^{jkz} + \Gamma_l e^{-jkz}}{e^{jkz} - \Gamma_l e^{-jkz}} \tag{12.1}$$

which is valid regardless of Z_g! (Note that the preceding statement is mathematically true, but from a practical point of view, a large reflection from the

[1] P. H. Smith, "Transmission-line Calculator," *Electronics*, vol. 12, pp. 29–31, January, 1939.

load may have undesirable effects on the generator.) The *normalized* input impedance is defined as

$$z_{\text{in}} \equiv \frac{Z_{\text{in}}}{R_0} = \frac{e^{jkz} + \Gamma_l e^{-jkz}}{e^{jkz} - \Gamma_l e^{-jkz}} \tag{12.2}$$

or

$$z_{\text{in}} = \frac{1 + |\Gamma_l| e^{j(\phi_l - 2kz)}}{1 - |\Gamma_l| e^{j(\phi_l - 2kz)}} \equiv r_a + jx_a \tag{12.3}$$

Equation 12.3 may be written as

$$|\Gamma_l| e^{j(\phi_l - 2kz)} = \frac{r_a - 1 + jx_a}{r_a + 1 + jx_a} \equiv U + jV \tag{12.4}$$

It is convenient at this point to recognize that

$$|U + jV| = (U^2 + V^2)^{1/2} \leq 1$$

since $|\Gamma_l| \leq 1$ for passive loads whose real parts are nonnegative. Tunnel diodes, for example, are exceptions and offer the possibility for $|\Gamma_l| > 1$, and hence amplification!

Solving for U and V in equation 12.4 gives

$$U = \frac{r_a^2 - 1 + x_a^2}{(r_a + 1)^2 + x_a^2} \tag{12.5}$$

and

$$V = \frac{2x_a}{(r_a + 1)^2 + x_a^2} \tag{12.6}$$

Solving for r_a and x_a in equation 12.3, then algebraically rearranging gives

$$\left(U - \frac{r_a}{r_a + 1}\right)^2 + (V)^2 = \left(\frac{1}{r_a + 1}\right)^2 \tag{12.7}$$

and

$$(U - 1)^2 + \left(V - \frac{1}{x_a}\right)^2 = \left(\frac{1}{x_a}\right)^2 \tag{12.8}$$

Equation 12.7 represents a family of constant r_a circles having centers on the U axis at $r_a/(r_a + 1)$ and radii $1/(r_a + 1)$ (the largest radius is *one*). Equation 12.8 represents a family of constant x_a circles having centers at 1, $1/x_a$ in the U, V plane and having radii $1/x_a$. We also note that

$$\Gamma_l e^{j(\phi_l - 2kz)} = U + jV = (U^2 + V^2)^{1/2} e^{j \tan^{-1}(V/U)}$$

or

$$\tan(\phi_l - 2kz) = \frac{V}{U} \tag{12.9}$$

The *normalized* input admittance, y_{in}, on the line at any point is the reciprocal of the normalized input impedance at the same point. It is a straightforward matter to begin with the reciprocal of equation 12.2 to represent y_{in}, and then show that y_{in} at a distance z from the load is identical with z_{in} at a distance $z \pm \lambda/4$ from the load. Since $|\Gamma_l| = (U^2 + V^2)^{1/2} \leq 1$ for passive loads, all possible values of z_{in} lie inside, or on, the *unit* circle in the U, V plane! Also, from equation 12.9, it can be seen that one revolution in the U, V plane (at constant radius) corresponds to a change in distance on the line of $\lambda/2$. This is not surprising, for we already know that impedances (and admittances) repeat every $\lambda/2$ on the line. We can now conclude that all possible values of normalized admittances also lie inside, or on, the unit circle in the U, V plane. In fact, since z_{in} and y_{in} are $\lambda/4$ apart, they must be *diametrically opposite* in the U, V plane. (See Problem 2 at the end of this chapter.)

Several important conclusions may now be reached concerning the Smith chart (the interior of the unit circle in the U, V plane).

1. Counterclockwise increasing angle measured from the U axis is $\phi_l - 2kz$. Thus, distance toward the load (z decreasing) is scaled linearly on the outside of the unit circle and measured in the counterclockwise direction. The actual location of $z = 0$ on the chart is arbitrary. On more expensive charts, the position of $z = 0$ can be set at will. As shown in Figure 12.2, the angle ϕ_l is scaled like $-z$, but its total excursion is 2π rad. It is easy to show that $\phi_l = 0$ if $x_l = 0$ and $R_l > R_0$. Values of z_{in} for these cases lie on the $+U$ axis and thus $\phi_l = 0$ also corresponds to the $+U$ axis.
2. Distance toward the generator is obviously measured in the clockwise direction (z increasing).
3. Since there is a one-to-one correspondence between Γ_l and z_{in} (or y_{in}), including $z_l = Z_l/R_0$ itself, rotation at constant radius, $|\Gamma_l|$, in the U, V plane consists of moving along a circle that uniquely specifies all possible values of z_{in} (or y_{in}) for a given $|\Gamma_l|$. (The effect of loss in the line might be pondered at this point in the development.)
4. Constant radius, $|\Gamma_l|$, in the U, V plane also uniquely determines such things as the VSWR, reflected power, and so forth. A large number of radially scaled parameters may be used with the chart if desired. A normalized impedance or admittance Smith chart is shown in Figure 12.4, which is used for Examples 1 and 2.

Figure 12.3(a) shows the loci of constant r (or constant g) circles on the normalized Smith chart. The largest circle represents $r = 0$, while the smallest circle (a point) represents $r = \infty$. The circles are labeled as they are for an actual chart (e.g., Figure 12.4). The loci of constant x (or constant b) circles are shown in Figure 12.3(b). They are incomplete circles. The largest circle represents $x = 0$ and is actually a straight line (a circle with an infinite radius), while the smallest circle (a point) represents $x = \pm\infty$. Notice

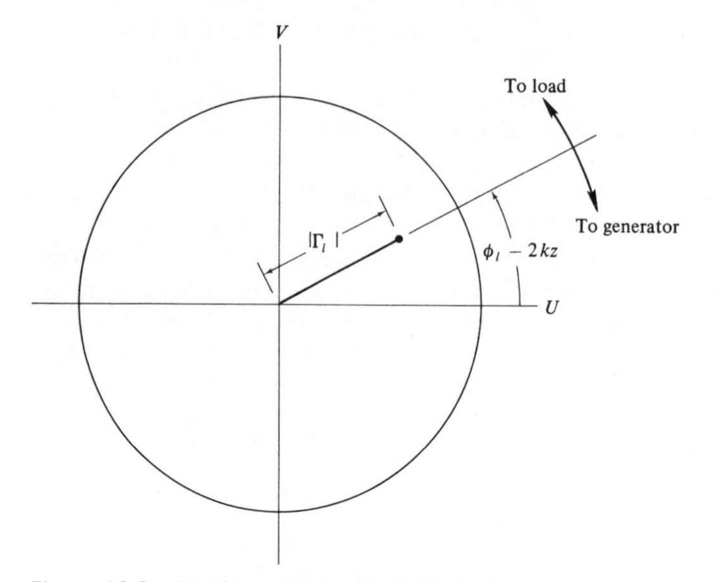

Figure 12.2. $|\Gamma_l|\,\underline{|\phi_l - 2kz}$ in the U-V plane.

Figure 12.3. Loci of constant r (or g) and constant x (or b) circles on the normalized Smith chart. (a) Constant r (or g) = 0, 0.3, 1.0, 3.0, ∞ circles. (b) Constant x (or b) = 0, 0.3, 1.0, 3.0, ∞ circles (incomplete). Positive x (or b) values lie above the horizontal axis, while negative values lie below. (c) The locus of z (or y) = 1 + j3 and z (or y) = 1 − j1.

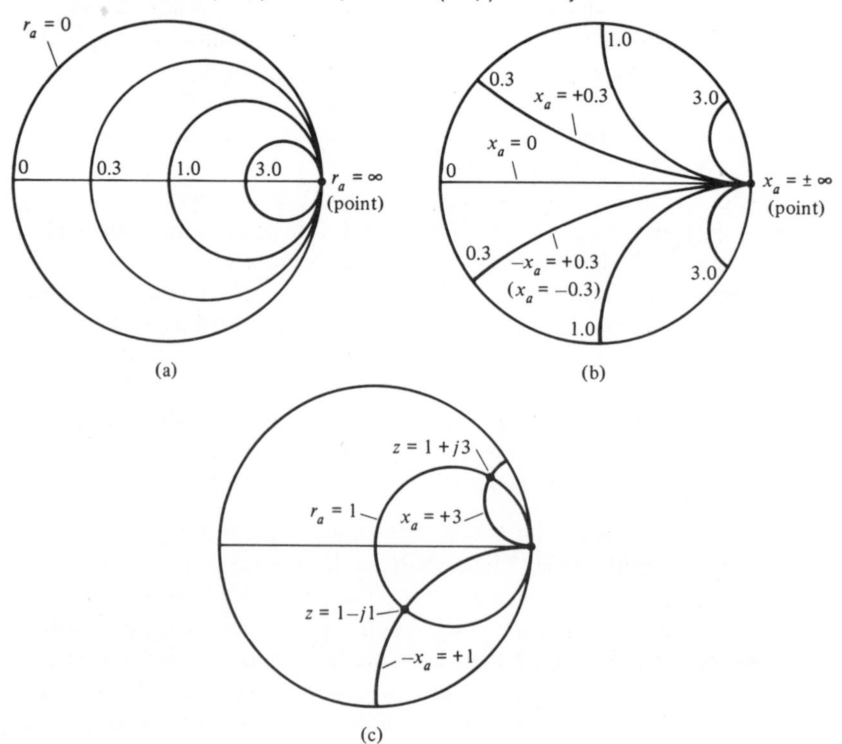

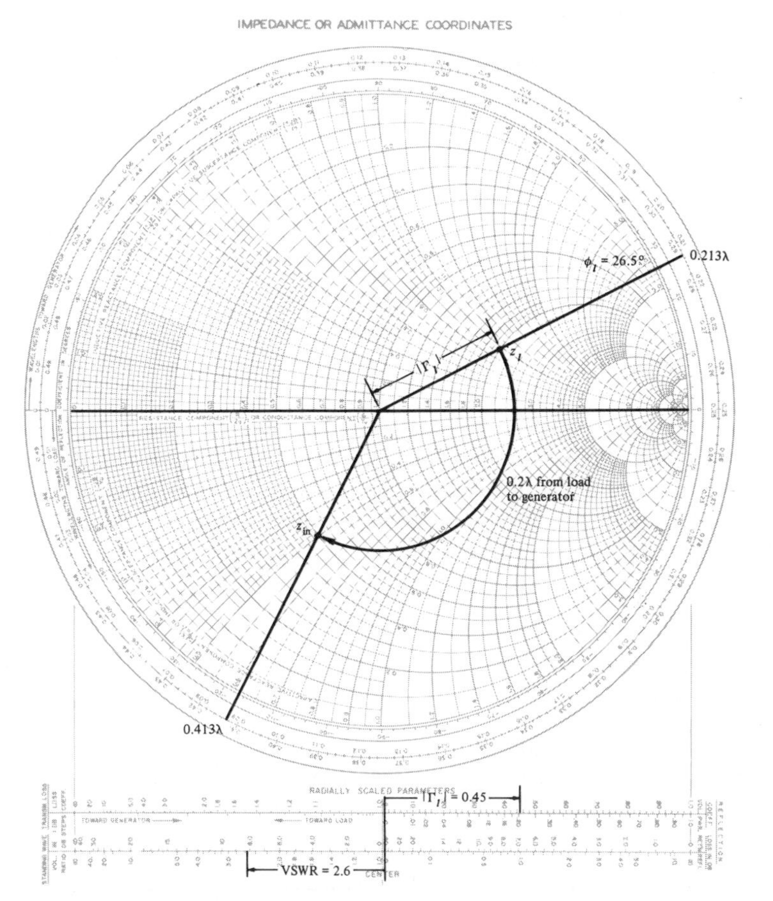

Figure 12.4. Smith chart data for Examples 1 and 2. *Source*: Copyright © 1949 by Kay Electric Company, Pine Brook, N.J. Renewal copyright © 1976 by P. H. Smith, New Providence, N.J.

carefully that positive reactances or susceptances lie *above* the horizontal axis, while negative reactances or susceptances lie *below* the horizontal axis. All numbers on the chart are labeled as positive, however. This can be confusing. Figure 12.3(c) demonstrates the loci of two points corresponding to $z = 1 + j3$ (or $y = 1 + j3$) and $z = 1 - j1$ (or $y = 1 - j1$). These points lie at the intersection of an r_a and an x_a circle. Some typical examples will help clarify any remaining questions about the use of the Smith chart.

EXAMPLE 1

If $Z_l = 100 + j50\ \Omega$ and $R_0 = 50\ \Omega$, find Γ_l and the VSWR. The normalized load impedance is $z_l = (100 + j50)/50 = 2 + j1$. This point is located on the chart as the intersection of the circles $r_a = 2$ and $x_a = +1$ as shown in Figure 12.4. From the previous discussion, it is obvious that the distance from the center of the circle to z_l is $|\Gamma_l|$. In Figure 12.4, this is indicated on the upper right scale below the chart labeled "refl. coef. vol." Thus, $|\Gamma_l| = 0.45$ as shown. The angle, ϕ_l, is measured from the $+U$ axis and is $+26.5°$. Thus

$$\Gamma_l = 0.45 \angle 26.5°$$

$|\Gamma_l|$ also uniquely determines the VSWR which is indicated on the lower left scale labeled "standing wave vol. ratio." Then,

$$\text{VSWR} = 2.6$$

EXAMPLE 2

Suppose we desire the input impedance 0.2λ from the load. Notice that the radial line through z_l intersects the "wavelength toward generator scale" at 0.213λ. This is only a reference reading and corresponds to $z = 0$. Consequently, the desired point where the input impedance is to be found is at $0.213\lambda + 0.200\lambda = 0.413\lambda$ on the same scale. At the same $|\Gamma_l|$, we find $z_{\text{in}} = 0.5 - j0.5$ or $Z_{\text{in}} = R_0 z_{\text{in}}$, so

$$Z_{\text{in}} = 25 - j25 \quad (\Omega)$$

Notice that both wavelength scales end at 0.5λ, and when rotating past this point in either direction, care must be taken to avoid losing the reference. This difficulty is not encountered with those charts having a rotatable wavelengths scale.

EXAMPLE 3

The following data are taken with an impedance bridge.

$$Z_{\text{in}}^{\text{sc}} = +j106 \quad (\Omega),$$

$$Z_{\text{in}}^{\text{oc}} = -j23.6 \quad (\Omega)$$

and

$$Z_{\text{in}} = 25 - j70 \quad (\Omega) \quad \text{(actual load in place)}$$

The actual load impedance, Z_l, is desired. From equation 11.55

$$R_0 = \sqrt{Z_{\text{in}}^{\text{oc}} Z_{\text{in}}^{\text{sc}}} = 50 \quad (\Omega)$$

The distance from the measuring point at the bridge to the actual location of the load can be found from $Z_{\text{in}}^{\text{sc}}$ (or $Z_{\text{in}}^{\text{oc}}$) as follows.

$$z_{\text{in}}^{\text{sc}} = \frac{Z_{\text{in}}^{\text{sc}}}{R_0} = +j2.12$$

is located on the chart. This point is located on the *unit* circle since the line is lossless. The short circuit $Z_l = z_l = 0$, in place when $Z_{\text{in}}^{\text{sc}}$ was measured, is located at $U = -1$, $V = 0$ as shown in Figure 12.5. Then

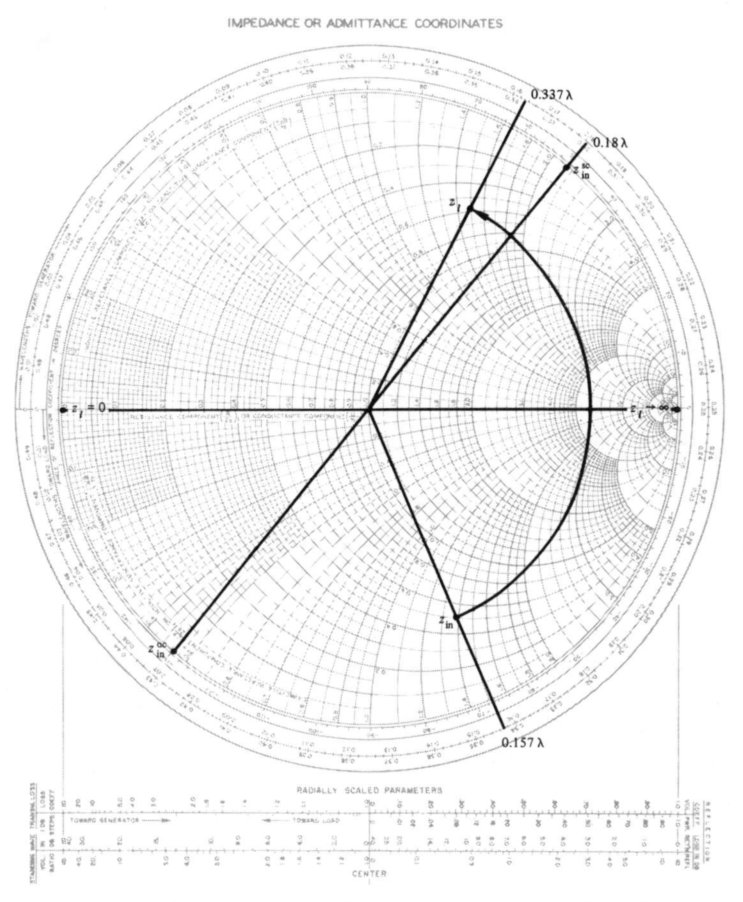

Figure 12.5. Smith chart data for Example 3. *Source*: Copyright © 1949 by Kay Electric Company, Pine Brook, N.J. Renewal copyright © 1976 by P. H. Smith, New Providence, N.J.

the distance desired is found as that distance from $z_l = 0$ toward $z_{\text{in}}^{\text{sc}}$, rotating in the "toward the generator" (clockwise) direction. Then $z = 0.18\lambda$ as shown. Notice that this information is also available from $z_{\text{in}}^{\text{oc}}$ and the location of $z_l \to \infty$ $(U = +1, V = 0)$.

The point $z_{\text{in}} = Z_{\text{in}}/R_0 = 0.5 - j1.4$ is next located on the chart as shown. A radial line through z_{in} gives a wavelength reference on the "toward load" scale of 0.157λ. The load then is located at $0.18\lambda + 0.157\lambda = 0.337\lambda$ as shown. Thus $z_l = 0.57 + j1.5$, or $Z_l = R_0 z_l$, or

$$Z_l = 28.5 + j75 \quad (\Omega)$$

EXAMPLE 4

The following data are taken with a 50 Ω slotted line and calibrated detector with an unknown load, Z_l in place.

$$|V|_{max} = 0 \quad (\text{dB})$$

$$|V|_{min} = -6 \quad (\text{dB})$$

and $|V|_{min}$ occurs at relative slotted line positions of 0.10 m, 0.35 m, 0.60 m, and 0.85 m measured toward the load. When the unknown load is replaced with a short circuit, the positions of $|V|_{min}$ are 0 m, 0.25 m, 0.50 m, and 0.75 m, measured *toward the load.* Determine Z_l and the frequency.

Frequency is normally determined by measuring $\lambda/2$ as the distance between adjacent voltage minima when the short circuit is in place. The VSWR is infinite in this case, and thus the "nulls" will be sharpest giving a very accurate position of $|V|_{min}$. Then, $\lambda/2 = 0.25$ m, or $\lambda = 0.5$ m. Since the dielectric is air,

$$f = \frac{3 \times 10^8}{0.5} = 600 \quad (\text{MHz})$$

The Smith chart may be used to determine Z_l in the following way. The VSWR is 2 (or 6 dB), and so z_l is somewhere on the Smith chart circle corresponding to VSWR = 2. Notice in Figure 12.6 that the VSWR scale corresponds to the r_a scale on the $+U$ axis and a 1/(VSWR) scale corresponds to the r_a scale on the $-U$ axis. This occurs because, according to equations 11.47 and 11.50,

$$V_{max} = C_1 e^{jkz}(1 + |\Gamma_l|)$$

while at the same location on the line (z_{max})

$$I_{min} = \frac{C_1}{R_0} e^{jkz}(1 - |\Gamma_l|)$$

using equations 11.48 and 11.50. Using equation 11.49

$$Z = Z_{max} = \frac{V_{max}}{I_{min}} = R_0(\text{VSWR})$$

Thus, the maximum *normalized* impedance equals the VSWR, and points of maximum impedance, corresponding to maximum voltage, are located on the $+U$ axis. In the same way, using equations 11.47, 11.48, 11.49, and 11.51, it is easy to show that

$$Z_{min} = \frac{V_{min}}{I_{max}} = \frac{R_0}{\text{VSWR}}$$

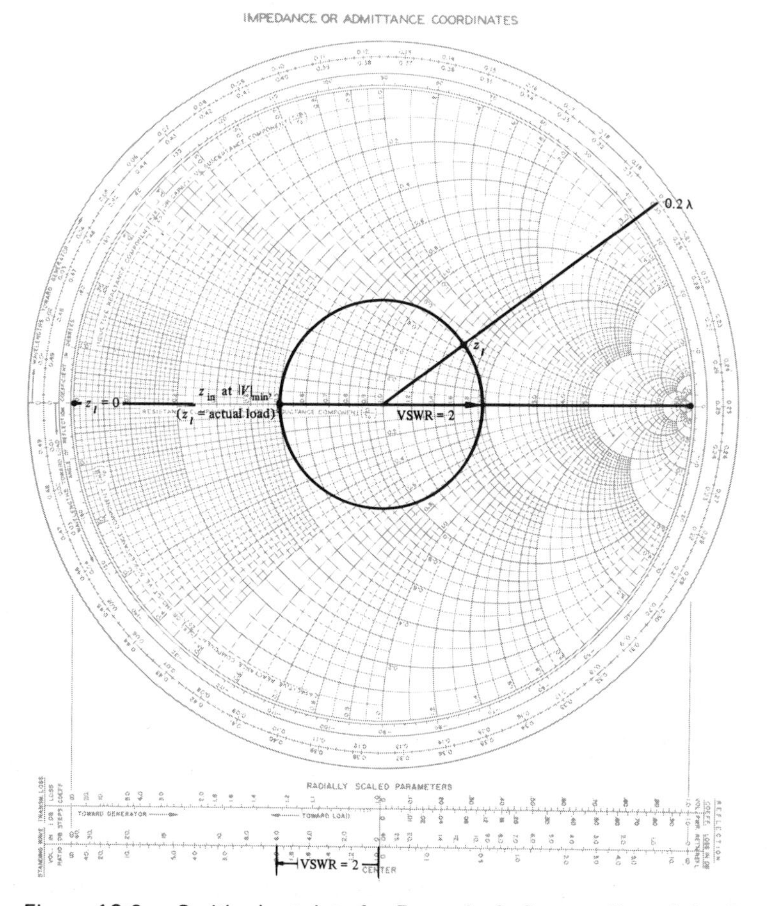

Figure 12.6. Smith chart data for Example 4. *Source*: Copyright © 1949 by Kay Electric Company, Pine Brook, N.J. Renewal copyright © 1976 by P. H. Smith, New Providence, N.J.

Thus, the minimum *normalized* impedance equals the reciprocal of the VSWR, and points of minimum impedance, corresponding to minimum voltage, are located on the $-U$ axis. With $Z_l = 0$ the voltage minimum or null occurs at 0λ "toward generator" or "toward load" since z_l itself is located at $U = -1$, $V = 0$. This is shown in Figure 12.6. The slotted line data indicate that $|V|_{\min}$ shifts $(0.10/0.50)\lambda = 0.2\lambda$ toward the load when the short circuit is replaced by the actual load. This is shown in Figure 12.7. Then z_l, on the chart, is 0.2λ toward the generator from the location of $|V|_{\min}$ with the actual load in place. This is shown in Figure 12.6 where it can be seen that $z_l = 1.56 + j0.69$ or $Z_l = 77.5 + j34.5\ \Omega$. The situation is clarified by Figure 12.7. Notice that only the "artificial" end of the line, or load location, is used.

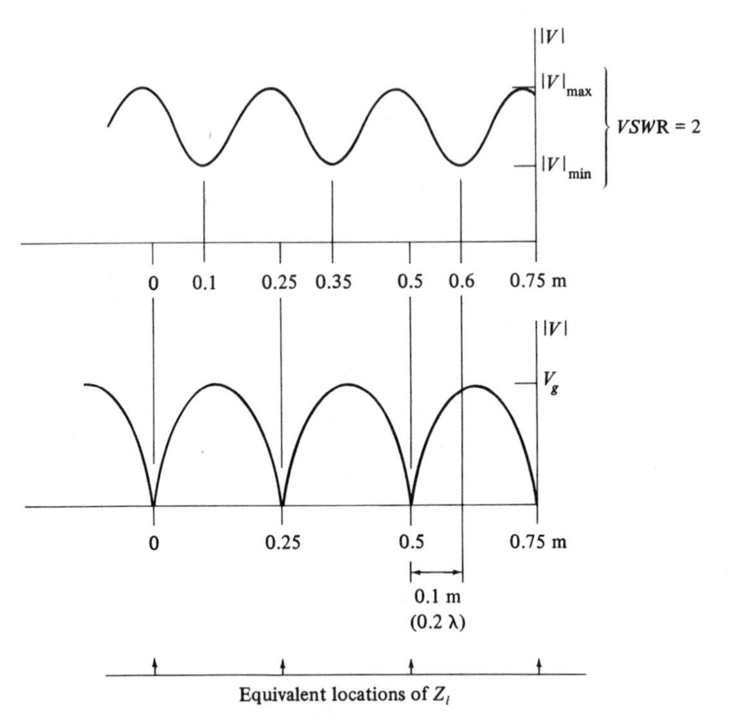

Figure 12.7. (a) $|V|$ versus z when Z_l = actual load, VSWR = 2. (b) $|V|$ versus z when $Z_l = 0$. (Example 4) $Z_g = R_0$.

Before proceeding to a new topic, some general statements concerning Smith charts and other transmission line calculators need to be made.

1. A *normalized* Smith chart applies to a line of *any* characteristic resistance and serves just as well for normalized admittances as for normalized impedances.[2]
2. If 50-Ω (or 20-m℧) lines are worked with almost exclusively, then a 50-Ω (or 20-m℧) Smith chart should be used.
3. Some impedance bridges read out in polar form. It is not difficult to show that a polar coordinate Smith chart contains circles of constant $|Z|$ and circles of constant $\angle Z$, and is available.
4. Quite often difficulty is encountered in accurately reading a normal Smith chart, especially in the neighborhood of $U = +1$, $V = 0$. This difficulty can be overcome by using an inverted circle (or Blanchard) chart.

12.3 TRANSMISSION LINE MATCHING

Transmission lines are used primarily to guide energy from a source to a load. It was previously shown that maximum power transfer occurs for $Z_l = R_0$

[2] Refer to the paragraph below equation 12.9.

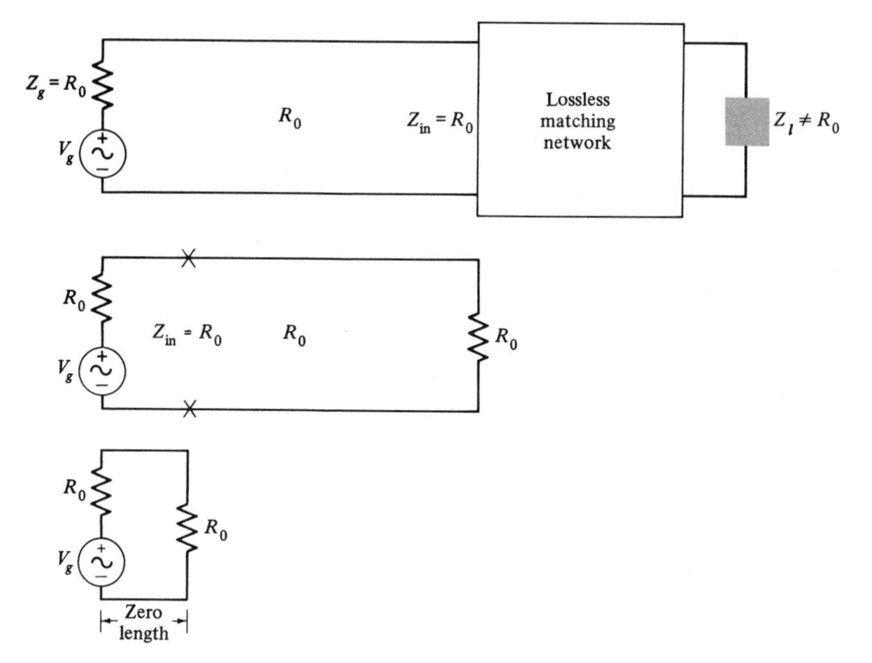

Figure 12.8. Successive equivalent circuits (neglecting time delays) for a lossless matching system.

(when $Z_g = R_0$). Quite often, the load on the line is such that its impedance cannot be easily altered. It is still desirable to dissipate maximum power *in* the load (or more correctly, *in* the *real* part of the load impedance). If we can design a lossless matching system, whose input impedance is R_0, and place this system on the generator side of the load, then we will have achieved maximum power transfer. The matching system and the line are lossless. There is no reflection back to the generator, and *all* of the power available in the *incident* wave *must* be ultimately dissipated in the *real* part of the load impedance. This matching scheme is indicated in Figure 12.8 by successive equivalent circuits (neglecting time delays).

Some examples of lossless elements, or at least the equivalents of lossless elements, suitable for use in a matching scheme, are the short- and open-circuited lossless lines. From equations 11.67 and 11.66, *normalized* values of input impedance and admittance for these elements are

$$z_{in}^{sc} = y_{in}^{oc} = +j \tan kz \tag{12.10}$$

and

$$z_{in}^{oc} = y_{in}^{sc} = -j \operatorname{ctn} kz \tag{12.11}$$

Equations 12.10 and 12.11 reveal that any reactance, $-\infty \leq x_{in} \leq +\infty$, is available from these line sections, called "stubs." Two cases where these stubs are utilized will be examined separately.

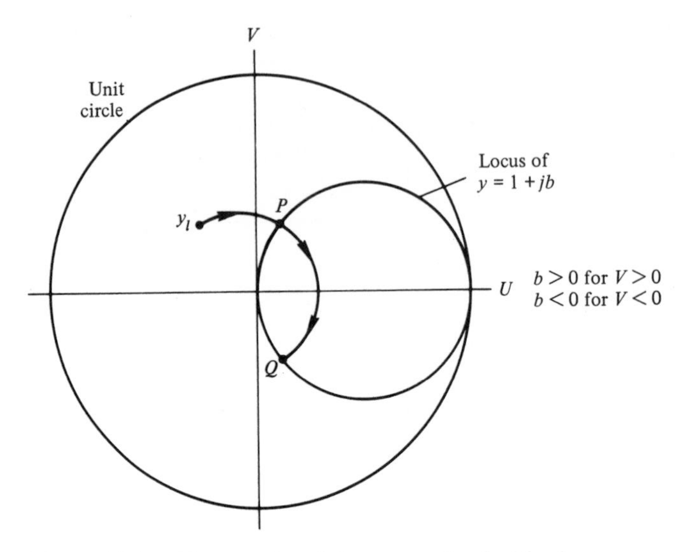

Figure 12.9. The $y = 1 + jb$ locus on the Smith chart.

12.4 SINGLE STUB MATCHING

The basic philosophy in any matching scheme is the same: we attempt to make $z_{\text{in}} = +1$, ($Z_{\text{in}} = R_0$) at some point on the line, or equivalently, we attempt to make $y_{\text{in}} = +1$ at some point on the line. Since the stubs will normally be *shunted across the line*, and admittances *add* in parallel, it is much easier to work with normalized admittances.

Since we are attempting to make $y_{\text{in}} = +1$ at some point on the line, the locus, $y = 1 + jb$ should be located on the Smith chart first. This locus is shown in Figure 12.9.[3] Inspection of Figure 12.9 reveals that for any y_l (with the exception of the special cases $y_l = +1$ and $y_l = +jb$), the $1 + jb$ circle is intersected twice (points P and Q) in rotating from y_l toward the generator. Thus, in any practical case, two values of z ($< \lambda/2$) can be found on the line where $y_{\text{in}} = 1 + jb$. Then, a shunt stub introduced at either of these points on the line will give $y_{\text{in}} = 1 + j0$ if the stub admittance is $y_s = -jb$. That is, $y_{\text{in}} = 1 + jb + y_s = 1 + jb - jb = 1 + j0$. If $y_{\text{in}} = 1 + j0$, then $z_{\text{in}} = 1 + j0$ and $Z_{\text{in}} = R_0$. This reproduces exactly the situation shown in Figure 12.8. An example is in order.

EXAMPLE 5

Suppose $R_0 = 50\ \Omega$ and $Z_l = 25 + j75\ \Omega$. Then $\Gamma_l = 0.74\,\underline{|64^\circ}$ and the VSWR $= 6.7$. The net load power, equation 11.73, is down from the matched case by $(1 - |\Gamma_l|^2) = 0.45(-3.45\ \text{dB})$. The standing wave

[3] Refer to Figure 12.3(b) for the loci of constant b (or x).

is excessively large, which will cause additional difficulties if the line has loss. It is desired to match the line to the load with a single shorted ($R_0 = 50\ \Omega$) stub. The steps in the procedure are numbered.

1. $$z_l = \frac{25 + j75}{50} = 0.5 + j1.5$$

   $$y_l = \frac{1}{0.5 + j1.5} = 0.2 - j0.6$$

   Note here that z_l and y_l are diametrically opposite on the Smith chart (Why?).
2. y_l is located on the chart and a wavelength reference of 0.412λ on the "toward generator" scale is obtained as shown in Figure 12.10.

Figure 12.10. Single stub matching problem, Example 5. *Source*: Copyright © 1949 by Kay Electric Company, Pine Brook, N.J. Renewal copyright © 1976 by P. H. Smith, New Providence, N.J.

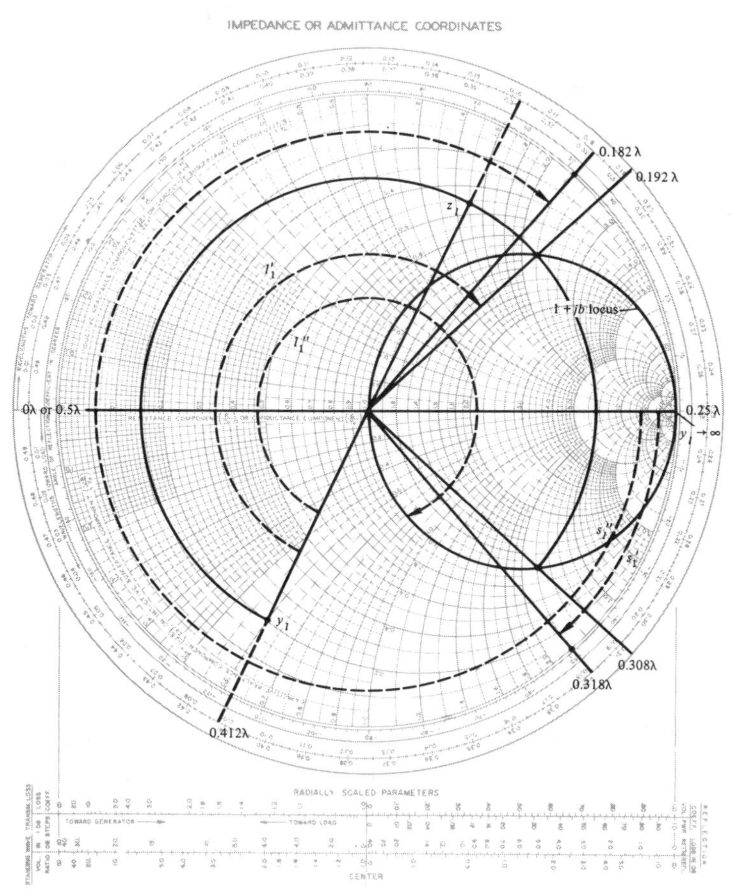

3. A constant radius circle ($|\Gamma_l| = 0.74$), is drawn through y_l and intersects the $1 + jb$ locus at two points.

$$\begin{bmatrix} y'_{\text{in}} = 1 + j2.2 \text{ at} \\ 0.192\lambda \text{ reference} \end{bmatrix} \quad \text{or} \quad \begin{bmatrix} y''_{\text{in}} = 1 - j2.2 \text{ at} \\ 0.308\lambda \text{ reference} \end{bmatrix}$$

4. The point of stub connection is the distance rotated, or

$$\begin{bmatrix} l'_1 = 0.5\lambda - 0.412\lambda + 0.192\lambda \\ l'_1 = 0.28\lambda \end{bmatrix}$$

or

$$\begin{bmatrix} l''_1 = 0.5\lambda - 0.412\lambda + 0.308\lambda \\ l''_1 = 0.396\lambda \end{bmatrix}$$

5. The required stub susceptance is

$$y'_s = -j2.2 \quad \text{or} \quad y''_s = +j2.2$$

6. The stub lengths are found by simply treating the stubs separately. The stub load is $y_l \to \infty$ (short circuit) which is located at $U = +1$, $V = 0$. Rotating from this point to the required susceptance gives the stub length. Thus,

$$\begin{bmatrix} s'_1 = 0.318\lambda - 0.25\lambda \\ s'_1 = 0.068\lambda \end{bmatrix} \quad \text{or} \quad \begin{bmatrix} s''_1 = 0.25\lambda + 0.182\lambda \\ s''_1 = 0.432\lambda \end{bmatrix}$$

This gives the matched system shown in Figure 12.11.

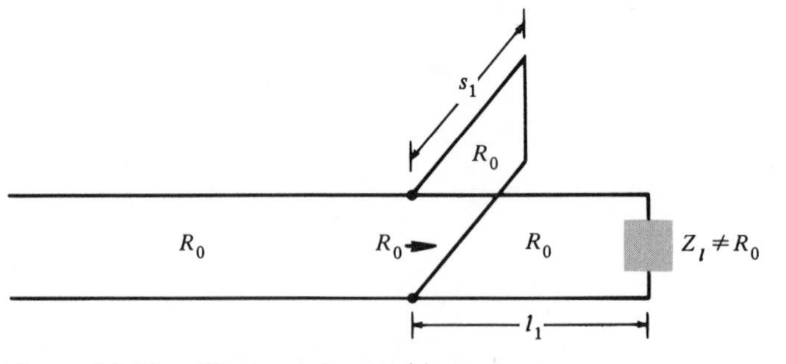

Figure 12.11. Single stub matching system.

The choice between the two possible solutions is not important unless the frequency is low enough so that the physical length of the line is an economic factor, or unless one solution has a better frequency bandwidth characteristic. See Problem 22. A shorted stub is usually chosen over the open stub, because (in coaxial systems at least) it has less tendency to radiate energy. In practice, l_1 and s_1 are adjusted to minimize reflected power.

In this regard, it is highly desirable to have some way to measure the reflected power *alone*. A device that accomplishes this is called a *directional*

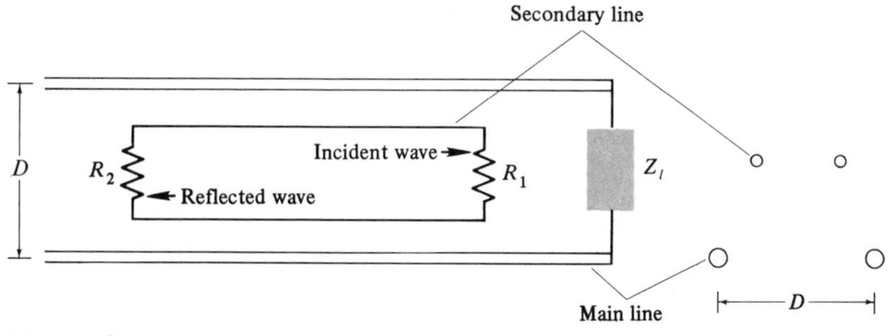

Figure 12.12. Elementary directional coupler.

coupler. The directional coupler takes many forms, one such form being shown in Figure 12.12. In simple terms, a small amount of energy is coupled from the main line to the secondary line. A signal, proportional to the incident voltage on the main line, can be measured at R_1, while a signal proportional to the reflected voltage on the main line can be measured at R_2. The directional coupler inherently takes a small amount of energy from the main line.

When l_1 and s_1 are adjusted for zero reflected power, all of the incident power is absorbed in the *load resistance*, R_l. Then

$$\langle P_l \rangle = \frac{V_g^2}{8R_0} = \frac{|I_l|^2}{2} R_l, \qquad I_l = \text{peak load current}$$

and

$$|I_l| = \frac{V_g}{2\sqrt{R_0 R_l}} = \frac{V_g}{2\sqrt{(25)(50)}} = \frac{V_g}{70.7} \quad \text{(A)}$$

peak value.

In order to gain some appreciation of the time saving features of the Smith chart, let us calculate the stub length, s_1, and point of connection, l_1, for the single stub system. From equation 12.3,

$$y_{\text{in}} = \frac{1 - |\Gamma_l| e^{j(\phi_l - 2kz)}}{1 + |\Gamma_l| e^{j(\phi_l - 2kz)}} = g_{\text{in}} + jb_{\text{in}} \tag{12.12}$$

or, rationalizing,

$$g_{\text{in}} = \frac{1 - |\Gamma_l|^2}{1 + |\Gamma_l|^2 + 2|\Gamma_l| \cos(\phi_l - 2kz)} \tag{12.13}$$

and

$$b_{\text{in}} = \frac{-2|\Gamma_l| \sin(\phi_l - 2kz)}{1 + |\Gamma_l|^2 + 2|\Gamma_l| \cos(\phi_l - 2kz)} \tag{12.14}$$

For a match, $g_{in} = 1$ at $z = l_1$, therefore, from equation 12.13,

$$\cos(\phi_l - 2kl_1) = -|\Gamma_l|$$

or

$$l_1 = \frac{\phi_l + \pi - \cos^{-1}|\Gamma_l|}{2k} \tag{12.15}$$

The susceptance b_{in} at l_1 is obtained by substituting equation 12.15 into 12.14, which gives

$$b_{in} = \frac{-2|\Gamma_l| \sin[\cos^{-1}(-|\Gamma_l|)]}{1 - |\Gamma_l|^2} \tag{12.16}$$

For a shorted stub having the same R_0 as the main line,

$$b_s = -\text{ctn } ks_1 \tag{12.17}$$

and a match occurs for $b_{in} = -b_s$. Then,

$$\text{ctn } ks_1 = \frac{2|\Gamma_l| \sin[\cos^{-1}(-|\Gamma_l|)]}{1 - |\Gamma_l|^2}$$

or

$$s_1 = \frac{1}{k} \tan^{-1} \frac{1 - |\Gamma_l|^2}{2|\Gamma_l| \sin[\cos^{-1}(-|\Gamma_l|)]} \tag{12.18}$$

Using numbers from the preceding example gives ($\Gamma_l = 0.74 \angle 64°$)

$$l_1 = \frac{(64/360)(2\pi) + \pi - \cos^{-1} 0.74}{4\pi} \lambda$$

or

$$l_1 = 0.339\lambda \pm 0.059\lambda$$

or

$$(l_1' = 0.28\lambda) \quad \text{or} \quad (l_1'' = 0.398\lambda)$$

Also,

$$s_1 = \frac{\lambda}{2\pi} \tan^{-1} \frac{0.453}{1.48 \sin[\cos^{-1}(-0.74)]}$$

or

$$s_1 = \frac{\lambda}{2\pi} \tan^{-1} \frac{0.306}{\sin(0.765\pi \quad \text{or} \quad 1.235\pi)}$$

Therefore,

$$s_1 = \frac{\lambda}{2\pi} \tan^{-1}(\pm 0.455)$$

and

$$(s_1' = 0.068\lambda) \quad \text{or} \quad (s_1'' = 0.431\lambda)$$

The calculated values of l_1 and s_1 agree very well with those taken from the Smith chart. This example shows that the use of the Smith chart is not only faster, it is much closer to what it is actually taking place on the line. This does not, of course, exclude the use of programmable calculators.

12.5 DOUBLE STUB MATCHING

A double stub matching system is shown in Figure 12.13. The spacings, l_1 and l_2, are normally fixed, while the stub lengths, s_1 and s_2, are normally adjustable. The normalized admittance at x-x with s_1 in place, but s_2 not in

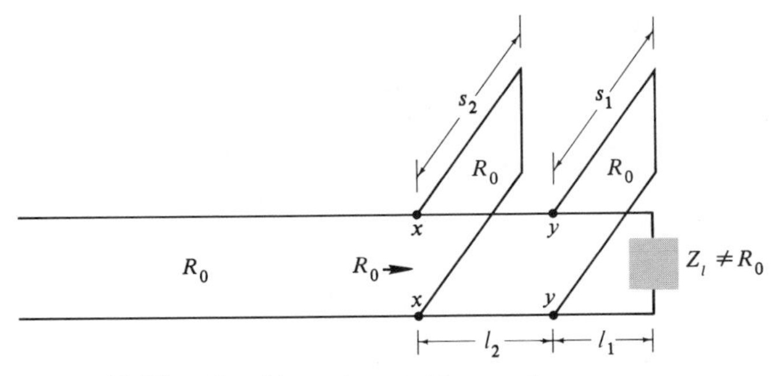

Figure 12.13. Double stub matching system.

place *must be* $1 + jb$. Then, s_2 is attached and adjusted so that its susceptance is $-jb$ (as in the single stub system), giving a matched system. Thus, the admittance at y-y, with s_1 in place, but s_2 not in place, must lie on the Smith chart locus of all points l_2 wavelengths *toward the load from the* $1 + jb$ locus (L_1). The new locus, L_2, is just a circle of the same size as the $1 + jb$ locus. It follows then that the stub s_1 has the function of moving the admittance at y-y *along a contour of constant* g to the locus L_2. An example will clarify the procedure.

EXAMPLE 6

Suppose $R_0 = 50\ \Omega$ and $Z_l = 25 + j75\ \Omega$ (as in the single stub example), so that $\Gamma_l = 0.74\,\underline{|64^\circ}$ and the VSWR $= 6.7$. Let $l_1 = 0.2\lambda$ and $l_2 = 0.375\lambda$. The steps in the design of this double stub system are as follows below for Figure 12.14.

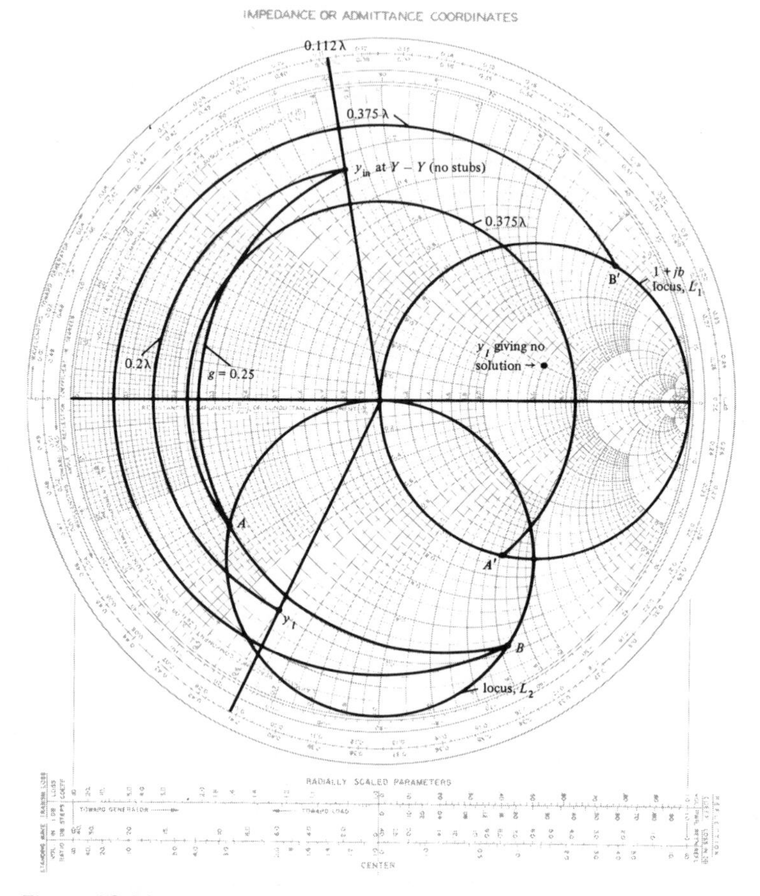

Figure 12.14. Double stub matching problem, Example 6. *Source*: Copyright © 1949 by Kay Electric Company, Pine Brook, N.J. Renewal copyright © 1976 by P. H. Smith, New Providence, N.J.

1. The first step in a double stub design is that of locating the locus L_2. This locus is just the L_1 $(1 + jb)$ locus rotated 0.375λ toward the load as shown in Figure 12.14. The locus L_2 is drawn on the chart.
2. The normalized load admittance is $y_l = 0.2 - j0.6$ and is located on the chart.
3. y_l is rotated 0.2λ toward the generator. This gives $y_{in} = 0.25 + j0.82$ (at y-y) at a reference of 0.112λ.
4. The stub, s_1, at y-y can move the admittance at y-y to *either* of the two points on the constructed locus L_2. These points are A, $y = 0.25 - j0.34$, and B, $y = 0.25 - j1.65$.

[It is worthwhile at this point to notice that if y_{in} at y-y without s_1 has a real part, g_{in}, greater than 2 (for example, $y_{in} = 3 + j1$), a solution does not exist! Thus, sometimes a solution is not possible for a double stub system, unless l_2 is altered.]

5. For point A, the stub s_1 must change the susceptance b from $+0.82$ to -0.34, while for point B, the stub s_1 must change the susceptance from $+0.82$ to -1.65. Therefore,

$$(b'_{s_1} = -1.16) \quad \text{or} \quad (b''_{s_1} = -2.47)$$

6. The required stub lengths are found as in the single stub system and are

$$(s'_1 = 0.114\lambda) \quad \text{or} \quad (s''_1 = 0.061\lambda)$$

7. Point A is rotated 0.375λ toward the generator to A' where $y_{in} = 1 - j1.65$ at x-x. Point B is rotated 0.375λ toward the generator to B' where $y_{in} = 1 + j3.6$.
8. The stub s_2 must cancel the susceptance at x-x, thus,

$$(y'_{s_1} = +j1.65) \quad \text{or} \quad (y''_{s_2} = -j3.6)$$

9. The stub lengths are found as before.

$$(s'_2 = 0.414\lambda) \quad \text{or} \quad (s''_2 = 0.044\lambda)$$

s'_1 and s'_2 must be used together, or s''_1 and s''_2 must be used together.

The same remarks, concerning a choice of one of the two possible solutions for the single stub system, apply here. The double stub system can also be treated analytically, but the effort is not worthwhile, unless many calculations using a programmable calculator or computer are contemplated.

12.6 CONCLUDING REMARKS

The Smith chart transmission line calculator enables rapid calculations to be made for parameters determining transmission line performance. The Smith chart is normally used with the slotted line or impedance bridge. Even if the line has loss, we will find that the Smith chart can be used. In most practical applications (where the loss is small) the loss can be ignored.

REFERENCES

Brown. (See references for Chapter 11.)
Chipman. (See references for Chapter 11.)
Moore. (See references for Chapter 11.)

PROBLEMS

1. Show that equations 12.7 and 12.8 can be derived from equation 12.3.
2. Prove that z_{in} and y_{in} are diametrically opposite on the Smith chart.
3. If $Z_l = 200 - j50$ on a 50-Ω lossless line, find Γ_l, VSWR, and Z_{in} at $z = 0.3\lambda$ from the load.
4. The following data are taken with an impedance bridge: $Z_{in}^{sc} = +j100\ \Omega$, $Z_{in}^{oc} = -j25\ \Omega$, and $Z_{in} = 100\underline{|30°}$ (actual load in place). Find Z_l.
5. The following data are taken with a slotted line: $|V|_{max} = 0$ dB, $|V|_{min} = -10$ dB, $z_{min} = 0, 0.375, 0.75, \ldots$ m (with short circuit as load) and positions of $|V|_{min} = 0.10, 0.475, 0.85, \ldots$ m (with load in place). Find the frequency. Find Z_l.
6. Design a single stub system to match a load, $Z_l = 10 - j50\ \Omega$ to a 50-Ω lossless line. Use a shunt shorted stub.
7. Repeat Problem 6 using a pair of series shorted stubs.
8. Repeat Problem 6 using double shunt shorted stubs spaced $\lambda/4$. One stub is placed directly across the load.
9. 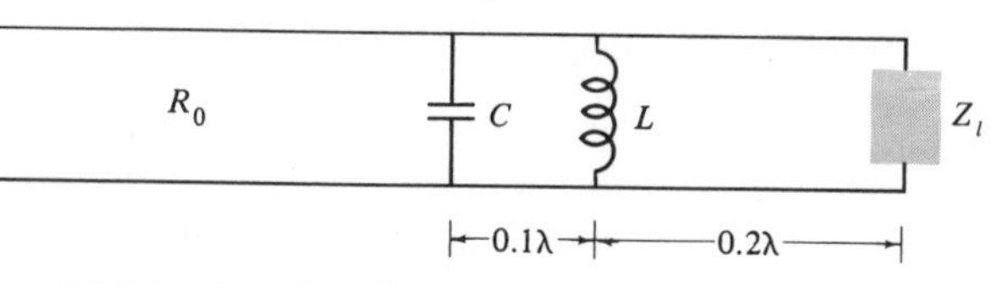

Figure 12.15. Lossless line.

If $Z_l = 100 + j200$, $L = 0.1\ \mu$H, $C = 20$ pF, $R_0 = 50\ \Omega$, and $f = 300$ MHz, find the VSWR to the left of the capacitor. See Figure 12.15.
10. Find $|I_l|$ (load current magnitude) in Problem 6 after the matching system is attached if $\langle P_{inc} \rangle$ is 10 W.
11. A load of 30 Ω in series with an inductor exists on the end of a lossless 50-Ω line and produces a VSWR of 3 at 300 MHz. Find L.
12. A matched system consists of a shunt stub (shorted) 0.1λ long located 0.2λ from the load. Find Z_l.
13. It was mentioned in Section 12.1 that amplitude modulation is often employed with the slotted line. Let

$$v_g(t) = V_0(1 + m_a \cos \omega_m t) \cos \omega_c t,$$

where ω_m is the modulation frequency, ω_c is the carrier frequency, and m_a is the percent modulation. Assuming a lossless line, $Z_g = R_0$, and a frequency dependent load impedance (e.g., $Z_l = R + j\omega L$), can the slotted line give exact results? See equation 11.77.
14. If $v_g(t)$ is an amplitude modulated signal as in Problem 13, use phasor concepts and find a time domain expression for the reflected voltage (generator matched). Use superposition.

15. Under what conditions could a frequency modulated voltage source be successfully employed with the slotted line?
16. An air-filled lossless coaxial cable has a nominal $R_0 = 50\ \Omega$. A lossless dielectric slug ($\varepsilon_R = 4$, filling the space between inner and outer conductors) 0.2 m long is inserted with one end 0.1 m from the load. If $Z_l = R_0$, what is the VSWR in the space from the generator to the slug? $f = 300$ MHz.
17. Show that "the electric field is distributed exactly like the line voltage" (as stated in Section 12.1) on the slotted coaxial line. See Problem 23, Chapter 11.
18. A load $Z_l = 100 + j50\ \Omega$ is connected to a line of length d and impedance R_0'. This line is connected to the generator through a 50-Ω line. Find R_0' and d for a matched system using the Smith chart.
19. If a $\lambda/4$ matching section is designed to match a 50-Ω line to $Z_l = 200\,\underline{|0^\circ}$ at 500 MHz, find the "bandwidth" of the lossless matching section. The bandwidth is (arbitrarily) measured at the frequencies where VSWR = 1.5. What is the maximum VSWR?
20. What VSWR corresponds to "half-power" ($\langle P_l \rangle / \langle P_{\text{inc}} \rangle = 0.5$)? Using the data of Problem 19, what can be said about the half-power bandwidth?
21. Repeat Problem 19 if two cascaded $\lambda/4$ transformers are used to match the 200-Ω load to the 50-Ω line. The intermediate impedance is 100 Ω.
22. Using the smaller of the two possible values for l_1, design a single shorted stub system at 500 MHz to match $Z_l = 200\ \Omega$ to a 50-Ω line. Use the smaller of the two possible values for s_1 and plot VSWR versus normalized frequency. What is the bandwidth in terms of VSWR = 1.5?

Chapter 13
Lossy Transmission Lines

All transmission lines have some loss, and ordinarily this is not desirable. In this chapter, we will be seeking methods to analyze the lossy transmission line. Fortunately, the most useful methods are merely extensions of those for the lossless case.

13.1 LOSSES IN GENERAL

The losses in a given material are extremely difficult to catagorize in the general case. A detailed mathematical treatment would necessarily be at the *microscopic* level, and, as such, is beyond the scope of this treatment. We can, however, give a qualitative description of the behavior of materials under the influence of a sinusoidal time-dependent force. In this way, we can obtain an understanding of the behavior of material parameters at the *macroscopic level.*

As an example, consider the *polarization* of a material (into a dielectric) under the influence of an external sinusoidal electric field. This effect was first considered in Section 4.3, with steady fields, where the resultant polarization was explained on the basis of the displacement of charges making up electric dipoles. These charged particles have mass and inertia. On this basis, we can immediately predict that the polarization will decrease with increasing frequency. In other words, as the frequency of the applied field increases, the inertia of the particles will tend to prevent the particle displacement from following (in time) the field. Thus, we can predict that the displacement will have in-phase and out-of-phase components.

Although there are several polarization mechanisms, we can consider *electronic polarization* as typical. In a somewhat oversimplified picture, we have the following forces acting on an electron when a field $E_x(t) = E_0 \cos \omega t$ is present.

1. The applied force; $-eE_0 \cos \omega t \; (Q = -e)$.
2. The Coulomb restoring force of attraction between the electron and (large mass) positive nucleus; proportional to displacement, x.
3. The frictional (damping) force; proportional to mass, m, and particle velocity, dx/dt.
4. The mass times acceleration, $m(d^2x/dt^2)$.

Newton's law gives the force equation,

$$m \frac{d^2x}{dt^2} = -eE_0 \cos \omega t - C_1 x - C_2 \frac{dx}{dt}$$

or in *phasor form*,

$$-m\omega^2 x = -eE_0 - C_1 x - j\omega C_2 x$$

The last equation should be familiar to electrical engineering students, because it is exactly the same as (dual to) the equation for the charge in an R, L, C series circuit when driven by $v = V_0 \cos \omega t$. Thus, a resonance will occur, and the displacement, x, will have real and imaginary parts.

The dipole moment is defined in equation 4.13, and is given here by $p_x = -ex$. The polarization, as defined by equation 4.16, is the dipole moment per unit volume, so it too is directly proportional to displacement, x. For linear, isotropic materials the susceptibility, χ_E, will also be directly proportional to x (see equation 4.26). Finally, if we combine equation 4.27 and equation 4.30, there results

$$\varepsilon = \varepsilon_0(1 + \chi_E)$$

We are now in a position to definitely conclude that, in general, the permittivity will be complex, because χ_E is complex. Using the analogy between the differential equation we have "derived" and that of the R, L, C circuit allows us to predict a resonance in the permittivity.

We can now simply define the complex permittivity[1] as

$$\varepsilon(\omega) = \varepsilon'(\omega) - j\varepsilon''(\omega)$$

[1] See Harrington, in the References at the end of Chapter 10.

It can be shown that $\varepsilon''(\omega)$ is positive. A similar development shows (with essentially the same assumptions) that for linear, isotropic ferromagnetic materials

$$\mu(\omega) = \mu'(\omega) - j\mu''(\omega)$$

The following definitions exist for these parameters.

$\varepsilon'(\omega)$ is called the *ac capacitivity*.
$\varepsilon''(\omega)$ is called the *dielectric loss factor*.
$\mu'(\omega)$ is called the *ac inductivity*.
$\mu''(\omega)$ is called the *magnetic loss factor*.

It is also common to use

$$\varepsilon(\omega) = |\varepsilon(\omega)| e^{-j\delta_d}$$

and

$$\mu(\omega) = |\mu(\omega)| e^{-j\delta_m}$$

where δ_d and δ_m are the *dielectric* and *magnetic loss angles*, respectively. Some tables of representative material parameters are presented in Appendix B.

For a linear, homogeneous, and isotropic material Maxwell's equations may be written

$$\nabla \times \mathbf{E} = -j\omega\mathbf{H} = -j\omega(\mu' - j\mu'')\mathbf{H}$$

and

$$\nabla \times \mathbf{H} = [\sigma + j\omega(\varepsilon' - j\varepsilon'')]\mathbf{E} = [(\sigma + \omega\varepsilon'') + j\omega\varepsilon']\mathbf{E}$$

These equations may be regarded, for our purposes, as representing the general case. Notice that there are three distinct parameters, σ, ε'', and μ'', which can contribute to the losses. Each of these may be frequency dependent. Notice particularly that σ and $\omega\varepsilon''$ play the same role and the *sum* $\sigma + \omega\varepsilon''$ may be "called" conductivity if we desire. From a practical point of view, it may be very difficult to distinguish σ and ε'', if one is measuring these parameters.

Equations were developed in Chapter 10 for uniform plane waves propagating at a single frequency in a *lossy* dielectric medium. We are not interested here in separating the terms that contribute to the loss, except to neglect μ''. Thus, in light of the preceding paragraph,

$$\mu = \mu'$$

and

$$\sigma + \omega\varepsilon'' + j\omega\varepsilon' = \sigma + j\omega\varepsilon$$

That is, we have eliminated the possibility of magnetic loss, and simply lumped $\sigma + \omega\varepsilon''$ into one term and called it σ. This is what we did in Chapters 9 and 10 without explicitly stating so. The quantity $\sigma/\omega\varepsilon$ is called the *loss tangent*, and in Chapter 9 we called $\hat{y} = \sigma + j\omega\varepsilon$ the admittivity.

13.2 FIELD EQUATIONS WITH LOSS

The beginning equations are now the same as those in Chapter 10 for a uniform plane wave propagating in a lossy dielectric

$$\frac{\partial E_x}{\partial z} = -j\omega\mu H_y \tag{13.1}$$

and

$$\frac{\partial H_y}{\partial z} = -(\sigma + j\omega\varepsilon)E_x \tag{13.2}$$

The solutions for propagation in the $+z$ direction were

$$E_x = E_0 e^{-\alpha z} e^{-j\beta z} \tag{13.3}$$

and

$$H_y = \frac{E_0}{\eta^+} e^{-\alpha z} e^{-j\beta z} \tag{13.4}$$

where

$$\eta^+ = \frac{j\omega\mu}{\gamma} \tag{13.5}$$

and

$$\gamma = \sqrt{j\omega\mu(\sigma + j\omega\varepsilon)} = \alpha + j\beta \tag{13.6}$$

It is still possible to introduce a pair of *perfectly conducting planes* parallel to the y axis without disturbing the field given by equations 13.3 and 13.4. If we proceed further, as in Chapter 11, and remove all but a finite width of the parallel plane system, we again expect fringing of the field at the edges, but no drastic changes. The field will still be TEM to z; that is, both E_z and H_z are zero. At this point we can parallel the rather lengthy rigorous development of Chapter 11 for the voltage and current differential equations of the line. This is really unnecessary, however, for inspection of equations 13.1 and 13.2 reveals that with dielectric loss only the term $\sigma + j\omega\varepsilon$ appears, and without this loss the same term is merely $j\omega\varepsilon$. Put a different way, a quadrature conduction current term is added to the (already present)

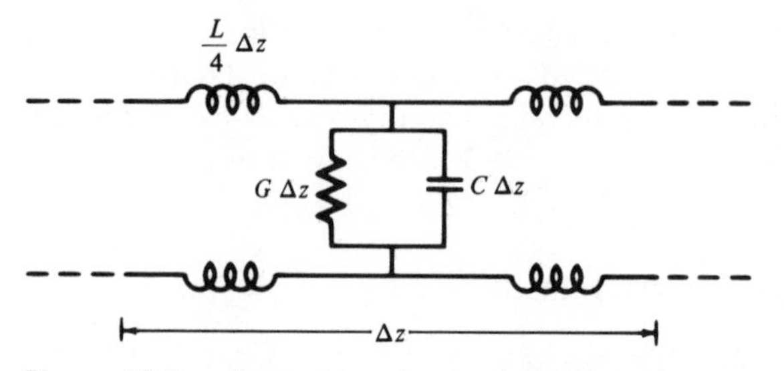

Figure 13.1. Equivalent circuit of distributed parameters for a line with perfect conductors embedded in a lossy dielectric.

displacement current. This means that the following replacements are in order.

$$j\omega\varepsilon \rightarrow \sigma + j\omega\varepsilon$$

$$j\omega C \rightarrow G + j\omega C$$

$$\frac{\sigma}{G} = \frac{\omega\varepsilon}{\omega C}, \qquad G = \frac{\sigma}{\varepsilon} C \quad \text{(conductance/length)} \tag{13.7}$$

$$LC = \mu\varepsilon$$

$$Z_0 = \sqrt{\frac{j\omega L}{G + j\omega C}} \quad \text{(characteristic impedance)} \tag{13.8}$$

$$\gamma = \sqrt{j\omega L(G + j\omega C)} \quad \text{(propagation constant)} \tag{13.9}$$

$$\frac{dV}{dz} = j\omega LI \tag{13.10}$$

$$\frac{dI}{dz} = (G + j\omega C)V \tag{13.11}$$

Remember that we are assuming no loss in the conductors. We will do nothing more with these equations at the present time, except to mention the equivalent circuit of distributed L, G, and C shown in Figure 13.1.

13.3 THE GENERAL LINE

Both the lossless line and the line with conductivity in the medium around the perfect conductors have been introduced by means of plane waves and the infinite parallel plane system. Exact results were possible in both cases, making it relatively simple to extend these results to lines of practical shape. The next logical step now would be that of allowing *finite* conductivity in the parallel plane conductors of the parallel plane system. If this is done, a TEM mode is no longer possible because a component of electric field must exist *inside* the conductor in the direction of the current ($\mathbf{J} = \sigma\mathbf{E}$). Since the boundary condition at the conductor-dielectric interface requires continuity

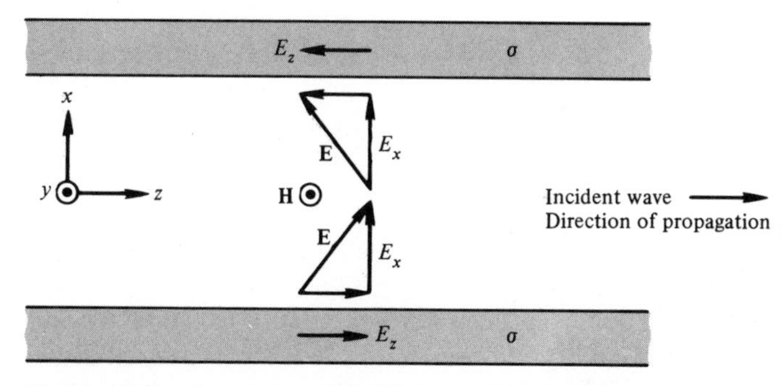

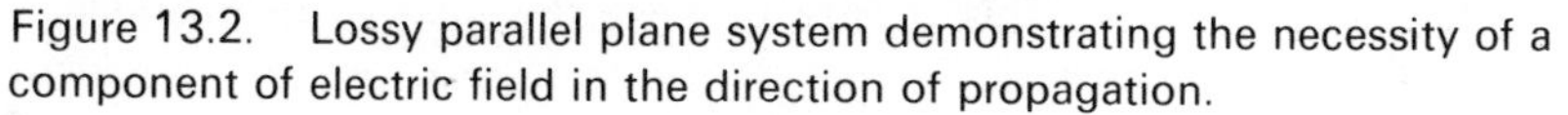

Figure 13.2. Lossy parallel plane system demonstrating the necessity of a component of electric field in the direction of propagation.

of electric field at the interface, a component of electric field exists outside the conductor in the current, or axial, direction. This, by definition, *is not a TEM to z mode.* The parallel plane system with finite conductivity is shown in Figure 13.2. Such a wave may be classified as a *T*ransverse *M*agnetic, or TM to z mode. That is, $E_z \neq 0$, $H_z = 0$, and the magnetic field lies in a plane transverse to the direction of net propagation.

Faced with this situation, we can either attempt an exact solution, which will not be easy because of the complicated boundary conditions, or we can make suitable approximations. The engineer is quite often willing to choose the latter course. In pursuing that course here, we will use a *perturbation* method. If the loss in the conductors is small, then the field external to the conductors is not altered, or perturbed, much from the field in the lossless case. An alternate statement is that if the losses are small, the power transmitted along the *system* is much greater than the power lost in the conductors of the *system.* (The use of the word *system* is meant to indicate that the technique being developed here applies to *any* electromagnetic guiding system.) If the magnetic field at the conductor surface is that of the perfect conductor case (as it almost is), then we can calculate the *power lost* as was done in Chapter 10. The *power flow* along the system may also be calculated using the lossless values of **E** and **H**. With this information, an attenuation constant, α, may be obtained in the following *general* way.

The time average power flow along the system is proportional to $e^{+2\alpha z}$, for a case where the power flow is in the $-z$ *direction.* Thus,

$$\langle P_f \rangle = P_0 e^{+2\alpha z} \quad \text{(W)} \tag{13.12}$$

The time average power dissipated per unit system length, $\langle P_d \rangle$, is the rate of decrease of $\langle P_f \rangle$ with $-z$, or

$$\langle P_d \rangle = -\frac{\partial \langle P_f \rangle}{\partial(-z)} = 2\alpha P_0 e^{+2\alpha z}$$

or

$$\langle P_d \rangle = 2\alpha \langle P_f \rangle \quad \text{(W/m)} \tag{13.13}$$

Therefore,

$$\boxed{\alpha = \frac{1}{2}\frac{\langle P_d \rangle}{\langle P_f \rangle}} \quad \text{(Np/m)} \tag{13.14}$$

Equation 13.14 is exact. It becomes approximate when we use approximate values for $\langle P_d \rangle$ and $\langle P_f \rangle$.

Besides the obvious effect of adding series resistance to the equivalent circuit of the line, another effect of finite conductor conductivity is the addition of an *internal inductance.* This occurs because the *internal* magnetic flux cannot link *all* of the current. Insofar as the transmission line equivalent circuit and equations are concerned, the internal and external inductances are merely added. Extending equations 13.8 and 13.9 to the general case gives

$$\boxed{Z_0 = \sqrt{\frac{Z}{Y}} = \sqrt{\frac{R + j\omega L}{G + j\omega C}}} \tag{13.15}$$

and

$$\boxed{\gamma = \sqrt{ZY} = \sqrt{(R + j\omega L)(G + j\omega C)}} \tag{13.16}$$

The equivalent circuit of the general line is as shown in Figure 13.3.

Following the same procedure as was used for the lossless line,

$$V(z, \omega) = C_1(e^{+(\alpha + j\beta)z} + \Gamma_l e^{-(\alpha + j\beta)z}) \tag{13.17}$$

and

$$I(z, \omega) = \frac{C_1}{Z_0}(e^{+(\alpha + j\beta)z} - \Gamma_l e^{-(\alpha + j\beta)z}) \tag{13.18}$$

where

$$\boxed{\Gamma_l = \frac{Z_l - Z_0}{Z_l + Z_0} = |\Gamma_l| e^{j\phi_l}} \tag{13.19}$$

$$\boxed{\alpha = \text{Re}\{\gamma\}} \tag{13.20}$$

and

$$\boxed{\beta = \text{Im}\{\gamma\}} \tag{13.21}$$

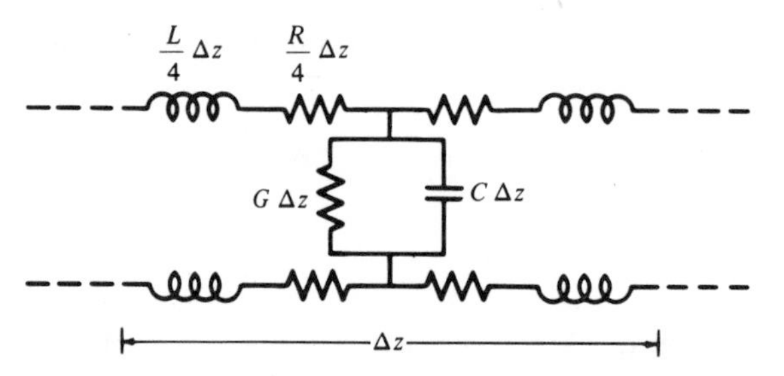

Figure 13.3. Equivalent circuit of distributed parameters for a general transmission line.

Taking the time domain value of the *incident voltage alone* from the phasor form (equation 13.17),

$$v(z, t) = \text{Re}\,\{C_1 e^{+j\omega t} e^{+\alpha z} e^{+j\beta z}\}$$

or

$$v(z, t) = C_1 e^{+\alpha z} \cos(\omega t + \beta z) \tag{13.22}$$

The phase velocity is obtained as in the lossless case and is

$$u_p = \frac{\omega}{\beta} = \frac{\omega}{\text{Im}\,\{\gamma\}} = \frac{\omega}{\text{Im}\,\{\sqrt{(R + j\omega L)(G + j\omega C)}\}} \tag{13.23}$$

Notice that, in general, u_p *is a function of* frequency, and thus, the line is not (in general) distortionless! This is another effect of the loss.

Taking the generator characteristics into account will eliminate the constant C_1 from equations 13.17 and 13.18. This is done, again, just like it was for the lossless case of Chapter 11, but rather than following that rather lengthy procedure, it is easier to simply replace R_0 by Z_0 and jk by $\alpha + j\beta$ in equations 11.61 and 11.62. This results in

$$V = \frac{V_g Z_0}{Z_0 + Z_g} e^{-(\alpha + j\beta)l} \frac{(e^{+(\alpha + j\beta)z} + \Gamma_l e^{-(\alpha + j\beta)z})}{1 - \Gamma_l \Gamma_g e^{-2(\alpha + j\beta)l}} \tag{13.24}$$

and

$$I = \frac{V_g}{Z_0 + Z_g} e^{-(\alpha + j\beta)l} \frac{(e^{+(\alpha + j\beta)z} - \Gamma_l e^{-(\alpha + j\beta)z})}{1 - \Gamma_l \Gamma_g e^{-2(\alpha + j\beta)l}} \tag{13.25}$$

Also, the input impedance is the ratio of V to I and is

$$Z_{\text{in}} = Z_0 \frac{e^{+(\alpha + j\beta)z} + \Gamma_l e^{-(\alpha + j\beta)z}}{e^{+(\alpha + j\beta)z} - \Gamma_l e^{-(\alpha + j\beta)z}} \tag{13.26}$$

If $Z_g = Z_0$, then $\Gamma_g = 0$ and the generator is matched to the line. In this case, equations 13.24 and 13.25 become

$$V = \frac{V_g}{2} e^{-(\alpha + j\beta)l}(e^{+(\alpha + j\beta)z} + \Gamma_l e^{-(\alpha + j\beta)z}) \tag{13.27}$$

and

$$I = \frac{V_g}{2Z_0} e^{-(\alpha + j\beta)l}(e^{+(\alpha + j\beta)z} - \Gamma_l e^{-(\alpha + j\beta)z}) \tag{13.28}$$

Some special cases with $Z_g = Z_0$ may now be investigated and compared to their lossless counterparts of Chapter 11.

1. *Open Circuit*

$$\begin{aligned} Z_l &\to \infty \\ \Gamma_l &= +1 \\ Z_{\text{in}}^{\text{oc}} &= Z_0 \operatorname{ctnh} \gamma z \\ |V| &= V_g e^{-\alpha l} |\cosh \gamma z| \\ |I| &= \frac{V_g}{|Z_0|} e^{-\alpha l} |\sinh \gamma z| \end{aligned} \tag{13.29}$$

It is worthwhile to examine $|V|$ plotted against z for various amounts of attenuation. This is shown in Figure 13.4. In these plots, an upper envelope may be determined by the tangent points where $\beta z = 0, \pi, 2\pi, \ldots$, so that $|\cosh \gamma z| = \cosh \alpha z$, and the upper envelope is $V_g e^{-\alpha l} \cosh \alpha z$. The lower envelope is determined similarly and is $V_g e^{-\alpha l} \sinh \alpha z$. The median line, $(V_g/2)e^{-\alpha(l-z)}$, actually represents the magnitude of the incident voltage, as will be seen in case (3). The current magnitude may be plotted the same way and is shown in Figure 13.5. It should be mentioned, again, that the VSWR is rather indefinite. If the loss is small, then the lossless value of the VSWR is usually quoted. That is, VSWR $= (1 + |\Gamma_l|)/(1 - |\Gamma_l|)$.

2. *Short Circuit*

$$\begin{aligned} Z_l &= 0 \\ \Gamma_l &= -1 \\ Z_{\text{in}}^{\text{sc}} &= Z_0 \tanh \gamma z \\ |V| &= V_g e^{-\alpha l} |\sinh \gamma z| \\ |I| &= \frac{V_g}{Z_0} e^{-\alpha l} |\cosh \gamma z| \end{aligned} \tag{13.30}$$

Because of the complementary relationship between equations 13.30 and 13.29, there is really no need to plot $|V|$ and $|I|$ for $Z_l = 0$.

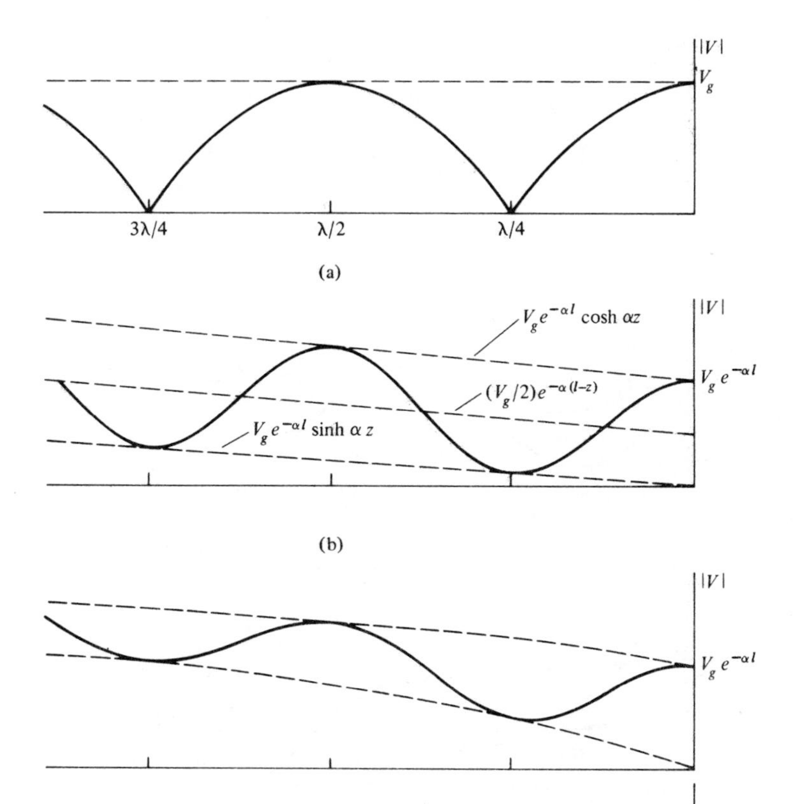

Figure 13.4. $|V|$ versus z, $Z_g = Z_0$, $Z_l \to \infty$. (a) No attenuation, $\alpha = 0$, (b) small attenuation, (c) larger attenuation.

3. *Matched Case*

$$Z_l = Z_0$$

$$\Gamma_l = 0$$

$$Z_{\text{in}} = Z_0$$

$$|V| = \frac{V_g}{2} e^{-\alpha(l-z)} \tag{13.31}$$

$$|I| = \frac{V_g}{2} e^{-\alpha(l-z)}$$

Maximum power transfer occurs because there is no reflection. $|V|$ and $|I|$ are shown in Figure 13.6 for moderate loss. The effect of the loss is rather obvious. Notice that $|V|$ and $|I|$ in Figure 13.6 are the median lines of Figure 13.4 and Figure 13.5. Why?

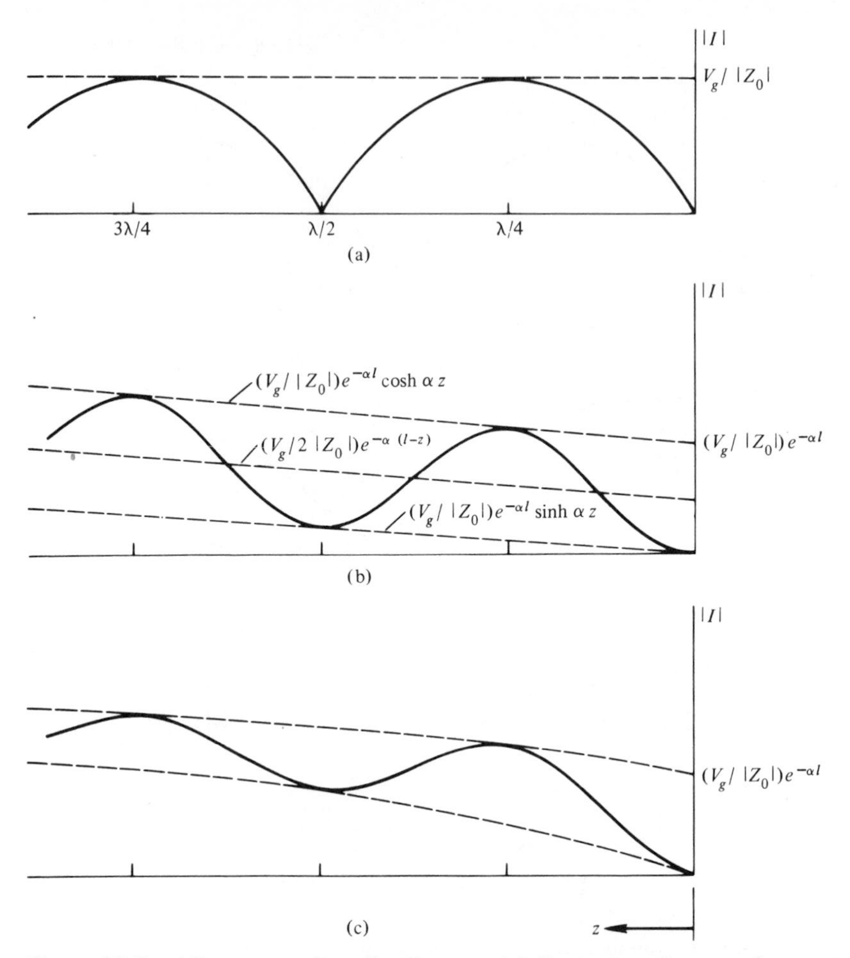

Figure 13.5. $|I|$ versus z, $Z_g = Z_0$, $Z_l \rightarrow \infty$. (a) No attenuation, $\alpha = 0$, (b) small attenuation, (c) larger attenuation.

Figure 13.6. (a) $|V|$ versus z, $Z_g = Z_0$, $Z_l = Z_0$, attenuation present. (b) $|I|$ versus z, $Z_g = Z_0$, $Z_l = Z_0$, attenuation present.

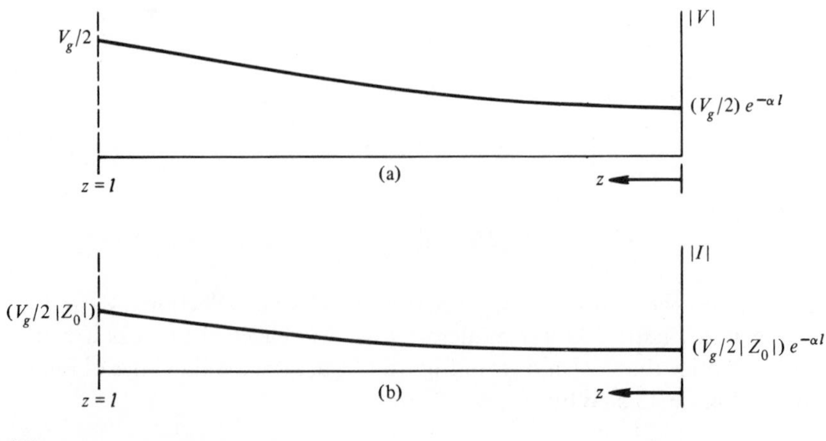

13.4 SPECIAL TWO-WIRE LINES

Some special values of R, L, G, and C, the distributed line constants, give special results and are summarized here.

(1) $R = G = 0$. This is the lossless case to which Chapter 11 was devoted.

(2) A distortionless line can be approximated in spite of the loss. Distortionless conditions are met if the time delay is frequency independent, or equivalently (see equation 11.76) if β is directly proportional to frequency and α is frequency independent. From equation 13.16

$$\gamma = \alpha + j\beta = \sqrt{ZY} = \sqrt{(R + j\omega L)(G + j\omega C)} \tag{13.32}$$

Therefore,

$$\alpha^2 + j2\alpha\beta - \beta^2 = RG - \omega^2 LC + j\omega(LG + RC)$$

or

$$\alpha^2 = \beta^2 + RG - \omega^2 LC$$

and

$$4\alpha^2\beta^2 = \omega^2(LG + RC)^2$$

β and α may be solved from the last two equations, giving the general equations,

$$\beta = \{[\omega^2 LC - RG + \sqrt{(RG - \omega^2 LC)^2 + \omega^2(LG + RC)^2}]/2\}^{1/2} \tag{13.33}$$

and

$$\alpha = \{[RG - \omega^2 LC + \sqrt{(RG - \omega^2 LC)^2 + \omega^2(LG + RC)^2}]/2\}^{1/2} \tag{13.34}$$

If it is possible to make

$$LG = RC \tag{13.35}$$

then equations 13.33 and 13.34 reduce to

$$\beta = \omega\sqrt{LC} \tag{13.36}$$

and

$$\alpha = \sqrt{RG} \tag{13.37}$$

and a distortionless line results. At frequencies where it is practical to "load" a line to satisfy equation 13.35, it is usually true that $R \gg G$. Then, L is usually increased by adding lumped inductance to the line at periodic intervals. Reducing R has the same effect, but is more expensive. Why?

Table 13.1 SPECIAL CASES—TRANSMISSION LINE LOSSES.

CASE	Z_0	$\gamma = \alpha + j\beta$
Lossless	$\sqrt{\frac{L}{C}} \equiv R_0$	$j\omega\sqrt{LC} = j\beta,\ \alpha = 0$
Distortionless $\frac{L}{C} = \frac{R}{G}$	$\sqrt{\frac{L}{C}} = \sqrt{\frac{R}{G}} = R_0$	$\sqrt{RG} + j\omega\sqrt{LC}$
Small loss $R \ll \omega L,\ G \ll \omega C$	$R_0\left[1 + j\left(\frac{G}{2\omega C} - \frac{R}{2\omega L}\right)\right]$	$\frac{R}{2R_0} + \frac{GR_0}{2} + j\omega\sqrt{LC}$
General	$\sqrt{\frac{R + j\omega L}{G + j\omega C}}$	$\sqrt{(R + j\omega L)(G + j\omega C)}$

(3) If the line loss is small, that is, $R \ll \omega L$ and $G \ll \omega C$, then a binomial expansion may be advantageously employed. From equation 13.10

$$Z_0 = \sqrt{\frac{L}{C}}\sqrt{\frac{1 + R/j\omega L}{1 + G/j\omega C}}$$

or

$$Z_0 = R_0\left(1 + \frac{R}{j2\omega L} + \cdots\right)\left(1 - \frac{G}{j2\omega C} - \cdots\right)$$

Retaining only the first order terms,

$$Z_0 \approx R_0\left[1 + j\left(\frac{G}{2\omega C} - \frac{R}{2\omega L}\right)\right] \tag{13.38}$$

In the same way,

$$\gamma \approx \frac{R}{2R_0} + \frac{GR_0}{2} + j\omega\sqrt{LC} \tag{13.39}$$

The results of these cases are summarized in Table 13.1.

According to equation 13.24, the voltage at some point z on the line (measured from the load) is

$$V = \frac{V_g Z_0}{Z_0 + Z_g} e^{-(\alpha + j\beta)(l - z)}$$

or

$$\frac{V(z, \omega)}{V_g(\omega)} = H(\omega) = \frac{Z_0(\omega)}{Z_0(\omega) + Z_g(\omega)} e^{-[\alpha(\omega) + j\beta(\omega)](l - z)} \tag{13.40}$$

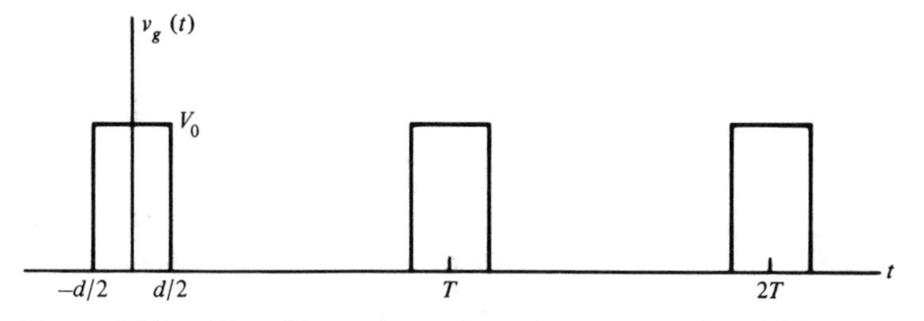

Figure 13.7. Repetitive rectangular pulse representing $v_g(t)$.

if the line is terminated with $Z_l = Z_0$, so that $\Gamma_l = 0$. $H(\omega)$ is the *transfer function* for the line when $v(z, t)$ is the response or "output" and $v_g(t)$ is the excitation or "input." It is simply the ratio of the *phasor* output to the phasor input. It is also the *Fourier transform* of the impulse response, $h(t)$.

We might ponder the question as to which of the cases in Table 13.1 gives distortionless performance.[2] The preceding equation shows that *amplitude* distortion as well as time delay distortion may cause difficulties. Amplitude distortion arises because of the presence of Z_0 in this equation. In Table 13.1, it can be seen that the lossless line has no distortion and no attenuation. The distortionless line has no distortion but does have attenuation (independent of frequency, however). The "small loss" case has no time delay distortion, but does have amplitude distortion as well as attenuation. The general case has both types of distortion as well as attenuation. A further complication which does not appear explicitly in Table 13.1 is that due to R, L, G, and C being *frequency* dependent parameters rather than constants.

As an example, suppose that we have a pulse generator connected to a lossy line which is matched at the load end. Let v_g be the repetitive pulse shown in Figure 13.7 (an even function). Using Appendix D for determining the Fourier coefficients, the Fourier series representation of v_g is

$$v_g = \sum_{n=0}^{\infty} \varepsilon_n a_n \cos n\omega_T t, \qquad \omega_T = \frac{2\pi}{T} \tag{13.41}$$

where

$$a_n = \frac{V_0 \omega_T d}{\pi} \frac{\sin n\omega_T d/2}{n\omega_T d/2} \quad \text{and} \quad \varepsilon_n = \begin{cases} 1/2, & n = 0 \\ 1, & n = 1, 2, 3, \ldots \end{cases}$$

Since each term in equation 13.41 represents a sinusoidal excitation, the phasor concept applies to each term. That is, the input $\varepsilon_n a_n \cos n\omega_T t$ has the phasor form $\varepsilon_n a_n \underline{/0^\circ}$, and for this input the phasor output is $\varepsilon_n a_n \underline{/0^\circ}\, |H(n\omega_T)| \underline{/H(n\omega_T)} = \varepsilon_n a_n |H(n\omega_T)| \underline{/H(n\omega_T)}$ whose time domain form is simply $\varepsilon_n a_n |H(n\omega_T)| \cos(n\omega_T t + \underline{/H(n\omega_T)})$. This means we may effectively let each term represent a v_g which is affected differently through $H(n\omega_T)$ as a traveling wave from the source end to the point z. We then simply sum the

[2] It has already been pointed out in Chapter 10 that the lossy line is anomalously dispersive.

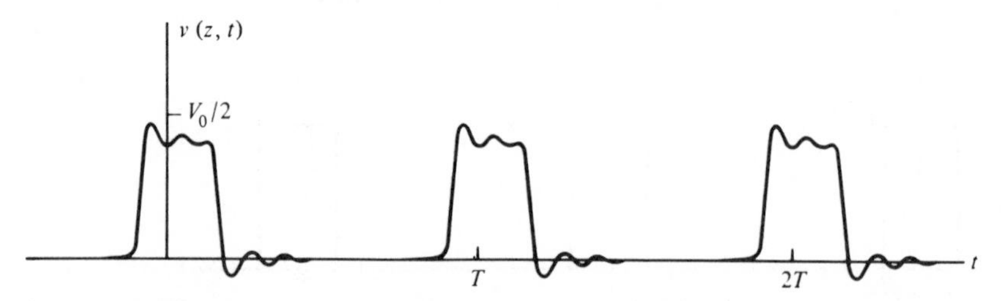

Figure 13.8. $v(z, t)$ versus t when loss is present and $v_g(t)$ is the rectangular pulse of Figure 13.7.

results (superposition) at z for each of these traveling waves. In order to simplify the results, let $Z_g = R_g$, a pure resistance, and assume that $R, L, G,$ and C *are constant*. Using equations 13.40 and 13.41 gives

$$v(z, t) = \sum_{n=0}^{\infty} \varepsilon_n a_n |H(n\omega_T)| \cos (n\omega_T t + \underline{|H(n\omega_T)}) \tag{13.42}$$

where

$$H(n\omega_T) = \frac{Z_0(n\omega_T)}{Z_0(n\omega_T) + Z_G(n\omega_T)} e^{-[\alpha(n\omega_T) + j\beta(n\omega_T)](l-z)} \tag{13.43}$$

$$Z_0(n\omega_T) = \sqrt{\frac{R + jn\omega_T L}{G + jn\omega_T C}} \tag{13.44}$$

and

$$\alpha(n\omega_T) + j\beta(n\omega_T) = \gamma(n\omega_T) = \sqrt{(R + jn\omega_T L)(G + jn\omega_T C)} \tag{13.45}$$

Equation 13.42 represents a *discrete* set of frequency terms; the fundamental and all of its harmonics. The results of a general case where the losses are moderate is shown in Figure 13.8. The degradation of the input pulse because of the attenuation and distortion is obvious.

13.5 QUALITY FACTOR

It is desirable at this time to derive an expression for Q, or quality factor, of a lossy line. Actually, most of the formal labor required for this has already been performed. A general definition for the Q is

$$\boxed{Q \equiv \omega \frac{\langle W \rangle}{\langle P_d \rangle}} \tag{13.46}$$

where $\langle W \rangle$ is the time average total energy stored per unit length of system (joules per meter) and $\langle P_d \rangle$ is the time average power dissipated per unit length of the system (watts per meter). Equation 13.13 may now be used in equation 13.46, giving

$$Q = \frac{1}{2\alpha} \frac{\omega\langle W\rangle}{\langle P_f\rangle} \tag{13.47}$$

where $\langle P_f\rangle$ is the time average power flow in the system (watts). The velocity of energy flow as defined in equation 10.37 is $u_e \doteq \langle P_f\rangle/\langle W\rangle$, so that $\langle P_f\rangle = u_e\langle W\rangle$ and thus

$$\boxed{Q = \frac{1}{2\alpha} \frac{\omega}{u_e}} \tag{13.48}$$

Equation 13.48 is exact[3] and applies to *any propagating system.*

We now assume that the losses are small and the mode of propagation on the line is *essentially TEM. In this case,* $u_e \approx \omega/\beta$, and equation 13.48 becomes

$$\boxed{Q = \frac{\beta}{2\alpha}} \tag{13.49}$$

which is *independent of length.* Since we have already assumed that α is small, the Q may be large; perhaps in the neighborhood of 10^3.

If we use the circuit theory result,

$$Q = \frac{\text{resonant frequency}}{\text{half-power bandwidth}} \tag{13.50}$$

instead of equation 13.46, and apply this to a resonant line length, the same result as that given by equation 13.49 is obtained.

The Q of a transmission line, along with Z_0, may be determined by a measurement similar to the one used to obtain R_0. In fact, this measurement tells us immediately whether the line has loss or not. (The line will always have loss, and whether it is neglected or not, is a matter of judgment.) This measurement consists of determining the input impedance with short-circuit and open-circuit load impedances. If these impedances are complex, then the line is lossy. Using equations 13.29 and 13.30,

$$Z_{\text{in}}^{\text{oc}} = Z_0 \operatorname{ctnh} \gamma z \tag{13.51}$$

and

$$Z_{\text{in}}^{\text{sc}} = Z_0 \tanh \gamma z \tag{13.52}$$

so that

$$\boxed{Z_0 = \sqrt{Z_{\text{in}}^{\text{sc}} Z_{\text{in}}^{\text{oc}}}} \tag{13.53}$$

[3] Notice that both $\langle P_f\rangle$ and $\langle W\rangle$ depend on z, so u_e is somewhat nebulous.

The quantity tanh γz can be found as Z_{in}^{sc}/Z_0 or as the square root of the ratio of equation 13.52 to equation 13.51.

$$\tanh(\alpha z + j\beta z) \equiv A + jB = \sqrt{\frac{Z_{in}^{sc}}{Z_{in}^{oc}}} = \frac{Z_{in}^{sc}}{Z_0} \tag{13.54}$$

It is not too difficult to show that

$$2\alpha z = \tanh^{-1}\left(\frac{2A}{1 + A^2 + B^2}\right) \tag{13.55}$$

and

$$2\beta z = \tan^{-1}\left(\frac{2B}{1 - A^2 - B^2}\right) + m\pi, \qquad m = 0, 1, 2, \ldots \tag{13.56}$$

Equation 13.56 has many solutions, one of which is correct. That is, it may be difficult to tell the difference between a short, but very lossy line, and a long line with little loss. An observation of the *physical* length will usually determine the correct solution. If this cannot be done, a low-frequency measurement of the total shunt capacitance of the line section will determine the correct solution. Using equations 13.49, 13.55, and 13.56,

$$Q = \frac{1}{2} \frac{\tan^{-1}[2B/(1 - A^2 - B^2)] + m\pi}{\tanh^{-1}[2A/(1 + A^2 + B^2)]} \tag{13.57}$$

13.6 LOSSY LINE ANALYSIS WITH THE SMITH CHART

Most quantities may be determined more rapidly by using the Smith chart than by straightforward calculation. The question now arises as to what modifications must be made in the use of the chart to account for the losses. The chart was originally derived using equation 12.3, or

$$z_{in} = \frac{1 + |\Gamma_l| e^{j(\phi_l - 2kz)}}{1 - |\Gamma_l| e^{j(\phi_l - 2kz)}} = r_a + jx_a \tag{13.58}$$

for the lossless line. For the *general* line equation 13.26 gives

$$z_{in} = \frac{Z_{in}}{Z_0} = \frac{1 + \Gamma_l e^{-2(\alpha + j\beta)z}}{1 - \Gamma_l e^{-2(\alpha + j\beta)z}} \tag{13.59}$$

Equation 13.59 may be written

$$z_{in} = \frac{1 + (|\Gamma_l| e^{-2\alpha z}) e^{j(\phi_l - 2\beta z)}}{1 - (|\Gamma_l| e^{-2\alpha z}) e^{j(\phi_l - 2\beta z)}} \tag{13.60}$$

A comparison of equations 11.58 and 11.60 reveals that k has been replaced by β and $|\Gamma_l|$ has been replaced by $|\Gamma_l| e^{-2\alpha z}$. The effects of these changes, insofar as the Smith chart is concerned, are rather obvious. Motion on the chart, *toward the generator*, must now be accompanied by an *inward* spiral,

($|\Gamma_l|e^{-2\alpha z}$), rather than constant radius motion, ($|\Gamma_l|$), as in the lossless case. The effect of the damping term, $e^{-2\alpha z}$, can be radially scaled, but usually transmission loss, $e^{-\alpha z}$, is what is actually scaled.

When making impedance measurements on a line using a bridge, the quantities measured are usually

$$Z_{\text{in}}^{\text{sc}} = Z_0 \tanh \gamma z$$

$$Z_{\text{in}}^{\text{oc}} = Z_0 \operatorname{ctnh} \gamma z$$

and

$$Z_{\text{in}} = Z_0 \frac{e^{\gamma z} + \Gamma_l e^{-\gamma z}}{e^{\gamma z} - \Gamma_l e^{-\gamma z}} \quad \text{(actual load in place)}$$

where

$$\Gamma_l = \frac{Z_L - Z_0}{Z_L + Z_0}$$

Substituting for Γ_l gives

$$Z_{\text{in}} = Z_0 \frac{Z_l + Z_0 \tanh \gamma z}{Z_0 + Z_l \tanh \gamma z} \tag{13.61}$$

Solving for Z_l in equation 13.61 in terms of the measured quantities gives

$$\boxed{Z_l = \frac{Z_{\text{in}}^{\text{oc}}(Z_{\text{in}} - Z_{\text{in}}^{\text{sc}})}{Z_{\text{in}}^{\text{oc}} - Z_{\text{in}}}} \tag{13.62}$$

EXAMPLE 1

The following measurements are made on a transmission line whose length is known to be less than $\lambda/4$.

$$Z_{\text{in}}^{\text{oc}} = 28.8\angle{-75^\circ} \quad (\Omega)$$

$$Z_{\text{in}}^{\text{sc}} = 80\angle{+85^\circ} \quad (\Omega)$$

and

$$Z_{\text{in}} = 96\angle{50^\circ} \quad (\Omega) \quad \text{(actual load in place)}$$

We want to find Z_l and Q. Substituting the measured values into equation 13.62 gives

$$Z_l = \frac{28.8\angle{-75^\circ}(96\angle{50^\circ} - 80\angle{85^\circ})}{28.8\angle{-75^\circ} - 96\angle{85^\circ}}$$

or

$$Z_l = 13.8\angle{36.74^\circ} \quad (\Omega)$$

Using equation 13.54

$$A + jB = \sqrt{\frac{80\angle 85^\circ}{28.8\angle -75^\circ}} = 1.667\angle 80^\circ$$

so

$$A = 0.29, \qquad B = 1.641$$

Using equation 13.55,

$$2\alpha z = \tanh^{-1}\left(\frac{0.58}{3.78}\right) = 0.154$$

and using equation 13.56,

$$2\beta z = \tan^{-1}\left(\frac{3.282}{-1.78}\right) + m\pi = \tan^{-1}(-1.846) + m\pi$$
$$= 2.067, 5.21, 8.35, \ldots$$

We know, however, that $z < \lambda/4$ or $2\beta z < (4\pi/\lambda)(\lambda/4) = \pi$, so we must choose

$$2\beta z = 2.067$$

Then, equation 13.49 gives

$$Q = \frac{\beta}{2\alpha} = \frac{\beta z}{2\alpha z} = \frac{1.034}{0.154} = 6.72$$

Notice that if the line had been known to be between $\lambda/2$ and $3\lambda/4$ in length, then

$$\frac{4\pi}{\lambda}\left(\frac{\lambda}{2}\right) = 2\pi < 2\beta z < \frac{4\pi}{\lambda}\left(\frac{3\lambda}{4}\right) = 3\pi$$

so

$$2\beta z = 8.35, \qquad \beta z = 4.175$$

and

$$Q = \frac{\beta z}{2\alpha z} = \frac{4.175}{0.154} = 27.1$$

The same quantities can be found using the Smith chart. From equation 13.53

$$Z_0 = 48\angle +5^\circ$$

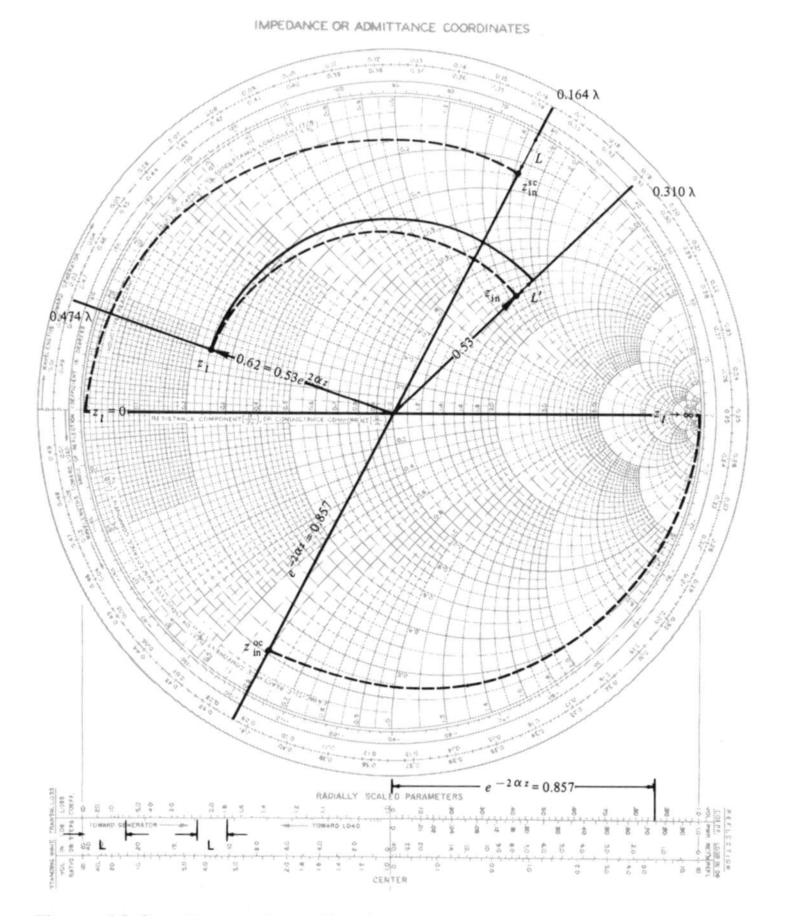

Figure 13.9. Smith chart data for Example 1. *Source*: Copyright © 1949 by Kay Electric Company, Pine Brook, N.J. Renewal copyright © 1976 by P. H. Smith, New Providence, N.J.

The normalized measurements are

$$z_{\text{in}}^{\text{oc}} = \frac{Z_{\text{in}}^{\text{oc}}}{Z_0} = 0.6\underline{|-80^\circ} = 0.1042 - j0.591$$

$$z_{\text{in}}^{\text{sc}} = \frac{Z_{\text{in}}^{\text{sc}}}{Z_0} = 1.667\underline{|+80^\circ} = 0.29 + j1.641$$

and

$$z_{\text{in}} = \frac{Z_{\text{in}}}{Z_0} = 2\underline{|45^\circ} = 1.414 + j1.414$$

These points are located on the Smith chart in Figure 13.9. Both $z_{\text{in}}^{\text{oc}}$ and $z_{\text{in}}^{\text{sc}}$ show that $z = 0.164\lambda$ ($<\lambda/4$). Motion from the short or open circuit has followed the exponential spiral on the chart. Using the reflection coefficient scale shows that

$$|\Gamma_l|e^{-2\alpha z} = 0.857$$

We know $|\Gamma_l| = 1$ for either the open or short circuit case, so

$$e^{-2\alpha z} = 0.857$$

or

$$\alpha z = 0.0772 \quad \text{(Np)}$$

Also,

$$\beta z = \frac{2\pi z}{\lambda} = \frac{2\pi}{\lambda}(0.164\lambda) = 1.03 \text{ rad}$$

Thus,

$$Q = \frac{\beta}{2\alpha} = \frac{\beta z}{2\alpha z} = 6.68$$

z_l is determined by rotating toward the load an amount 0.164λ from z_{in} along an *outward* spiraling exponential curve. The coefficient of reflection magnitude corresponding to z_{in} is 0.53, so we must spiral out to a radius $0.53e^{+2\alpha z} = 0.53/0.857 = 0.62$. Here we find

$$z_l = 0.24 + j0.155 = 0.286\underline{|32.86^\circ}$$
$$Z_l = Z_0 z_l = 13.7\underline{|37.9^\circ} \quad (\Omega)$$

These answers compare favorably to those obtained by direct calculation. The *transmission loss radial scale* may be used instead of the $|\Gamma_l|$ scale. This scale is calibrated in terms of $e^{-\alpha z}$ (not $e^{-2\alpha z}$) in 1-dB steps. The distance L in Figure 13.9 gives a transmission loss of about 0.67 dB, so

$$e^{-\alpha z} = 0.924, \quad (-0.67 \text{ dB})$$

or

$$\alpha z = 0.077 \quad \text{(Np)}$$

This checks the previous result. Note that the distances L and L' are both 0.67 dB, but are *not* the same physical length! Also, α and β have not been determined, but αz, and total phase shift, βz, have been determined.

In the previous example, the value of Q which was determined was rather low. In fact, it is so low that its accuracy may be questioned, since the formula, $Q = \beta/2\alpha$, was based on the assumption that the losses are small, or the Q is high. In this example, the short-circuit and open-circuit impedances were purposely chosen to make the Q rather low so that the Smith chart could be easily used. Normally, the Q will be of the order of 10^3.

It should be emphasized that the Q we have calculated in the previous example is that for a *propagating system.* Quite often, as has already been mentioned, sections of lines are used as equivalent lumped elements in certain applications.

EXAMPLE 2

Suppose that the same line that we just considered is to be used (with a short circuit at its load end) to represent an inductive reactance. What is the "inductor" Q in this application? We have

$$\begin{aligned} Z_{\text{in}}^{\text{sc}} &= 80\underline{|85^\circ} = R_{\text{in}}^{\text{sc}} + jX_{\text{in}}^{\text{sc}} \\ &= 6.97 + j79.7 \quad (\Omega) \end{aligned}$$

The "inductor" Q is defined here as being

$$Q_L = \frac{\omega L_{\text{in}}^{\text{sc}}}{R_{\text{in}}^{\text{sc}}} = \frac{X_{\text{in}}^{\text{sc}}}{R_{\text{in}}^{\text{sc}}} = \tan 85^\circ = 11.43$$

The Q we have just calculated is a *nonresonant* value for an equivalent lumped inductor ($z < \lambda/4$) which the line represents at its input terminals. $Z_{\text{in}}^{\text{sc}}$ can be found from the Smith chart by taking values of $Z_{\text{in}}^{\text{sc}}$ from the short-circuit spiral and multiplying by Z_0. From $Z_{\text{in}}^{\text{sc}}$ the Q_L for any length may be found. Notice particularly that this Q will depend on the line length!

We have also assumed that the short circuit truly represents zero impedance in the preceding examples. The loss in the actual "short circuit" can be calculated with previously used techniques, but it is obvious that this loss will be small compared to the other losses unless the line length is very small. An example that considers all losses will be presented in Section 13.8. A much more practical example is now considered.

EXAMPLE 3

Suppose we want to determine the input characteristics for a resonant shorted line with "small loss." We again assume that "shorted" means $Z_l = 0$, but we know from our study of the skin effect that Z_l is not zero unless the short is accomplished with a perfect conductor. We have

$$Z_{\text{in}}^{\text{sc}} = Z_0 \tanh \gamma z \tag{13.30}$$

where Z_0 and γ are given by equations 13.38 and 13.39, respectively. Given R, L, G, C, and ω it is a straightforward calculation to find the input impedance. It is also a straightforward, although lengthy, matter to find the shortest line length that gives resonance (line length where Z_{in} is real). Instead of making general calculations explicitly, let us consider two cases separately. First, assume that the dielectric is air or some other (essentially) lossless material. Then $G = 0$,

$$Z_0 \approx R_0 - jR_0 \frac{R}{2\omega L}$$

$$\gamma_c \approx \frac{R}{2R_0} + j\omega\sqrt{LC} = \alpha_c + j\beta$$

Solving equation 13.30 for the resonant length gives

$$z_0 \approx \frac{\lambda/4}{1 + (\alpha_c/\beta)^2} \tag{13.63}$$

which is slightly smaller than $\lambda/4$. Small argument formulas for the trigonometric and hyperbolic functions have been used.[4] At this resonant frequency

$$R_{\text{in}}^{\text{sc}} \approx R_0 \frac{2}{\pi} \frac{\beta}{\alpha_c} \tag{13.64}$$

which is very large ($\beta \gg \alpha$) and indicative of the high Q. The Q is

$$Q_c \approx \frac{\beta}{2\alpha_c} = \frac{\omega\sqrt{LC}}{2R/2R_0} = \frac{\omega L}{R} \tag{13.65}$$

Subscripts c refer to *conductor* losses.

If $R = 0$ (or losses due to R are small compared to those due to G), then

$$Z_0 \approx R_0 + jR_0 \frac{G}{2\omega C}$$

$$\gamma \approx \frac{GR_0}{2} + j\omega\sqrt{LC} = \alpha_d + j\beta$$

Solving equation 13.30 for the resonant length gives

$$z_0 \approx \frac{\lambda/4}{1 - (\alpha_d/\beta)^2} \tag{13.66}$$

which is slightly larger[5] than $\lambda/4$. At this resonant length

$$R_{\text{in}}^{\text{sc}} \approx R_0 \frac{2}{\pi} \frac{\beta}{\alpha_d} \tag{13.67}$$

which is large and the same form as for the other case. The Q is

$$Q_d = \frac{\beta}{2\alpha_d} = \frac{\omega\sqrt{LC}}{2GR_0/2} = \frac{\omega C}{G} = \frac{\omega\varepsilon}{\sigma} \tag{13.68}$$

[4] See Problem 12.

[5] If the line is "distortionless," the resonant length is exactly $\lambda/4$.

The last result shows that the dielectric Q is just the Q of the dielectric! When both losses are present,

$$Q_{cd} = \frac{\beta}{2\alpha} = \frac{\omega\sqrt{LC}}{R/R_0 + GR_0}$$

or, simply

$$\boxed{\frac{1}{Q_{cd}} = \frac{1}{Q_c} + \frac{1}{Q_d}} \qquad (13.69)$$

The last examples have involved problems that were independent of the type of transmission line being employed (coax, two-wire, etc.). We would like to treat one problem completely, including the effect of the imperfect short circuit. This will require the explicit use of some field theory (from which we seem to have strayed) and the specification of a particular type line. This problem is considered in Section 13.8. We first must consider particular types of transmission lines.

13.7 LINE PARAMETERS

We have seen that G and C are relatively simple to calculate, whereas R and L are not so simple to find because of partial flux linkage. In the following we will make two calculations for R and L: one for low frequencies where the current density is essentially uniform throughout the conductor, and another for high frequencies where the current density is assumed to be uniform throughout a layer of depth δ, the skin depth. The resistance is proportional to $1/\delta$, and therefore proportional to $f^{1/2}$. The Q_c due to conductor losses, is $\omega L/R$, so Q_c is proportional to $f^{1/2}$. This result indicates that Q_c increases with frequency. The formula for δ was derived on the basis that $\sigma \gg \omega\varepsilon$, and obviously if ω increases indefinitely, the formula is not applicable. On the other hand, the frequency would be something like 10^{17} Hz for most conductors before the formula is not valid. At these frequencies two-conductor systems do not behave as we have described them in the preceding chapters, and are not useful. The lossless case is also presented for completeness.

The geometries to be considered are the parallel plane, coax and two-wire line shown in Figure 13.10. There are three points to keep in mind when making the high frequency calculation.

(1) The external inductance per unit length can be easily obtained from C. This was demonstrated in equation 8.57.

$$\boxed{L_{\text{ext}} = \frac{\mu\varepsilon}{C}} \quad \text{(H/m)} \qquad (13.70)$$

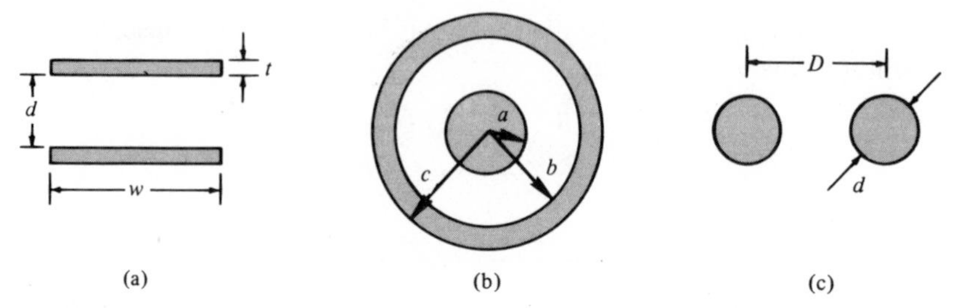

Figure 13.10. Common two-wire transmission lines. (a) Planar, (b) coaxial (c) two-wire (round).

(2) The conductance per unit length can also be easily obtained from C. See equation 13.7.

$$\boxed{G = \frac{\sigma_d}{\varepsilon} C} \quad (℧/\text{m}) \tag{13.71}$$

(3) Because of the skin effect and surface impedance concept, $R_s \approx X_s \approx \omega L_s \approx \omega L_{\text{int}}$, as demonstrated in Section 10.12. Thus,

$$\boxed{L_{\text{int}} = \frac{R_s}{\omega} = \frac{R}{\omega}} \quad (\text{H/m}) \tag{13.72}$$

where

$$\boxed{R_s = R = \frac{1}{\sigma_c\, \delta w}} \quad (\Omega/\text{m}) \tag{13.73}$$

and w is the "width" through which the skin current flows. The low-frequency inductance for the parallel plane system is calculated most easily by means of equation 8.62. The low-frequency inductance for the coax was found as equation 8.35, while the low-frequency inductance for the two-wire line is equation 8.39. Results for the geometries in Figure 13.10 are listed in Table 13.2.

13.8 THE COAXIAL CAVITY[6]

A very well shielded resonant cavity results from a shorted (nominal) quarter-wave section of rigid coaxial line. This device may be used for many things, such as measuring σ_d or ε. It is shown in Figure 13.11. Our first task will be to find Q_d. The Q was defined in equation 13.46,

$$Q = \omega \frac{\langle W \rangle}{\langle P_d \rangle}$$

[6] See Hayt in the References at the end of the chapter.

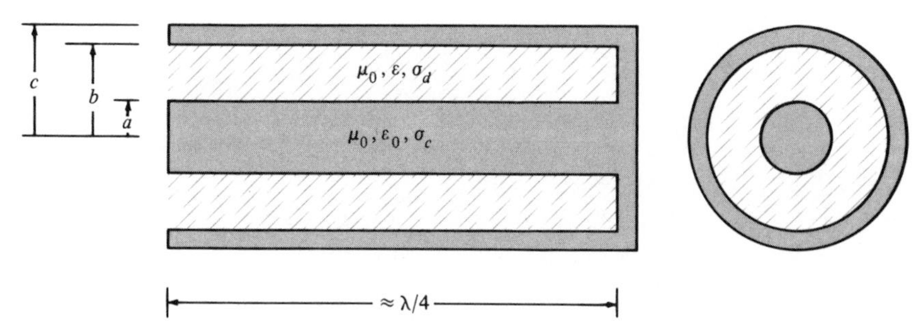

Figure 13.11. The coaxial resonant cavity.

and this definition will apply here, although it makes more sense to call $\langle W \rangle$ the average total energy stored in the cavity and to call $\langle P_d \rangle$ the average total power dissipated in the cavity. We must find the fields before $\langle W \rangle$ and $\langle P_d \rangle$ can be found. We will use the lossless field equations here (perturbation, mentioned earlier), and these can easily be found. We demonstrated conclusively in Section 11.2 that both **E** and **H** are distributed (for the lossless case) in transverse planes precisely the same way as they are statically. We know, furthermore, that they are quadrature standing waves longitudinally. In other words.

$$E_\rho(\rho, z, \omega) = \frac{E_0}{\rho} \sin kz \tag{13.74}$$

$$H_\phi(\rho, z, \omega) = j\frac{E_0}{\eta\rho} \cos kz \tag{13.75}$$

as can easily be verified. Then the dielectric Q[7] is

$$Q_d = \omega \frac{\dfrac{\varepsilon}{2}\displaystyle\int_a^b\int_0^{2\pi}\int_0^{\lambda/4}\frac{1}{2}|E_\rho|^2\rho\,d\rho\,d\phi\,dz + \frac{\mu}{2}\int_a^b\int_0^{2\pi}\int_0^{\lambda/4}\frac{1}{2}|H_\phi|^2\rho\,d\rho\,d\phi\,dz}{\dfrac{1}{2}\sigma_d\displaystyle\int_a^b\int_0^{2\pi}\int_0^{\lambda/4}|E_\rho|^2\rho\,d\rho\,d\phi\,dz} \tag{13.76}$$

where the lossless resonant $\lambda/4$ length has been used. We already know from Section 13.6, however, that the dielectric Q is just the Q of the dielectric.

[7] Do not forget to use "$\frac{1}{2}$ real part of." For example,

$$\langle W_E \rangle = \tfrac{1}{2}\text{Re}\,\{\tfrac{1}{2}\textstyle\iiint \mathbf{D}\cdot\mathbf{E}^*\,dv\}$$

Table 13.2 COMMON TRANSMISSION LINE PARAMETERS

CASE	PARALLEL PLANE SYSTEM	COAXIAL CABLE SYSTEM	TWO-WIRE LINE SYSTEM
Lossless	$C = \frac{\varepsilon w}{d}$	$C = \frac{2\pi\varepsilon}{\ln (b/a)}$	$C = \frac{\pi\varepsilon}{\cosh^{-1} D/d} \approx \frac{\pi\varepsilon}{\ln (2D/d)}, D \gg d$
	$L = \frac{\mu d}{w}$	$L = \frac{\mu}{2\pi} \ln \frac{b}{a}$	$L = \frac{\mu}{\pi} \cosh^{-1} \frac{D}{d} \approx \frac{\mu}{\pi} \ln \frac{2D}{d}, D \gg d$
High-frequency	$C = \frac{\varepsilon w}{d}$	$C = \frac{2\pi\varepsilon}{\ln (b/a)}$	$C = \frac{\pi\varepsilon}{\cosh^{-1} D/d}$
	$G = \frac{\sigma_d w}{d}$	$G = \frac{2\pi\sigma_d}{\ln (b/a)}$	$G = \frac{\pi\sigma_d}{\cosh^{-1} D/d}$
	$R = \frac{2}{\sigma_c \delta w}$	$R = \frac{1}{\sigma_c 2\pi\delta} \left(\frac{1}{a} + \frac{1}{b}\right)$	$R = \frac{2}{\pi\sigma_c \delta(d - \delta)} \approx \frac{2}{\pi\sigma_c \delta d}$
	$L = \frac{\mu d}{w} + \frac{2}{\omega\sigma_c \delta w} = \frac{\mu}{w}(d + \delta)$	$L = \frac{\mu}{2\pi} \left[\ln \frac{b}{a} + \frac{\delta}{2}\left(\frac{1}{a} + \frac{1}{b}\right)\right]$	$L = \frac{\mu}{\pi}\left(\cosh^{-1} \frac{D}{d} + \frac{\delta}{d}\right)$ $\approx \frac{\mu}{\pi}\left(\ln \frac{2D}{d} + \frac{\delta}{d}\right), D \gg d$
	$C = \frac{\varepsilon w}{d}$	$C = \frac{2\pi\varepsilon}{\ln (b/a)}$	$C = \frac{\pi\varepsilon}{\cosh^{-1} D/d}$
	$G = \frac{\sigma_d w}{d}$	$G = \frac{2\pi\sigma_d}{\ln (b/a)}$	$G = \frac{\pi\sigma_d}{\cosh^{-1} D/d}$

Low-frequency	$R = \dfrac{2}{\sigma_c t w}$ $L = \dfrac{\mu}{w}\left(d - \dfrac{t}{3}\right)$	$R = \dfrac{1}{\pi\sigma_c}\left(\dfrac{1}{a^2} + \dfrac{1}{b^2}\right)$ $L = \dfrac{\mu}{2\pi}\left[\ln\dfrac{b}{a} + \dfrac{1}{4} + \dfrac{1}{4(c^2 - b^2)} \times \left(b^2 - 3c^2 + \dfrac{4c^4}{c^2 - b^2}\ln\dfrac{c}{b}\right)\right]$	$R = \dfrac{8}{\pi d^2 \sigma_c}$ $L = \dfrac{\mu}{\pi}\left(\ln\dfrac{2D}{d} + \dfrac{1}{4}\right)$

Eliminating H_ϕ with equation 13.75 gives

$$Q_d = \omega \frac{\frac{\varepsilon}{2}\iiint \cdots + \frac{\mu}{2\eta^2}\iiint \cdots}{\sigma_d \iiint \cdots} = \omega \frac{\varepsilon \iiint \cdots}{\sigma_d \iiint \cdots}$$

or

$$\boxed{Q_d = \frac{\omega\varepsilon}{\sigma_d}} \tag{13.77}$$

We next find the Q due to conductor losses, Q_c. Using our skin effect concepts, we have a "uniform" current density

$$J_{za} = \left.\frac{H_\phi}{\delta}\right|_{\rho=a} = j\frac{E_0}{\eta a\delta}\cos kz$$

flowing in the inner conductor, while in the outer conductor, the current density is

$$J_{zb} = -\left.\frac{H_\phi}{\delta}\right|_{\rho=b} = -j\frac{E_0}{\eta b\delta}\cos kz$$

The shorting plug at $z = 0$ has a current density

$$J_{\phi 0} = \frac{H_\phi|_{z=0}}{\delta} = j\frac{E_0}{\eta\rho\delta}$$

Then, the average power loss for these three conductors will be, respectively, (see equation 10.86)

$$\langle P_d\rangle_{\rho=a} = \frac{1}{2}\left\{\frac{1}{\sigma_c}\int_{a-\delta}^{a}\int_{0}^{2\pi}\int_{0}^{\lambda/4}|J_{za}|^2\rho\, d\rho\, d\phi\, dz\right\}$$

$$\langle P_d\rangle_{\rho=b} = \frac{1}{2}\left\{\frac{1}{\sigma_c}\int_{b}^{b+\delta}\int_{0}^{2\pi}\int_{0}^{\lambda/4}|J_{zb}|^2\rho\, d\rho\, d\phi\, dz\right\}$$

$$\langle P_d\rangle_{z=0} = \frac{1}{2}\left\{\frac{1}{\sigma_c}\int_{a}^{b}\int_{0}^{2\pi}\int_{-\delta}^{0}|J_{\phi 0}|^2\rho\, d\rho\, d\phi\, dz\right\}$$

Using the total average power dissipated and the same total average energy stored as in finding Q_d we have

$$Q_c = \frac{2/\delta \ln b/a}{(1/a)(1 - \delta/2a) + (1/b)(1 + \delta/2b) + (8/\lambda) \ln (b/a)} \tag{13.78}$$

We recall from equation 13.69

$$\frac{1}{Q_{cd}} = \frac{1}{Q_c} + \frac{1}{Q_d}$$

This resonant line must have an equivalent in terms of parallel lumped parameters R_e, L_e, and C_e which we can find. The voltage at the input to the line can be found in terms of equation 13.74 as

$$V_{\text{in}} = E_0 \ln \frac{b}{a} \tag{13.79}$$

Now, the *maximum* energy stored in the capacitor (when the energy stored in the inductor is zero) is

$$\frac{1}{2} C_e V_{\text{in}}^2 = \frac{1}{2} C_e E_0^2 \left(\ln \frac{b}{a} \right)^2 = \frac{\pi^2 \varepsilon E_0^2}{4k} \ln \frac{b}{a}$$

The right-hand term in the preceding equation was obtained from the numerator of equation 13.76. Solving for C_e,

$$C_e = \frac{\pi \varepsilon \lambda}{4 \ln (b/a)} \tag{13.80}$$

Also,

$$Q_{cd} = \omega_0 C_e R_e$$

or

$$R_e = \frac{Q_{cd}}{\omega_0 C_e} \tag{13.81}$$

and finally,

$$\omega_0^2 = \frac{1}{L_e C_e}$$

or

$$L_e = \frac{1}{\omega_0^2 C_e} \tag{13.82}$$

A numerical example is in order.

EXAMPLE 4

Let us determine the important quantities for a resonant quarter-wave coaxial resonator if $f_0 = 300$ MHz, $\mu = \mu_0$, $a = 1$ cm, $b = 2.3$ cm. In the first part we assume that the dielectric is air so that $\sigma_d \approx 0$, $\varepsilon = \varepsilon_0$. The walls are silver ($\sigma_c = 6.17 \times 10^7$) plated. Then,

$$\lambda = \frac{u}{f} = \frac{300 \times 10^6}{300 \times 10^6} = 1 \quad \text{(m)}, \qquad l = \frac{\lambda}{4} = \frac{1}{4} = 25 \quad \text{(cm)}$$

$$\delta = (\pi f \mu \sigma_c)^{-1/2} = 3.7 \times 10^{-6} \quad \text{(m)}$$

$$Q_c = 3000$$

$$C_e = 8.34 \quad \text{(pF)}$$

$$R_e = 191{,}000 \quad (\Omega)$$

$$L_e = 0.0337 \quad (\mu\text{H})$$

Circuit theory tells us that the bandwidth, f_0/Q_c, is 0.1 MHz.

Now suppose the cavity interior is filled with a material we are testing which has $\varepsilon_R = 9$, $\sigma_d = 150 \times 10^{-6}$. Then,

$$\lambda = \frac{u}{f} = \frac{100 \times 10^6}{300 \times 10^6} = 0.333 \quad \text{(m)}, \qquad l = \frac{\lambda}{4} = 8.33 \quad \text{(cm)}$$

$$\delta = 3.7 \times 10^{-6} \quad \text{(m)}$$

$$Q_c = 2753$$

$$Q_d = \frac{\omega\varepsilon}{\sigma_d} = 1000$$

$$Q_{cd} = 734$$

$$C_e = 25 \quad \text{(pF)}$$

$$R_e = 15{,}600 \quad (\Omega)$$

$$L_e = 0.0113 \quad (\mu\text{H})$$

bandwidth = 0.409 MHz

Now, with l fixed the resonant frequency is proportional to $(\varepsilon_R)^{-1/2}$, so by means of a simple binomial expansion.

$$\frac{\Delta f_0}{f_0} = -\frac{1}{2}\frac{\Delta\varepsilon_R}{\varepsilon_R}$$

That is, a 0.1% decrease in dielectric constant gives a 0.2% increase in resonant frequency, or 0.6 MHz. Such a change is easily detected. What uses could be made of this behavior?

Before leaving this example (second part), it is worthwhile to consider the distributed parameters and make the same calculations. Using Table 13.2,

$$R = 0.100 \quad (\Omega/\text{m})$$

$$L = 0.1666 \quad (\mu\text{H/m})$$

$$C = 600 \quad (\text{pF/m})$$

$$G = 1.13 \times 10^{-3} \quad (\mho/\text{m})$$

Using equations 13.65 and 13.68,

$$Q_c = \frac{\omega L}{R} = 3140, \qquad Q_d = \frac{\omega C}{G} = \frac{\omega \varepsilon}{\sigma_d} = 1000$$

Therefore,

$$Q_{cd} = 758$$

Using equation 13.64 with the total attenuation (see Problem 12)

$$R_{\text{in}}^{\text{sc}} = R_e = R_0 \frac{2}{\pi} \frac{\beta}{\alpha} = R_0 \frac{4}{\pi} \frac{\beta}{2\alpha}$$

or

$$R_e = R_0 \frac{4}{\pi} Q_{cd} = \omega L_e Q_{cd} = \frac{Q_{cd}}{\omega C_e}$$

Thus,

$$R_e = 16{,}100 \quad (\Omega)$$

$$L_e = 0.0113 \quad (\mu\text{H})$$

$$C_e = 25 \quad (\text{pF})$$

The latter results differ slightly from the former because the approximations are not the same, and the imperfect short circuit has been neglected. The latter results are, however, much easier to obtain.

From a practical point of view it may be more likely that we are interested in *measuring* σ_d and ε_R within a line of resonant length $l = (2p - 1)\lambda'/4$. This measurement may be difficult to obtain with a slotted line because of the losses in the line itself, the losses in nonideal short and open circuits, frequency modulation in the source, and so forth all tend to mask the large VSWR expected with a low-loss dielectric. This difficulty is not insurmountable.

If $|\Gamma_l| \approx 1$, as for short- or open-circuit loads, or for essentially purely reactive loads, then $|V|_{\max}$ and $|V|_{\min}$ will occur for all practical purposes at positions where the lossless theory predicts they will occur. That is, with $Z_l \approx 0, |\Gamma_l| \approx 1, \phi_l \approx \pm\pi$ (and since $\phi_l - 2\beta z \approx -2n\pi$ for $|V|_{max}$), we have

$$z_{max} \approx \frac{(2n-1)\pi}{2\beta}, \qquad n = 1, 2, 3, \ldots \tag{13.83}$$

For $n = 1$, z_{max} is the location of the *first* maximum of $|V|$. See Figure 13.5. In the same way

$$z_{min} \approx \frac{m\pi}{\beta}, \qquad m = 1, 2, 3, \ldots \tag{13.84}$$

Now consider Figure 13.12. The input impedance looking into the resonant section is

$$Z_{in} = Z_0 \frac{1 + \Gamma_l^s}{1 - \Gamma_l^s} = Z_0' \frac{e^{\gamma' l} + \Gamma_l' e^{-\gamma' l}}{e^{\gamma' l} - \Gamma_l' e^{-\gamma' l}} \tag{13.85}$$

which is exact. Superscripts (s) refer to short circuit. But since $l = (2p-1)\lambda'/4$,

$$\gamma' l = \alpha' l + j\frac{\pi}{2}(2p-1)$$

and

$$e^{\gamma' l} = e^{\alpha' l}[j(-1)^{p+1}]$$

Therefore,

$$Z_0 \frac{1 + \Gamma_l^s}{1 - \Gamma_l^s} = Z_0' \frac{e^{\alpha' l} - \Gamma_l' e^{-\alpha' l}}{e^{\alpha' l} + \Gamma_l' e^{-\alpha' l}} \tag{13.86}$$

Next, assume that Z_0 and Z_0' are essentially real and μ is the same everywhere so that $Z_0' \approx Z_0 \lambda'/\lambda$. Assume $Z_l' \approx 0$ so that $\Gamma_l' \approx -1$. Assume that we have nearly an open circuit at the input to the resonant section so that $\phi_l^s \approx 0$. With these assumptions equation 13.86 becomes

$$\frac{1 + |\Gamma_l^s|}{1 - |\Gamma_l^s|} \approx \frac{\lambda'}{\lambda} \operatorname{ctnh} \alpha' l \tag{13.87}$$

or, since $\alpha' l \ll 1$, $\operatorname{ctnh} \alpha' l \approx 1/\alpha' l$, and

$$\frac{1 + |\Gamma_l^s|}{1 - |\Gamma_l^s|} \approx \frac{4}{\alpha'(2p-1)\lambda} \tag{13.88}$$

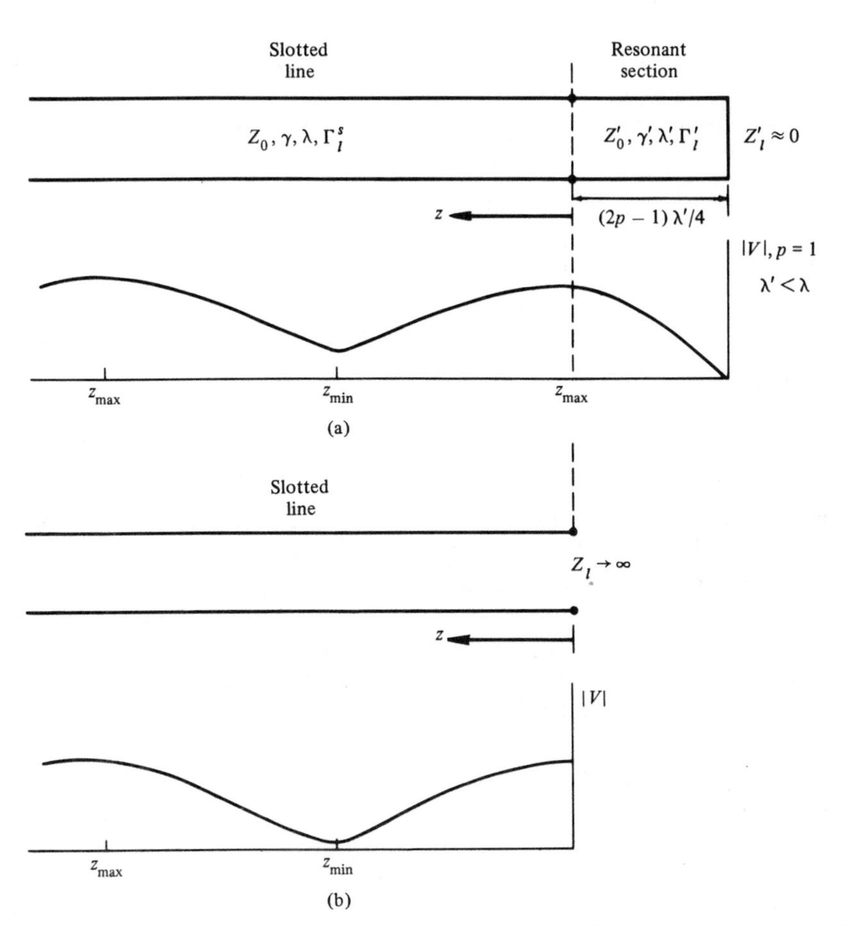

Figure 13.12. Measurement of α'. (a) Resonant section connected to slotted line (VSWR = s). (b) Open circuited slotted line (VSWR = s^o), $s^o > s$.

Now, on the slotted line the VSWR ($\equiv s$) is

$$s = \frac{e^{\alpha z_{\max}} + |\Gamma_l^s| e^{-\alpha z_{\max}}}{e^{\alpha z_{\min}} - |\Gamma_l^s| e^{-\alpha z_{\min}}} \tag{13.89}$$

which is not unique until m and n are specified. That is, we must specify *which* maximum or minimum along the line we intend to use. Solving equation 13.89 for $|\Gamma_l^s|$ and substituting into equation 13.88, we have

$$\frac{s \cosh(\alpha z_{\min}) - \sinh(\alpha z_{\max})}{\cosh(\alpha z_{\max}) - s \sinh(\alpha z_{\min})} \approx \frac{4}{\alpha'(2p-1)\lambda} \tag{13.90}$$

Since $\alpha z \ll 1$, this becomes

$$\frac{s}{1 - s\alpha z_{\min}} \approx \frac{4}{\alpha'(2p-1)\lambda} \tag{13.91}$$

If an open circuit is placed on the end of the slotted line, it nearly duplicates conditions in Figure 13.12. That is, $z_{min} \approx (2n - 1)\pi/2\beta$ (see equation 13.83), and, furthermore, since $\Gamma_l \approx 1$ in this situation, the VSWR ($\equiv s^o$) is

$$s^o = \frac{\cosh(\alpha z_{max})}{\sinh(\alpha z_{min})} \approx \frac{1}{\alpha z_{min}}$$

or

$$\alpha z_{min} \approx \frac{1}{s^o}$$

Thus, equation 13.91 becomes

$$\alpha' \approx \frac{4}{(2p - 1)\lambda} \frac{1 - s/s^o}{s} \tag{13.92}$$

Notice that s and s^o must be measured[8] at the *same* z_{max} and z_{min} points on the slotted line! The two VSWR measurements, s and s^o, correspond to the two configurations in Figure 13.12, (a) and (b), respectively.

When the resonant frequency is found, λ can be measured, then s^o and s ($s^o > s$) can be measured, and α' can be calculated with equation 13.92. α' is the attenuation constant of the resonant section for *both* conductor and dielectric losses. The dielectric loss is

$$\alpha_d = \alpha' - \alpha_c = \alpha' - \frac{\beta'}{2Q_c} = \alpha' - \frac{\pi}{\lambda' Q_c} \tag{13.93}$$

and Q_c can be calculated with equation 13.78. Finally,

$$\varepsilon_R = \frac{\varepsilon}{\varepsilon_0} = \left(\frac{\lambda}{\lambda'}\right)^2 \tag{13.94}$$

and equations 13.77 and 13.49 can be combined to give

$$\sigma_d = 2\alpha_d f \lambda' \varepsilon_0 \varepsilon_R \tag{13.95}$$

or

$$\frac{\sigma_d}{\omega\varepsilon} = \frac{\alpha_d \lambda'}{\pi} \quad \text{(dielectric loss tangent)} \tag{13.96}$$

EXAMPLE 5

Suppose that a certain $\lambda'/4$ section is 10 cm long and is filled with a dielectric whose conductivity and permittivity we desire to measure. The lowest resonant frequency is determined to be 300 MHz. The VSWR is measured as 38.9 dB with the resonant section in place, but

[8] See Problem 22.

when it is replaced with an open circuit (such that z_{max} and z_{min} are unaltered), the VSWR increases to 51.3 dB. The Q_c due to conductor losses has been calculated as 2500. We have

$$s = 88.10$$

$$s^o = 367.28$$

and

$$\lambda' = 0.4 \quad (\text{m})$$

Using equation 13.92,

$$\alpha' = \frac{4}{1}\frac{1 - 0.240}{88.10} = 0.035 \qquad (p = 1)$$

Using equation 13.93,

$$\alpha_d = 0.035 - \frac{\pi}{(0.4)(2500)} = 0.031$$

Using equation 13.94,

$$\varepsilon_R = \left(\frac{1}{0.4}\right)^2 = 6.25$$

Finally, equation 13.95 gives

$$\sigma_d = 2(0.031)(3 \times 10^8)(0.4)\left(\frac{1}{36\pi}\right) \times 10^{-9}(6.25)$$

$$\sigma_d = 4.16 \times 10^{-4}$$

or

$$\frac{\sigma_d}{\omega\varepsilon} = 0.0039$$

13.9 CONCLUDING REMARKS

In this chapter, we have seen that the analysis of the lossy line is very similar to that for the lossless line. In most cases, the modifications required are minor. In stub matching, for example, the best method would be that of designing the system on the basis of no losses, and then readjusting the stub lengths or spacings experimentally to account for the losses. The values of R, L, C, and G, the distributed parameters of a transmission line, were found. It must be remembered that these parameters, as well as the perturbation technique we developed, are only valid for low-loss or, high-Q cases. These are the cases of most interest to the engineer. Finally, time domain analysis could be pursued, but we did not choose to do this because the problem is, at best, difficult to treat, and when we recall that most material "constants" are actually frequency dependent, the problem is even more complicated.

REFERENCES

Chipman. (See references for Chapter 11.)
Hayt. (See references for Chapter 2.) The $\lambda/4$ resonant line is discussed in Chapter 13.

PROBLEMS

1. Derive equations 13.33 and 13.34.
2. Show that $Z_0 = \sqrt{L/C} = \sqrt{R/G}$ for a "distortionless" line.
3. A telephone line has $R = 6.3 \times 10^{-3}\ \Omega/\text{m}$, $L = 2.27 \times 10^{-6}$ H/m, $G = 180\ \mu\mu\mho/\text{m}$, and $C = 5.2\ \mu\mu\text{F/m}$. Find α, β, Z_0, and u_p at 1 KHz.
4. How much inductance should be added to make the line of Problem 3 distortionless? What is the new value of α?
5. If $Z_g = Z_l = Z_0$, then from equation 13.24

$$V(z, \omega) = V_g(\omega)e^{-\gamma(\omega)(l-z)}, \qquad \gamma = \alpha + j\beta$$

   If $v_g(t) = V_0\,\delta(t)$, outline the steps required to find $v(z, t)$. Find $v(z, t)$ explicitly for (a) the lossless line, (b) the distortionless line, and (c) the small loss case (equation 13.39).
6. What is the Q of the line in Problem 3? Is this a valid calculation?
7. Show that the Q of the distortionless line as given by equation 13.49 is $\omega L/2R$. What is the Q of the line in Problem 4?
8. Repeat Example 1 if the line is known to be between $5\lambda/2$ and $11\lambda/4$ in length.
9. Use equation 13.57 to find the Q in Example 1.
10. A line with small loss is resonant ($z_0 \approx \lambda/4$) and its Q is 100. Find its length if the dielectric is lossless.
11. Calculate the high-frequency parameters for a two-wire line with copper conductors and polyethylene dielectric: $\sigma_c = 5.8 \times 10^7$, $\varepsilon_R = 2.26$, $\sigma_d = 5 \times 10^{-6}$, $f = 100$ MHz, $D = 10^{-2}$, $d = 10^{-3}$.
12. Starting with equation 13.30 and using Table 13.1 show that for the small loss case, the resonant length is

$$z_0 \approx \frac{\lambda/4}{1 - (X_0/R_0)(\alpha/\beta)} \quad \text{and} \quad R_{\text{in}}^{\text{sc}} \approx \frac{R_0}{\alpha z_0} = R_0 \frac{4}{\pi} Q,$$

$$Z_0 = R_0 + jX_0$$

13. The voltage on a general line for a $+z$ traveling wave can be written

$$V(z, \omega) = V_0(\omega)e^{-\gamma(\omega)z}$$

Assume that $R, L, G,$ and C are constant. Use Table E.1 (Appendix E) and show that the impulse response $[V_0(\omega) = 1]$ is

(a) $$v(z, t) = e^{-1/2[(R/L)+(G/C)]t}\,\delta(t - \sqrt{LC}z) + \frac{1}{4}\left(\frac{R}{L} - \frac{G}{C}\right)^2 \sqrt{LC}ze^{-1/2[(R/L)+(G/C)]t} \times \frac{I_1\left[\frac{1}{2}\left(\frac{R}{L} - \frac{G}{C}\right)\sqrt{t^2 - LCz^2}\right]}{\frac{1}{2}\left(\frac{R}{L} - \frac{G}{C}\right)\sqrt{t^2 - LCz^2}}\,u(t - \sqrt{LC}z)$$

and

(b) $$i(z, t) = \frac{e^{-1/2[(R/L)+(G/C)]t}\,\delta(t - \sqrt{LC}z)}{R_0} + \frac{1}{4}\left(\frac{R}{L} - \frac{G}{C}\right)^2 t\,\frac{e^{-1/2[(R/L)+(G/C)]t}}{R_0} \times \frac{I_1\left[\frac{1}{2}\left(\frac{R}{L} - \frac{G}{C}\right)\sqrt{t^2 - LCz^2}\right]}{\frac{1}{2}\left(\frac{R}{L} - \frac{G}{C}\right)\sqrt{t^2 - LCz^2}}\,u(t - \sqrt{LC}z) + \frac{1}{2}\left(\frac{R}{L} - \frac{G}{C}\right)\frac{e^{-1/2[(R/L)+(G/C)]t}}{R_0} \times I_0\left[\frac{1}{2}\left(\frac{R}{L} - \frac{G}{C}\right)\sqrt{t^2 - LCz^2}\right]u(t - \sqrt{LC}z)$$

$$R_0 = \sqrt{\frac{L}{C}}$$

(c) Interpret the results of (a) and (b).

14. Using the results of Problem 13 show that a distortionless line $(R/L = G/C)$ is distortionless. Notice that

$$\lim_{x\to 0} \frac{I_1(x)}{x} = \frac{1}{2} \quad \text{and} \quad \lim_{x\to 0} I_0(x) = 1$$

15. Find the voltage impulse response at the middle of a 600-m-long distortionless line if $Z_g = R_0 = \sqrt{L/C}$ and $Z_l = 0$. Use equation 13.24 and the same transform as used in Problems 13 and 14. $R = 6 \times 10^{-3}\,\Omega/\text{m}$, $L = 3 \times 10^{-6}$ H/m, and $C = 5 \times 10^{-12}$ F/m. Compare this result to that of Example 2, Chapter 11.

16. (a) Calculate Z_0 and γ for the line of Problem 11.
    (b) Suppose it is assumed that this line is lossless in order to expedite a single stub design to match $Z_l = 400\ \Omega$ at 100 MHz. Using $R_0 = \text{Re}\{Z_0\}$, design such a system using the minimum value of l_1. Next, calculate the impedance at the stub junction and compare this to Z_0.
    (c) Comment on results.

17. (a) Calculate α for the rigid coaxial line whose dimensions are those of Example 4 (second part, $f = 300$ MHz).
    (b) Repeat if a and b are both halved.

18. The incident fields in the vicinity of a receiving antenna are such that the antenna can deliver 10^{-12} W into a matched load. If the transmission line between the antenna and receiver has $\alpha = 3$ dB/100 ft and is 100 ft long, what is the maximum power delivered to the receiver?
19. Why is the "velocity factor" for most commercial two-wire line with polyethylene dielectric 0.85 rather than $0.67 = 1/\sqrt{\varepsilon_R}$?
20. Derive an expression for wavelength as a function of frequency in a good conductor if the material parameters are constants.
21. Following the procedure of Section 13.7, and making reasonable assumptions, find the high-frequency parameters for a microstrip transmission line consisting of a thin (t) conducting strip of width w located a height h ($h \ll w$) above a large conducting ground plane. A dielectric is sandwiched between the two conducting planes as part of the manufacturing process.

22. (a) For a lossless transmission line show that

$$\frac{|V|}{|V|_{\max}} = \frac{1}{s}\sqrt{1 + (s^2 - 1)\sin^2[\beta(z - z_{\min})]}, \qquad (s = \text{VSWR})$$

    (b) For a line with small loss, $s > 10$ and $|V|/|V|_{\max} = \sqrt{2}/s$ (3 dB above $|V|_{\min}$) show that

$$s \approx \frac{\lambda}{\pi \Delta z}$$

    where Δz is the distance between the two points, which are 3 dB above $|V|_{\min}$ (*width of minimum method for determining* VSWR).

Chapter 14
Lossless Waveguides and Cavities

In the microwave region of frequencies, beginning roughly at 1 GHz, ordinary two-conductor systems, commonly called transmission lines, become inefficient, so that another guiding system is sought. A hollow pipe or cylinder of some arbitrary, but uniform, cross section might be expected to have less loss than the ordinary line. This pipe is commonly called a *waveguide*, but all guiding systems are really waveguides. In this chapter, we will investigate the propagation characteristics of hollow waveguides.

Much useful information was derived in preceding chapters from the TEM mode propagating between infinite parallel planes. We can now re-examine this rather unique guiding system, searching for other modes of propagation. It was shown in Chapter 13 that a mode other than the TEM mode was required (strictly speaking) if the conducting planes were not perfectly conducting. Is it possible for the perfectly conducting parallel plane system to support these higher order modes?

There are many methods for solving boundary value problems of the type posed here, but they are all essentially the same in that what they accomplish is a solution to Maxwell's equations fitting the boundary conditions for the particular problem. The first method used here, which is sufficiently general to accommodate all of the problems in this text, plus many more, is that of deriving the fields from *potentials.* It involves the separation of variables technique which was used in Chapter 5, but does not involve any new mathematics. Since much time and effort was expended in Section 9.5 to set up a magnetic vector potential, and it was certainly stated that many times its use is the most efficient way to obtain **E** and **H**, it seems only natural to use it when the opportunity arises. On the other hand, in many cases it is sufficient to present a simplified treatment which produces only the most important case, that of the dominant TE_{10} mode for the rectangular waveguide. This is done in Section 14.6. Those who are primarily interested in this case can begin this chapter with Section 14.6, and save considerable time.

14.1 POTENTIALS[1]

In a region free of isolated electric charges and conduction current densities, such as the interior of a hollow one conductor system, Maxwell's equations are

$$\nabla \times \mathbf{E} = -j\omega\mu\mathbf{H} \tag{14.1}$$

and

$$\nabla \times \mathbf{H} = j\omega\varepsilon\mathbf{E} \tag{14.2}$$

Taking the divergence of both sides of equations 14.1 and 14.2 gives

$$\nabla \cdot \nabla \times \mathbf{E} = -j\omega\mu(\nabla \cdot \mathbf{H}) \equiv 0 \tag{14.3}$$

and

$$\nabla \cdot \nabla \times \mathbf{H} = j\omega\varepsilon(\nabla \cdot \mathbf{E}) \equiv 0 \tag{14.4}$$

by vector identity. We have assumed that μ and ε are scalar quantities, independent of position and direction. Then,

$$\nabla \cdot \mathbf{H} = 0 = \nabla \cdot \mathbf{B} \tag{14.5}$$

and

$$\nabla \cdot \mathbf{E} = 0 = \nabla \cdot \mathbf{D} \tag{14.6}$$

or, *both* **E** and **H** are *solenoidal* fields! In this case, it is possible to define

$$\mathbf{E} \equiv -\nabla \times \mathbf{F} \tag{14.7}$$

and

$$\mathbf{B} \equiv \nabla \times \mathbf{A} \tag{14.8}$$

[1] Those who are interested primarily in the TE_{10} dominant mode for the rectangular guide can proceed to Section 14.6, second paragraph, for a simplified treatment.

and equations 14.5 and 14.6 will always be satisfied. $\mathbf{F}$ is called *electric vector potential* (volts), and $\mathbf{A}$ is, of course, called *magnetic vector potential* (webers per meter). Note that it would be possible to define $\mathbf{D} = \pm\nabla \times \mathbf{F}$ and $\mathbf{H} = \pm\nabla \times \mathbf{A}$, and these potentials would only differ from those of equations 14.7 and 14.8 by constants. We will use $\mathbf{B} = \nabla \times \mathbf{A}$ since we have already done so in Chapter 9, although the use of $\mathbf{H} = \nabla \times \mathbf{A}$ retains more symmetry in the equations. It is also important to note that *magnetic charges* do not exist in nature, so for the μ and ε used here, $\mathbf{B}$ and $\mathbf{H}$ are always *solenoidal* and $\mathbf{B}$ or $\mathbf{H} = \nabla \times \mathbf{A}$ is always possible. On the other hand, equation 14.7 was possible *only* because we assumed a region free of isolated *electric charges.*

Equations 14.7 and 14.8 do not completely define $\mathbf{F}$ and $\mathbf{A}$, for *both* the *divergence and the curl* of a vector are required to completely specify it. We can specify the divergence of $\mathbf{F}$ and $\mathbf{A}$ at a more convenient point in the development. It is convenient here to remember that the equations are still *linear*, so part of the field may be determined from $\mathbf{A}$, with $\mathbf{F} = 0$; then the rest of the field is determined from $\mathbf{F}$ with $\mathbf{A} = 0$. By superposition, a total field may be found as the sum of the two parts *if necessary*. Quite often, it is not necessary to add the two partial fields. For this reason, it is desirable to consider the two cases separately.

(1) $\mathbf{F} = 0$. This case was treated in considerable detail for the time domain (t shown explicitly) with sources (ρ_v and $\mathbf{J}$) present. All we need for present purposes is to use phasor notation and set all the sources equal to zero. From equation 9.65, the Lorentz condition is

$$\nabla \cdot \mathbf{A} = -j\omega\mu\varepsilon\Phi_a \tag{14.9}$$

where Φ_a is the ordinary *electric scalar* potential, and the added subscript merely reminds us to associate it with $\mathbf{A}$. Using equations 9.67, 9.78, and 9.77, we have

$$\begin{aligned} \mathbf{F} &= 0 \\ \nabla^2\mathbf{A} + k^2\mathbf{A} &= 0 \quad \text{(wave equation)} \\ \mathbf{E} &= -j\omega\mathbf{A} + \frac{1}{j\omega\mu\varepsilon}\nabla(\nabla \cdot \mathbf{A}) \\ \mathbf{B} &= \nabla \times \mathbf{A} \end{aligned} \tag{14.10}$$

(2) $\mathbf{A} \equiv 0$. From equations 14.7 and 14.2,

$$\nabla \times \mathbf{H} = -j\omega\varepsilon(\nabla \times \mathbf{F})$$

or

$$\nabla \times (\mathbf{H} + j\omega\varepsilon\mathbf{F}) = 0 \tag{14.11}$$

and $\mathbf{H} + j\omega\varepsilon\mathbf{F}$ is a *lamellar* field. As in case (1), we define

$$\mathbf{H} + j\omega\varepsilon\mathbf{F} \equiv -\nabla\Phi_f \tag{14.12}$$

where Φ_f is *magnetic scalar potential* (amperes). Now substitute equations 14.7 and 14.12 into 14.1, so that

$$\nabla \times \nabla \times \mathbf{F} = j\omega\mu(-j\omega\varepsilon\mathbf{F} - \nabla\Phi_f)$$

or

$$\nabla \times \nabla \times \mathbf{F} - k^2\mathbf{F} + j\omega\mu\nabla\Phi_f = 0$$

Expanding by vector identity (again) gives

$$\nabla(\nabla \cdot \mathbf{F}) - \nabla^2\mathbf{F} - k^2\mathbf{F} + j\omega\mu\nabla\Phi_f = 0$$

or

$$\nabla(\nabla \cdot \mathbf{F} + j\omega\mu\Phi_f) - \nabla^2\mathbf{F} - k^2\mathbf{F} = 0 \tag{14.13}$$

An obvious choice for $\nabla \cdot \mathbf{F}$ is (again) the Lorentz condition,

$$\nabla \cdot \mathbf{F} \equiv -j\omega\mu\Phi_f \tag{14.14}$$

so that

$$\nabla^2\mathbf{F} + k^2\mathbf{F} = 0 \tag{14.15}$$

Substituting equation 14.14 into 14.12 gives

$$\mathbf{H} = -j\omega\varepsilon\mathbf{F} + \frac{1}{j\omega\mu}\nabla(\nabla \cdot \mathbf{F}) \tag{14.16}$$

Case (2) may be summarized

$$\begin{aligned} \mathbf{A} &= 0 \\ \nabla^2\mathbf{F} + \mathrm{k}^2\mathbf{F} &= 0 \quad \text{(wave equation)} \\ \mathbf{H} &= -j\omega\varepsilon\mathbf{F} + \frac{1}{j\omega\mu}\nabla(\nabla \cdot \mathbf{F}) \\ \mathbf{E} &= -\nabla \times \mathbf{F} \end{aligned} \tag{14.17}$$

Note that these results could have been written down directly from equation 14.10 after comparing equations 14.1, 14.2, 14.7, and 14.8 (principle of duality).

As previously mentioned, the sum of the fields given by equations 14.10 and 14.17 is a general solution. We will find that equations 14.10 and 14.17 will *separately* accommodate all the waveguide problems of interest here.

14.2 THE PARALLEL PLANE SYSTEM

Consider the parallel plane system shown in Figure 14.1. Mode propagation other than TEM in such a system was predicted in Chapter 10. It is advantageous to consider two cases separately.

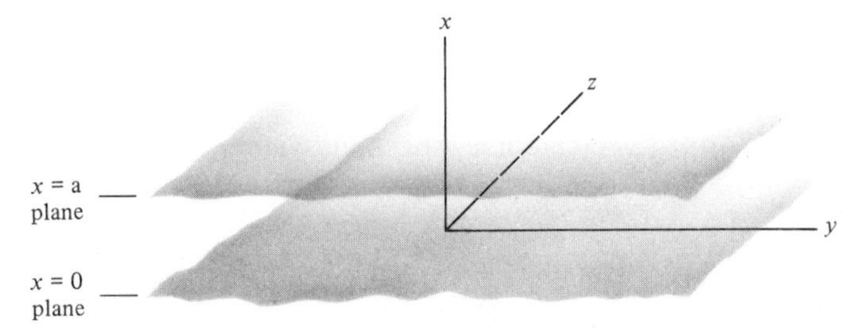

Figure 14.1. Parallel plane waveguide system.

(1) The choice $\mathbf{A} = \mathbf{a}_z A_z$, $A_x = A_y = 0$ will give $H_z = 0$, which is one of the two cases we wish to examine. Substituting this value of $\mathbf{A}$ into equations 14.10, and then expanding in Cartesian coordinates gives

$$\frac{\partial^2 A_z}{\partial x^2} + \frac{\partial^2 A_z}{\partial y^2} + \frac{\partial^2 A_z}{\partial z^2} + k^2 A_z = 0 \tag{14.18}$$

and

$$E_x = \frac{1}{j\omega\mu\varepsilon}\frac{\partial^2 A_z}{\partial x \partial z} \qquad H_x = \frac{1}{\mu}\frac{\partial A_z}{\partial y}$$

$$E_y = \frac{1}{j\omega\mu\varepsilon}\frac{\partial^2 A_z}{\partial y \partial z} \qquad H_y = -\frac{1}{\mu}\frac{\partial A_z}{\partial x} \tag{14.19}$$

$$E_z = \frac{1}{j\omega\mu\varepsilon}\left(k^2 + \frac{\partial^2}{\partial z^2}\right)A_z \qquad \boxed{H_z \equiv 0}$$

This is always a *TM to z mode*[2] since the magnetic field is entirely transverse ($H_z = 0$) to the z direction.

We now assume that there is no variation with y and look for propagation in the $+z$ direction. It immediately follows that $H_x = E_y = 0$ since $\partial/\partial y = 0$ by assumption. The boundary conditions on the electric field require that the tangential electric field vanish on the surface of the perfect conductors, or

$$E_z = 0\begin{cases} x = 0 \\ x = a \end{cases} \tag{14.20}$$

The boundary condition on the magnetic field will be automatically accounted for. A general solution to equation 14.18, as derived in Appendix H by separation of variables (we used this technique for a two-dimensional problem in Section 5.3.) is

$$A_z(x, y, z) = h(k_x x)h(k_y y)h(k_z z) \tag{14.21}$$

[2] This designation was first mentioned in Section 13.3.

where the *separation equation* is

$$k_z^2 = k^2 - k_x^2 - k_y^2 \tag{14.22}$$

k_x, k_y, and k_z are called *eigenvalues* (characteristic values) and h is a *harmonic* function, or

$$h(k_x x) = A \cos k_x x + B \sin k_x x \tag{14.23}$$

$$h(k_y y) = C \cos k_y y + D \sin k_y y \tag{14.24}$$

and

$$h(k_z z) = E \cos k_z z + F \sin k_z z \tag{14.25}$$

We expect propagation in the $+z$ direction, so we choose $F = -jE$ to make

$$h(k_z z) = Ee^{-jk_z z} \tag{14.26}$$

Notice that we could just as well choose propagation in the $-z$ direction. We have specified that $\partial/\partial y = 0$, so $k_y = 0$ and

$$h(k_y y) = C \tag{14.27}$$

Finally, notice from equation 14.19 that equation 14.20 is satisfied if $A = 0$ and $k_x = m\pi/a$. We have now determined that (with $BCE \equiv 1$)

$$A_z = \sin \frac{m\pi x}{a} e^{-jk_z z} \tag{14.28}$$

and

$$k_z^2 = k^2 - \left(\frac{m\pi}{a}\right)^2, \qquad m = 1, 2, 3, \ldots \tag{14.29}$$

The field is given by equation 14.19, or

$$\begin{aligned} E_x &= -\frac{k_z m\pi}{\omega\mu\varepsilon a} \cos \frac{m\pi x}{a} e^{-jk_z z} \\ E_z &= \frac{1}{j\omega\mu\varepsilon}\left(\frac{m\pi}{a}\right)^2 \sin \frac{m\pi x}{a} e^{-jk_z z} \\ H_y &= -\frac{m\pi}{a\mu} \cos \frac{m\pi x}{a} e^{-jk_z z} \end{aligned} \tag{14.30}$$

From equation 14.29, we have

$$k_z = \sqrt{k^2 - \left(\frac{m\pi}{a}\right)^2} \tag{14.31}$$

Inspection of equation 14.31 reveals that propagation will occur if $k^2 > (m\pi/a)^2$, or if $k > m\pi/a$, for in this case, k_z is *real*. On the other hand, if

$k < m\pi/a$, k_z is imaginary, and we *do not have wave propagation*, but merely exponentially damped ($e^{-\alpha z}$), *evanescent*, fields. This situation may be summarized as follows.

$$k_z = \begin{cases} \sqrt{k^2 - \left(\dfrac{m\pi}{a}\right)^2} = \beta, & k > \dfrac{m\pi}{a} \\ \sqrt{\left(\dfrac{m\pi}{a}\right)^2 - k^2} = -j\alpha, & k < \dfrac{m\pi}{a} \end{cases} \tag{14.32}$$

Notice the choice of signs for k_z For $k > m\pi/a$ we choose $k_z = +\beta$ to give a $+z$ traveling wave, $e^{-j\beta z}$. For $k < m\pi/a$ we choose $k_z = -j\alpha$ to give $e^{-j(-j\alpha)z} = e^{-\alpha z}$, an exponentially increasing term being impossible since z is not bounded. The *transition* occurs where $k_z = 0$, or $k = m\pi/a$. Solving for the *particular* frequency at this transition, called the *cutoff frequency*,

$$k_c = \frac{m\pi}{a} = \omega_c\sqrt{\mu\varepsilon}$$

or

$$\boxed{\omega_c = \frac{m\pi}{a\sqrt{\mu\varepsilon}}} \tag{14.33}$$

Substituting equation 14.33 into 14.32 gives a better form.

$$k_z = \begin{cases} k\sqrt{1 - \left(\dfrac{\omega_c}{\omega}\right)^2} = \beta, & \omega > \omega_c \\ -jk_c\sqrt{1 - \left(\dfrac{\omega}{\omega_c}\right)^2} = -j\alpha, & \omega < \omega_c \end{cases} \tag{14.34}$$

(2) The choice $\mathbf{F} = \mathbf{a}_z F_z$, $F_x = F_y = 0$ will give $E_z = 0$, which is the other case we wish to examine. Substituting this value of $\mathbf{F}$ into equations 14.17, and expanding into Cartesian coordinates:

$$\frac{\partial^2 F_z}{\partial x^2} + \frac{\partial^2 F_z}{\partial y^2} + \frac{\partial^2 F_z}{\partial z^2} + k^2 F_z = 0 \tag{14.35}$$

and

$$\begin{aligned} E_x &= -\frac{\partial F_z}{\partial y} & H_x &= \frac{1}{j\omega\mu}\frac{\partial^2 F_z}{\partial x\,\partial z} \\ E_y &= \frac{\partial F_z}{\partial x} & H_y &= \frac{1}{j\omega\mu}\frac{\partial^2 F_z}{\partial y\,\partial z} \\ \Aboxed{E_z &\equiv 0} & H_z &= \frac{1}{j\omega\mu}\left(k^2 + \frac{\partial^2}{\partial z^2}\right)F_z \end{aligned} \tag{14.36}$$

This is always a TE to z mode since the electric field is transverse to $z(E_z = 0)$.

We now assume that there is no variation with y and again look for propagation in the $+z$ direction. It immediately follows that $E_x = H_y = 0$ since $\partial/\partial y = 0$ by assumption. The partial differential equation, equation 14.35, is the same as equation 14.18 whose solution was equation 14.21. Then the same possibilities as given by equations 14.23, 14.24, and 14.25 exist. Again, we choose $F = -jE$, so that

$$h(k_z z) = Ee^{-jk_z z} \tag{14.37}$$

and, again, $k_y = 0$, since $\partial/\partial y = 0$. Then,

$$h(k_y y) = C \tag{14.38}$$

Inspection of equation 14.36 reveals that the boundary conditions,

$$E_y = 0 \begin{cases} x = 0 \\ x = a \end{cases} \tag{14.39}$$

are satisfied if $B = 0$ and $k_x = m\pi/a$. We now have $(ACE \equiv 1)$

$$F_z = \cos \frac{m\pi x}{a} e^{-jk_z z} \tag{14.40}$$

and, as before,

$$k_z^2 = k^2 - \left(\frac{m\pi}{a}\right)^2, \qquad m = 1, 2, 3, \ldots \tag{14.41}$$

The field is given by equations 14.36, or

$$\begin{aligned} E_y &= -\frac{m\pi}{a} \sin \frac{m\pi x}{a} e^{-jk_z z} \\ H_x &= \frac{k_z m\pi}{\omega\mu a} \sin \frac{m\pi x}{a} e^{-jk_z z} \\ H_z &= \frac{1}{j\omega\mu} \left(\frac{m\pi}{a}\right)^2 \cos \frac{m\pi x}{a} e^{-jk_z z} \end{aligned} \tag{14.42}$$

Again,

$$k_z = \begin{cases} k\sqrt{1 - \left(\dfrac{\omega_c}{\omega}\right)^2} = \beta, & \omega > \omega_c \\ -jk_c\sqrt{1 - \left(\dfrac{\omega}{\omega_c}\right)^2} = -j\alpha, & \omega < \omega_c \end{cases} \tag{14.43}$$

where

$$\omega_c = \frac{m\pi}{a\sqrt{\mu\varepsilon}} \tag{14.44}$$

(3) In equation 14.10, let $\mathbf{A} = \mathbf{a}_z A_z$ and $A_x = A_y = 0$. We now desire to find a *uniform* field; that is, a field independent of y *and* x. This should be similar to the TEM to z mode, or uniform plane wave, which has been discussed in previous chapters.

If A_z is independent of y, then the field (equation 14.19) is certainly independent of y. If A_z is *also* independent of x, then the only solution is the trivial null field. Let us reexamine equation 14.18 with $\partial A_z/\partial y \equiv 0$. This is

$$\frac{\partial^2 A_z}{\partial x^2} + \frac{\partial^2 A_z}{\partial z^2} + k^2 A_z = 0 \tag{14.45}$$

A solution to equation 14.45 is certainly equation 14.28,

$$A_z = \sin \frac{m\pi x}{a} e^{-jk_z z}$$

but this does not give a field independent of x! Apparently, another solution to equation 14.45 must be found. Inspection of equation 14.19 reveals that A_z can be a linear function of x and the field will be independent of x. Let us see if this type of solution will satisfy equation 14.45. (Actually, we can be more formal and use the separation of variables technique as in the Appendix.) We then assume

$$A_z(x, z) = (Ax + B)Z(z) \tag{14.46}$$

where $Z(z)$ is a function of z only. Then equation 14.45 becomes

$$(Ax + B)\frac{d^2 Z}{dz^2} + k^2(Ax + B)z = 0$$

or

$$\frac{d^2 Z}{dz^2} + k^2 Z = 0 \tag{14.47}$$

We want a wave propagating in the $+z$ direction, so an appropriate solution to equation 14.47 is

$$Z = Ce^{-jkz} \tag{14.48}$$

Note that the phase constant is the *intrinsic* phase constant! Then equation 14.46 becomes

$$A_z = (Dx + E)e^{-jkz} \tag{14.49}$$

where $D \equiv AC$ and $E \equiv BC$. The constant E contributes nothing to the field so we let $E = 0$ and $D = -\mu$, or

$$A_z = -\mu x e^{-jkz} \tag{14.50}$$

Table 14.1 PARALLEL PLANE WAVEGUIDE FIELD EQUATIONS WITH PROPAGATION IN THE $+z$ DIRECTION, $k_z = \sqrt{k^2 - (m\pi/a)^2}$, $k = \omega\sqrt{\mu\varepsilon}$

TM TO z	TE TO z	TEM TO z
$E_x = E_0 \cos\frac{m\pi x}{a} e^{-jk_z z}$	$E_y = E_0 \sin\frac{m\pi x}{a} e^{-jk_z z}$	$E_x = E_0 e^{-jkz}$
$E_z = j\frac{E_0}{k_z}\frac{m\pi}{a} \sin\frac{m\pi x}{a} e^{-jk_z z}$	$H_x = -\frac{k_z E_0}{\omega\mu} \sin\frac{m\pi x}{a} e^{-jk_z z}$	$H_y = \frac{E_0}{\eta} e^{-jkz}$
$H_y = \frac{\omega\varepsilon E_0}{k_z} \cos\frac{m\pi x}{a} e^{-jk_z z}$	$H_z = j\frac{E_0}{\omega\mu}\frac{m\pi}{a} \cos\frac{m\pi x}{a} e^{-jk_z z}$	
$m = 1, 2, 3, \ldots$	$m = 1, 2, 3, \ldots$	

The field, from equation 14.19, is

$$E_x = \frac{jk\mu}{j\omega\mu\varepsilon} e^{-jkz} \tag{14.51}$$

and

$$H_y = e^{-jkz} \tag{14.52}$$

or

$$E_x = \eta e^{-jkz} \tag{14.53}$$

and

$$H_y = e^{-jkz} \tag{14.54}$$

This *is* our familiar friend, the TEM to z mode, or uniform plane wave.

Before attempting to make any comparisons between the modes, let us summarize the results. It is convenient to normalize by letting the peak value of E_x (or E_y) be E_0 in each case. This has been done in Table 14.1.

The phase velocity is the velocity of a point of constant phase. Taking the TM mode as an example,

$$E_x(x, z, \omega) = E_0 \cos\frac{m\pi x}{a} e^{-jk_z z}$$

or using the time domain form,

$$E_x(x, z, t) = E_0 \cos\frac{m\pi x}{a} \cos(\omega t - k_z z) \tag{14.55}$$

For a point of constant phase,

$$\omega t - k_z z = \text{constant}$$

or

$$\omega - k_z \frac{dz}{dt} = 0$$

or

$$u_p = \frac{\omega}{k_z} \tag{14.56}$$

The phase velocity is real for $\omega > \omega_c$, or

$$u_p = \frac{\omega}{k\sqrt{1 - (\omega_c/\omega)^2}}$$

or

$$\boxed{u_p = \frac{1}{\sqrt{\mu\varepsilon}\sqrt{1 - (\omega_c/\omega)^2}} \geq \frac{1}{\sqrt{\mu\varepsilon}}} \tag{14.57}$$

Recall the similar situation for oblique incidence in Chapter 10. Equation 14.57 states that the phase velocity is greater than the (plane wave) velocity of light in the same medium! We have given this phenomenon some attention in Chapter 10, and will again do so shortly.

Also, the system or guide wavelength, λ_g, is the distance in the z direction for 2π rad of phase shift. Therefore,

$$k_z(z_1 - z_2) = 2\pi = k_z \lambda_g$$

or

$$\boxed{\lambda_g = \frac{2\pi}{k_z} = \frac{1}{f\sqrt{\mu\varepsilon}\sqrt{1 - (\omega_c/\omega)^2}} \geq \lambda} \tag{14.58}$$

The velocity of energy flow is defined in equation 10.37 as

$$u_e \equiv \frac{\langle P_f \rangle}{\langle W \rangle} \tag{14.59}$$

or

$$u_e = \frac{\displaystyle\int_0^1 \int_0^a \frac{1}{2} \operatorname{Re}\{\mathbf{E} \times \mathbf{H}^*\} \cdot \mathbf{a}_z \, dx \, dy}{\displaystyle\int_0^1 \int_0^1 \int_0^a \frac{1}{2} \operatorname{Re}\left\{\frac{\mathbf{D}}{2} \cdot \mathbf{E}^* + \frac{\mathbf{B}}{2} \cdot \mathbf{H}^*\right\} dx \, dy \, dz} \tag{14.60}$$

for a section of the parallel plane system which is 1 m wide. Taking the TM mode as an example again, equation 14.60 becomes

$$u_e = \frac{\dfrac{\omega\varepsilon E_0^2}{2k_z} \displaystyle\int_0^a\int_0^1 \cos^2 \frac{m\pi x}{a}\, dx\, dy}{\dfrac{\varepsilon}{4} E_0^2 \displaystyle\int_0^a\int_0^1\int_0^1 \left(\frac{k^2}{k_z^2} + \cos \frac{2m\pi x}{a}\right) dx\, dy\, dz} \tag{14.61}$$

or

$$u_e = \frac{2\omega}{k_z} \frac{\displaystyle\int_0^a \cos^2 \frac{m\pi x}{a}\, dx}{\displaystyle\int_0^a \left(\frac{k^2}{k_z^2} + \cos \frac{2m\pi x}{a}\right) dx} \tag{14.62}$$

This becomes

$$u_e = \frac{2\omega}{k_z} \frac{(a/2)}{(k^2/k_z^2)(a)} = \frac{\omega k_z}{k^2} \tag{14.63}$$

or

$$\boxed{u_e = \frac{\sqrt{1 - (\omega_c/\omega)^2}}{\sqrt{\mu\varepsilon}} \leq \frac{1}{\sqrt{\mu\varepsilon}}} \tag{14.64}$$

Equation 14.64 states that the velocity of energy flow is less than the plane wave velocity of light in the same medium. It follows from equations 14.57 and 14.64 that

$$\boxed{u_p u_e = \frac{1}{\mu\varepsilon}} \tag{14.65}$$

The same result is obtained for TE to z modes. The wave impedance equations follow from the definitions in Chapter 10.

It is interesting and informative to plot some of the important quantities against frequency or normalized frequency, ω/ω_c. Figure 14.2 shows that for $\omega > \omega_c$ the TE and TM wave impedances are real, indicating power flow. For $\omega < \omega_c$, the wave impedances are reactive, indicating the lack of power flow. For $\omega \gg \omega_c$, the TE and TM wave impedances approach the *intrinsic* impedance, $\sqrt{\mu/\varepsilon}$.

The phase velocity of both the TE and TM modes is infinite at the cutoff frequency, and approaches the intrinsic (TEM) phase velocity,

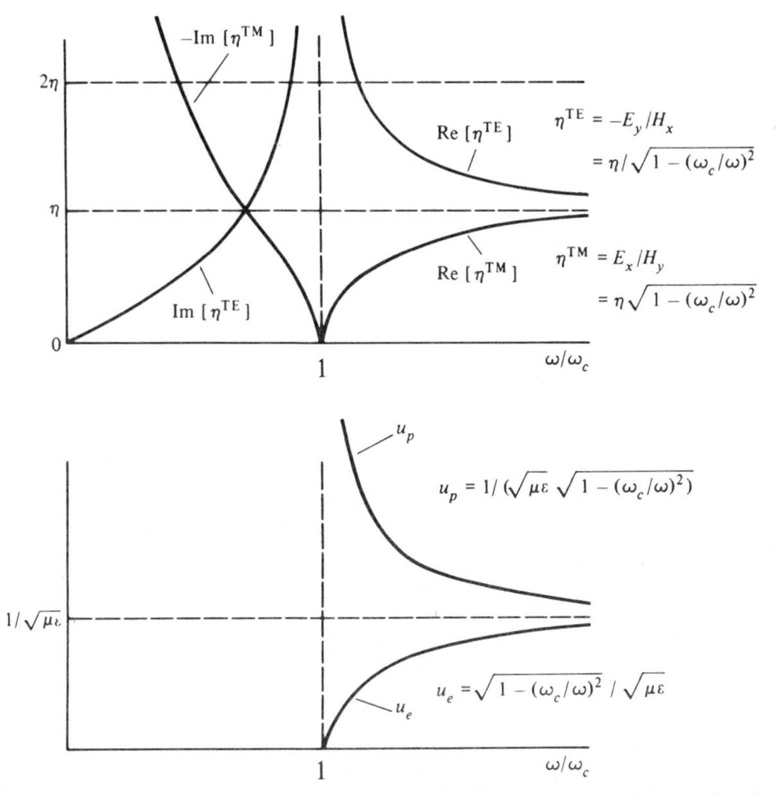

Figure 14.2. (a) TM (E_x/H_y) and TE ($-E_y/H_x$) wave impedances for the parallel plane system. (b) Phase velocity and velocity of energy flow for the parallel plane system.

$1/\sqrt{\mu\varepsilon}$, for $\omega \gg \omega_c$. The velocity of energy flow for both modes is zero at the cutoff frequency and approaches the intrinsic value for $\omega \gg \omega_c$. The three parts of Figure 14.2 are just different ways to view the same phenomenon: propagation without attenuation (no losses) above the cutoff frequency and exponential damping without propagation below cutoff. Thus, for the TE and TM modes the parallel plane system acts like a high-pass filter.

Each value of $m = 1, 2, 3, \ldots$ gives a different mode (a "normal" mode in a more general sense), and more than one mode can exist at the same time (Why?). The mode with lowest cutoff frequency is called the *dominant mode*, and for the parallel plane system, this is the TEM mode. It can be seen from Table 14.1 that the TEM mode can be obtained from the TM mode *field equations* (not potential) by setting $m = 0$. This was avoided previously to prevent confusion. If we wish to list the modes in order of cutoff frequency, we should list them showing m explicitly; that is, TE_m and TM_m should be the mode designations. Note that in all cases except the TEM, the cutoff frequency depends on the dimension a.

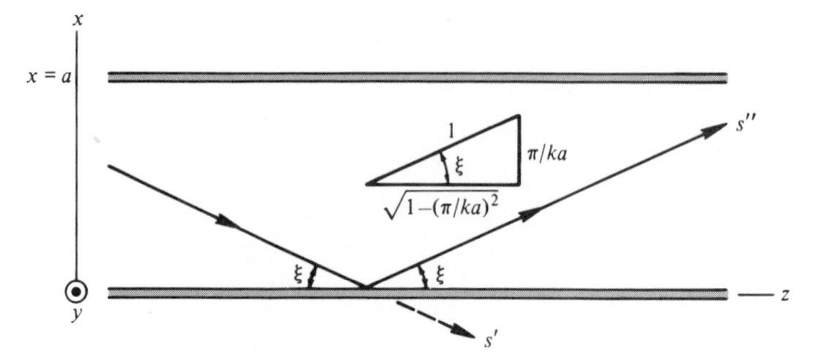

Figure 14.3. Plane wave directions s' and s'' for the TE_1 mode.

14.3 TE AND TM MODE PHASE VELOCITY

From the previous section, there remains the apparently puzzling situation of having a velocity, the phase velocity, greater than the velocity of light or intrinsic velocity. It is also puzzling that the velocity of energy flow is *less* than the *intrinsic* velocity. Let us reexamine the TE_1 field equations to gain some insight into this situation. From Table 14.1 (for $m = 1$) we have

$$E_y = E_0 \sin \frac{\pi x}{a} e^{-jk_z z} \tag{14.66}$$

which can be written

$$E_y = \frac{E_0}{2j} \left[e^{j(\pi x/a)} - e^{-j(\pi x/a)}\right] e^{-j\sqrt{k^2 - (\pi/a)^2} z} \tag{14.67}$$

This may be rearranged to

$$E_y = \frac{E_0}{2j} e^{jk(\pi x/ka)} e^{-jk\sqrt{1-(\pi/ka)^2} z} - \frac{E_0}{2j} e^{-jk(\pi x/ka)} e^{-jk\sqrt{1-(\pi/ka)^2} z} \tag{14.68}$$

or

$$E_y = \frac{E_0}{2j} e^{-jks'} - \frac{E_0}{2j} e^{-jks''} \tag{14.69}$$

where s' and s'' are directions as shown in Figure 14.3. The first term is a *uniform plane wave* traveling in the s' direction, while the second is a *uniform plane wave* traveling in the s'' direction. In fact, one plane wave is just a reflection of the other plane wave. Notice that the angle ξ is

$$\xi = \cos^{-1} \sqrt{1 - \left(\frac{\pi}{ka}\right)^2} \tag{14.70}$$

or

$$\xi = \cos^{-1} \sqrt{1 - \left(\frac{\omega_c}{\omega}\right)^2} \tag{14.71}$$

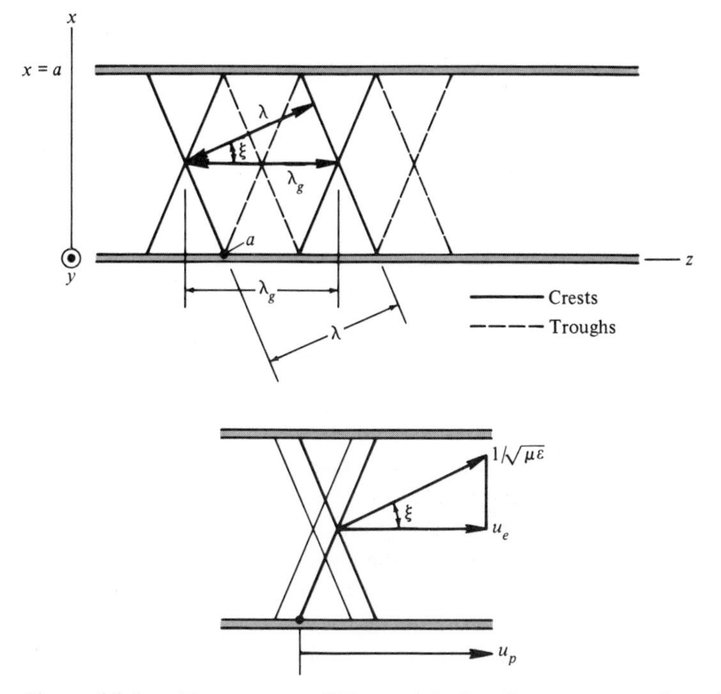

Figure 14.4. Plane waves (TE_1 mode) showing u_p, u_e, u, λ, and λ_g for the parallel plane system.

A very simple interpretation of what happens at the cutoff frequency is now apparent. When $\omega = \omega_c$, then $\xi = \pi/2$, and there is no *net* plane wave propagation in the $+z$ direction (and thus no power flow). When $\omega > \omega_c$, the component of *plane wave* velocity in the $+z$ direction is $1/\sqrt{\mu\varepsilon}$ $\cos\xi$ since the plane wave velocities in the s' and s'' directions are $1/\sqrt{\mu\varepsilon}$. This component velocity can be written

$$\frac{1}{\sqrt{\mu\varepsilon}}\cos\xi = \frac{1}{\sqrt{\mu\varepsilon}}\sqrt{1 - \left(\frac{\omega_c}{\omega}\right)^2} \tag{14.72}$$

Thus, the reason for the smaller velocity is now obvious.

What about the phase velocity? A better picture of what this phase velocity actually is may be obtained by adding lines of constant phase onto the plane waves of Figure 14.3. These are shown in Figure 14.4. The phase velocity is the velocity of a point of constant phase for the composite wave. Point a (the point of intersection of the plane wave equiphase trough and the perfectly conducting plane at $x = 0$) is such a point. The plane wave travels a distance λ while the intersection point, a, travels a distance λ_g. Geometrically, $\lambda = \lambda_g \cos\xi$. Therefore, the velocities are related by

$$\frac{1/\sqrt{\mu\varepsilon}}{u_p} = \frac{\lambda_g}{\lambda} = \frac{\lambda}{\lambda\cos\xi} \tag{14.73}$$

Again, recall the similar discussion for oblique incidence in Section 10.16. Equations 14.73 and 14.72 can be combined to give

$$u_p = \frac{1}{\sqrt{\mu\varepsilon}\sqrt{1 - (\omega_c/\omega)^2}} > \frac{1}{\sqrt{\mu\varepsilon}} \tag{14.74}$$

We have also shown that

$$\lambda_g = \frac{\lambda}{\sqrt{1 - (\omega_c/\omega)^2}} \tag{14.75}$$

It is helpful in viewing Figure 14.4 to consider the analogous picture of an ocean wave striking a straight shore line at an angle slightly off the normal. Visualize the TE_m and TM_m modes between infinite, perfectly conducting planes as being nothing more than plane waves reflected back and forth between the planes, while having a component velocity, u_e, in the direction of propagation. It is also helpful at this point to compare Figure 14.3 to Figure 10.13. Notice that the angle ξ is the complement of θ_i. We merely needed to add another perfectly conducting plane parallel to the first and located where the resultant tangential electric field was zero. The boundary conditions would then have been satisfied, and after all of the possible electromagnetic field orientations and spacings between conductors had beem examined, the TEM, TE_m, and TMs modes of this section would have resulted.

Finally, notice that the time delay encountered by a sinusoidal signal in traveling a distance l in the parallel plane waveguide, because of the zig-zag paths will be given by (see Problem 6 at the end of this chapter).

$$\boxed{\tau_D = \frac{l}{u_e}}$$

for TM or TE modes, while for the TEM mode it is given by

$$\tau_D = l\sqrt{\mu\varepsilon}$$

Thus, a signal is delayed more by the zig-zag path. Moreover, a guiding system of this type, in the TE and TM modes, cannot be distortionless even if it is lossless, because u_e (or u_p) is *frequency dependent*. The same thing is true for the more important hollow enclosed waveguide to be considered in Section 14.5.

It is interesting that the parallel plane waveguide in the TE or TM modes is a *normally* dispersive[3] system even in the absence of losses. The

[3] ($u_g < 1/\sqrt{\mu\varepsilon}$), see Section 10.18 and Figure 10.16.

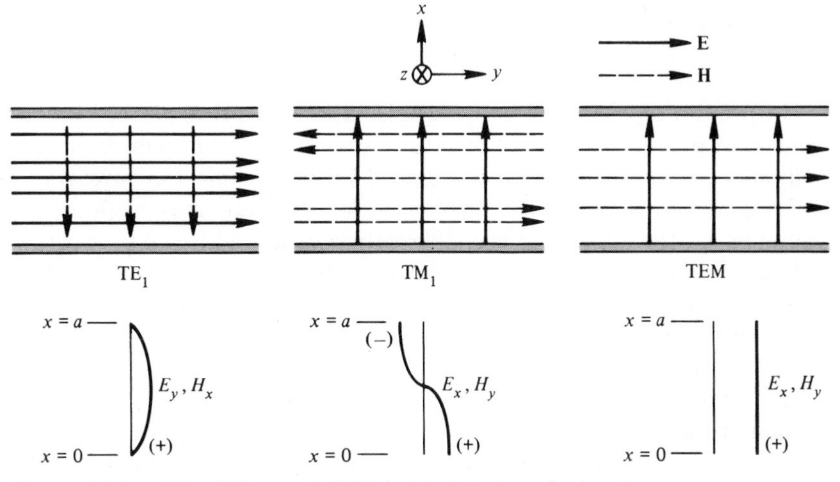

Figure 14.5. TE_1, TM_1, and TEM field plots ($z = 0$ plane).

group velocity, introduced in Section 10.7, is $u_g = d\omega/d\beta$, and it is easy to show that this becomes

$$u_g = \frac{\sqrt{1 - (\omega_c/\omega)^2}}{\sqrt{\mu\varepsilon}} = u_e$$

so that the group time delay is

$$\tau_D = \frac{l}{u_g} = \frac{l\sqrt{\mu\varepsilon}}{\sqrt{1 - (\omega_c/\omega)^2}}$$

Additional light is shed on the concept of time delay in Problems 6 and 12 at the end of the chapter, and on the concept of dispersion in Problems 13 and 14.

14.4 TE AND TM FIELD PLOTS

Electric and magnetic field plots for the TE_1 and TM_1 modes for fixed time, t, may be sketched from the field equations in Table 14.1. These, along with the TEM mode, are shown in Figure 14.5. The spacing of the field lines indicates the intensity ($\cos \pi x/a$ or $\sin \pi x/a$) of the field quantities. The finite width of a practical parallel plane system limits its usefulness for all modes except the TEM mode (stripline).

14.5 ONE-CONDUCTOR HOLLOW RECTANGULAR WAVEGUIDE

With considerable background available now, let us proceed to form an enclosed, hollow waveguide by inserting perfectly conducting planes at $y = 0$ and $y = b$ into the parallel plane system. The new configuration is shown in Figure 14.6. We want to know if the TEM, TM_m, and TE_m modes can still exist (in the hollow region) subject to the additional boundary conditions.

Inspection of Figure 14.5(a) or Table 14.1 reveals that the TEM mode *cannot* exist because E_x will not be zero at $y = 0$ and $y = b$ as the boundary conditions demand. Inspection of Figure 14.5(c) or Table 14.1 reveals that the TM_1 mode also cannot be supported for the same reason. Boundary conditions for the TE_1 mode are inherently satisfied, and so it can exist in the hollow structure of Figure 14.6. We now know that *at least* one mode of propagation can be supported by our hollow, one-conductor waveguide system. We might expect that higher order modes can exist also. The general case will be treated shortly.

It is not difficult to show that a hollow conducting system of *any* constant cross section will not support a TEM mode inside. If z is the axial direction for such a waveguide, then $E_z = H_z = 0$ for the TEM mode (by definition). Then, the magnetic field lines that form closed loops in the $x - y$ plane must encircle current. This cannot be conduction current for there is no internal conductor. It also cannot be displacement current because $D_z = E_z = 0$. Hence, the TEM mode cannot exist inside a hollow one conductor system!

General solutions for the hollow rectangular waveguide of Figure 14.6 should be easy to obtain, since we have gained much valuable experience in the technique from the simpler parallel plane (two-dimensional) system. The TM and TE modes will be treated separately.

(1) We already know that the choice $\mathbf{A} = \mathbf{a}_z A_z$, $A_x = A_y = 0$ and $\mathbf{F} = 0$ will give a TM to z mode. In fact, equation 14.19 represents the field. The boundary conditions of Figure 14.6 are

$$E_x = 0\begin{cases} y = 0 \\ y = b \end{cases}$$

$$E_y = 0\begin{cases} x = 0 \\ x = a \end{cases} \tag{14.76}$$

and

$$E_z = 0\begin{cases} x = 0, & x = a \\ y = 0, & y = b \end{cases}$$

Equations 14.21 through 14.25 are general and still apply. These equations are repeated here for convenience.

$$A_z = h(k_x x)h(k_y y)h(k_z z) \tag{14.21}$$

where

$$k_z^2 = k^2 - k_x^2 - k_y^2 \tag{14.22}$$

$$h(k_x x) = A \cos k_x x + B \sin k_x x \tag{14.23}$$

$$h(k_y y) = C \cos k_y y + D \sin k_y y \tag{14.24}$$

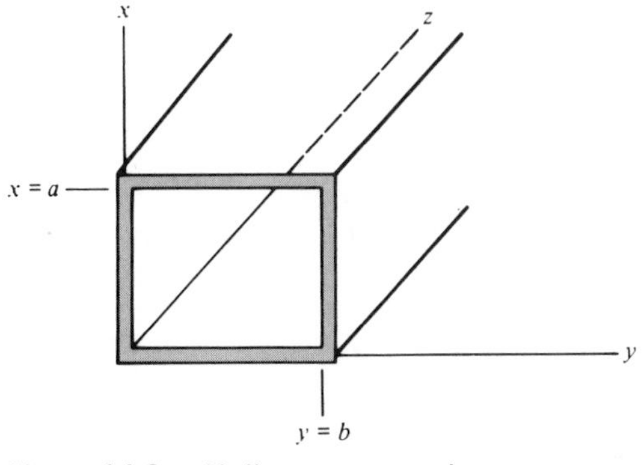

Figure 14.6. Hollow, one-conductor, rectangular wave guide.

and

$$h(k_z z) = E \cos k_z z + F \sin k_z z \tag{14.25}$$

We immediately choose $F = -jE$ so that

$$h(k_z z) = Ee^{-jk_z z}$$

Boundary conditions on E_z will be satisfied if they are satisfied for A_z itself. This requires that $A = C = 0$,

$$k_x = \frac{m\pi}{a}, \qquad m = 1, 2, 3, \ldots \tag{14.77}$$

and

$$k_y = \frac{n\pi}{b}, \qquad n = 1, 2, 3, \ldots \tag{14.78}$$

The use of *either* m or $n = 0$ gives a trivial solution. Thus,

$$A_z = BDE \sin k_x x \sin k_y y \, e^{-jk_z z} \tag{14.79}$$

and boundary conditions on E_x and E_y are also satisfied. It is convenient to let $BDE = 1$, so that

$$A_z = \sin k_x x \sin k_y y \, e^{-jk_z z} \tag{14.80}$$

Substituting equations 14.77 and 14.78 into 14.22 gives

$$k_z^2 = k^2 - \left(\frac{m\pi}{a}\right)^2 - \left(\frac{n\pi}{b}\right)^2 \tag{14.81}$$

The field is given by substitution of equation 14.80 into 14.19 and will be presented in tabular form.

Following the procedure used for the parallel plane system, we have

$$k_z = \begin{cases} \sqrt{k^2 - \left(\frac{m\pi}{a}\right)^2 - \left(\frac{n\pi}{b}\right)^2} = \beta, & k^2 > \left(\frac{m\pi}{a}\right)^2 + \left(\frac{n\pi}{b}\right)^2 \\ \sqrt{\left(\frac{m\pi}{a}\right)^2 + \left(\frac{n\pi}{b}\right)^2 - k^2} = -j\alpha, & k^2 < \left(\frac{m\pi}{a}\right)^2 + \left(\frac{n\pi}{b}\right)^2 \end{cases} \tag{14.82}$$

The transition or cutoff frequency occurs for $k_z = 0$, or

$$\boxed{\omega_c = \frac{\pi}{\sqrt{\mu\varepsilon}} \sqrt{\left(\frac{m}{a}\right)^2 + \left(\frac{n}{b}\right)^2}} \tag{14.83}$$

Substituting equation 14.83 into 14.82 gives

$$k_z = \begin{cases} k\sqrt{1 - \left(\frac{\omega_c}{\omega}\right)^2} = \beta, & \omega > \omega_c \\ -jk_c\sqrt{1 - \left(\frac{\omega}{\omega_c}\right)^2} = -j\alpha, & \omega < \omega_c \end{cases} \tag{14.84}$$

(2) The choice of $\mathbf{F} = \mathbf{a}_z F_z$, $F_x = F_y = 0$, and $\mathbf{A} = 0$ gives a TE to z mode. For reasons already given, the solution to F_z must be of the form

$$F_z = h(k_x x)h(k_y y)e^{-jk_z z} \tag{14.85}$$

The boundary conditions are, of course, the same as those given by equation 14.76, while the TE to z field is given by equation 14.36. Boundary conditions on E_z are automatically satisfied since $E_z = 0$ for any TE to z mode. Inspection of equation 14.36 reveals that boundary conditions on E_x and E_y are satisfied if $h(k_x x)$ and $h(k_y y)$ are such that

$$F_z = \cos\frac{m\pi x}{a}\cos\frac{n\pi y}{b}\,e^{-jk_z z} \tag{14.86}$$

In this case, $m = 0, 1, 2, 3, \ldots$, and $n = 0, 1, 2, 3, \ldots$, but *both* m and n cannot be zero, or a trivial solution will result.

The separation equation is the same as equation 14.22 since the eigenvalues (k_x, k_y, and k_z) are the same. For this reason the cutoff frequencies are again given by equation 14.83. The field is obtained by substituting equation 14.86 into 14.36. The general results for both the TE to z and TM to z mode are shown in Table 14.2. The peak values of E_x, in the TM case, and E_y, in the TE case, have been normalized to E_0 for convenience.

Table 14.2 RECTANGULAR WAVEGUIDE FIELD EQUATIONS WITH PROPAGATION IN THE $+z$ DIRECTION,

$$k_z = \sqrt{k^2 - (m\pi/a)^2 - (n\pi/b)^2},\ k = \omega\sqrt{\mu\varepsilon}$$

TM TO z	TE TO z
$E_x = E_0 \cos\frac{m\pi x}{a} \sin\frac{n\pi y}{b} e^{-jk_z z}$	$E_x = -\frac{na}{mb} E_0 \cos\frac{m\pi x}{a} \sin\frac{n\pi y}{b} e^{-jk_z z}$
$E_y = \frac{na}{mb} E_0 \sin\frac{m\pi x}{a} \cos\frac{n\pi y}{b} e^{-jk_z z}$	$E_y = E_0 \sin\frac{m\pi x}{a} \cos\frac{n\pi y}{b} e^{-jk_z z}$
$E_z = ja\frac{k^2 - k_z^2}{m\pi k_z} E_0 \sin\frac{m\pi x}{a} \sin\frac{n\pi y}{b} e^{-jk_z z}$	$E_z \equiv 0$ $m = 0, 1, 2, \ldots$; $n = 0, 1, 2, \ldots$; m, n both not zero
$H_x = -\frac{na}{mb}\frac{\omega\varepsilon}{k_z} E_0 \sin\frac{m\pi x}{a} \cos\frac{n\pi y}{b} e^{-jk_z z}$	$H_x = -\frac{k_z}{\omega\mu} E_0 \sin\frac{m\pi x}{a} \cos\frac{n\pi y}{b} e^{-jk_z z}$
$H_y = \frac{\omega\varepsilon}{k_z} E_0 \cos\frac{m\pi x}{a} \sin\frac{n\pi y}{b} e^{-jk_z z}$	$H_y = -\frac{na}{mb}\frac{k_z}{\omega\mu} E_0 \cos\frac{m\pi x}{a} \sin\frac{n\pi y}{b} e^{-jk_z z}$
$H_z \equiv 0$ $m = 1, 2, \ldots$; $n = 1, 2, \ldots$	$H_z = ja\frac{k^2 - k_z^2}{m\pi\omega\mu} E_0 \cos\frac{m\pi x}{a} \cos\frac{n\pi y}{b} e^{-jk_z z}$

The equations for the wave impedance, phase velocity, energy velocity, group velocity, and guide wavelength are identical in form to those for the parallel plane system (TE and TM modes). $\langle S_Z \rangle$ is now more complicated, but may always be determined from

$$\langle S_z \rangle = \tfrac{1}{2}\,\mathrm{Re}\,\{\mathbf{E} \times \mathbf{H}^*\}_z \tag{14.87}$$

Plots of γ, η^+, u_p, and u_e will be identical to those of Figure 14.2.

Modes are designated TE_{mn} or TM_{mn}. The dominant mode (lowest ω_c) for the rectangular waveguide is the TE_{10} for $a > b$, or the TE_{01} for $b > a$. This is determined by inspection of equation 14.83. Field plots for the TE_{10}, TE_{20}, and TM_{11} modes are shown in Figure 14.7 using the field equations in Table 14.2. These field distributions are obtained from the time domain forms at $t = 0$, and are, therefore, like "snapshots." Higher order field plots may be obtained in the same manner, if desired.

14.6 THE DOMINANT TE_{10} MODE

We now assume that $a > b$, so that the TE_{10} mode is dominant. Waveguides are normally operated in this mode. The field is determined from Table 14.2

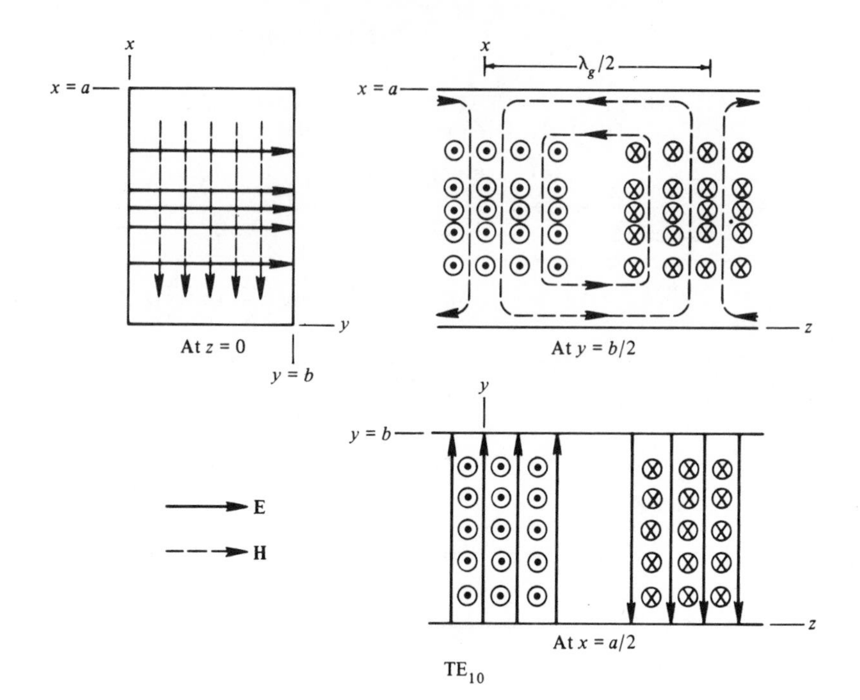

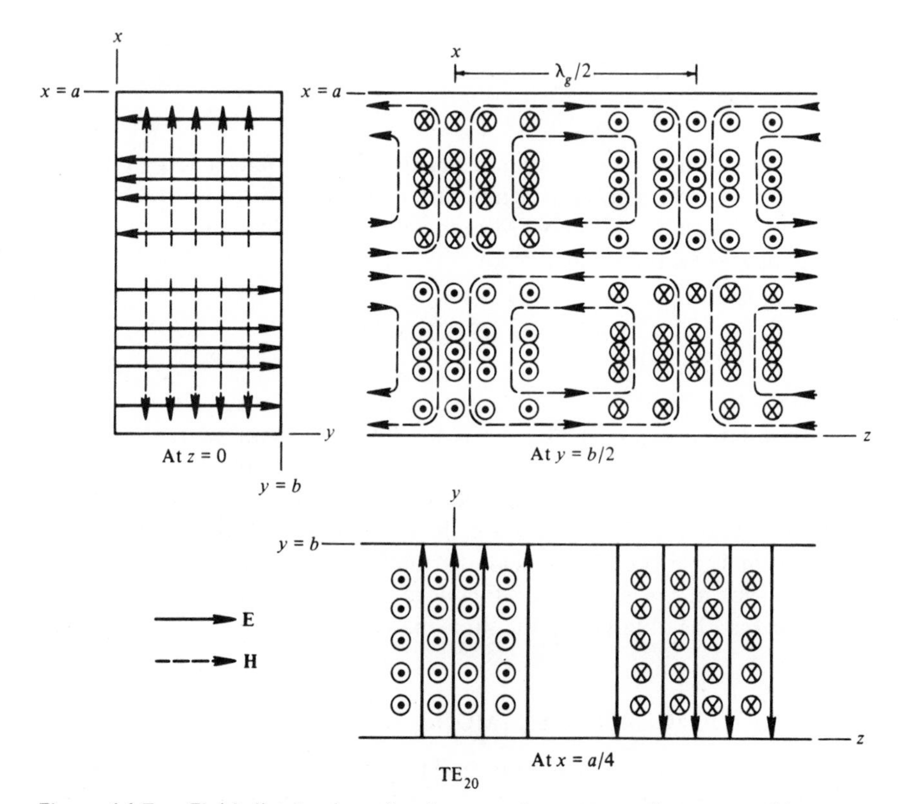

Figure 14.7. Field distributions for three modes, rectangular waveguide.

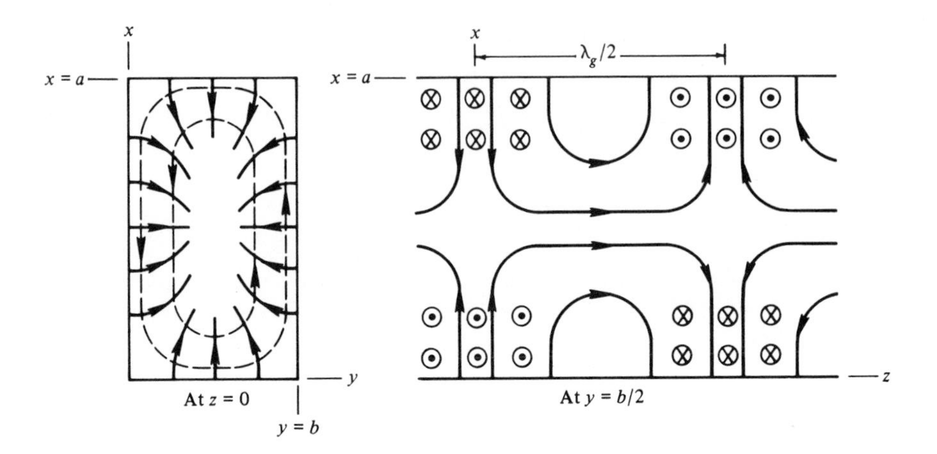

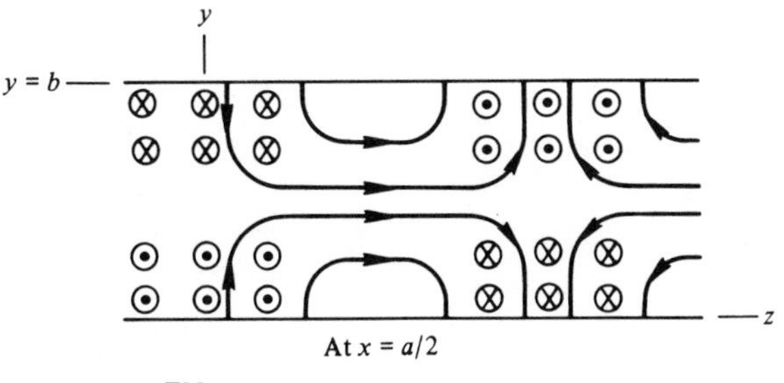

TM_{11}

Figure 14.7. (*Continued*)

with $m = 1$ and $n = 0$. It is

$$
\boxed{
\begin{aligned}
E_y &= E_0 \sin \frac{\pi x}{a} e^{-jk_z z} \\
H_x &= -\frac{k_z E_0}{\omega\mu} \sin \frac{\pi x}{a} e^{-jk_z z} \\
H_z &= j \frac{E_0 \pi}{\omega\mu a} \cos \frac{\pi x}{a} e^{-jk_z z}
\end{aligned}
}
\tag{14.88}
$$

The cutoff frequency is given by

$$
\omega_c = \frac{\pi}{a\sqrt{\mu\varepsilon}} \tag{14.89}
$$

while the propagation constant is

$$\gamma = jk_z = \begin{cases} j\beta = jk\sqrt{1 - \left(\dfrac{\omega_c}{\omega}\right)^2}, & \omega > \omega_c \\ \alpha = k_c\sqrt{1 - \left(\dfrac{\omega}{\omega_c}\right)^2}, & \omega < \omega_c \end{cases} \tag{14.90}$$

Equations 14.88 are identical to those for the TE_1 parallel plane mode listed in Table 14.1. The time average power density is found from equation 14.87 as

$$\langle S_z \rangle = \frac{E_0^2 \sin^2 \dfrac{\pi x}{a}}{2\eta^+} \tag{14.91}$$

In the same way,

$$\eta^+ = -\frac{E_y}{H_x} = \frac{\eta}{\sqrt{1 - (\omega_c/\omega)^2}} \tag{14.92}$$

$$u_p = \frac{1}{\sqrt{\mu\varepsilon}\sqrt{1 - (\omega_c/\omega)^2}} \tag{14.93}$$

$$u_e = \frac{\sqrt{1 - (\omega_c/\omega)^2}}{\sqrt{\mu\varepsilon}} = u_g \tag{14.94}$$

and

$$\lambda_g = \frac{\lambda}{\sqrt{1 - (\omega_c/\omega)^2}} \tag{14.95}$$

Previous remarks concerning the composition of the fields in the TE_1 parallel plane waveguide apply here also. That is, propagation occurs by means of plane waves taking a zig-zag path along the waveguide.

Before proceeding with a discussion of the properties of the TE_{10} mode, it is worthwhile at this point to digress and obtain a solution for the field equations for this mode in the *simplest* possible manner without the use of vector potential. We will allow ourselves to utilize the results from the oblique incidence case studied in Chapter 10. The geometry is that of Figure 14.6. We are looking for propagation in the $+z$ direction, and from the oblique incidence problem, we know that the phase constant will not be $k = \omega\sqrt{\mu\varepsilon}$. Therefore, we assume a z variation of $e^{-jk_z z}$. Since this is a TE mode, **E** is entirely transverse to the direction of propagation, and $E_z \equiv 0$. In Figure 10.13, $\mathbf{E} = \mathbf{a}_x E_x$, and the reflection takes place on the $z = 0$ plane. In Figure 14.6, however, we wish for the reflection to take place on the $x = 0$ and the $x = a$ planes, with no variation in the y direction. That is, we require that $\partial/\partial y \equiv 0$. The boundary conditions for the tangential components of **E** are

$$E_x = 0\begin{cases} y = 0 \\ y = b \end{cases} \quad \text{and} \quad E_y = 0\begin{cases} x = 0 \\ x = a \end{cases}$$

Maxwell's equations are

$$\mathbf{E} = \frac{1}{j\omega\varepsilon}\nabla \times \mathbf{H} \quad \text{and} \quad \mathbf{H} = -\frac{1}{j\omega\mu}\nabla \times \mathbf{E}$$

Expanding Maxwell's equations in Cartesian coordinates,

$$E_x = \frac{1}{j\omega\varepsilon}\left(\frac{\partial H_z}{\partial y} - \frac{\partial H_y}{\partial z}\right) \qquad H_x = -\frac{1}{j\omega\mu}\left(\frac{\partial E_z}{\partial y} - \frac{\partial E_y}{\partial z}\right)$$

$$E_y = \frac{1}{j\omega\varepsilon}\left(\frac{\partial H_x}{\partial z} - \frac{\partial H_z}{\partial x}\right) \qquad H_y = -\frac{1}{j\omega\mu}\left(\frac{\partial E_x}{\partial z} - \frac{\partial E_z}{\partial x}\right)$$

$$E_z = \frac{1}{j\omega\varepsilon}\left(\frac{\partial H_y}{\partial x} - \frac{\partial H_x}{\partial y}\right) \qquad H_z = -\frac{1}{j\omega\mu}\left(\frac{\partial E_y}{\partial x} - \frac{\partial E_x}{\partial y}\right)$$

Because of the fact that $E_z \equiv 0$ and $\partial/\partial y \equiv 0$, the preceding set of equations reduces to

$$E_x = -\frac{1}{j\omega\varepsilon}\frac{\partial H_y}{\partial z} \quad \text{(a)} \qquad H_x = \frac{1}{j\omega\mu}\frac{\partial E_y}{\partial z} \quad \text{(d)}$$

$$E_y = \frac{1}{j\omega\varepsilon}\left(\frac{\partial H_x}{\partial z} - \frac{\partial H_z}{\partial x}\right) \quad \text{(b)} \qquad H_y = -\frac{1}{j\omega\mu}\frac{\partial E_x}{\partial z} \quad \text{(e)}$$

$$0 = \frac{1}{j\omega\varepsilon}\frac{\partial H_y}{\partial x} \quad \text{(c)} \qquad H_z = -\frac{1}{j\omega\mu}\frac{\partial E_y}{\partial x} \quad \text{(f)}$$

The simplest solution to equation (c) is $H_y \equiv 0$, and, with this, the simplest solution to equation (e) is $E_x \equiv 0$. In this case the boundary conditions on E_x are automatically satisfied. Equation (a) is automatically satisfied also. We are left with

$$E_y = \frac{1}{j\omega\varepsilon}\left(\frac{\partial H_x}{\partial z} - \frac{\partial H_z}{\partial x}\right) \tag{g}$$

$$H_x = \frac{1}{j\omega\mu}\frac{\partial E_y}{\partial z} \tag{h}$$

$$H_z = -\frac{1}{j\omega\mu}\frac{\partial E_y}{\partial x} \tag{i}$$

Substituting equations (h) and (i) into (g) gives

$$E_y = -\frac{1}{k^2}\left(\frac{\partial^2 E_y}{\partial z^2} + \frac{\partial^2 E_y}{\partial x^2}\right) \tag{j}$$

We have already chosen the z variation to be of the form $e^{-jk_z z}$. Comparing the first and second terms on the right side of equation (j) indicates that the x variation should be a form which, when differentiated twice with respect to x, gives the same form. A sine or cosine form is in order, and when the boundary conditions on E_y are recalled, the cosine is eliminated. We are led to try[4] a solution of the form

$$E_y = E_0 \sin k_x x e^{-jk_z z} \tag{k}$$

Substituting this into equation (j) and reducing gives

$$1 = \frac{1}{k^2}(k_z^2 + k_x^2)$$

or

$$k^2 = k_z^2 + k_x^2 \tag{l}$$

The boundary conditions on E_y [equation (k)] are satisfied if $k_x = \pi/a$, $2\pi/a$, $3\pi/a, \ldots$. Propagation obviously ceases, according to equation (k), if $k_z = 0$, and this occurs when $k_x = k = k_c$ according to equation (l). Thus,

$$k_c = \omega_c\sqrt{\mu\varepsilon} = \frac{\pi}{a}, \frac{2\pi}{a}, \frac{3\pi}{a}, \ldots$$

The dominant mode has the lowest cutoff frequency, ω_c, so for this case

$$\omega_c = \frac{\pi}{a\sqrt{\mu\varepsilon}} \tag{14.89}$$

Equation (l) gives $k_z = \pm k\sqrt{1 - (\omega_c/\omega)^2}$. The choice of sign is important. For $\omega > \omega_c$, k_z must be positive to give propagation in the positive z direction. On the other hand, it is also true that $k_z = \pm k_c\sqrt{(\omega/\omega_c)^2 - 1} = \pm jk_c\sqrt{1 - (\omega/\omega_c)^2}$, and for $\omega < \omega_c$ we must choose the sign to give a damped or decreasing field, an exponentially increasing field being impossible. Summarizing,

$$jk_z = \begin{cases} j\beta = jk\sqrt{1 - \left(\dfrac{\omega_c}{\omega}\right)^2}, & \omega > \omega_c \\ \alpha = k_c\sqrt{1 - \left(\dfrac{\omega}{\omega_c}\right)^2}, & \omega < \omega_c \end{cases} \tag{14.90}$$

Substituting equation (k) into equations (h) and (i), using $k_x = \pi/a$, we obtain the field equations for a mode with *one variation* with x ($\sin \pi x/a$,

[4] An educated guess.

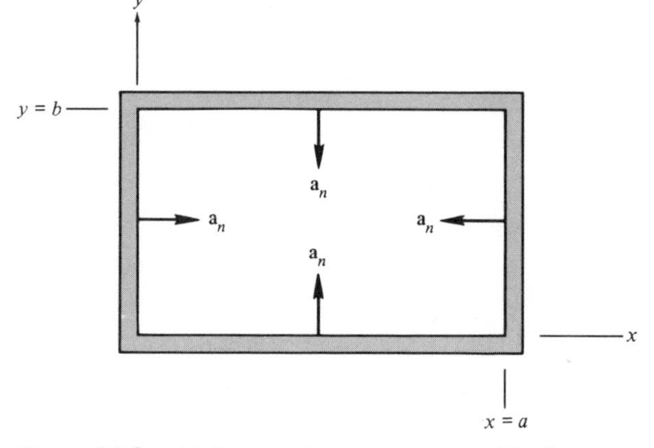

Figure 14.8. Unit normal vectors, $\mathbf{a}_n$, inside the rectangular guide.

$0 \leq x \leq a$) and *no variation* with y, called the TE_{10} mode.

$$E_y = E_0 \sin \frac{\pi x}{a} e^{-jk_z z}$$

$$H_x = -\frac{k_z E_0}{\omega\mu} \sin \frac{\pi x}{a} e^{-jk_z z} \qquad (14.88)$$

$$H_z = j\frac{E_0 \pi}{\omega\mu a} \cos \frac{\pi x}{a} e^{-jk_z z}$$

If we now calculate $\langle S_z \rangle, \eta^+, u_p, u_e$, and λ_g, we will obtain equations identical with equations 14.91 through 14.95, respectively. Thus, this simpler approach gives the identical results for the TE_{10} mode as the more general approach used earlier.

It was found in the case of the coaxial line that an axial slot could be cut in the outer conductor for probing the line. Can an axial slot be cut in the rectangular waveguide to examine the TE_{10} field? To answer this question, we must examine the current flow on the inside of the waveguide walls. Since the waveguide has been assumed lossless (the walls are perfect conductors), this current will be a surface current according to

$$\mathbf{J}_s = \mathbf{a}_n \times \mathbf{H}\big|_{\text{on walls}} \qquad (14.96)$$

where $\mathbf{a}_n$ is a unit normal vector as shown in Figure 14.8. From equations 14.88 and 14.96,

$$J_{sz} = -H_x\big|_{y=b} = \frac{k_z E_0}{\omega\mu} \sin \frac{\pi x}{a} e^{-jk_z z} \qquad (14.97)$$

and

$$J_{sx} = H_z\big|_{y=b} = j\frac{E_0 \pi}{\omega\mu a} \cos \frac{\pi x}{a} e^{-jk_z z} \qquad (14.98)$$

In particular, at the top of the waveguide in the center ($x = a/2$, $y = b$)

$$J_{sz} = \frac{k_z E_0}{\omega\mu} e^{-jk_z z} \tag{14.99}$$

and

$$J_{sx} = 0 \tag{14.100}$$

Therefore, a narrow axial slot may be cut in the top (or the bottom) of the waveguide at $x = a/2$ without appreciably disturbing the flow of current (or the field).

A slotted waveguide (TE_{10} mode) may be probed in much the same way as the slotted coaxial line. If a standing wave is set up in the slotted section, the distance between minima *is not* $\lambda/2$, but is $\lambda_g/2$ (see Figure 14.4). Measurement of λ_g and the dimension a are sufficient to determine the operating frequency by simply using equations 14.89 and 14.95. An example at this point is useful.

EXAMPLE 1

Suppose that an air filled x-band waveguide ($a = 0.9$ inches, $b = 0.4$ inches) is probed with a slotted section, and it is found that the distance between minima is 2 cm. The minimum position does not shift when the actual load is replaced by a shorting plane (short circuit). The standing wave ratio is 2. What is the load impedance and what is the operating frequency?

$$\frac{\lambda_g}{2} = 2 \times 10^{-2} \quad \text{or} \quad \lambda_g = 4 \times 10^{-2} \quad \text{(m)}$$

$$a = 0.9(2.54 \times 10^{-2}) = 2.286 \times 10^{-2} \quad \text{(m)}$$

From equation 14.89,

$$\omega_c = \frac{3\pi \times 10^8}{2.286 \times 10^{-2}} = 41.5 \times 10^9$$

or

$$f_c = 6.56 \quad \text{(GHz)}$$

From equation 14.95,

$$\lambda_g = \frac{\lambda}{\sqrt{1 - (\omega_c/\omega)^2}} = \frac{1}{f\sqrt{\mu\varepsilon}\sqrt{1 - (\omega_c/\omega)^2}}$$

or solving for f,

$$f = \sqrt{f_c^2 + \frac{1}{\mu\varepsilon(\lambda_g)^2}} = \sqrt{(6.56)^2 + (7.5)^2} \times 10^9$$

or

$$f = 9.96 \quad (\text{GHz})$$

The *normalized* load impedance is on the Smith chart circle corresponding to VSWR = 2 as shown in Figure 14.9. Since the minimum position

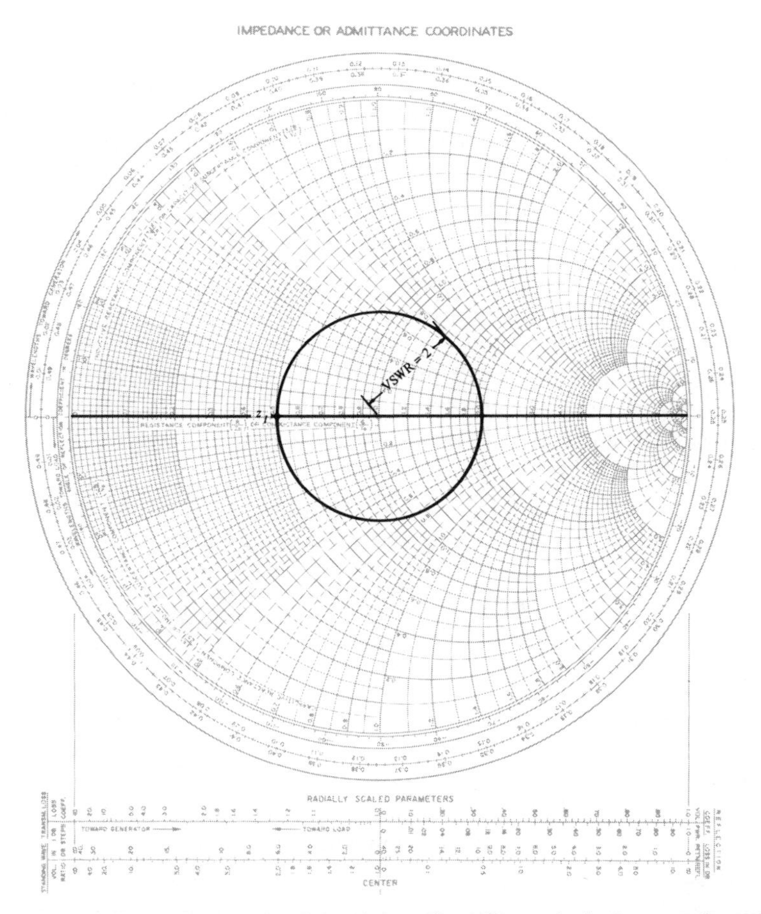

Figure 14.9. Waveguide impedance data, Example 1. *Source*: Copyright © 1949 by Kay Electric Company, Pine Brook, N.J. Renewal copyright © 1976 by P. H. Smith, New Providence, N.J.

did not shift when the short was replaced by the actual load, we have the *equivalent* of a voltage minimum at the load. This means the impedance at the load is minimum, or $z_l = \frac{1}{2}$. This is a normalized impedance and must be related to a waveguide characteristic resistance. The measurements in the slotted section were made at $x = a/2$ and, in this case,

$$\langle S_z \rangle = \frac{E_0^2}{2\eta^+}$$

from equation 14.91. A logical decision then is to call η^+ the waveguide impedance. That is,

$$(Z_0)_{\mathrm{TE}_{10}} = \eta^+ = \frac{\eta}{\sqrt{1 - (\omega_c/\omega)^2}}$$

or

$$(Z_0)_{\mathrm{TE}_{10}} = \frac{377}{\sqrt{1 - (6.56/9.96)^2}} = \frac{377}{0.752} = 501 \quad (\Omega)$$

Therefore, the equivalent load impedance is

$$Z_L = z_L (Z_0)_{\mathrm{TE}_{10}} = \tfrac{1}{2}(501) = 250.5 \quad (\Omega)$$

EXAMPLE 2

Suppose we attempt to operate this waveguide at a frequency of 5 GHz. What happens? Since the cutoff frequency is 6.56 GHz, an evanescent wave results. Using equation 14.90,

$$\alpha = k_c \sqrt{1 - \left(\frac{\omega}{\omega_c}\right)^2} = \omega_c \sqrt{\mu\varepsilon} \sqrt{1 - \left(\frac{\omega}{\omega_c}\right)^2}$$

or

$$\alpha = \frac{(41.2 \times 10^9)}{3 \times 10^8} \sqrt{1 - \left(\frac{5}{6.56}\right)^2} = (137.4)(0.65)$$

or

$$\alpha = 89.4 \quad (\mathrm{Np/m})$$

or

$$\alpha = 775 \quad (\mathrm{dB/m})$$

This is indeed a very large attenuation constant!

EXAMPLE 3

What are the cutoff frequencies of the higher order modes? From equation 14.83,

$$f_c = \frac{1}{2\sqrt{\mu\varepsilon}} \sqrt{\left(\frac{m}{a}\right)^2 + \left(\frac{n}{b}\right)^2}$$

Then, for the TE_{01} mode

$$f_c = \frac{1.5 \times 10^8}{b} = \frac{1.5 \times 10^8}{(0.4)(2.54 \times 10^{-2})} = 14.72 \quad \text{(GHz)}$$

For the TE_{20} mode

$$f_c = \frac{3 \times 10^8}{a} = \frac{3 \times 10^8}{(0.9)(2.54 \times 10^{-2})} = 13.12 \quad \text{(GHz)}$$

Other higher order modes will have higher cutoff frequencies. A good operating frequency for the dominant mode (TE_{10}) would lie above its own cutoff frequency and below the next highest cutoff frequency, or

$$6.56 \times 10^9 < f < 13.12 \times 10^9$$

Then 10 GHz is a good operating frequency for this waveguide. We (theoretically) have a two-to-one frequency range of operation.

It is interesting to plot the currents on the waveguide walls for the TE_{10} mode. Utilizing Figure 14.8 and equations 14.97 and 14.98,

$$J_{sz} = \frac{k_z E_0}{\omega\mu} \sin\frac{\pi x}{a} e^{-jk_z z} \tag{14.97}$$

and

$$J_{sx} = j\frac{E_0 \pi}{\omega\mu a} \cos\frac{\pi x}{a} e^{-jk_z z} \tag{14.98}$$

on the top of the waveguide at $y = b$. On the sidewall at $x = a$, we have

$$J_{sy} = H_z\big|_{x=a} = -j\frac{E_0 \pi}{\omega\mu a} e^{-jk_z z}$$

The surface current is plotted in Figure 14.10. What appear to be charge sinks and sources, $\lambda_g/2$ apart, are really not regions of current discontinuity. The current is continuous by means of the displacement current, $j\omega\varepsilon E_y$, inside the waveguide according to the continuity equation

$$\nabla \cdot \mathbf{J} = -j\omega\rho_v \tag{14.101}$$

which for surface current densities becomes

$$\nabla \cdot \mathbf{J}_s = -j\omega\rho_s \tag{14.102}$$

The surface charge density is ρ_s. The sidewall current obviously flows without divergence. It is relatively simple to verify that equation 14.102 holds

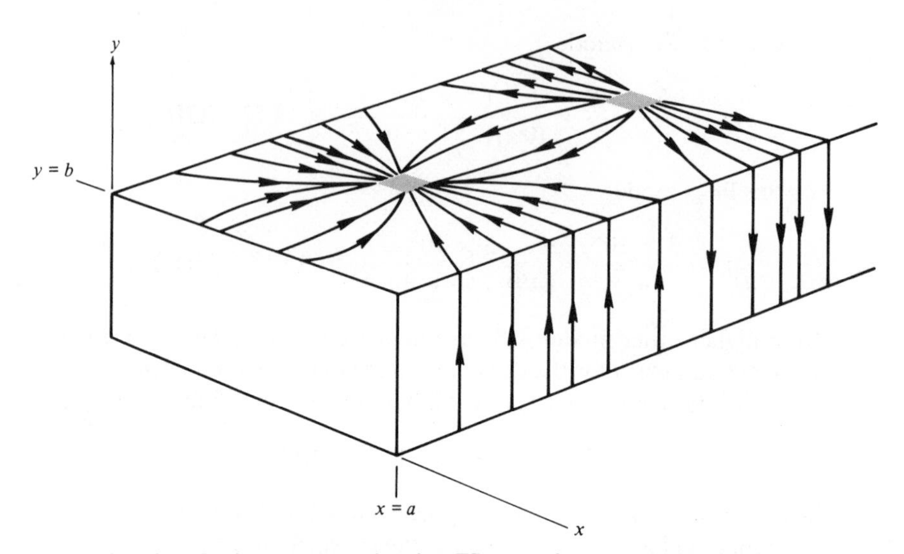

Figure 14.10. Surface current density, TE_{10} mode, rectangular guide.

for the top (or bottom) of the waveguide. The left side of equation 14.102 may be written (for the top)

$$\begin{aligned}\nabla \cdot \mathbf{J}_s &= \nabla \cdot (\mathbf{a}_n \times \mathbf{H}) = \nabla \cdot [-\mathbf{a}_y \times (\mathbf{a}_x H_x + \mathbf{a}_z H_z)] \\ &= \nabla \cdot [\mathbf{a}_z H_x - \mathbf{a}_x H_z] \\ &= -\frac{\partial H_z}{\partial x} + \frac{\partial H_x}{\partial z}\end{aligned}$$

while the right side may be written (for the top)

$$-j\omega\rho_s = -j\omega D_n = j\omega D_y$$

Thus, we have

$$\left(\frac{\partial H_x}{\partial z} - \frac{\partial H_z}{\partial x}\right)_{y=b} = j\omega D_y\big|_{y=b} \tag{14.103}$$

or

$$|(\nabla \times \mathbf{H})_y|_{y=b} = j\omega\varepsilon E_y|_{y=b} \tag{14.104}$$

Equation 14.104 is just the y component of one of Maxwell's equations (evaluated at $y = b$), and this was satisfied initially!

Other waveguide geometries are possible, of course. For example, if it is necessary for mechanical reasons to have a rotatable joint, a circular waveguide is called for. The general characteristics of a circular waveguide are very similar to those for the rectangular waveguide. It will possess cutoff frequencies, and so forth. It is different mathematically in that higher functions, called Bessel functions are encountered. For this reason an analysis of the circular guide is not given here, but is presented in Appendix H for the interested reader.

14.7 LOSSLESS CAVITY RESONATORS

Energy storage elements are required for engineering purposes in the microwave range of frequencies. It was found in Chapter 13 that certain resonant lengths of transmission line behave as resonant circuits at certain resonant frequencies. We spent a good deal of time and effort analyzing the resonant quarter-wave coaxial cavity. Another example might be a parallel plane system shorted at both ends. We can visualize the two shorting ends as capacitor plates and the planes as inductances. In extending the resonant frequency upwards, we would naturally attempt to reduce the inductance by paralleling more conductors between the ends (like a cage). Ultimately, this leads to a closed box or cavity. It is this geometry, called a rectangular cavity, that we will investigate in this section.

The geometry of the rectangular cavity is shown in Figure 14.11. Notice that this cavity is merely a section of rectangular waveguide shorted at both ends. Instead of a traveling wave in the axial (z) direction, we expect two traveling waves in *opposite* directions, or a *standing wave.*

(1) For a TM to z mode, equations 14.21 and 14.22 are still valid. They are

$$A_z(x, y, z) = h(k_x x)h(k_y y)h(k_z z) \tag{14.105}$$

and

$$k_z^2 = k^2 - k_x^2 - k_y^2 \tag{14.106}$$

The field is given by equation 14.19. As before, the boundary conditions at $x = 0, a$ and $y = 0, b$ are satisfied by the choice

$$h(k_x x) = \sin \frac{m\pi x}{a}, \qquad m = 1, 2, 3, \ldots \tag{14.107}$$

and

$$h(k_y y) = \sin \frac{n\pi y}{b}, \qquad n = 1, 2, 3, \ldots \tag{14.108}$$

We now have the additional boundary condition

$$E_x = E_y = 0 \begin{cases} z = 0 \\ z = c \end{cases} \tag{14.109}$$

Inspection of equation 14.19 reveals that equation 14.109 is satisfied if

$$h(k_z z) = \cos \frac{p\pi z}{c}, \qquad p = 0, 1, 2 \ldots \tag{14.110}$$

Therefore, equation 14.105 becomes

$$\boxed{A_z = \sin \frac{m\pi x}{a} \sin \frac{n\pi y}{b} \cos \frac{p\pi z}{c}} \tag{14.111}$$

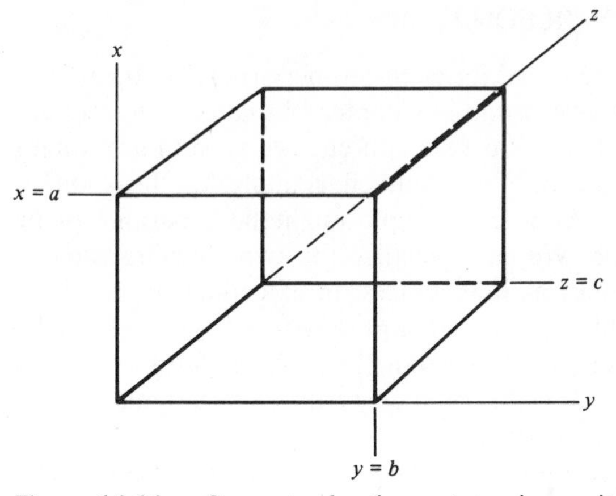

Figure 14.11. Geometry for the rectangular cavity.

The separation equation becomes

$$\left(\frac{p\pi}{c}\right)^2 = k^2 - \left(\frac{m\pi}{a}\right)^2 - \left(\frac{n\pi}{b}\right)^2 \tag{14.112}$$

which can *only* be satisfied now at *discrete* frequencies called the *resonant* frequencies. This is a typical characteristic of lossless, infinite Q, systems. Solving equation 14.112 for the resonant frequencies ($k = \omega\sqrt{\mu\varepsilon}$),

$$\boxed{f_r = \frac{1}{2\sqrt{\mu\varepsilon}}\sqrt{\left(\frac{m}{a}\right)^2 + \left(\frac{n}{b}\right)^2 + \left(\frac{p}{c}\right)^2}} \tag{14.113}$$

The dominant TM_{mnp} mode is the TM_{110}.

(2) For a TE to z mode, we follow the same procedure and find that

$$\boxed{F_z = \cos\frac{m\pi x}{a}\cos\frac{n\pi y}{b}\sin\frac{p\pi z}{c}} \tag{14.114}$$

is the potential required to satisfy both the wave equation and boundary conditions. In this case, $m = 0, 1, 2, \ldots, n = 0, 1, 2, \ldots$, and $p = 1, 2, 3, \ldots$, but *both* m and n cannot be zero. The field is given by equation 14.36, and the separation equation is identical to equation 14.112. Then, the resonant frequencies are the same and are given by equation 14.113. If $a < b < c$, the dominant TE_{mnp} mode is the TE_{011} mode. It is easy to show that more than one mode may have the same resonant frequency. If this occurs, one mode will usually dominate the other because of the manner in which the cavity is excited. For example, an electric loop may excite one mode while being oriented in such a position as to make it impossible to couple to the other mode.

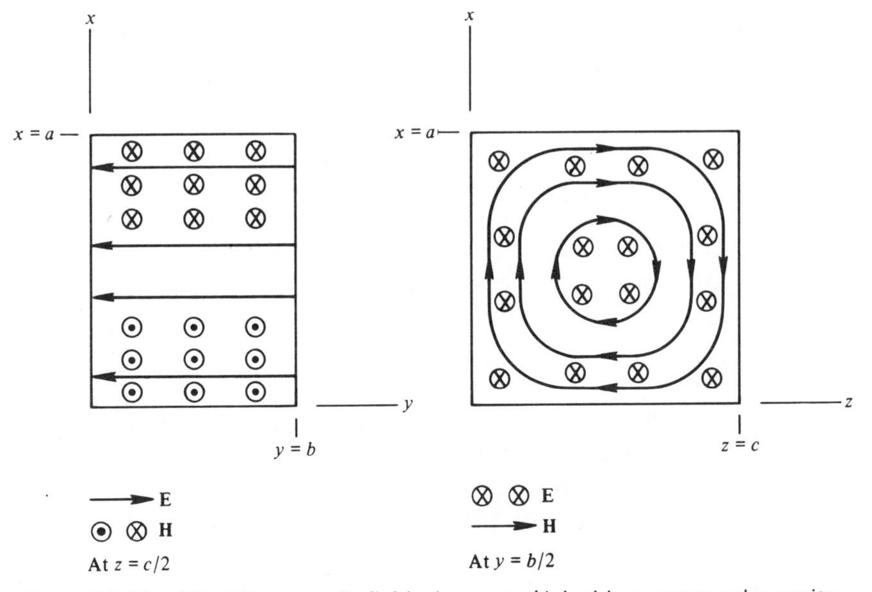

Figure 14.12. The TE_{101} mode fields ($a = c > b$) inside a rectangular cavity.

For $a = c > b$, the dominant mode is the TE_{101} whose field distribution is shown in Figure 14.12.

If we need a tunable cavity (for measurement purposes, for example), then the circular cavity should be considered because its axial dimension can easily be altered by means of threads on one end. The circular cavity is analyzed in Appendix H.

14.8 CONCLUDING REMARKS

In this chapter, we have examined the propagation characteristics of the lossless hollow rectangular waveguide. The technique that was followed employs the vector potential in a very general way. Unfortunately, a large number of rather complicated field equations resulted. It was not surprising to find that a simpler technique easily produced similar results for the TE_{10} dominant mode, but with less *general* usefulness. The more important characteristics of hollow waveguides such as the cutoff frequency, phase velocity, velocity of energy flow, and so on, should be carefully considered. At this introductory level, the details of the derivation of the field equations are not so important.

Most of the characteristics of the hollow waveguide are similar to those of two-wire systems, with the notable exception that a TEM mode can never propagate inside a hollow one-conductor system. Many modes can be generated unintentionally in a waveguide around the source (a probe, for example), or around some other discontinuity. In a uniform section of the waveguide, only those modes "above cutoff" can propagate energy. It is worth noting that many systems in nature can support "normal modes."[5] The familiar vibrating string satisfies a wave equation, and can certainly vibrate in many normal modes.

Practical cavities are only sections of shorted waveguide (the spherical cavity is a notable exception), and are therefore usually easy to treat mathematically. We can expect the Q of resonant cavities to be rather high. The actual mode that is excited in a system not only depends on the *geometry* of the particular cavity or waveguide being used, but also on the *geometry* of the exciting source (loop, monopole, etc.).

REFERENCES

Harrington. (See references for Chapter 10.)

Jordan. (See references for Chapter 10.)

Marcuvitz, N. *Waveguide Handbook*, MIT Radiation Laboratory Series, vol. 10. New York: McGraw-Hill, 1951. A standard reference for guiding systems.

Ramo. (See references for Chapter 4.)

Skilling. (See references for Chapter 3.)

Thomassen, K. I. *Introduction to Microwave Fields and Circuits*. New Jersey: Prentice-Hall, 1971. The impulse response of guiding systems is investigated on page 89.

PROBLEMS

1. Show that the parallel plane TM_0 ($m = 0$) mode is actually the TEM mode.
2. For the TE_{10} rectangular waveguide mode define a voltage as the line integral of **E** across the center of the waveguide, and define a current as the total current in the wall at $y = 0$. Show that $V = bE_0 e^{-j\beta z}$ and $I = (2aE_0)/(\pi Z_0)e^{-j\beta z}$, where $Z_0 = \eta^+$.
3. A lossless waveguide is 1 cm by 2 cm in cross section and is operating at 10 GHz. What effective load impedance should it have in order to avoid reflections (*dominant mode in air*)?
4. Design a rectangular waveguide to have a center frequency of 3000 MHz (*s*-band), with a 2:1 frequency range of single mode operation and maximum power handling capability ($\mu = \mu_0, \varepsilon = \varepsilon_0$).
5. A waveguide is 0.9 by 0.4 inches in cross section and operating at 10 GHz. What is the (evanescent) attenuation constant for the closest (to the dominant propagating mode) nonpropagating mode?
6. The effective load impedance on the waveguide of Problem 5 is 300 Ω.

   (a) What is the SWR?
   (b) Design a quarter-wave matching section using a rectangular dielectric plug.

[5] It may be surprising to learn that one can find other modes in the rectangular guide. See Harrington, in the references at the end of the chapter.

7. For the waveguide of Problem 5, what is the difference in time delay per unit length for two signals; one at 10,000 MHz and one at 10,006 MHz? Six MHz is a television channel width. What phase shift does this represent?
8. Find the resonant frequency for the TE_{101} rectangular cavity mode if $a = c = 2b = 4$ cm and the dielectric is air.
9. Calculate new dimensions for the cavity of Problem 8 if the dielectric is teflon and the resonant frequency is unchanged.
10. Show that $u_g = u_e$ for a lossless rectangular waveguide.
11. What maximum time average power can be handled by an x-band rectangular waveguide (0.9 × 0.4 inches) operating at 10 GHz in the TE_{10} mode? The dielectric is air which breaks down at 3×10^6 V/m.
12. Consider the TEM mode of operation for the waveguide of Fig. 14.1. The field at $y = 0$ is

$$E_x(z, \omega) = E_0(\omega)e^{-jkz} = E_0(\omega)e^{-j\omega\sqrt{\mu\varepsilon}\,z}$$

(a) If $E_0(t) = E_0\,\delta(t)$, show that $E_x(z, t) = E_0\,\delta(t - z\sqrt{\mu\varepsilon})$.
(b) Interpret this result.

13. Consider the TE modes of operation for the waveguide of Figs. 14.1 and 14.6. Assume that these modes are excited in such a way that only E_y exists. Then at $x = a/2$, $y = 0$,

$$E_y(z, \omega) = E_0(\omega)e^{-\gamma z} = E_0(\omega)e^{-\sqrt{\omega_c^2 - \omega^2}\sqrt{\mu\varepsilon}\,z}$$

(a) If $E_0(t) = E_0\,\delta(t)$, show that

$$E_y(z, t) = E_0\,\delta(t - z\sqrt{\mu\varepsilon}) - E_0\,\frac{\omega_c z\sqrt{\mu\varepsilon}}{\sqrt{t^2 - z^2\mu\varepsilon}} \times J_1(\omega_c\sqrt{t^2 - z^2\mu\varepsilon})u(t - z\sqrt{\mu\varepsilon})$$

Hint: use Table E.1 (Appendix E).
(b) Interpret this result and compare it to that in Problem 12.

14. A signal is *suddenly* applied to a lossless two-conductor guiding system A and the mode is unknown. The same signal is simultaneously applied to lossless two-conductor guiding system B. Both systems have uniform (but different) cross sections and identical lengths. Observers with appropriate equipment are located at the load end of each system. Which observer first detects a signal, and what is the total time delay?

15. (a) Find the electromagnetic field for the TE_{101} rectangular cavity mode. Let the peak value of E_y be E_0, and remember that $\omega = \omega_r$ (only).
(b) Show that $\nabla \cdot \mathbf{H} = 0$.
(c) What determines the dimension b?

16. (a) If the TE_{101} mode of Problem 15 is to be excited by a small monopole antenna extending from the center conductor of a coaxial cable, where should it be placed? See Figure 14.12.
    (b) Suppose the voltage on the cable at the input to the cavity is of the form

$$v(t) = \sum_{q=0}^{\infty} a_q \cos q\omega_r t$$

where f_r is the resonant frequency for the TE_{101} mode. How will the field distribution compare to the case where $v(t) = a_1 \cos \omega_r t$ only? Consider only TE_{mnp} modes.

17. Calculate $\langle \mathbf{S} \rangle$ for the field of Problem 15.
18. Find the time domain forms for the field of Problem 15.
19. (a) Refer to Appendix H and find the electromagnetic field for the TM_{010} cylindrical cavity mode.
    (b) Compare this field to that of Problem 15 or 18.
20. (a) What radius must the cavity of Problem 19 have if its resonant frequency (TM_{010} mode) is to be the same as that (TE_{101} mode) of the rectangular cavity of Problem 8?
    (b) Calculate d for the cylindrical cavity so that it has the same volume as the cavity of Problem 8. Compare the surface areas.
    (c) Predict which cavity will have the higher Q.
    (d) Which cavity can handle the greatest electric field?

Chapter 15
Waveguide and Cavity Losses

If waveguide and cavity losses are small, as will generally be the case, the perturbation method, introduced in Chapter 13, will give good engineering results. This method assumes that the field quantities are essentially unchanged from the lossless case when a small loss is present. The loss is then calculated from the lossless field equations and the known parameters of the surrounding media. In this chapter we will consider the most important cases and derive the attenuation constants and quality factors for them.

15.1 GENERAL REMARKS

An exact equation, equation 13.14, was derived in Chapter 13 for the attenuation constant for waves propagating with damping or loss. This equation is

$$\alpha = \frac{1}{2}\frac{\langle P_d \rangle}{\langle P_f \rangle} \quad \text{(Np/m)} \tag{15.1}$$

where $\langle P_d \rangle$ is the time average power dissipated per unit length and $\langle P_f \rangle$ is the time average power flow in watts. The dissipation arises from several sources. The most important of these is the finite conductivity of the waveguide or cavity walls. The interior of the hollow waveguide or cavity may be a lossy dielectric. The interior medium may also have magnetic loss. Thus, it might be difficult to identify the source of the loss. Another factor that complicates the situation is that σ, μ, and ε *all* depend on frequency in general.

An equation suitable for finding the Q of *any propagating system* of uniform cross section is equation 13.48,

$$Q = \frac{\omega}{2\alpha u_e} \tag{15.2}$$

In equation 15.2, α is the attenuation constant due to losses in the propagating system *during* propagation and is not to be confused with the attenuation constant for a (lossless) evanescent mode below cutoff (therefore nonpropagating). An approximate formula for the Q may be found by noting that for any of the lossless systems we have studied

$$u_e = \frac{\sqrt{1 - (\omega_c/\omega)^2}}{\sqrt{\mu\varepsilon}}, \qquad f > f_c \tag{15.3}$$

so that equation 15.2 may be written

$$\boxed{Q \approx \frac{\omega\sqrt{\mu\varepsilon}}{2\alpha\sqrt{1 - (\omega_c/\omega)^2}}} \tag{15.4}$$

Equation 15.4 is approximate since equation 15.3 applies only to lossless systems. Notice that equation 15.4 reduces to equation 13.49 for transmission lines, $Q \approx \beta/2\alpha$, because in the "small loss" TEM mode case $\beta \approx \omega\sqrt{\mu\varepsilon} = k$ and $\omega_c = 0$! In order to use equation 15.4, we must calculate ω_c using lossless formulas. It should be pointed out here that if there is any loss, then there is no true cutoff frequency, since γ can never be zero. The procedure we will follow will be that of calculating the attenuation constant due to the waveguide walls separately from the attenuation constant due to the interior medium. When finished, we may use

$$\alpha = \alpha_c + \alpha_d \tag{15.5}$$

and

$$\frac{1}{Q_{cd}} = \frac{1}{Q_c} + \frac{1}{Q_d} \tag{15.6}$$

We must use the general form, equation 15.1, when calculating the attenuation constant due to the finite conductivity of the waveguide walls, α_c. The result will depend greatly on the waveguide geometry and the particular mode being employed. A calculation of α_c, then, will be made in Section 15.2 for a specific example.

Suppose now the interior region of the waveguide has loss. In Chapter 13 we found for the resonant quarter-wave line that

$$Q_d = \frac{\omega\varepsilon}{\sigma_d} \tag{13.86}$$

This is a general result, being independent of mode and geometry, so we will not repeat its calculation for the rectangular waveguide.

15.2 RECTANGULAR WAVEGUIDE—TE$_{10}$ MODE

The field equations for the TE$_{10}$ to z dominant mode propagating in a lossless waveguide are given by equation 14.88. They are

$$E_y = E_0 \sin \frac{\pi x}{a} e^{-jk_z z}$$

$$H_x = -\frac{k_z E_0}{\omega\mu} \sin \frac{\pi x}{a} e^{-jk_z z} \tag{15.7}$$

and

$$H_z = j\frac{E_0}{\omega\mu a} \cos \frac{\pi x}{a} e^{-jk_z z}$$

where

$$k_z = \beta = k\sqrt{1 - \left(\frac{\omega_c}{\omega}\right)^2}, \qquad f > f_c \tag{15.8}$$

and

$$\omega_c = \frac{\pi}{a\sqrt{\mu\varepsilon}} \tag{15.9}$$

If losses in the internal medium are neglected for the time being, we need consider only the losses in the conductor. Then, we use equation 15.1,

$$\alpha_c = \frac{1}{2}\frac{\langle P_d \rangle}{\langle P_f \rangle}$$

Now,

$$\langle P_f \rangle = -\tfrac{1}{2} \operatorname{Re} \left\{ \int_0^a \int_0^b E_y H_x^* \, dx \, dy \right\}$$

$$= \tfrac{1}{2} \operatorname{Re} \left\{ \int_0^a \int_0^b \frac{E_0^2}{\omega\mu} k_z \sin^2 \frac{\pi x}{a} \, dx \, dy \right\} \tag{15.10}$$

$$\langle P_f \rangle = \frac{ab}{4\eta^+} E_0^2 \quad \text{(W)} \tag{15.11}$$

a *lossless* value. Also

$$\langle P_d \rangle = \tfrac{1}{2} R_s \oint\!\!\oint_{\text{walls}} |H_{\tan}|^2 \, ds \tag{15.12}$$

where R_s is the surface resistance (see equation 10.92) and $|H_{\tan}|$ is the *lossless* value of the magnetic field tangent to the conducting surface. Thus,

$$\begin{aligned}\langle P_d \rangle = \tfrac{1}{2} R_s \Bigg\{ & \int_0^1 \int_0^b |H_z|^2_{x=0} \, dy \, dz + \int_0^1 \int_0^b |H_z|^2_{x=a} \, dy \, dz \\ & + \int_0^1 \int_0^a (|H_x|^2 + |H_z|^2)_{y=0} \, dx \, dz \\ & + \int_0^1 \int_0^a (|H_x|^2 + |H_z|^2)_{y=b} \, dx \, dz \Bigg\} \\ = \frac{R_s}{2} \Bigg[& 2\left(\frac{\pi E_0}{\omega\mu a}\right)^2 b + a\left(\frac{\beta E_0}{\omega\mu}\right)^2 + a\left(\frac{\pi E_0}{\omega\mu a}\right)^2 \Bigg]\end{aligned}$$

or

$$\langle P_d \rangle = \frac{R_s E_0^2}{2} \left[\left(\frac{\pi}{\omega\mu a}\right)^2 (a + 2b) + \left(\frac{\beta}{\omega\mu}\right)^2 a \right] \quad \text{(W/m)} \tag{15.13}$$

Substituting equations 15.11 and 15.13 into equation 15.1 and rearranging,

$$\alpha_c = \frac{R_s \eta^+}{ab} \left[\left(\frac{f_c}{\eta f}\right)^2 (a + 2b) + \frac{a}{(\eta^+)^2} \right] \tag{15.14}$$

Remember that R_s and η^+ are both frequency dependent.

EXAMPLE 1

As an example, suppose a silver plated waveguide is operated in the TE_{10} mode at 10 GHz with $a = 1.25 \times 10^{-2}$ m and $b = 0.625 \times 10^{-2}$ m. The waveguide is "loaded" with polystyrene.

From equation 10.90,

$$R_s = \sqrt{\frac{\pi f \mu}{\sigma_c}} = 2.52 \times 10^{-2} \quad (\Omega)$$

for silver, a "good conductor" at 10 GHz. Also, the (lossless) intrinsic impedance is ($\varepsilon_R = 2.53$)

$$\sqrt{\frac{\mu}{\varepsilon}} = 237 \quad (\Omega)$$

(see Appendix B). The cutoff frequency is

$$f_c = \frac{1}{2a\sqrt{\mu\varepsilon}} = 7.53 \quad \text{(GHz)}$$

and

$$\eta^+ = \frac{\eta}{\sqrt{1 - (f_c/f)^2}} = 360 \quad (\Omega)$$

Substituting these values into equation 15.14 gives

$$\alpha_c \approx 3.04 \times 10^{-2} \quad \text{(Np/m)}$$

Using equation 15.4,

$$Q_c \approx 8360$$

Using equation 13.77, with $\sigma_d/\omega\varepsilon \approx 4.33 \times 10^{-4}$ (Appendix C),

$$Q_d = 2309$$

Then, from equation 15.6,

$$Q = \frac{Q_c Q_d}{Q_c + Q_d} = 1810$$

A plot of α and β against frequency for a general case is shown in Figure 15.1.

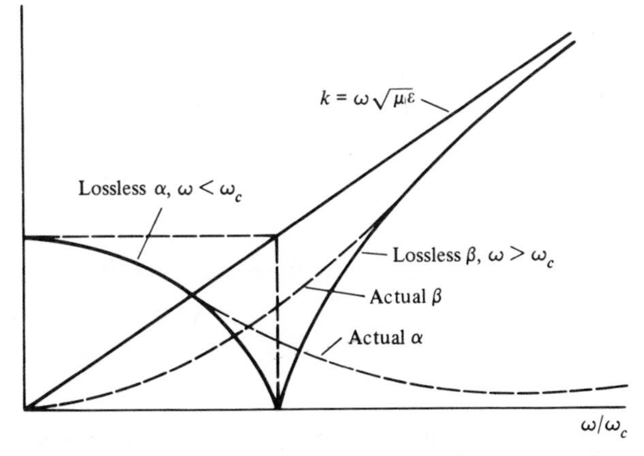

Figure 15.1. α, β, k versus ω/ω_c for the rectangular waveguide.

15.3 RECTANGULAR CAVITY—TE_{101} MODE

The general expression for cavity Q is

$$\boxed{Q \equiv \omega \frac{\langle W_T \rangle}{\langle P_{dT} \rangle}} \tag{15.15}$$

where $\langle W_T \rangle$ is the time average *total* energy stored in the cavity (joules) and $\langle P_{dT} \rangle$ is the time average total power dissipated in the cavity (watts). The procedure followed here is that of examining one of the more practical cases, rather than attempting to solve the general cases.

If $a > b > c$, the lowest resonant frequency for TE to z modes in the rectangular cavity occurs when $m = 1$, $n = 0$, and $p = 1$, giving the TE_{101} mode. In this case, the lossless field is obtained by substituting equation 14.35,

$$F_z = \cos\frac{\pi x}{a}\sin\frac{\pi z}{c}$$

into equation 14.36, giving

$$\begin{aligned} E_x &= 0 & H_x &= -\frac{\pi^2}{j\omega\mu ac}\sin\frac{\pi x}{a}\cos\frac{\pi z}{c} \\ E_y &= -\frac{\pi}{a}\sin\frac{\pi x}{a}\sin\frac{\pi z}{c} & H_y &= 0 \\ E_z &\equiv 0 & H_z &= \frac{\pi^2}{j\omega\mu a^2}\cos\frac{\pi x}{a}\sin\frac{\pi z}{c} \end{aligned} \tag{15.16}$$

Equations 15.16 may be normalized to give

$$E_y = E_0 \sin\frac{\pi x}{a}\sin\frac{\pi z}{c} \tag{15.17}$$

$$H_x = \frac{E_0}{j2f_r\mu c}\sin\frac{\pi x}{a}\cos\frac{\pi z}{c} \tag{15.18}$$

and

$$H_z = -\frac{E_0}{j2f_r\mu a}\cos\frac{\pi x}{a}\sin\frac{\pi z}{c} \tag{15.19}$$

See Figure 14.12 for the distribution of this field in the cavity. The resonant frequency,

$$f_r = \frac{I}{2\sqrt{\mu\varepsilon}}\sqrt{\frac{1}{a^2}+\frac{1}{c^2}} \tag{15.20}$$

is obtained from equation 14.136. Notice that because these are lossless equations, the *only* applicable frequency is f_r.

The electric and magnetic fields are 90° out-of-phase, indicating no loss. When the magnetic field is zero (in time), the electric field is maximum, and *all* of the stored energy is in the electric field. When the electric field is zero, the reverse is true. The total energy is conserved, or constant, and independent of time. The time average energy in the electric field and magnetic field are equal with sinusoidal excitation. Thus

$$W_T = \langle W_T \rangle = 2\langle W_e \rangle = 2\langle W_m \rangle \tag{15.21}$$

and

$$\langle W_e \rangle = \tfrac{1}{2}\,\mathrm{Re} \iiint\limits_{\text{vol}} \frac{\mathbf{D}}{2} \cdot \mathbf{E}^* \, dv \tag{15.22}$$

For the cavity, the *lossless* equations give

$$\langle W_e \rangle = \frac{\varepsilon}{4} \int_0^a \int_0^b \int_0^c |E_y|^2 \, dx \, dy \, dz$$

or

$$\langle W_e \rangle = \langle W_m \rangle = \frac{\varepsilon}{16} E_0^2 \, abc \tag{15.23}$$

and

$$\langle W_T \rangle = \frac{\varepsilon}{8} E_0^2 \, abc \quad \text{(J)} \tag{15.24}$$

For the time being, we neglect any loss in the internal medium and consider only conductor loss in the cavity walls. We now need to calculate the *total* time average power dissipated in the walls. This is done in a manner similar to that for determining the time average power dissipated per unit length, $\langle P_d \rangle$, for the waveguide. That is,

$$\begin{aligned}
\langle P_{dT} \rangle &= \tfrac{1}{2} R_s \oint\!\!\oint\limits_{\text{walls}} |H_{\tan}|^2 \, ds \\
&= \tfrac{1}{2} R_s \int_0^c \int_0^b \left| H_z \right|^2_{x=0} dy \, dz + \int_0^c \int_0^b \left| H_z \right|^2_{x=a} dy \, dz \\
&\quad + \int_0^c \int_0^a (|H_x|^2 + |H_z|^2)_{y=0} \, dx \, dz \\
&\quad + \int_0^c \int_0^a (|H_x|^2 + |H_z|^2)_{y=b} \, dx \, dz \\
&\quad + \int_0^b \int_0^a \left| H_x \right|^2_{x=0} dx \, dy + \int_0^b \int_0^a \left| H_x \right|^2_{z=c} dx \, dy \\
&= \tfrac{1}{2} R_s \left(\frac{E_0}{2 f_r \mu} \right)^2 \left(\frac{bc}{a^2} + \frac{c}{2a} + \frac{a}{2c} + \frac{ab}{c^2} \right)
\end{aligned}$$

$$\langle P_{dT} \rangle = \tfrac{1}{2} R_s \frac{E_0^2}{\eta^2} \frac{a^2 c^2}{a^2 + c^2} \left(\frac{bc}{a^2} + \frac{c}{2a} + \frac{a}{2c} + \frac{ab}{c^2} \right) \quad \text{(W)} \tag{15.25}$$

From equation 15.15

$$Q_c = 2\pi f_r \frac{\langle W_T \rangle}{\langle P_{dT} \rangle}$$

$$= \frac{\pi \varepsilon f_r \eta^2 b(a^2 + c^2)}{2R_s\left(\frac{bc^2}{a} + \frac{c^2}{2} + \frac{a^2}{2} + \frac{a^2 b}{c}\right)}$$

$$Q_c = \frac{\pi\eta}{4R_s} \frac{b(a^2 + c^2)^{3/2}}{\left(bc^3 + \frac{ac^3}{2} + \frac{ca^3}{2} + ba^3\right)} \tag{15.26}$$

If $a = c$,

$$Q_c = \frac{1.11\eta}{R_s\left(1 + \frac{a}{2b}\right)} \tag{15.27}$$

EXAMPLE 2

Suppose a copper rectangular cavity has $a = c = 2 \times 10^{-2}$ m and $b = 10^{-2}$ m and is operated in the TE_{101} mode. If the interior is a vacuum, find Q_c.

The resonant frequency is

$$f_r = \frac{1}{2\sqrt{\mu\varepsilon}} \sqrt{\frac{1}{a^2} + \frac{1}{c^2}} = 10.62 \quad \text{(GHz)}$$

so

$$R_s = \sqrt{\frac{\pi f_r \mu}{\sigma_c}} = 2.69 \times 10^{-2} \quad (\Omega)$$

Therefore,

$$Q_c = 7780$$

Quite often it is desirable, or even necessary, to resort to *dielectric loading* of a cavity to reduce its physical size for a given resonant frequency. If the frequency is high enough so that the dielectric loss must be considered, then

$$Q_d = \frac{\omega\varepsilon}{\sigma_d} \tag{15.28}$$

EXAMPLE 3

Repeat Example 2 if the cavity is loaded with polystyrene ($\varepsilon_R = 2.53$, $\sigma_d/\omega\varepsilon \approx 4.33 \times 10^{-4}$).

$$f_r = \frac{1}{2\sqrt{\mu\varepsilon}}\left(\frac{1}{a^2} + \frac{1}{c^2}\right) = 6.68 \quad \text{(GHz)}$$

$$R_s = \sqrt{\frac{\pi f_r \mu}{\sigma_c}} = 2.13 \times 10^{-2} \quad (\Omega)$$

Therefore,

$$Q_c = \frac{1.11\eta}{R_s(1 + a/2b)} = 6150, \qquad Q_d = \frac{\omega\varepsilon}{\sigma_d} = 2309, \quad \text{and} \quad Q_{cd} = 1680$$

The bandwidth is $f_r/Q_{cd} = 3.98$ MHz. The Q is an optimistic value because no consideration was given to the fact that the cavity cannot be constructed exactly as the theory requires, and because the loss in the networks required to couple energy into and out of the cavity were not considered. The actual bandwidth will be larger than that calculated. If modulation frequencies up to 5 MHz (video, for example) are present, then the wider bandwidth is desirable.

Other cavity resonators are treated similarly, but we can predict here that for similar field distributions (they cannot be identical) the circular cavity will have a higher Q than the rectangular cavity, and the spherical cavity will have a higher Q than the cylindrical cavity. This follows from the smaller surface areas (same volume) and reduced joule heating for the spherical compared to the circular cavity, and the circular compared to the rectangular cavity. An example of this is considered in the problems for the ambitious student.

15.4 CONCLUDING REMARKS

The quality factors, Q_c and Q_d, as calculated in this chapter are intrinsic values, depending in general on mode, material, and geometry. We can expect in practice to obtain smaller values of Q and larger values of α than predicted by the theory. For example, coupling energy into, or out of, a waveguide requires some kind of network. If the effects of the coupling network are included and lumped into an external Q, then

$$\frac{1}{Q_{\text{eff}}} = \frac{1}{Q_c} + \frac{1}{Q_d} + \frac{1}{Q_{\text{ext}}}$$

where Q_{eff} is the overall or effective Q. Q_{eff} will obviously be smaller than any of the Q's which contribute to it. It should also be remembered that the perturbation technique we have employed here is approximate. Further considerations are beyond the scope of this material.

REFERENCES

Harrington. (See references for Chapter 10.)
Jordan. (See references for Chapter 10.)
Ramo. (See references for Chapter 4.)

PROBLEMS

1. Show that the attenuation constant for the parallel plane waveguide with a perfect dielectric is (see Table 14.1)

$$\alpha_c = \frac{2R_s(\omega_c/\omega)^2}{\eta a} \frac{1}{\sqrt{1-(\omega_c/\omega)^2}}$$

and

$$\alpha_c = \frac{2R_s}{\eta a} \frac{1}{\sqrt{1-(\omega_c/\omega)^2}}$$

for the TE_m and TM_m modes, respectively.

2. Calculate the Q of a rectangular waveguide (air filled) with copper walls. The dimensions are 0.4 by 0.9 inches and the dominant mode is being employed. The operating frequency is 10 GHz. Repeat for 7 GHz. The *nominal* cutoff frequency is 6.56 GHz (x-band).
3. Calculate Q_c of the cavity of Problem 8, Chapter 14, if the walls are silver. Repeat for brass.
4. Calculate the Q_c of the cavity of Problem 9, Chapter, 14, if the walls are silver.
5. Plot α_c versus frequency for the TE_{10} rectangular waveguide mode with copper walls and an air dielectric, if $a = 2b = 5.08$ cm.
6. Calculate the bandwidth of the cavities of Problems 3 and 4.
7. Refer to Problems 19 and 20 in Chapter 14 where the cylindrical cavity TM_{010} mode is discussed. The field is (see Appendix H)

$$E_z = E_0 J_0\left(2.405 \frac{\rho}{a}\right)$$

$$H_\phi = j \frac{E_0}{\eta} J_1\left(2.405 \frac{\rho}{a}\right)$$

and

$$f_r = \frac{2.405}{2\pi a\sqrt{\mu\varepsilon}}$$

Calculate the Q_c of this cavity if the walls are silver plated. the dielectric is air, and $a = 2.16$ cm (requested in Problem 20a, Chapter 14). Does this calculation agree with the prediction of Problem 20c, Chapter 14?

8. (a) Plot E_z/E_0 versus ρ for Problem 7.
   (b) Plot $H_\phi/(jE_0/\eta)$ versus ρ for Problem 7.

9. Calculate the bandwidth of the cavity of Problem 7.
10. Compare α_c for the waveguide of Problem 5 at 5 GHz with that for a two-wire line with $D = 10^{-2}$ m, $d = 10^{-3}$ m, copper conductors, and air dielectric. Comment on the results.
11. Repeat Problem 10 for a coaxial cable with $a = 0.3 \times 10^{-2}$ m, $b = 0.6 \times 10^{-2}$ m, copper conductors, and air dielectric.
12. Following the method of Section 13.8, find an equivalent lumped circuit for the cavity of Problems 3, 4, and 7.

Chapter 16
Radiation

An antenna is the interface, or connecting link, between some guiding system and (usually) free space. As such, its function is either to *radiate* electromagnetic energy (in the transmitting case), or *receive* electromagnetic energy (in the receiving case). In the former case, a transmitting system feeds the guiding system to the antenna, and in the latter case, the guiding system feeds a receiving system from the antenna. Most antenna characteristics apply in a receiving case *or* a transmitting case, and can be determined, one from the other, by a generalized reciprocity relation.

In this chapter we will briefly examine the basic fundamentals of radiation and reception. The elementary and classic *Hertzian dipole* will be utilized as an example because it can be regarded as a basic building block from which all radiation studies can begin. Perhaps the most important antenna from a practical point of view is the half-wave dipole. This antenna is not only commonly used singly, but frequently used in multiple arrangements (antenna arrays) where directional radiation characteristics are desired. In this chapter, an approximate solution for the distribution of current on an isolated half-wave dipole antenna will be obtained. This procedure will demonstrate the fact that the real problem in radiation studies (and, indeed, in most of field theory) is not in determining the fields, but, rather, in determining the source distributions which, in turn, produce the correct fields to satisfy Maxwell's equations and the boundary conditions. If we can somehow find the source distribution, then, as we shall shortly see, finding the radiation field is a straightforward proposition.

16.1 THE RADIATION FIELD

Figure 16.1 shows a source (a time-dependent current or charge) producing an electromagnetic field. We already know that this will occur. We also know that the phasor field will, in general, be given by

$$\mathbf{B} = \nabla \times \mathbf{A} \tag{16.1}$$

and

$$\mathbf{E} = \frac{1}{j\omega\mu\varepsilon}\,\nabla(\nabla \cdot \mathbf{A}) - j\omega\mathbf{A} \tag{16.2}$$

as shown in Chapter 9. The magnetic vector potential, $\mathbf{A}$, must satisfy the partial differential equation

$$\nabla^2\mathbf{A} + k^2\mathbf{A} = \begin{cases} -\mu\mathbf{J}, & \mathbf{r} \text{ inside vol}' \\ 0, & \mathbf{r} \text{ outside vol}' \end{cases} \tag{16.3}$$

A solution to equation 16.3 is certainly the Helmholtz or superposition integral which we obtained in Chapter 9. The phasor form was equation 9.76.

$$\mathbf{A}(\mathbf{r}, \omega) = \frac{\mu}{4\pi} \iiint\limits_{\text{vol}'} \frac{\mathbf{J}(\mathbf{r}', \omega)e^{-jkR}}{R}\,dv' \tag{16.4}$$

where

$$R = |\mathbf{r} - \mathbf{r}'|$$

and the unit impulse response for the rectangular components of the phasor form is by inspection

$$h(|\mathbf{r} - \mathbf{r}'|, \omega) = \frac{\mu}{4\pi}\frac{e^{-jk|\mathbf{r}-\mathbf{r}'|}}{|\mathbf{r} - \mathbf{r}'|} = \frac{\mu}{4\pi}\frac{e^{-j\omega\sqrt{\mu\varepsilon}\,R}}{R} \tag{16.5}$$

In other words, a unit impulsive point source at $\mathbf{r}'$ produces the potential given by equation 16.5 at $\mathbf{r}$. We have assumed that μ and ε are scalar constants everywhere.

We will now point out explicitly those features which are special to the *radiation* zone or far-field. The radiation zone is characterized by $|\mathbf{r}| \gg |\mathbf{r}'|$, in which case $\mathbf{r}$ and $\mathbf{r} - \mathbf{r}'$ are essentially parallel, or

$$|\mathbf{r} - \mathbf{r}'| \approx r - r' \cos \xi \tag{16.6}$$

as Figure 16.1 clearly shows. It is convenient at this point to list for future use the term $r' \cos \xi$ where the source coordinates are rectangular, cylindrical,

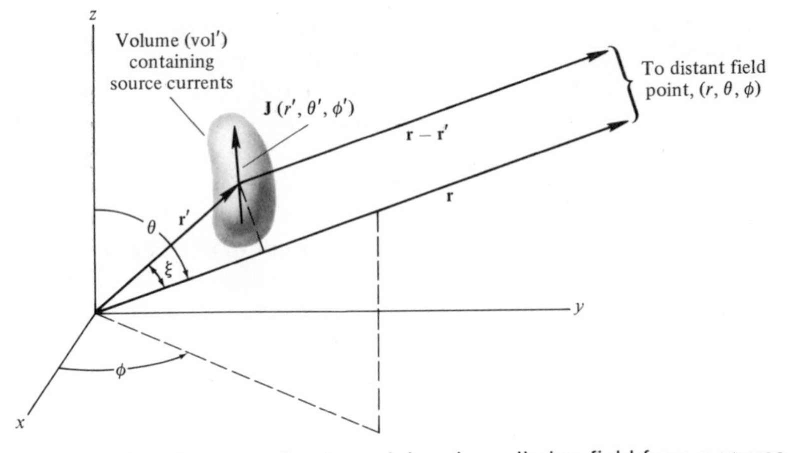

Figure 16.1. Geometry for determining the radiation field from a source current.

and spherical, while the field coordinates are spherical in each case. In order to do this, we use

$$\frac{\mathbf{r} \cdot \mathbf{r}'}{r} = r' \cos \xi = \frac{xx' + yy' + zz'}{r} \tag{16.7}$$

with the coordinate transformations found in Appendix A.1. The result is

$$\boxed{r' \cos \xi = (x' \cos \phi + y' \sin \phi) \sin \theta + z' \cos \theta} \tag{16.8}$$

or

$$\boxed{r' \cos \xi = \rho' \sin \theta \cos (\phi - \phi') + z' \cos \theta} \tag{16.9}$$

or

$$\boxed{r' \cos \xi = r'[\cos \theta \cos \theta' + \sin \theta \sin \theta' \cos (\phi - \phi')]} \tag{16.10}$$

Insofar as the denominator of equation 16.4 (a magnitude only) is concerned, we may go one step further and use

$$|\mathbf{r} - \mathbf{r}'| \approx r \quad \text{(denominator only)} \tag{16.11}$$

The last approximation is not possible for the phase term in the numerator of equation 16.4 because the difference between **r** and **r**′ becomes extremely important if it amounts to an appreciable part of a wavelength. In other words, $(k\mathbf{r} - k\mathbf{r}')$ may represent a large and important phase angle. Since the denominator of equation 16.4 is a positive quantity, the difference between $|\mathbf{r} - \mathbf{r}'|$ and r is unimportant, so long as $r \gg r'$. Equation 16.4 is then represented accurately by

$$\mathbf{A}(\mathbf{r}, \omega) = \frac{\mu}{4\pi} \frac{e^{-jkr}}{r} \iiint_{\text{vol}'} \mathbf{J}(\mathbf{r}', \omega) e^{jkr' \cos \xi} \, dv' \tag{16.12}$$

for the radiation zone, or far-field.

Now, since $\mathbf{J}(\mathbf{r}', \omega)$ is independent of r, and, as equations 16.8, 16.9, and 16.10 show, $\cos \xi$ is independent of r, we are able to conclude that the triple integral in equation 16.12 is independent of r! In spherical coordinates, equation 16.12 will always have the general form

$$\mathbf{A}(\mathbf{r}, \omega) = \frac{\mu}{4\pi} \frac{e^{-jkr}}{r} [f_r(\theta, \phi)\mathbf{a}_r + f_\theta(\theta, \phi)\mathbf{a}_\theta + f_\phi(\theta, \phi)\mathbf{a}_\phi] \tag{16.13}$$

The special feature of equation 16.12 is that r is *outside* the integral sign, and, therefore, many of the operations required by equation 16.1 and 16.2 can be performed with no difficulty, regardless of whether $\mathbf{J}(\mathbf{r}')$ is known or not.

In spherical coordinates, equations 16.1 and 16.2 may be written

$$E_r = \frac{1}{j\omega\mu\varepsilon} [\nabla(\nabla \cdot \mathbf{A})]_r - j\omega A_r \tag{16.14}$$

$$E_\theta = \frac{1}{j\omega\mu\varepsilon} [\nabla(\nabla \cdot \mathbf{A})]_\theta - j\omega A_\theta \tag{16.15}$$

$$E_\phi = \frac{1}{j\omega\mu\varepsilon} [\nabla(\nabla \cdot \mathbf{A})]_\phi - j\omega A_\phi \tag{16.16}$$

$$H_r = \frac{1}{\mu} (\nabla \times \mathbf{A})_r \tag{16.17}$$

$$H_\theta = \frac{1}{\mu} (\nabla \times \mathbf{A})_\theta \tag{16.18}$$

$$H_\phi = \frac{1}{\mu} (\nabla \times \mathbf{A})_\phi \tag{16.19}$$

As another specialization of the radiation zone, we retain *only* the terms that vary with $1/r$ from the preceding set of equations. The terms that vary as $1/r^2$ or $1/r^3$ are negligibly small for large r, and make up what is called the "near-field." The near-field will be mentioned again in more detail in the next section, but it can be stated now that it, as opposed to the fields in the radiation zone, or far-field, will not contribute to the time average power flow away from the source. When the indicated expansions are carried out (a rather lengthy proposition) using equation 16.13, the result is

$$\boxed{\begin{array}{ll} E_r = 0 & H_r = 0 \\ E_\theta = -j\omega A_\theta & H_\theta = \dfrac{-E_\phi}{\eta} \\ E_\phi = -j\omega A_\phi & H_\phi = \dfrac{E_\theta}{\eta} \end{array}} \tag{16.20}$$

for the radiation zone. Therefore, with equations 16.20, the radiation field may be obtained without differentiating the vector potential! That is, this operation has now been performed once and for all! Then, the radiation field consists of outward traveling spherical waves which decrease[1] as $1/r$. In order to avoid difficulties in integrating nonconstant unit vectors, it is always advisable to perform the integration using the Cartesian components (where the unit vectors are constant) of the current. In this case equation 16.12 together with equations 16.20 give the most general form of the radiation field.

$$\boxed{\begin{aligned} E_\theta(r, \theta, \phi) &= \frac{-j\omega\mu}{4\pi}\frac{e^{-jkr}}{r}\left[\cos\theta\cos\phi \iiint_{\text{vol}'} J_x(\mathbf{r}')e^{jkr'\cos\xi}\,dv' \right. \\ &\quad + \cos\theta\sin\phi \iiint_{\text{vol}'} J_y(\mathbf{r}')e^{jkr'\cos\xi}\,dv' \\ &\quad \left. - \sin\theta \iiint_{\text{vol}'} J_z(\mathbf{r}')e^{jkr'\cos\xi}\,dv' \right] \\ E_\phi(r, \theta, \phi) &= \frac{-j\omega\mu}{4\pi}\frac{e^{-jkr}}{r}\left[-\sin\phi \iiint_{\text{vol}'} J_x(\mathbf{r}')e^{jkr'\cos\xi}\,dv' \right. \\ &\quad \left. + \cos\phi \iiint_{\text{vol}'} J_y(\mathbf{r}')e^{jkr'\cos\xi}\,dv' \right] \\ H_\theta &= -\frac{E_\phi}{\eta} \\ H_\phi &= \frac{E_\theta}{\eta} \end{aligned}} \tag{16.21}$$

Equations 16.21 indicate that if the current density **J** is known, then, in a formal sense at least, the solution is complete. This is true regardless of the

[1] This is a spreading effect.

type antenna one is analyzing. For most antennas made of conductors in the form of wires, such as the dipole antenna, we can approximate the current density reasonably well. On the other hand, slot antennas and aperture type antennas are such that it is very difficult to estimate the current densities on the conducting surfaces. These antennas are usually analyzed in terms of the known (or approximated) fields over the aperture, for which equations other than those developed here will be needed.

The time average power density is

$$\langle \mathbf{S} \rangle = \tfrac{1}{2}\,\mathrm{Re}\{\mathbf{E} \times \mathbf{H}^*\}$$

or, when equation 16.20 is substituted,

$$\langle \mathbf{S} \rangle = \frac{\mathbf{a}_r}{2\eta}\left(|E_\theta|^2 + |E_\phi|^2\right) \quad (\mathrm{W/m^2}) \tag{16.22}$$

That is, the time average power density is apparently directed *radially outward*. The outward power flow, or *radiated power* is obtained by integrating the normal component of $\langle \mathbf{S} \rangle$ over a sphere of (large) radius r, or

$$\langle P_r \rangle = \int_0^{2\pi}\int_0^{\pi} \langle \mathbf{S} \rangle \cdot \mathbf{a}_r r^2 \sin\theta \, d\theta \, d\phi \quad (\mathrm{W}) \tag{16.23}$$

Substituting equation 16.22 into equation 16.23 gives

$$\boxed{\langle P_r \rangle = \frac{r^2}{2\eta}\int_0^{2\pi}\int_0^{\pi} \left(|E_\theta|^2 + |E_\phi|^2\right) \sin\theta \, d\theta \, d\phi} \tag{16.24}$$

If the source distribution (current density, surface current density, or current) is known explicitly, then the formal solution to the radiation problem is complete, being given by the preceding equations. Unfortunately, this is rarely the case, especially for those antenna geometries of practical importance. An *exact* solution for the source distribution requires the solution to a boundary value problem, and is almost never tractable. Quite often, we are willing to approximate this source distribution.

16.2 THE HERTZIAN DIPOLE

Now that we have (rather suddenly) armed ourselves with the basic equations governing radiation, let us take a specific example. The Hertzian dipole is a filamentary current I extending over an *incremental* length l. Since the length is incremental, the Hertzian dipole may be regarded as the basic building block for radiation problems. This point is further emphasized if we simply recognize that the Hertzian dipole is, in fact, the impulsive point source

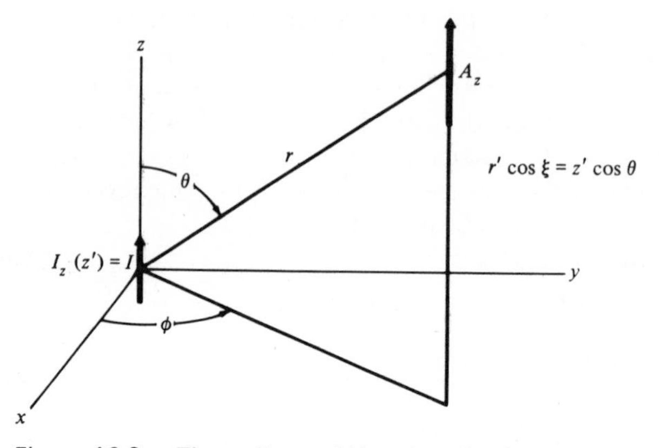

Figure 16.2. The z-directed Hertzian dipole at the origin.

mentioned earlier. Consider a z-directed Hertzian dipole at the origin as shown in Figure 16.2. It has a moment or strength Il and its current density is

$$\boxed{\mathbf{J}(\mathbf{r}') = \mathbf{a}_z Il\,\delta(x')\,\delta(y')\,\delta(z')} \quad (\text{A/m}^2) \tag{16.25}$$

When equation 16.25 is substituted into equation 16.21, we immediately obtain

$$\boxed{E_\theta(r, \theta) = j\,\frac{\eta(kl)I}{4\pi}\,\frac{e^{-jkr}}{r}\sin\theta, \qquad H_\phi = \frac{E_\theta}{\eta}} \tag{16.26}$$

and when equation 16.25 is substituted into equation 16.4 or 16.12, we obtain

$$\boxed{A_z = \frac{\mu Il}{4\pi}\,\frac{e^{-jkr}}{r}} \tag{16.27}$$

The geometry is shown in Figure 16.2.

This result can also be obtained directly from the differential equation, equation 16.3. Equation 16.3 becomes

$$\nabla^2 A_z + k^2 A_z = 0, \qquad r > 0 \tag{16.28}$$

or since A_z can depend only on r (the Hertzian dipole is a *point* source for A_z),

$$\frac{1}{r^2}\frac{d}{dr}\left(r^2\frac{dA_z}{dr}\right) + k^2 A_z = 0 \tag{16.29}$$

Equation 16.29 may be rearranged to

$$\frac{d^2 A_z}{dr^2} + \frac{2}{r}\frac{dA_z}{dr} + k^2 A_z = 0 \tag{16.30}$$

which has the solution

$$A_z = C_1 \frac{e^{-jkr}}{r} + C_2 \frac{e^{jkr}}{r} \tag{16.31}$$

We already know from equation 16.27 that the first solution is the correct one. Even though the second solution is mathematically correct, it is non-physical. It is an inward spherical traveling wave, and if the surrounding medium is lossy ($\varepsilon = \varepsilon' - j\varepsilon''$, for example), then it takes the form

$$C_2 \frac{e^{\alpha r + j\beta r}}{r}$$

which tends to infinity for large r. Therefore,

$$A_z = C_1 \frac{e^{-jkr}}{r} \tag{16.32}$$

The static solution ($k = 0$) to this problem is

$$A_z = \frac{\mu I l}{4\pi r} \tag{16.33}$$

as given in Problem 11, Chapter 7. Therefore, the phasor solution is

$$A_z = \frac{\mu I l}{4\pi} \frac{e^{-jkr}}{r} \tag{16.34}$$

We have not used equation 16.12 explicitly to obtain this solution. It is worthwhile to actually do this. For the filamentary current that we actually have, equation 16.12, with equation 16.8, becomes

$$A_z = \frac{\mu}{4\pi} \frac{e^{-jkr}}{r} \int_{-l/2}^{l/2} I_z(z') e^{jkz' \cos\theta}\, dz'$$

$$= \frac{\mu I}{4\pi} \frac{e^{-jkr}}{r} \int_{-l/2}^{l/2} e^{jkz' \cos\theta}\, dz'$$

or

$$A_z = \frac{\mu I l}{4\pi} \frac{e^{-jkr}}{r} \left\{ \frac{\sin\left[(kl/2)\cos\theta\right]}{(kl/2)\cos\theta} \right\} \tag{16.35}$$

It was originally stipulated that l was incremental, so the braced term is approximately unity. Thus,

$$A_z = \frac{\mu I l}{4\pi} \frac{e^{-jkr}}{r} \qquad (l \to 0) \tag{16.36}$$

which is our previous result.

The radiation field is given by equation 16.20

$$E_\theta = j\,\frac{\eta(kl)I}{4\pi}\,\frac{e^{-jkr}}{r}\sin\theta,\; H_\phi = \frac{E_\theta}{\eta} \tag{16.37}$$

since $A_\theta = -A_z \sin\theta$. The radiated power is given by equation 16.37 and 16.24. It is

$$\langle P_r \rangle = \frac{\eta(kl)^2 I^2}{32\pi^2}\int_0^{2\pi}\int_0^{\pi} \sin^3\theta \, d\theta \, d\phi \tag{16.38}$$

where

$$|E_\theta| = \frac{\eta k l I}{4\pi r}\sin\theta \tag{16.39}$$

The double integral is $8\pi/3$, and equation 16.38 becomes

$$\langle P_r \rangle = 10(kl)^2 |I|^2 \quad \text{(W)} \tag{16.40}$$

Remember that $|I|$ is a *peak* value. If the same power is dissipated in a resistor, then

$$\langle P_r \rangle = \tfrac{1}{2}|I|^2 R \tag{16.41}$$

Equating equations 16.40 and 16.41, and calling this resistance the *radiation* resistance we have

$$\boxed{R_{\text{rad}} = 20(kl)^2} \quad (\Omega) \tag{16.42}$$

Recalling that we originally stipulated that l was incremental, or equivalently, $kl \ll 1$, we see that R_{rad} is very small. The Hertzian dipole is not efficient in a practical situation because power loss due to ordinary conductor resistance (ohmic loss) will normally be many times greater than that due to radiation. In other words, the total resistance in any transmission system connected to a Hertzian dipole will be many times greater than the radiation resistance. It can be shown that the reactance of the Hertzian dipole is large and negative.

EXAMPLE 1

It is informative to consider, as a numerical example, a "short circuit" at the end of an open wire transmission line. Let this short circuit be a piece of straight wire 1 cm long. If the operating frequency is 600 MHz, and the rest of the line is neglected, then

$$R_{\text{rad}} = 20(0.04\pi)^2 = 0.316 \quad (\Omega) \tag{16.43}$$

On the other hand, a straight piece of wire carrying a uniform current at 60 Hz would need to be 10^5 m long to possess the same radiation resistance!

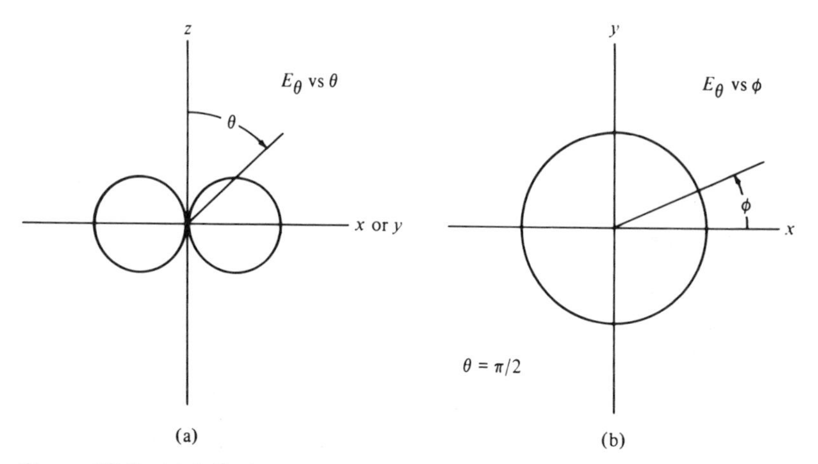

Figure 16.3. (a) E-plane pattern of a Hertzian dipole (E_θ, normalized, versus θ). (b) H-plane pattern of a Hertzian dipole (E_θ, normalized, versus ϕ, $\theta = \pi/2$).

Problem 2 requests the complete field (near and far) for the Hertzian dipole. It is given by

$$E_r = \frac{Il}{j2\pi\omega\varepsilon} \cos\theta \frac{e^{-jkr}}{r}\left(jk + \frac{1}{r}\right)$$

$$E_\theta = \frac{Il}{j4\pi\omega\varepsilon} \sin\theta \frac{e^{-jkr}}{r}\left(-k^2 + \frac{jk}{r} + \frac{1}{r^2}\right)$$

$$H_\phi = \frac{Il}{4\pi} \sin\theta \frac{e^{-jkr}}{r}\left(jk + \frac{1}{r}\right)$$

The radiation pattern of an antenna is usually a plot of the normalized electric field against the angle θ or ϕ for fixed r. This pattern displays the directional characteristics of the antenna as a radiator. In other words, it shows how well the antenna radiates in a given direction compared to other directions. It is customary to present two patterns. The first, called the E-plane pattern, is a plot of normalized $|\mathbf{E}|$ versus angle in the plane containing $\mathbf{E}$ and the principal beam (if one exists) of the antenna. The second, called the H-plane pattern, is a plot of normalized $|\mathbf{E}|$ versus angle in the plane containing $\mathbf{H}$ and the principal beam of the antenna. Thus, for the Hertzian dipole, the E-plane is the x-z plane or the y-z plane (or any plane between these two) since E_θ is independent of ϕ. The H-plane is the x-y plane. These patterns are shown in Figure 16.3. We will see later that the receiving pattern of a receiving antenna is the same as its transmitting pattern.

16.3 THE HALF-WAVE ($\lambda/2$) DIPOLE

The cylindrical half-wave dipole antenna has been in use for many years as either a single radiator, or as one of the radiating elements in an antenna array

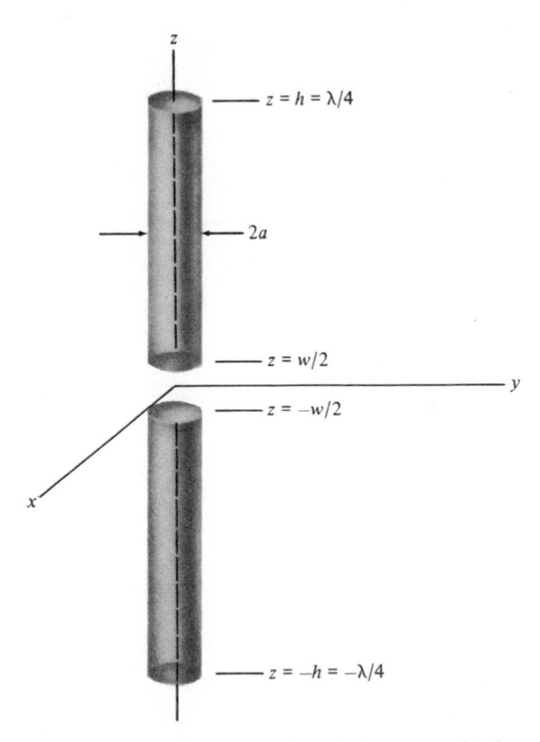

Figure 16.4. The cylindrical half-wave dipole, coaxial with the z axis.

where more pronounced directional characteristics are desired.[2] It consists of two collinear legs separated by a very small gap of width w. Each leg has a radius a and a length $h \approx \lambda/4$, so that the total length is $\lambda/2$. The dipole is shown in Figure 16.4. Normally, the dipole is driven by a balanced transmission line connected across the gap. Since the transmission line itself becomes an inherent part of the radiating system, we will simply remove the line and assume that there is an applied emf from some source which results in a uniform z-directed electric field around the circumference of the gap. In this way, we can theoretically remove the effects of a feeding system. Furthermore, if $h \gg a$, we can simply ignore the effects of any current that might exist on the end caps of a solid dipole and concern ourselves only with the currents on the lateral surfaces of the dipole. With the preceding assumptions, plus assuming that the dipole is made of perfect conductors, we have only z-directed surface currents resulting from the applied emf at the gap. In addition, we have complete azimuthal independence ($\partial/\partial\phi \equiv 0$).

There is one remaining problem. What is $J_{sz}(z') = I_z(z')/2\pi a$? There are two approaches to this problem. First, we can make a reasonable guess as to what the current distribution $I_z(z')$ is, or we can set up a boundary value

[2] Problem 31 requests the radiation patterns for two simple cases for a two-element $\lambda/2$ dipole array.

problem with $I_z(z')$ as the unknown. We will discuss the latter possibility first.

An integral equation can be formulated in terms of the vector potential on the antenna surface with $I_z(z')$ as the unknown. It is based on the fact that the tangential electric field, $E_z(z)$, is zero on the perfectly conducting lateral surface of the dipole. Many methods have been devised to solve this integral equation. We will mention only one, the *method of moments*, because we used it in Chapter 5 to solve a very similar problem (the charged cylinder). In fact the same technique can be employed: namely that of assuming that the half-wave dipole consists of a chain of N semi-Hertzian dipoles, each with an unknown but constant (with z') current over a small but finite length ($2h/N$). The integral equation is then converted into a set of N simultaneous equations for the N unknown Hertzian dipole current amplitudes. Results of this approach (for large N) agree very well with measured current distributions. An exact solution to the problem has not been found even though it has received much attention for approximately 50 years.

Perhaps the simplest accurate solution to the boundary value problem is obtained by merely observing the measured current distribution and then using simple analytic forms to approximate it. For example, excellent results are obtained if we assume a two-term current,[3]

$$I_z(z') = I_t \cos kz' + b \sin 2k|z'| \qquad \left(h \approx \frac{\lambda}{4}\right) \tag{16.44}$$

The unknown I_t and b are then determined with the method of moments by forcing the above mentioned integral equation to hold at two points in its interval of validity. This generates two simultaneous equations in two unknowns, from which I_t and b may easily be found. Results from this method are very good.

Fortunately, the (phasor) radiation field is relatively insensitive to small errors in $I_z(z')$, so a reasonable guess will give useful results. On the other hand, the impedance of the antenna at the driving point (which a designer needs to know) does in general depend strongly on the exact form of $I_z(z')$. Classically, the usual assumed current (see Figure 16.5, $h = \lambda/4$),

$$I_z(z') = \frac{I_t \sin k(h - |z'|)}{\sin kh} \tag{16.45}$$

results from taking an open-circuited two-wire lossless transmission line, bending the legs 90° to form the dipole, and then insisting that the current has not changed from the transmission line form. This cannot be correct because, in the first place, if $h = n\lambda/2$ ($n = 1, 2, \ldots$), the terminal or driving point current, I_t, in equation 16.45 must be zero (to keep the current finite at other points) indicating that the driving point impedance (V_t/I_t) is infinite!

[3] See Neff, Siller, and Tillman in the references at the end of the chapter.

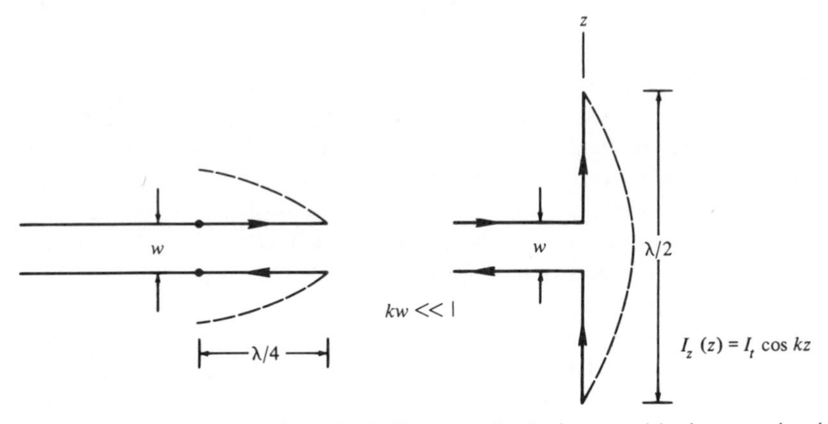

Figure 16.5. The evolution of a half-wave dipole from an ideal open-circuited transmission line. Currents shown dashed.

Secondly, if we use equation 16.45, calculate the radiation field, and then its inverse Fourier transform or impulse response, we find that the field is a train of *undamped* impulses[4] repeating forever! This cannot be correct. Nevertheless, equation 16.45 gives fairly good answers for thin antennas ($ka \ll \lambda$), and particularly for the special case of $h = \lambda/4$, where equation 16.45 becomes

$$I_z(z') = I_t \cos kz' \tag{16.46}$$

Equation 16.44 then represents a (known) good approximation, equation 16.46, plus a first order correction.

The half-wave dipole impedance calculated using the current given by equation 16.46 is dependent on the dipole radius a, but for a reasonable radius ($h/a = 75$),

$$Z_r = 73.1 + j42.5 \quad (\Omega) \tag{16.47}$$

A more accurate impedance from the integral equation approach using equation 16.44 is ($h/a = 75$)

$$Z_r = 87.8 + j39.1 \quad (\Omega) \tag{16.48}$$

We will next calculate the radiation field resulting from equation 16.46. In equation 16.9, $\rho' = a$, and we choose $\phi = 0$ since the fields are obviously independent of ϕ. Thus.

$$r' \cos \xi = a \sin \theta \cos \phi' + z' \cos \theta \tag{16.49}$$

In the present problem the current is a surface current density, $J_{sz}(z')$, so

[4] See Problems 11 and 12.

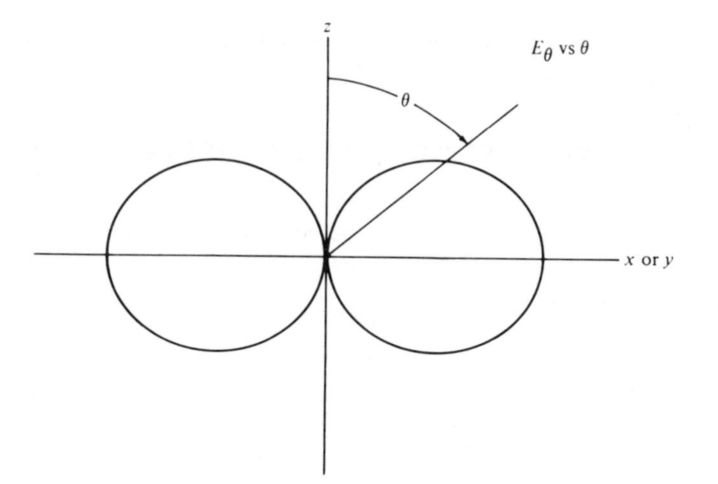

Figure 16.6. E-plane pattern of a half-wave dipole antenna with sinusoidal current.

equation 16.21 becomes

$$E_\theta(r, \theta) = \frac{j\omega\mu}{4\pi} \frac{e^{-jkr}}{r} \sin\theta \int_{-h}^{h} \int_{0}^{2\pi} J_{sz}(z') e^{jk(a \sin\theta \cos\phi' + z' \cos\theta)} a \, d\phi' \, dz' \tag{16.50}$$

or

$$E_\theta(r, \theta) = \frac{j\omega\mu}{4\pi} \frac{e^{-jkr}}{r} \sin\theta \int_{-\lambda/4}^{\lambda/4} \frac{I_t \cos kz'}{2\pi a} e^{jkz' \cos\theta} \int_{0}^{2\pi} e^{jka \sin\theta \cos\phi'} a \, d\phi' \, dz' \tag{16.51}$$

We have omitted the base gap width, w, as being negligibly small. The second integral of equation 16.51 is actually well known, but it is also approximately equal to $2\pi a$ since a is small and $e^{jka \sin\theta \cos\phi} \approx 1$. Thus,

$$E_\theta(r, \theta) = \frac{j\omega\mu}{4\pi} \frac{e^{-jkr}}{r} \sin\theta \int_{-\lambda/4}^{\lambda/4} I_t \cos kz' e^{jkz' \cos\theta} \, dz' \tag{16.52}$$

which is easy to integrate or simply extract from integral tables, so

$$\boxed{E_\theta(r, \theta) = j60 I_t \frac{e^{-jkr}}{r} \frac{\cos(\pi/2 \cos\theta)}{\sin\theta}} \tag{16.53}$$

and

$$\boxed{H_\phi(r, \theta) = E_\theta(r, \theta)/\eta} \tag{16.54}$$

The E-plane pattern is shown in Figure 16.6. Notice that it is remarkably similar to the pattern of a Hertzian dipole.

16.4 GAIN, DIRECTIVITY, APERTURE, AND EFFECTIVE LENGTH

The directive properties of an antenna may be conveniently described in terms of *power per unit solid angle* (watts per sterradian), $P_s(\theta, \phi)$. This quantity is related to the radiated power by

$$P_r = \iint P_s(\theta, \phi)\, d\Omega = \int_0^{2\pi}\int_0^{\pi} P_s(\theta, \phi) \sin\theta\, d\theta\, d\phi \quad \text{(W)} \qquad (16.55)$$

where $d\Omega$ is the differential solid angle. $P_s(\theta, \phi)$ is related to the Poynting vector, $\mathbf{S}$, through equation 16.23.

$$P_s(\theta, \phi) = r^2 \mathbf{a}_r \cdot \mathbf{S} = r^2 S_r \qquad (16.56)$$

Now, if we imagine an isotropic source which radiates equally in all directions, then $P_s(\theta, \phi) = P_{si}$ (constant) and in this case

$$P_r = P_{si} \int_0^{2\pi}\int_0^{\pi} \sin\theta\, d\theta\, d\phi = 4\pi P_{si} \qquad (16.57)$$

An equivalent statement is that the spatial average value of $P_s(\theta, \phi)$ for any antenna is $P_r/4\pi$. The isotropic source is a very convenient reference, but is physically unrealizable.

The *gain function* of an antenna is defined as

$$\boxed{G(\theta_0, \phi_0) = \frac{P_s(\theta_0, \phi_0)}{P_r/4\pi} = \frac{P_s(\theta_0, \phi_0)}{P_{si}} = 4\pi \frac{P_s(\theta_0, \phi_0)}{P_r}} \qquad (16.58)$$

It is the power per unit solid angle in a specified direction (θ_0, ϕ_0) divided by the average power per solid angle. Alternatively, it is the power per unit solid angle divided by the power per unit solid angle for an isotropic source (reference) when both radiate the same total power. In terms of time average quantities,

$$\boxed{G(\theta_0, \phi_0) = 4\pi \frac{\langle P_s(\theta_0, \phi_0)\rangle}{\langle P_r\rangle}} \qquad (16.59)$$

Substituting equations 16.22, 16.24, and 16.56,

$$G(\theta_0, \phi_0) = 4\pi \frac{|E_\theta(\theta_0, \phi_0)|^2 + |E_\phi(\theta_0, \phi_0)|^2}{\int_0^{2\pi}\int_0^{\pi} (|E_\theta(\theta, \phi)|^2 + |E_\phi(\theta, \phi)|^2) \sin\theta \, d\theta \, d\phi} \tag{16.60}$$

or substituting equations 16.22 and 16.56 and replacing $\langle P_r \rangle$ with $|I_t|^2 R_r/2$ (for those antennas where the identification of a terminal current is possible),

$$G(\theta_0, \phi_0) = \frac{r^2}{30R_r} \frac{|E_\theta|^2 + |E_\phi|^2}{|I_t|^2} \tag{16.61}$$

The maximum value of the gain function [obtained in the direction of the "main beam," or the direction of maximum field, $(|E_\theta(\theta, \phi)|^2 + |E_\phi(\theta, \phi)|^2)^{1/2}$] is the directivity, G, of the antenna. That is,

$$G = G(\theta_0, \phi_0)_{\text{max}} \tag{16.62}$$

This is also commonly called the gain of the antenna. Equation 16.62 will give an optimistic figure because it neglects ohmic losses, $\langle P_d \rangle$, in the antenna. The total power "loss" consists of the sum of radiation and ohmic losses, so the antenna efficiency is $\langle P_r \rangle / (\langle P_r \rangle + \langle P_d \rangle)$. The power gain, G_p, is then defined as the directivity times the efficiency, or

$$G_p = G \frac{\langle P_r \rangle}{\langle P_r \rangle + \langle P_d \rangle} \tag{16.63}$$

The efficiency is very high for most antennas, with the exception of those that are electrically small, and those that are *superdirective* or *supergain*. The latter require currents which may drastically reduce the efficiency.

EXAMPLE 2

Let us calculate the gain (G) of a Hertzian dipole. From equation 16.26

$$\left| E_\theta \right|^2_{\text{max}} = \left| \frac{\eta k l I}{4\pi r} \right|^2, \qquad \theta = \frac{\pi}{2}$$

and

$$|E_\phi|^2 = 0$$

Therefore,

$$G = 4\pi \frac{\left|\frac{\eta k l I}{4\pi r}\right|^2}{\left|\frac{\eta k l I}{4\pi r}\right|^2 \int_0^{2\pi}\int_0^{\pi} \sin^3\theta\, d\theta\, d\phi}$$

$$G = \frac{2}{\int_0^{\pi} \sin^3\theta\, d\theta} = \frac{2}{4/3} = 1.5 \quad (1.76 \text{ dB})$$

Thus, the power per unit solid angle in the direction $\theta = \pi/2$ (any ϕ) is 1.5 times as much as if the power were radiated equally in all directions. For the Hertzian monopole (see Problem 7), which is half of the dipole oriented vertically and driven against a flat (perfectly conducting) ground, radiation occurs for $0 \leq \theta \leq \pi/2$ and so

$$G = \frac{2}{\int_0^{\pi/2} \sin^3\theta\, d\theta} = \frac{2}{2/3} = 3 \quad (4.77 \text{ dB})$$

Television broadcast antennas often use an array of vertically stacked antennas to increase the field (and therefore the power per unit solid angle) along $\theta \approx \pi/2$ at the expense of higher angle (nearer $\theta = 0$) radiation. The radiation pattern is usually omnidirectional with respect to ϕ. Thus, a gain is achieved. The transmitter power may be effectively increased.

EXAMPLE 3

A television transmitter produces 10 kW. The antenna has a gain of 10. Neglecting losses in the feedline and ohmic losses in the antenna, what is the effective radiated power? The effective radiated power is 10×10 kW $= 100$ kW, and, furthermore, this antenna produces an electric field which is $\sqrt{10}$ times as large at a given point (in the direction where G is specified) as it would be if $G = 1$. The 10-kW transmitter is just as effective as a 100-kW transmitter with an antenna gain of 1.

The gain function, radiation pattern and impedance of an antenna are all the same in the transmitting mode or the receiving mode. We have thus far only discussed transmitting antennas.

An optimistic "rule-of-thumb" formula for gain calculations can be obtained from equation 16.58. Imagine an antenna with only one beam forming

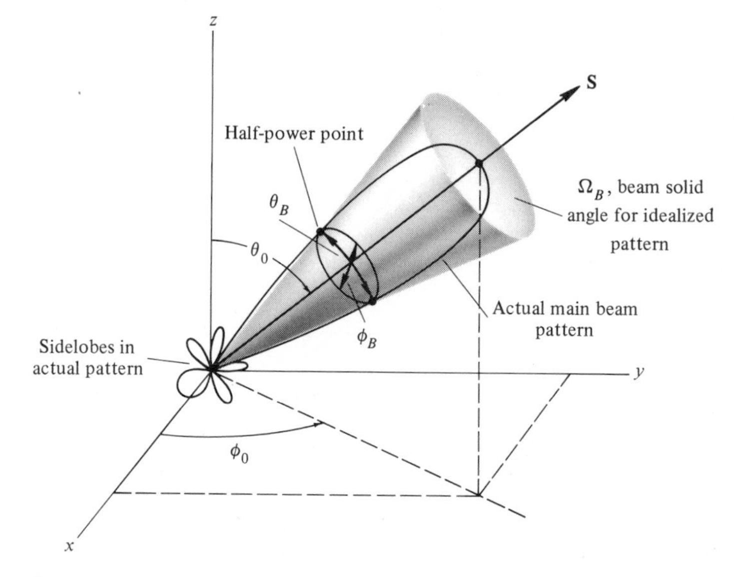

Figure 16.7. Actual and idealized power patterns showing the approximate equivalence at the half-power points.

its radiation pattern (no sidelobes) such that *all* the power flows through the solid angle defined by the beam, and, in addition, the power per unit solid angle is constant over the beam. This picture is truly optimistic. It is shown in Figure 16.7. For the idealized power pattern shown equation 16.58 gives

$$G(\theta_0, \phi_0) = G(\theta_0, \phi_0)_{\max} = G = 4\pi \frac{P_s(\theta_0, \phi_0)}{P_r}$$

or

$$G = 4\pi \frac{P_s(\theta_0, \phi_0)}{\iint P_s(\theta_0, \phi_0)\, d\Omega} = 4\pi \frac{P_s(\theta_0, \phi_0)}{P_s(\theta_0, \phi_0) \iint d\Omega} = \frac{4\pi}{\Omega_B} \tag{16.64}$$

where Ω_B is the beam solid angle as shown in Figure 16.7. For many actual antennas, particularly those with high-gain, small sidelobes and a single main beam, it turns out that $\Omega_B \approx \theta_B \phi_B$, where θ_B and ϕ_B are the *beamwidths* (in radians) at the *half-power points* measured in the $\mathbf{a}_\theta$ and $\mathbf{a}_\phi$ directions, respectively. That is, θ_B and ϕ_B are measured at the points where $P_s(\theta, \phi) = 0.5P_s(\theta_0, \phi_0)_{\max}$. Therefore,

$$\boxed{G \approx \frac{4\pi}{\theta_B \phi_B}} \tag{16.65}$$

Notice that an isotropic antenna has no sidelobes, one "main beam" and furthermore, $\Omega_B = 4\pi$ (a sphere), so equation 16.64 correctly gives $G = 1$!

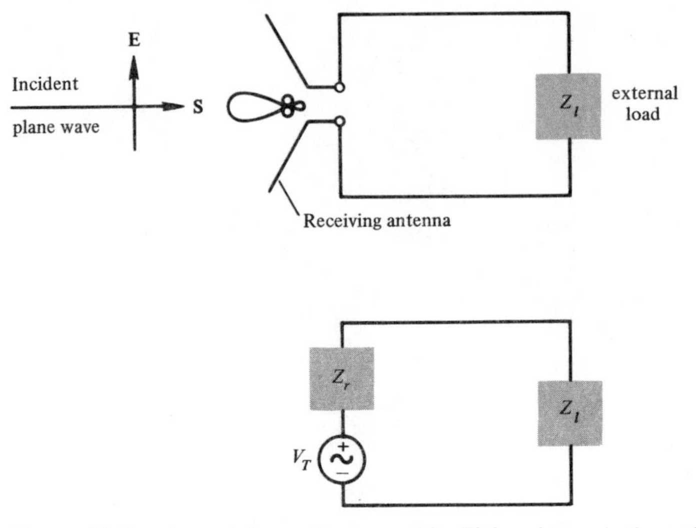

Figure 16.8. A receiving antenna and its Thévenin equivalent circuit when subjected to a uniform plane wave.

It should be obvious by now that an alternate and perhaps more meaningful definition for antenna gain exists. It was alluded to in Example 3. The gain is the ratio of the power required for an isotropic source to the power required for the actual antenna when the antenna and isotropic source produce the same electric field at a given distance. Conversely, it is the square of the ratio of the electric fields when the powers are the same. Thus, if a Hertzian dipole produces 132 mV/m (rms) at 1 mile for 1 kW, an isotropic antenna would produce $132/\sqrt{1.5} = 108$ mV/m (rms) at 1 mile for 1 kW.

A receiving antenna possesses an effective aperture, A, or equivalent area from which it extracts power from the power density of an incoming (essentially) plane wave. It is logical to define the aperture as the maximum received power in a load connected to the antenna divided by the power density of the plane wave. That is,

$$\boxed{A = \frac{\langle P_L \rangle}{\langle |\mathbf{S}| \rangle}} \quad (\text{m}^2) \tag{16.66}$$

The receiving antenna is equivalent to a Thévenin generator in series with its own impedance, $R_r + jX_r$. This is shown in Figure 16.8. The external load impedance is $Z_l = R_l + jX_l$, which might be the input impedance to a transmission line feeding a receiver. In any case, maximum power transfer occurs for a conjugate match (recall circuit theory?), or when $Z_r = Z_l^*$ and $X_r = -X_l$. In this situation it is easy to show that $\langle P_L \rangle = |V_T|^2/8R_r$. ($|V_T|$ is a peak value.)

Let us assume for simplicity that the receiving antenna is correctly polarized for the incoming plane wave, and is oriented for maximum reception. (Its main beam is aligned toward **S** and the antenna is rotated so that the electric field it would produce if transmitting is parallel to the electric field of the plane wave.) If this is not the case, it is not difficult to correctly degrade V_T later. If this is the case, $\langle |\mathbf{S}| \rangle = |E|^2/2\eta$ for the plane wave, and equation 16.66 becomes

$$A = \frac{\eta |V_T|^2}{4R_r |E|^2} \tag{16.67}$$

It is reasonable from Figure 16.8 and equation 16.67 that V_T and E are related, and V_T is E multiplied by some kind of length in order to be dimensionally correct. This length is called the effective length of the receiving antenna. Thus,

$$V_T = l_E E$$

or

$$\boxed{\frac{V_T}{E} = l_E} \tag{16.68}$$

and equation 16.67 becomes

$$A = \frac{\eta}{4R_r} |l_E|^2 \tag{16.69}$$

Notice that A is a real number only (equation 16.69), but it can be shown that l_E is, in general, a complex number.

It is somewhat tedius and beyond the scope of this treatment to derive a general equation for l_E, but conclusive results can be obtained in a simple manner anyway. There is one antenna we have studied whose actual length equals its effective length, and this occurs because it has the *same* current over its entire "length." It is the elementary Hertzian dipole. Substituting equation 16.42 for R_r in equation 16.69 we obtain

$$A = \frac{\lambda^2}{4\pi}(1.5)$$

and since $G = 1.5$ for the Hertzian dipole,

$$\boxed{A = \frac{\lambda^2 G}{4\pi}} \tag{16.70}$$

Equation 16.70 was derived for a special case, but the result applies to any antenna. For large aperture type antennas, such as parabolic dishes (fed at the focal point by a smaller antenna), the aperture is approximately equal to the frontal area of the parabola, and equation 16.70 affords an easy way to approximate the gain.

Equations 16.69 and 16.70 give us a way to calculate the effective length of any antenna. Substituting equation 16.69 into 16.70 and solving for $|l_E|$,

$$\boxed{|l_E| = \sqrt{\frac{GR_r}{\pi\eta}}\,\lambda \quad \text{(m)}} \tag{16.71}$$

For a given antenna *type* would you rather operate at a low frequency or a high frequency? Equation 16.71 gives an obvious answer.

A small dipole or monopole, which is half of a dipole driven against a large ground plane, can be made to perform much like the Hertzian dipole. That is, the current that normally would be zero at the ends of the dipole, or at the end of the monopole, can be increased by "end loading." One way to accomplish this is to connect radial conductors at the dipole ends and extend these conductors in a direction perpendicular to the dipole somewhat like the top of an umbrella. This is usually done at very low frequencies where a half-wave dipole or quarter-wave monopole cannot be built because of space limitations. The radiation resistance is very small for the Hertzian dipole or end loaded dipole, as has already been mentioned. The gain of a half-wave dipole is only slightly greater than that of the Hertzian dipole, but its radiation resistance is much greater, and hence its effective length according to equation 16.71 is greater. In addition, the ohmic resistance contributing to heat loss may be large compared to the radiation resistance for the shorter antennas, but for half-wave dipoles, the situation is reversed. Coupling to and from the half-wave dipole with transmission lines is simple since line impedances and dipole impedances are close to the same values, and matching is not difficult.

For those antennas where the identification of a terminal current is possible, we can use equations 16.41 (with $I = I_t$), 16.22, 16.56, and 16.59 in 16.71 to obtain

$$|l_E| = \frac{2r}{\eta|I_t|}\sqrt{(|E_{\theta_0}|^2 + |E_{\phi 0}|^2)}\,\lambda \tag{16.72}$$

EXAMPLE 4

What is the effective length of a half-wave dipole? Using equations 16.72 and 16.53 ($\theta = \pi/2$),

$$|l_E| = \frac{2r}{\eta|I_t|}|E_{\theta_0}|\lambda = \frac{2r}{\eta|I_t|}\frac{60|I_t|}{r}\lambda = \frac{\lambda}{\pi} = \frac{2}{k} \tag{16.73}$$

It is worth mentioning that for *any* angle θ equations 16.72 and 16.53 give

$$|l_E(\theta)| = \frac{2}{k}\frac{\cos(\pi/2 \cos\theta)}{\sin\theta} \tag{16.74}$$

Since $V_T = l_E E$, the "pattern" of a receiving antenna is the same as its transmitting pattern. Compare equation 16.74 to equation 16.53. The pattern factor $\cos(\pi/2 \cos\theta)/\sin\theta$, is the "degrading" or reduction factor for l_E mentioned earlier.

The behavior of an antenna above a perfectly conducting infinite ground plane can be understood with the use of image theory, which was introduced for the static case in Section 5.6. Proceeding here along the same lines as in Section 5.6, we want to find the field of vertical current element in the presence of a horizontal ground plane. We next want to find the field of a horizontal current element in the presence of the same ground plane. An arbitrarily directed current can be resolved into vertical and horizontal components, and the resultant field for the arbitrary current can be obtained as a superposition of the fields from the vertical and horizontal current elements. We will employ the Hertzian dipole since any current can be resolved into Hertzian dipoles because of the impulsive nature of this source.

Consider Figure 16.9(a). We want to find the tangential electrical field in the $z = 0$ plane due to the two symmetrically located Hertzian dipoles with codirected currents. The tangential field is E_ρ, and using A.2 in Appendix A $E_\rho = E_r \sin\theta + E_\theta \cos\theta$. The complete field of a single Hertzian dipole is listed below Example 1. Because of the symmetry in Fig. 16.9(a), the field for both dipoles can be put in the form

$$\begin{aligned} E_\rho &= A\cos\theta_1 \sin\theta_1 + B\sin\theta_1\cos\theta_1 \\ &\quad + A\cos\theta_2\sin\theta_2 + B\sin\theta_2\cos\theta_2, \qquad A = A(r), \qquad B = B(r) \\ E_\rho &= \frac{A+B}{2}(\sin 2\theta_1 + \sin 2\theta_2) \end{aligned}$$

Now, $\theta_1 = \pi - \theta_2$, so $\sin 2\theta_1 = -\sin 2\theta_2$, and $E_\rho = 0$, $z = 0$. In this case, the problem in Figure 16.9(a) may be replaced with the equivalent problem (for $z > 0$) of Figure 16.9(b), consisting of the original dipole at $z = h$ and an infinite conducting plane at $z = 0$, where $E_\rho = 0$. That is, in Figures 16.9(a) and (b) the sources are the same, the media are the same, and the boundary conditions are the same for $z > 0$; therefore the fields are the same for $z > 0$. There is no field for $z < 0$ in Figure 16.9(b). Thus, we say that the image of a vertical current above a perfectly conducting horizontal ground plane is a codirected current symmetrically located below the ground plane.

In order to use the dipole equations without modification for the case of currents parallel to the ground plane (which we called horizontal earlier), we place the ground plane in the $x = 0$ plane as shown in Figure 16.9(d), and

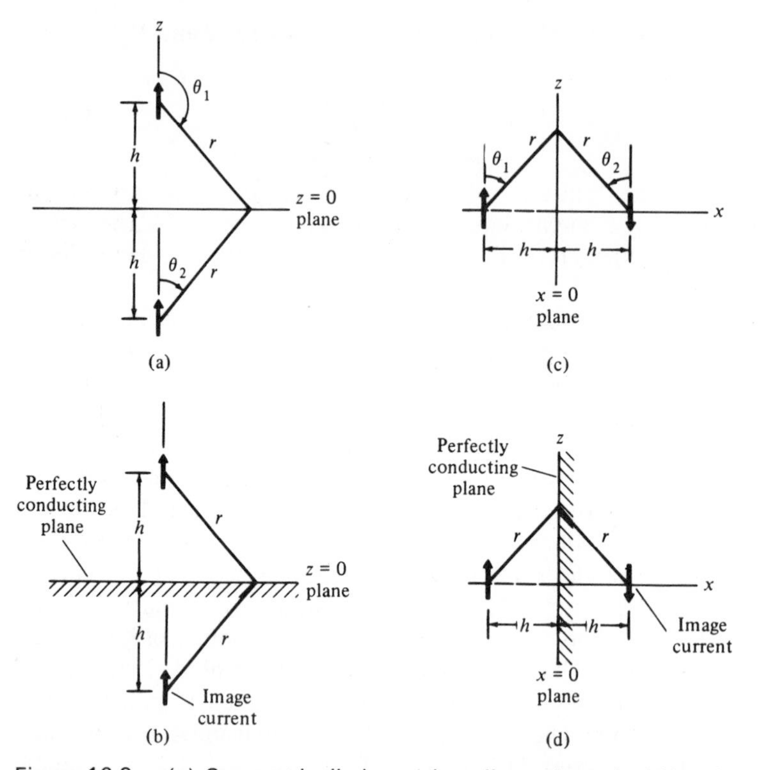

Figure 16.9. (a) Symmetrically located, codirected, vertical Hertzian dipoles perpendicular to the $z = 0$ plane. (b) Single Hertzian dipole located above an infinite, perfectly conducting ground plane at $z = 0$. This source is equivalent to those in (a) for $z > 0$. (c) Symmetrically located, oppositely directed Hertzian dipoles parallel to the $x = 0$ plane. (d) Single Hertzian dipole located to the left of the $x = 0$ plane. This source is equivalent to those in (c) for $x < 0$.

proceed as before. This time the tangential electric field is E_z, and $E_z = E_r \cos\theta - E_\theta \sin\theta$. Notice that the currents are oppositely directed. Thus,

$$\begin{aligned} E_z &= A\cos\theta_1\cos\theta_1 - B_1\sin\theta_1\sin\theta_1 \\ &\qquad - A\cos\theta_2\cos\theta_2 + B\sin\theta_2\sin\theta_2 \\ E_z &= 0, x = 0 \end{aligned}$$

since $\theta_1 = \theta_2$. The same remarks can now be made concerning equivalent problems in Figures 16.9(c) and (d). Thus, we say that the image of a horizontal current above a horizontal ground plane is an oppositely directed current symmetrically located below the ground plane.

A vertical monopole driven against a perfectly conducting horizontal ground plane has a vertical image current below, so that for the same terminal current it produces the same field ($z > 0$), but half the radiated power, and thus has half the radiation resistance as its isolated dipole counterpart.

The important quantities for the simple antennas we have discussed plus other types mentioned in the problems are summarized in Table 16.1. Notice that **E** as listed in the table is calculated assuming that the wire radius $a \ll \lambda$ or $ka \ll 1$. Thus, if the gain is calculated on the basis of this value of **E** it will not quite agree with the value of gain calculated by some other method (using equation 16.71, for example) if that method includes a finite value for a in every term. Numerical integration was employed in the two-term (equation 16.44) method, and this is another source for small errors. A description of the basic wire antennas is given below.

1. The dipole is oriented vertically along the z axis. l is the dipole length. Results apply for $kl \ll 1$, and if the current is essentially uniform across the dipole length and filamentary.
2. The monopole is oriented vertically and driven against a perfectly conducting ground plane at $z = 0$. The monopole height is $h = l/2$. Results apply for $kh \ll 1$, and if the current is essentially uniform across the monopole height and filamentary.
3. The dipole is oriented vertically along the z axis. l is the dipole length. The dipole current is $I_z(z) = I_t(1 - |z|/h)$ and is filamentary. See Problem 8. ($h = l/2$.)
4. The monopole is oriented vertically and driven against a perfectly conducting ground plane at $z = 0$. The monopole height is $h = l/2$. The monopole current is $I_z(z) = I_t(1 - z/h)$ and is filamentary.
5. The magnetic dipole here is a filamentary circular current loop of radius a (not the wire radius) lying in the $z = 0$ plane and centered at the origin with one turn. See Problems 4 and 5.
6. Same as 5 with N turns.
7. The half-wave dipole is oriented vertically along the z axis. The length is $\lambda/2$, and the current is $I_z(z) = I_t \cos kz$. $h = l/2 = \lambda/4$ and $h/a = 75$, where a is the wire radius. This value of a was used to calculate R_r, but a filamentary wire was used with negligible difference in results to calculate the other listed quantities.
8. The quarter-wave monopole is oriented vertically and driven against a perfectly conducting ground plane at $z = 0$. The height is $h = \lambda/4$ and $h/a = 75$, where a is the wire radius. This value of a was used to calculate R_r, but a filamentary wire was used with negligible difference in results to calculate the other listed quantities. $I_z(z) = I_t \cos kz$.
9. The folded half-wave dipole with sinusoidal current is a regular half-wave dipole with an extra wire folded back from one end to the other as in Figure 16.12. The rest of the description is the same as in 7.
10. Same as 8, except that an extra wire is connected to the end of the monopole and folded back parallel to the original wire and connected to the ground plane.

Table 16.1 BASIC WIRE ANTENNA PARAMETERS

ANTENNA	$\dfrac{\mathbf{E}}{(\eta/\pi)(e^{-jkr}/r)}$	$R_r\ (\Omega)$	$\lvert\mathbf{E}\rvert_{\text{rms}}$ (V/m) $\theta = \pi/2$ $\langle P_r\rangle = 10^3$ $r = 1$ mile	G	$\lvert l_E\rvert$ (m)
Isotropic	—	—	0.108	1.0	—
1. Hertzian dipole	$\dfrac{jklI}{4}\sin\theta\mathbf{a}_\theta$	$20(kl)^2$	0.132	1.5	l
2. Hertzian monopole	$\dfrac{jkhI}{2}\sin\theta\mathbf{a}_\theta$	$40(kh)^2$	0.186	3.0	$2h$
3. Short dipole	$\dfrac{jklI_t}{8}\sin\theta\mathbf{a}_\theta$	$5(kl)^2$	0.132	1.5	$\dfrac{l}{2}$
4. Short monopole	$\dfrac{jkhI_t}{4}\sin\theta\mathbf{a}_\theta$	$10(kh)^2$	0.186	3.0	h
5. Magnetic dipole	$\dfrac{\pi(ka)^2 I}{4}\sin\theta\mathbf{a}_\phi$	$20(\pi)^2(ka)^4$	0.132	1.5	πka^2

6. Magnetic dipole N-turn	$\frac{\pi(ka)^2 NI}{4} \sin\theta \mathbf{a}_\phi$	$20(N\pi)^2(ka)^4$	0.132	1.5	$N\pi ka^2$
7. $\lambda/2$ dipole sine-current	$\frac{jI_t}{2} \frac{\cos(\pi/2 \cos\theta)}{\sin\theta} \mathbf{a}_\theta$	73.08	0.138	1.64	$\frac{2}{k}$
8. $\lambda/4$ Mono. sine-current	$\frac{jI_t}{2} \frac{\cos(\pi/2 \cos\theta)}{\sin\theta} \mathbf{a}_\theta$	36.54	0.195	3.28	$\frac{2}{k}$
9. Folded $\lambda/2$ dipole sine-current	$jI_t \frac{\cos(\pi/2 \cos\theta)}{\sin\theta} \mathbf{a}_\theta$	292.3	0.138	1.64	$\frac{4}{k}$
10. Folded $\lambda/4$ mono. sine-current	$jI_t \frac{\cos(\pi/2 \cos\theta)}{\sin\theta} \mathbf{a}_\theta$	146.2	0.195	3.28	$\frac{4}{k}$
11. $\lambda/2$ dipole 2-term current	$\frac{jI_t}{2} \left[\frac{\cos(\pi/2 \cos\theta)}{\sin\theta} + \alpha \frac{1 + \cos(\pi/2 \cos\theta)}{4 - \cos^2\theta}\right] \mathbf{a}_\theta$	87.8	0.138	1.64	$\frac{2.19}{k}$
12. $\lambda/4$ mono. 2-term current	$\frac{jI_t}{2} \left[\frac{\cos(\pi/2 \cos\theta)}{\sin\theta} + \alpha \frac{1 + \cos(\pi/2 \cos\theta)}{4 - \cos^2\theta}\right] \mathbf{a}_\theta$	43.9	0.195	3.28	$\frac{2.19}{k}$

11. Same as 7, except that the current is $I_z(z) = I_t \cos kz + b \sin 2k|z|$, $\alpha = (2b/I_t) \sin\theta$, $(2b/I_t) = 0.178 - j0.236$.
12. Same as 8, except that the current is $I_z(z) = I_t \cos k_z + b \sin 2kz$, $\alpha = (2b/I_t) \sin\theta$, $(2b/I_t) = 0.178 - j0.236$.

16.5 THE FRIIS TRANSMISSION FORMULA

Having determined a number of useful antenna parameters, it is a relatively simple task to apply these to an elementary communications system. Consider Figure 16.10 where such a system is shown. Let us assume for the sake of

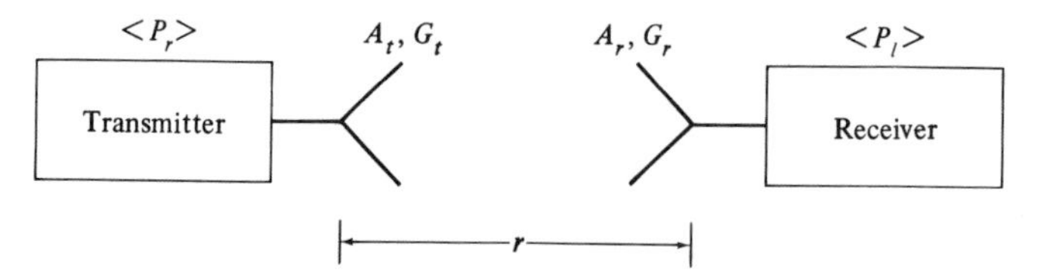

Figure 16.10. Elementary communication system.

simplicity that it is ideal. That is, the antennas and feedlines are lossless, the antennas are aligned in an optimum manner, and the surrounding medium is a lossless dielectric. These restrictions can be removed later without too much difficulty. In fact, we already know how to remove most of them.

As we have seen, the time average power density at the receiving antenna is given by

$$\langle|\mathbf{S}|\rangle = \frac{\langle P_r\rangle}{4\pi r^2} G_t \tag{16.75}$$

where $\langle P_r\rangle$, as before, is the time average transmitter power or power radiated (with no losses) and G_t is the transmitting antenna gain. The time average power delivered to a matched load, $\langle P_l\rangle$, is $\langle|\mathbf{S}|\rangle$ times the receiving antenna aperture A_r. Thus,

$$\langle P_l\rangle = \langle P_r\rangle \frac{A_r G_t}{4\pi r^2}\,. \tag{16.76}$$

Eliminating G_t by means of equation 16.70 gives

$$\boxed{\langle P_l\rangle = \langle P_r\rangle \frac{A_t A_r}{\lambda^2 r^2}} \tag{16.77}$$

while eliminating A_r by equation 16.70 gives

$$\boxed{\langle P_l\rangle = \langle P_r\rangle \frac{\lambda^2 G_t G_r}{16\pi^2 r^2}} \tag{16.78}$$

Either equation 16.77 or 16.78 is known as the *Friis transmission formula.*

EXAMPLE 5

Assume that a signal of 1 μV (rms) across a matched receiver load is the minimum value for an acceptable signal-to-noise ratio and bandwidth at 27×10^6 Hz (citizen's band). Assume 75-Ω lines and half-wave dipoles in free space on both ends with $\langle P_r \rangle = 5$ W. Find r_{max}.

$$\langle P_l \rangle = \frac{(10^{-6})^2}{75} = \frac{10^{-12}}{75} \quad \text{(W)}$$

Solving for r in equation 16.78

$$r_{max} = \sqrt{\frac{\langle P_r \rangle \lambda^2 G_t G_r}{\langle P_l \rangle 16\pi^2}}$$

or since $G_t = G_r = 1.64$,

$$r_{max} = \sqrt{\frac{\langle P_r \rangle (1.64\lambda)^2}{\langle P_l \rangle 16\pi^2}} = \sqrt{\frac{75(5)(300 \times 10^6 \times 1.64)^2}{10^{-12} 16(27 \times 10^6)^2 \pi^2}}$$

$$r_{max} = 28.08 \times 10^6 \quad \text{(m)}$$

This is an optimistic number for applications near the earth's surface because free space conditions were assumed, but it does demonstrate, for example, that communications in this mode could be easily established between the earth and a satellite.

16.6 CONCLUDING REMARKS

The basic concepts of radiation were introduced in this chapter. A general set of equations which governs the whole of radiation was derived. There are antennas, such as slot antennas and aperture antennas, for which other forms may be more useful, but the equations of Section 16.1 still form the basic theory. The elementary Hertzian dipole was considered as an example and was found to be very easy to analyze.

The popular half-wave dipole was partially analyzed for two reasons. First of all, it is widely used and understanding its behavior is important. Secondly, the distribution of current over the antenna surface is not actually known and until it can be found, or at least approximated, no antenna problem is simple. An approximate solution for the dipole current using an integral equation was mentioned. When the current distribution has been found, other antenna characteristics can be obtained in a straightforward manner.

The Hertzian dipole, as we have seen, is a mathematical convenience, and cannot be constructed in practice. *Short* dipoles are constructed in one form or another in practice for very low frequency (VLF) or long-wave radiation. What are the characteristics of a short dipole? This question is partly answered in Problem 8 at the end of this chapter.

REFERENCES

Harrington. (See references for Chapter 10.)
Jordan. (See references for Chapter 10.)
Neff, H. P., Siller, C. A., and Tillman, J. D. "A Trigonometric Approximation to the Current in the Solution of Hallén's Equation," *IEEE Trans. on Antennas and Propagation* AP 17, No. 6 (Nov. 1969): 805–806.
Weeks, W. L. *Antenna Engineering*. New York: McGraw-Hill, 1968.

PROBLEMS

1. Verify equations 16.20.
2. Find the complete field (not just the radiation field) of the Hertzian dipole. Start with equation 16.27.
3. Verify equation 16.40.
4. Find the radiation field of a magnetic dipole which is a small electric current loop of radius a in the $z = 0$ plane centered at the origin. Note that this field is dual to that of the Hertzian dipole. Use equation 16.21.
5. Use the results of Problem 4 to show that

$$R_{\text{rad}} = \eta \frac{k^4 s^2}{6\pi} \quad (\Omega)$$

for the magnetic dipole.
6. Sketch the normalized radiation field for the magnetic dipole.
7. (a) Find the radiation resistance of a Hertzian monopole which is half of the dipole driven against ground.
   (b) Calculate the value R_{rad} if $h(=l/2) = 15.24$ m and $f = 25$ kHZ.
   (c) Calculate the field for $\theta = \pi/2$ and a radiated power of 1 W.
8. A short dipole that is not top loaded has a current, which is given closely by $I_z(z) = I_t(1 - |z|/h)$, $h = l/2$. Show that $R_r = 20(kh)^2 = 5(kl)^2$, which is one-fourth that of the Hertzian dipole. See Figure 16.11. ($h \ll \lambda$)

Figure 16.11. The short dipole and its current.

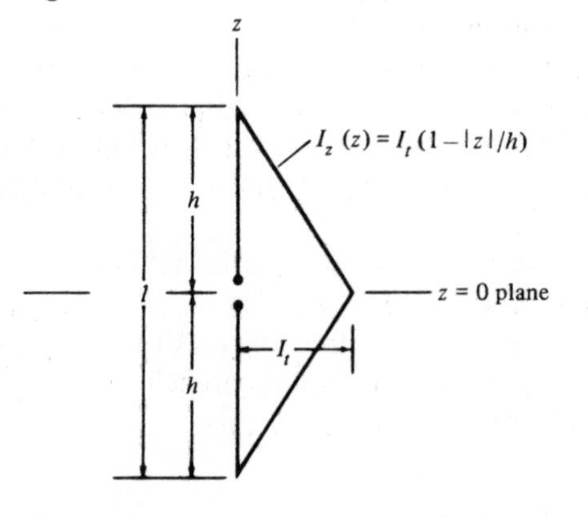

9. Show that equation 16.45 reduces to the current given in Problem 8 if $kh \ll 1$.
10. Starting with equation 16.21 show that

$$E_\theta = \frac{j\omega\mu}{4\pi}\frac{e^{-jkr}}{r}\sin\theta\int_{-h}^{h} I_z(z')e^{jkz'\cos\theta}\,dz'$$

for a filamentary dipole on the z axis.
11. If the current is assumed to be given by equation 16.45 show that E_θ in Problem 10, for $\theta = \pi/2$, is given by

$$E_\theta(r, \omega) = \frac{60}{r}e^{-j\omega r/c}I_t(\omega)j\tan\frac{\omega h}{2c}, \qquad c = \frac{1}{\sqrt{\mu\varepsilon}}$$

12. Find the unit impulse response (in time) for the field in Problem 11. That is, let $i_t(t) = \delta(t)$ so that $I_t(\omega) = 1$. Hint:

$$j\tan\frac{\omega h}{2c} = 1 + 2\sum_{n=1}^{\infty}(-1)^n e^{-jn\omega h/c}$$

Then use Appendix E. Interpret the result.
13. A $\lambda/2$ dipole is connected to a lossless 75-Ω balanced line.
   (a) What is the VSWR? Use equation 16.48.
   (b) If the incident power on the line is 1 kW, what is the maximum field at 1 mile? Assume that the field can be calculated accurately enough by equation 16.53.
   (c) Repeat (b) if the dipole is matched to the line.
14. Find R_r for the thin half-wave dipole with sinusoidal current if

$$\int_0^{\pi/2}\frac{\cos^2(\pi/2\cos\theta)}{\sin\theta}\,d\theta = 0.609$$

15. Derive equation 16.60 as indicated in the text.
16. Derive equation 16.7 as indicated in the text.
17. (a) Show that the rms field of a thin half-wave dipole with sinusoidal current is 138 mV/m at 1 mile for 1 kW and $\theta = \pi/2$.
   (b) Repeat (a) for 4 kW.
18. A thin half-wave dipole receiving antenna is connected to a 75-Ω line. A correctly polarized plane wave with $E = 100$ mV/m (peak) is incident on the dipole with **E** parallel to the dipole. Find $\langle P_l \rangle$.
19. Repeat Problem 18 if **E** is polarized at a 45° angle from the dipole axis but the plane of **E** and **H** is still parallel to the dipole axis.
20. Repeat Problem 18 if **E** and the dipole axis are in the same plane but inclined at 45°.
21. Repeat Problem 18 if $R_0 = 300$ Ω (R_l is still 75 Ω).
   (a) The line length is 5λ.
   (b) The line length is 4.25λ.

22. A folded half-wave dipole shown in Figure 16.12 produces essentially

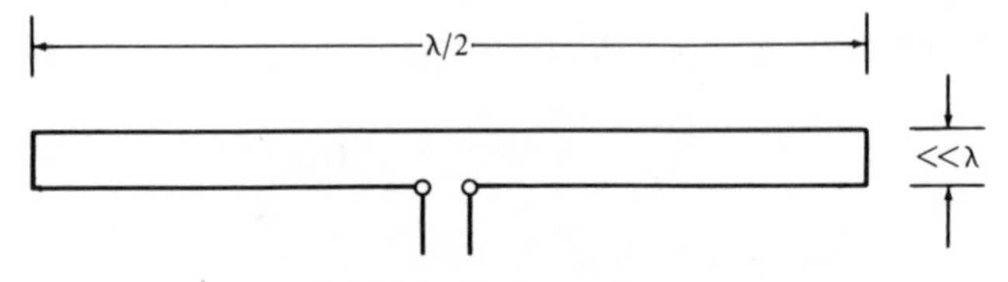

Figure 16.12. A folded half-wave dipole.

the same field as the ordinary half-wave dipole for the same power, but has approximately four times the impedance. Assuming sinusoidal current, find:

(a) The impedance.
(b) The gain.
(c) The effective length.
(d) $\langle P_l \rangle$ in a matched load.
(e) Compare the answer in (d) to that for the ordinary half-wave dipole.

Hint: The folded dipole is essentially two coincident half-wave dipoles as far as the radiated field is concerned.

23. What happens to the power in Z_r of Figure 16.8?
24. The effective radiated video power for a channel 6 television station (83.25×10^6 Hz) is 100 kW. Calculate $\langle P_l \rangle$ for a receiving antenna with a gain of 5 if the range is 100 km.
25. A certain antenna has an effective length $|l_E| = 2.5/k$ and a radiation resistance of 100 Ω. What is its gain? What rms field does it produce for 1 kW at 1 mile?
26. The television transmitting antenna of Example 3 produces omnidirectional coverage in azimuth. What is its beamwidth in the vertical plane?

27. (a) Starting with the results of Problem 10, use equation 16.45 and find the field for a thin dipole of any length carrying sinusoidal current.

$$E_\theta = j60 \frac{e^{-jkr}}{r} \frac{I_t}{\sin kh} \frac{\cos(kh\cos\theta) - \cos kh}{\sin\theta}, \quad \eta = 120\pi$$

(b) Plot $|E_\theta|$ (normalized) versus θ for $h = \lambda/2$ (full-wave dipole).

28. Starting with equation 16.59 show that the gain of a dipole carrying sinusoidal current (Problem 27) is ($\theta = \pi/2$)

$$G = \frac{120}{R_r} \tan^2 \frac{kh}{2}$$

29. The following data are obtained by numerical integration for a thin dipole with sinusoidal current.

h/λ	0.20	0.25	0.30	0.35	0.40	0.45
R_r	40.9	73.1	132.7	253.6	578.9	2241

h/λ	0.50	0.55	0.60	0.65	0.70	0.75
R_r	∞	1728	376.3	140.6	92.9	104

Plot $|E_\theta|$ at 1 mile for 1 kW and $\theta = \pi/2$ versus h/λ. Why is a $5\lambda/8$ monopole often used in practice?

30. Given that

$$\int_0^{2\pi} e^{jka \sin \theta \cos \phi'}\, d\phi' = 2\pi a J_0(ka \sin \theta)$$

and assuming that R_r is still given by equation 16.47, what is $|E_\theta|$ for a $\lambda/2$ dipole at 1 mile for 1 kW, $\theta = \pi/2$ and $h/a = 75$? $J_0(x) = 1 - x^2/4 + x^4/64 - \cdots$.

31. As an example of improving the directional characteristics of an antenna, consider first a pair of isotropic sources located at $(0, 0, d/2)$ and $(0, 0, -d/2)$ with currents $1\angle 0°$ and $1\angle 0°$ as shown in Figure 16.13. Use the origin as a phase reference.

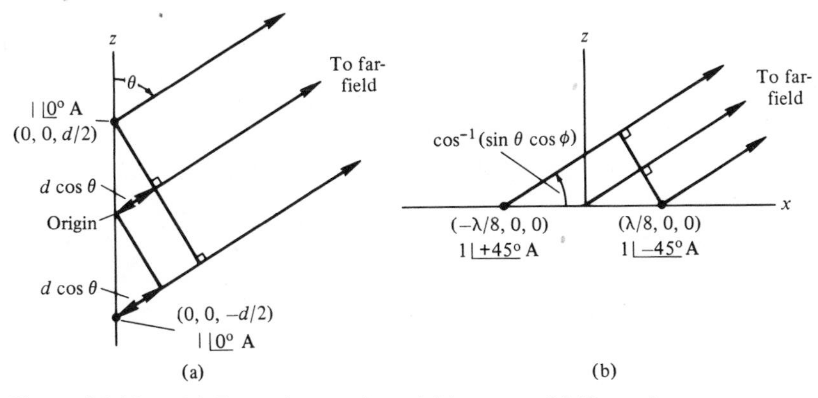

Figure 16.13. (a) Two-element broadside array. (b) Two-element endfire array. Hypothetical isotropic sources shown.

(a) Show that the normalized field, called the "array factor" will be $2 \cos (kd \cos \theta/2)$ in the far-field.

(b) Replace the isotropic sources with half-wave dipoles lying along the z axis and show that the field is

$$E_\theta = j120\, \frac{e^{-jkr}}{r}\, \frac{\cos^2 (\pi/2 \cos \theta)}{\sin \theta}$$

if $d = \lambda/2$. This is a two-element "broadside" array.

(c) Plot E_θ versus θ.

(d) Repeat (a), (b), and (c) for a pair of half-wave dipoles at $(\lambda/8, 0, 0)$ and $(-\lambda/8, 0, 0)$ with current $1\underline{|-45^\circ}$ and $1\underline{|+45^\circ}$, respectively, and alignment with the z axis. This is a two-element "endfire" array. Make the plot in the $y = 0$ plane.

Appendix A
Vector Relations

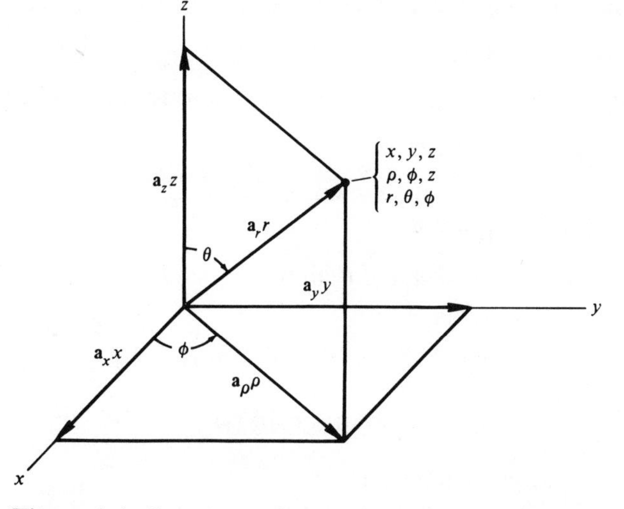

Figure A.1. Geometry of a standard coordinate system: rectangular (x, y, z), cylindrical (ρ, ϕ, z) and spherical (r, θ, ϕ).

UNIT VECTORS

$\mathbf{a}_x, \mathbf{a}_y, \mathbf{a}_z$—rectangular (constant)
$\mathbf{a}_\rho, \mathbf{a}_\phi, \mathbf{a}_z$—cylindrical (nonconstant)
$\mathbf{a}_r, \mathbf{a}_\theta, \mathbf{a}_\phi$—spherical (nonconstant)

COORDINATE TRANSFORMATIONS

$$\begin{aligned}
x &= \rho \cos \phi = r \sin \theta \cos \phi \\
y &= \rho \sin \phi = r \sin \theta \sin \phi \\
z &= r \cos \theta \\
\rho &= \sqrt{x^2 + y^2} = r \sin \theta \\
\phi &= \tan^{-1} y/x \\
r &= \sqrt{x^2 + y^2 + z^2} = \sqrt{\rho^2 + z^2} \\
\theta &= \tan^{-1} \sqrt{x^2 + y^2}/z = \tan^{-1} \rho/z
\end{aligned} \tag{A.1}$$

COORDINATE COMPONENT TRANSFORMATIONS

$$\begin{aligned}
A_x &= A_\rho \cos \phi - A_\phi \sin \phi \\
&= A_r \sin \theta \cos \phi + A_\theta \cos \theta \cos \phi - A_\phi \sin \phi \\
A_y &= A_\rho \sin \phi + A_\phi \cos \phi \\
&= A_r \sin \theta \sin \phi + A_\theta \cos \theta \sin \phi + A_\phi \cos \phi \\
A_z &= A_r \cos \theta - A_\theta \sin \theta \\
A_\rho &= A_x \cos \phi + A_y \sin \phi = A_r \sin \theta + A_\theta \cos \theta \\
A_\phi &= -A_x \sin \phi + A_y \cos \phi \\
A_r &= A_x \sin \theta \cos \phi + A_y \sin \theta \sin \phi + A_z \cos \theta \\
&= A_\rho \sin \theta + A_z \cos \theta \\
A_\theta &= A_x \cos \theta \cos \phi + A_y \cos \theta \sin \phi - A_z \sin \theta \\
&= A_\rho \cos \theta - A_z \sin \theta
\end{aligned} \tag{A.2}$$

DIFFERENTIAL ELEMENTS OF VECTOR LENGTH

$$d\mathbf{l} = \begin{cases} \mathbf{a}_x\, dx + \mathbf{a}_y\, dy + \mathbf{a}_z\, dz \\ \mathbf{a}_\rho\, d\rho + \mathbf{a}_\phi \rho\, d\phi + \mathbf{a}_z\, dz \\ \mathbf{a}_r\, dr + \mathbf{a}_\theta\, r\, d\theta + \mathbf{a}_\phi r \sin \theta\, d\phi \end{cases} \tag{A.3}$$

DIFFERENTIAL ELEMENTS OF VECTOR AREA

$$d\mathbf{s} = \begin{cases} \mathbf{a}_x\, dy\, dz + \mathbf{a}_y\, dx\, dz + \mathbf{a}_z\, dx\, dy \\ \mathbf{a}_\rho \rho\, d\phi\, dz + \mathbf{a}_\phi\, d\rho\, dz + \mathbf{a}_z \rho\, d\rho\, d\phi \\ \mathbf{a}_r r^2 \sin\theta\, d\theta\, d\phi + \mathbf{a}_\theta r \sin\theta\, dr\, d\phi + \mathbf{a}_\phi r\, dr\, d\theta \end{cases} \tag{A.4}$$

DIFFERENTIAL ELEMENTS OF VOLUME

$$dv = \begin{cases} dx\, dy\, dz \\ \rho\, d\rho\, d\phi\, dz \\ r^2 \sin\theta\, dr\, d\theta\, d\phi \end{cases} \tag{A.5}$$

VECTOR OPERATIONS—RECTANGULAR COORDINATES

$$\nabla\alpha = \mathbf{a}_x \frac{\partial\alpha}{\partial x} + \mathbf{a}_y \frac{\partial\alpha}{\partial y} + \mathbf{a}_z \frac{\partial\alpha}{\partial z}$$

$$\nabla \cdot \mathbf{A} = \frac{\partial A_x}{\partial x} + \frac{\partial A_y}{\partial y} + \frac{\partial A_z}{\partial z}$$

$$\nabla \times \mathbf{A} = \mathbf{a}_x\left(\frac{\partial A_z}{\partial y} - \frac{\partial A_y}{\partial z}\right) + \mathbf{a}_y\left(\frac{\partial A_x}{\partial z} - \frac{\partial A_z}{\partial x}\right) + \mathbf{a}_z\left(\frac{\partial A_y}{\partial x} - \frac{\partial A_x}{\partial y}\right) \tag{A.6}$$

$$\nabla^2\alpha = \frac{\partial^2\alpha}{\partial x^2} + \frac{\partial^2\alpha}{\partial y^2} + \frac{\partial^2\alpha}{\partial z^2} \equiv \nabla \cdot \nabla\alpha$$

$$\nabla^2\mathbf{A} = \mathbf{a}_x\nabla^2 A_x + \mathbf{a}_y\nabla^2 A_y + \mathbf{a}_z\nabla^2 A_z \equiv \nabla(\nabla \cdot \mathbf{A}) - \nabla \times (\nabla \times \mathbf{A})$$

VECTOR OPERATIONS—CYLINDRICAL COORDINATES

$$\nabla\alpha = \mathbf{a}_\rho \frac{\partial\alpha}{\partial\rho} + \mathbf{a}_\phi \frac{1}{\rho}\frac{\partial\alpha}{\partial\phi} + \mathbf{a}_z \frac{\partial\alpha}{\partial z}$$

$$\nabla \cdot \mathbf{A} = \frac{1}{\rho}\frac{\partial}{\partial\rho}(\rho A_\rho) + \frac{1}{\rho}\frac{\partial A_\phi}{\partial\phi} + \frac{\partial A_z}{\partial z}$$

$$\nabla \times \mathbf{A} = \mathbf{a}_\rho\left(\frac{1}{\rho}\frac{\partial A_z}{\partial\phi} - \frac{\partial A_\phi}{\partial z}\right) + \mathbf{a}_\phi\left(\frac{\partial A_\rho}{\partial z} - \frac{\partial A_z}{\partial\rho}\right) + \mathbf{a}_z \frac{1}{\rho}\left[\frac{\partial}{\partial\rho}(\rho A_\phi) - \frac{\partial A_\rho}{\partial\phi}\right]$$

$$\nabla^2\alpha = \frac{1}{\rho}\frac{\partial}{\partial\rho}\left(\rho\frac{\partial\alpha}{\partial\rho}\right) + \frac{1}{\rho^2}\frac{\partial^2\alpha}{\partial\phi^2} + \frac{\partial^2\alpha}{\partial z^2}$$

$$\nabla^2\mathbf{A} = \mathbf{a}_\rho\left(\nabla^2 A_\rho - \frac{2}{\rho^2}\frac{\partial A_\phi}{\partial\phi} - \frac{A_\rho}{\rho^2}\right) + \mathbf{a}_\phi\left(\nabla^2 A_\phi + \frac{2}{\rho^2}\frac{\partial A_\rho}{\partial\phi} - \frac{A_\phi}{\rho^2}\right) + \mathbf{a}_z\nabla^2 A_z \tag{A.7}$$

VECTOR OPERATIONS—SPHERICAL COORDINATES

$$\nabla\alpha = \mathbf{a}_r \frac{\partial\alpha}{\partial r} + \mathbf{a}_\theta \frac{1}{r}\frac{\partial\alpha}{\partial\theta} + \mathbf{a}_\phi \frac{1}{r\sin\theta}\frac{\partial\alpha}{\partial\phi}$$

$$\nabla\cdot\mathbf{A} = \frac{1}{r^2}\frac{\partial}{\partial r}(r^2 A_r) + \frac{1}{r\sin\theta}\frac{\partial}{\partial\theta}(A_\theta\sin\theta) + \frac{1}{r\sin\theta}\frac{\partial A_\phi}{\partial\phi}$$

$$\nabla\times\mathbf{A} = \mathbf{a}_r \frac{1}{r\sin\theta}\left[\frac{\partial}{\partial\theta}(A_\phi\sin\theta) - \frac{\partial A_\theta}{\partial\phi}\right] + \mathbf{a}_\theta \frac{1}{r}\left[\frac{1}{\sin\theta}\frac{\partial A_r}{\partial\phi} - \frac{\partial}{\partial r}(rA_\phi)\right] + \mathbf{a}_\phi \frac{1}{r}\left[\frac{\partial}{\partial r}(rA_\theta) - \frac{\partial A_r}{\partial\theta}\right]$$

$$\nabla^2\alpha = \frac{1}{r^2}\frac{\partial}{\partial r}\left(r^2\frac{\partial\alpha}{\partial r}\right) + \frac{1}{r^2\sin\theta}\frac{\partial}{\partial\theta}\left(\sin\theta\frac{\partial\alpha}{\partial\theta}\right) + \frac{1}{r^2\sin^2\theta}\frac{\partial^2\alpha}{\partial\phi^2}$$

$$\begin{aligned}\nabla^2\mathbf{A} = {} & \mathbf{a}_r\left[\nabla^2 A_r - \frac{2}{r^2}\left(A_r + \text{ctn}\,\theta A_\theta + \csc\theta\frac{\partial A_\phi}{\partial\phi} + \frac{\partial A_\theta}{\partial\theta}\right)\right] \\ & + \mathbf{a}_\theta\left[\nabla^2 A_\theta - \frac{1}{r^2}\left(\csc^2\theta A_\theta - 2\frac{\partial A_r}{\partial\theta} + 2\,\text{ctn}\,\theta\csc\theta\frac{\partial A_\phi}{\partial\phi}\right)\right] \\ & + \mathbf{a}_\phi\left[\nabla^2 A_\phi - \frac{1}{r^2}\left(\csc^2\theta A_\phi - 2\csc\theta\frac{\partial A_r}{\partial\phi} - 2\,\text{ctn}\,\theta\csc\theta\frac{\partial A_\theta}{\partial\phi}\right)\right]\end{aligned} \tag{A.8}$$

ADDITION AND MULTIPLICATION

$$\begin{aligned} A^2 &= \mathbf{A}\cdot\mathbf{A} \\ |A|^2 &= \mathbf{A}\cdot\mathbf{A}^* \\ \mathbf{A}+\mathbf{B} &= \mathbf{B}+\mathbf{A} \\ \mathbf{A}\cdot\mathbf{B} &= \mathbf{B}\cdot\mathbf{A} \\ \mathbf{A}\times\mathbf{B} &= -\mathbf{B}\times\mathbf{A} \\ (\mathbf{A}+\mathbf{B})\cdot\mathbf{C} &= \mathbf{A}\cdot\mathbf{C}+\mathbf{B}\cdot\mathbf{C} \\ (\mathbf{A}+\mathbf{B})\times\mathbf{C} &= \mathbf{A}\times\mathbf{C}+\mathbf{B}\times\mathbf{C} \\ \mathbf{A}\times(\mathbf{B}\times\mathbf{C}) &= (\mathbf{A}\cdot\mathbf{C})\mathbf{B}-(\mathbf{A}\cdot\mathbf{B})\mathbf{C} \\ \mathbf{A}\cdot(\mathbf{B}\times\mathbf{C}) &= \mathbf{B}\cdot(\mathbf{C}\times\mathbf{A}) = \mathbf{C}\cdot(\mathbf{A}\times\mathbf{B}) \end{aligned} \tag{A.9}$$

DIFFERENTIATION

$$\nabla(\alpha + \beta) = \nabla\alpha + \nabla\beta$$

$$\nabla \cdot (\mathbf{A} + \mathbf{B}) = \nabla \cdot \mathbf{A} + \nabla \cdot \mathbf{B}$$

$$\nabla \times (\mathbf{A} + \mathbf{B}) = \nabla \times \mathbf{A} + \nabla \times \mathbf{B}$$

$$\nabla(\alpha\beta) = \alpha\nabla\beta + \beta\nabla\alpha$$

$$\nabla \cdot (\alpha\mathbf{A}) = \alpha\nabla \cdot \mathbf{A} + \mathbf{A} \cdot \nabla\alpha$$

$$\nabla \times (\alpha\mathbf{A}) = \alpha(\nabla \times \mathbf{A}) - \mathbf{A} \times \nabla\alpha \tag{A.10}$$

$$\nabla \cdot (\mathbf{A} \times \mathbf{B}) = \mathbf{B} \cdot (\nabla \times \mathbf{A}) - \mathbf{A} \cdot (\nabla \times \mathbf{B})$$

$$\nabla^2\mathbf{A} = \nabla(\nabla \cdot \mathbf{A}) - \nabla \times (\nabla \times \mathbf{A})$$

$$\nabla \times (\alpha\nabla\beta) = \nabla\alpha \times \nabla\beta$$

$$\nabla \times \nabla\alpha = 0$$

$$\nabla \cdot (\nabla \times \mathbf{A}) = 0$$

$$\nabla \times (\mathbf{A} \times \mathbf{B}) = \mathbf{A}(\nabla \cdot \mathbf{B}) + (\mathbf{B} \cdot \nabla)\mathbf{A} - \mathbf{B}(\nabla \cdot \mathbf{A}) - (\mathbf{A} \cdot \nabla)\mathbf{B}$$

INTEGRATION

$$\iiint_{\text{vol}} \nabla \cdot \mathbf{A}\, dv = \oiint_s \mathbf{A} \cdot d\mathbf{s} \quad \text{(divergence theorem)}$$

$$\iint_s \nabla \times \mathbf{A} \cdot d\mathbf{s} = \oint_l \mathbf{A} \cdot d\mathbf{l} \quad \text{(Stoke's theorem)}$$

$$\iiint_{\text{vol}} \nabla \times \mathbf{A}\, dv = -\oiint_s \mathbf{A} \times d\mathbf{s} \tag{A.11}$$

$$\iiint_{\text{vol}} \nabla\alpha\, dv = \oiint_s \alpha\, d\mathbf{s}$$

$$\iint_s \mathbf{a}_n \times \nabla\alpha\, ds = \oint_l \alpha\, d\mathbf{l}$$

Appendix B
Physical Constants

Permittivity of free space (vacuum), $\varepsilon_0 = 8.854 \times 10^{-12}$ F/m $\approx \dfrac{10^{-9}}{36\pi}$ F/m

Permeability of free space (vacuum), $\mu_0 \equiv 4\pi \times 10^{-7}$ H/m

Electron charge magnitude, $e = 1.602 \times 10^{-19}$ C

Electron rest mass, $m_e = 9.109 \times 10^{-31}$ kg

Proton rest mass, $m_p = 1.673 \times 10^{-27}$ kg

Speed of light in free space, $\dfrac{1}{\sqrt{\mu_0 \varepsilon_0}} = 2.998 \times 10^8$ m/s

Appendix C
Material Parameters

Representative values of conductivity, relative permittivity, and loss tangent for various materials are listed below. Most of the data have been taken from the references listed below.[1,2,3]

Table C.1 REPRESENTATIVE VALUES OF CONDUCTIVITY, σ_c, AT ROOM TEMPERATURE AND $f = 0$

MATERIAL	σ_c(℧/m)
Silver	6.17×10^7
Copper	5.80×10^7
Gold	4.10×10^7
Aluminum	3.82×10^7
Tungsten	1.82×10^7
Brass	1.5×10^7
Solder	0.7×10^7
Nichrome	0.1×10^7
Seawater	4×10^0
Ferrites	$\approx 10^{-2}$
Fresh water	$\approx 10^{-3}$
Porcelain	$\approx 10^{-10}$

[1] *Dielectric Materials and Applications*, Part V, Technology Press, M.I.T., Cambridge, Mass., 1954.

[2] D. G. Fink and J. M. Carroll, *Standard Handbook for Electrical Engineers*, 10th ed., McGraw-Hill, New York, 1968.

[3] International Telephone and Telegraph Company, Inc., *Reference Data for Radio Engineers*, 5th ed., Howard W. Sams & Co., Indianapolis, Ind., 1968.

Table C.2 REPRESENTATIVE VALUES OF RELATIVE PERMITTIVITY (DIELECTRIC CONSTANT), ε_R, AT ROOM TEMPERATURE AND f = 100 Hz

MATERIAL	ε_R
Bakelite	8.2
Clay soil (dry)	4.73
Ice	4.2
Lucite	3.2
Nylon	3.88
Plexiglas	3.40
Polyethylene	2.25
Polystyrene	2.56
Sandy soil (dry)	3.42
Styrafoam	1.03
Teflon	2.1
Water	78.0

Table C.3 REPRESENTATIVE VALUES OF LOSS TANGENT, $\sigma_d/\omega\varepsilon$, FOR VARIOUS MATERIALS AT ROOM TEMPERATURE. FREQUENCY (Hz)

MATERIAL	10^2	10^4	10^6	10^8	10^{10}
Bakelite	0.134	0.0631	0.0593	0.0773	0.0369
Clay soil (dry)	0.121	0.119	0.0661		0.013
Loamy soil (dry)	0.0686	0.0353	0.0182		0.00139
Lucite	0.0625	0.0315	0.0144	0.00678	0.00319
Nylon	0.0144	0.0233	0.0258	0.0209	
Plexiglas	0.0603	0.0300	0.0139		0.00676
Polyethylene	4.89×10^{-4}	3.11×10^{-4}	4.00×10^{-4}		4.00×10^{-4}
Polystyrene	5.08×10^{-5}	5.08×10^{-5}	7.03×10^{-5}	11.8×10^{-5}	43.3×10^{-5}
Sandy soil (dry)	0.196	0.0342	0.0170		0.00364
Styrafoam	19.4×10^{-5}	10.3×10^{-5}	19.4×10^{-5}		14.6×10^{-5}
Teflon	52.4×10^{-5}	33.3×10^{-5}	19.0×10^{-5}	19.0×10^{-5}	38.5×10^{-5}
Water			0.0190	0.00701	1.03

Appendix D
Fourier Series

A periodic function $f(x)$ with period L, or a nonperiodic function defined over an interval of length L and periodically extended, which satisfies the Dirichlet conditions, can be expanded into a Fourier series.

$$f(x) = \frac{A_0}{2} + \sum_{n=1}^{\infty} \left[A_n \cos\left(\frac{2n\pi x}{L}\right) + B_n \sin\left(\frac{2n\pi x}{L}\right) \right] \tag{D.1}$$

where

$$A_n = \frac{2}{L} \int_0^L f(x) \cos\left(\frac{2n\pi x}{L}\right) dx \tag{D.2}$$

and

$$B_n = \frac{2}{L} \int_0^L f(x) \sin\left(\frac{2n\pi x}{L}\right) dx \tag{D.3}$$

A trigonometric series of the form of equation D.1 is a Fourier series with a least-mean-square error approximation to $f(x)$ if, and only if, the coefficients are given by equations D.2 and D.3 (Euler formulas). The independent variable x may be distance, angle, or time.

The exponential form of the Fourier series is

$$f(x) = \sum_{n=-\infty}^{\infty} C_n e^{j(2n\pi x/L)} \tag{D.4}$$

where

$$C_n = \frac{1}{L} \int_0^L f(x) e^{-j(2n\pi x/L)}\, dx \tag{D.5}$$

The Fourier series expresses a function as a superposition of *discrete* sinusoidal terms.

Appendix E
Fourier Transforms

A function $f(x)$ can also be represented as a superposition of sinusoidal functions in an infinite interval, forming a *continuous*, rather than discrete, spectrum. This representation is a logical extension of the exponential Fourier series representation of Appendix D. The infinite sum becomes an infinite integral.

$$f(x) = \frac{1}{2\pi} \int_{-\infty}^{\infty} F(k_x) e^{jxk_x} \, dk_x \tag{E.1}$$

where

$$F(k_x) = \int_{-\infty}^{\infty} f(x) e^{-jk_x x} \, dx \tag{E.2}$$

$F(k_x)$ as given by equation E.2 is called the *Fourier transform* of $f(x)$, and $f(x)$ as given by equation E.1 is the *inverse Fourier transform.* Together, equations E.1 and E.2 form a *Fourier transform* pair. If x is distance (meters), then k_x is wave number (meters^{-1}).

If time (t) is the independent variable, then radian frequency (ω) is the transform variable, and the transform pair is simply

$$f(t) = \frac{1}{2\pi} \int_{-\infty}^{\infty} F(\omega) e^{jt\omega} \, d\omega \tag{E.3}$$

$$F(\omega) = \int_{-\infty}^{\infty} f(t) e^{-j\omega t} \, dt \tag{E.4}$$

This usage is more frequently encountered in engineering problems than equations E.1 and E.2. A short list of Fourier transform pairs is given in Table E.1 (page 548).

Table E.1 FOURIER TRANSFORM PAIRS

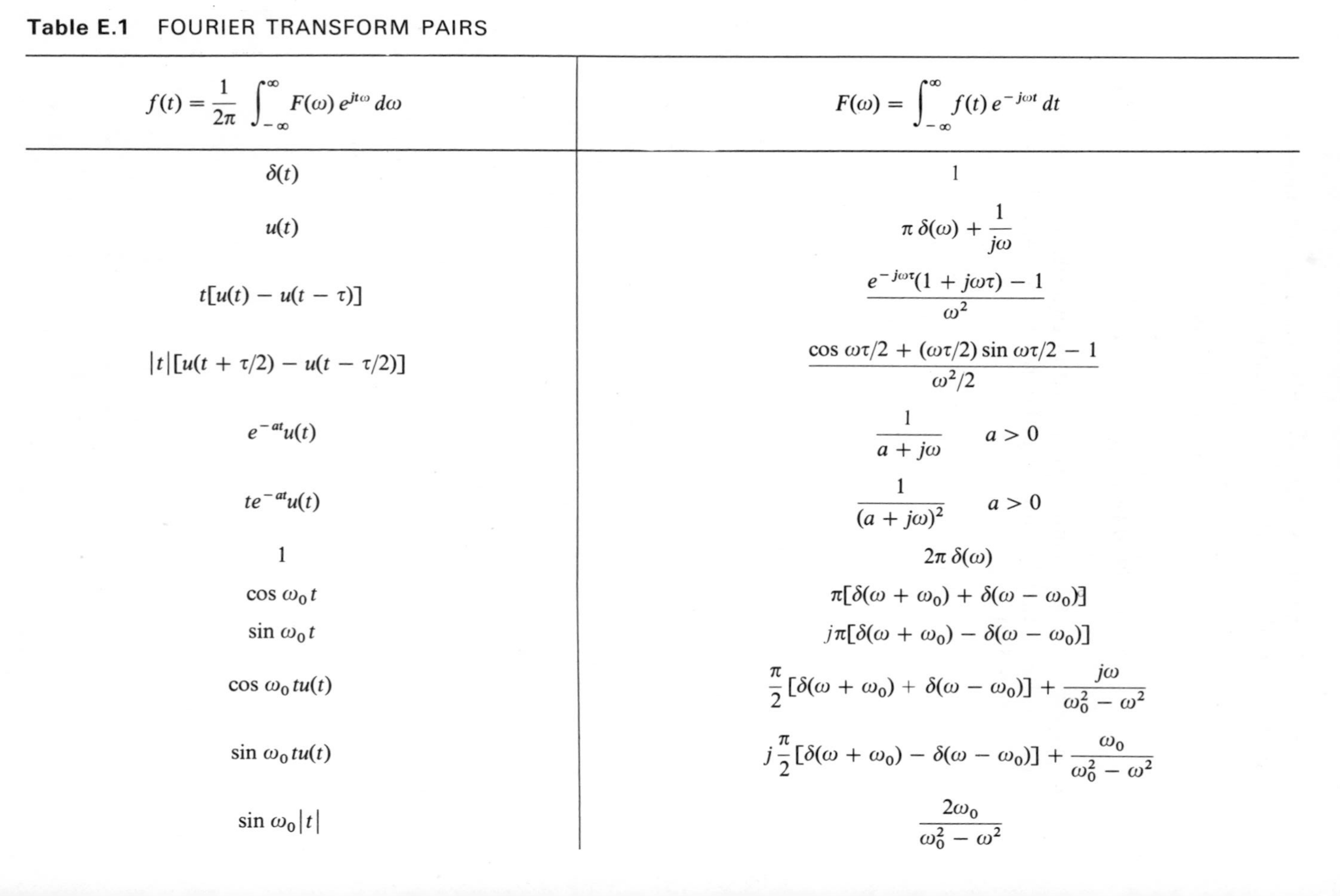

$f(t) = \frac{1}{2\pi} \int_{-\infty}^{\infty} F(\omega)\, e^{jt\omega}\, d\omega$	$F(\omega) = \int_{-\infty}^{\infty} f(t)\, e^{-j\omega t}\, dt$
$\delta(t)$	1
$u(t)$	$\pi\, \delta(\omega) + \frac{1}{j\omega}$
$t[u(t) - u(t - \tau)]$	$\frac{e^{-j\omega\tau}(1 + j\omega\tau) - 1}{\omega^2}$
$\|t\|[u(t + \tau/2) - u(t - \tau/2)]$	$\frac{\cos \omega\tau/2 + (\omega\tau/2) \sin \omega\tau/2 - 1}{\omega^2/2}$
$e^{-at}u(t)$	$\frac{1}{a + j\omega} \quad a > 0$
$te^{-at}u(t)$	$\frac{1}{(a + j\omega)^2} \quad a > 0$
1	$2\pi\, \delta(\omega)$
$\cos \omega_0 t$	$\pi[\delta(\omega + \omega_0) + \delta(\omega - \omega_0)]$
$\sin \omega_0 t$	$j\pi[\delta(\omega + \omega_0) - \delta(\omega - \omega_0)]$
$\cos \omega_0 t u(t)$	$\frac{\pi}{2}[\delta(\omega + \omega_0) + \delta(\omega - \omega_0)] + \frac{j\omega}{\omega_0^2 - \omega^2}$
$\sin \omega_0 t u(t)$	$j\frac{\pi}{2}[\delta(\omega + \omega_0) - \delta(\omega - \omega_0)] + \frac{\omega_0}{\omega_0^2 - \omega^2}$
$\sin \omega_0 \|t\|$	$\frac{2\omega_0}{\omega_0^2 - \omega^2}$

$f(t)$	$F(\omega)$
$F(t)$	$2\pi f(-\omega)$ symmetry
$f(at)$	$\frac{1}{\lvert a\rvert} F\left(\frac{\omega}{a}\right)$ scaling
$f(t-t_0)$	$F(\omega)e^{-j\omega t_0}$ time shift
$f(t)e^{j\omega_0 t}$	$F(\omega-\omega_0)$ frequency shift
$\frac{d^n f(t)}{dt^n}$	$(j\omega)^n F(\omega)$ time differentiation
$(-jt)^n f(t)$	$\frac{d^n F(\omega)}{d\omega^n}$ frequency differentiation
$\int_{-\infty}^{t} f(\tau)\,d\tau$	$\frac{F(\omega)}{j\omega} + \pi F(0)\,\delta(\omega)$ time integration
$f_1(t) * f_2(t) = \int_{-\infty}^{\infty} f_1(\tau) f_2(t-\tau)\,d\tau$	$F_1(\omega)F_2(\omega)$ time convolution
$f_1(t) f_2(t)$	$\frac{1}{2\pi}[F_1(\omega) * F_2(\omega)] = \frac{1}{2\pi}\int_{-\infty}^{\infty} F_1(\eta)F_2(\omega-\eta)\,d\eta$ frequency convolution
$\frac{1}{t} e^{-[(a+b)/2]t} I_1\left(\frac{a-b}{2} t\right)^*$	$\frac{\sqrt{j\omega+a}-\sqrt{j\omega+b}}{\sqrt{j\omega+a}+\sqrt{j\omega+b}}$
$\delta(t-t_0) - a^2 t_0 \frac{J_1(a\sqrt{t^2-t_0^2})^*}{\sqrt{t^2-t_0^2}} u(t-t_0)$	$e^{-t_0\sqrt{a^2-\omega^2}}$

Table E.1 *(continued)*

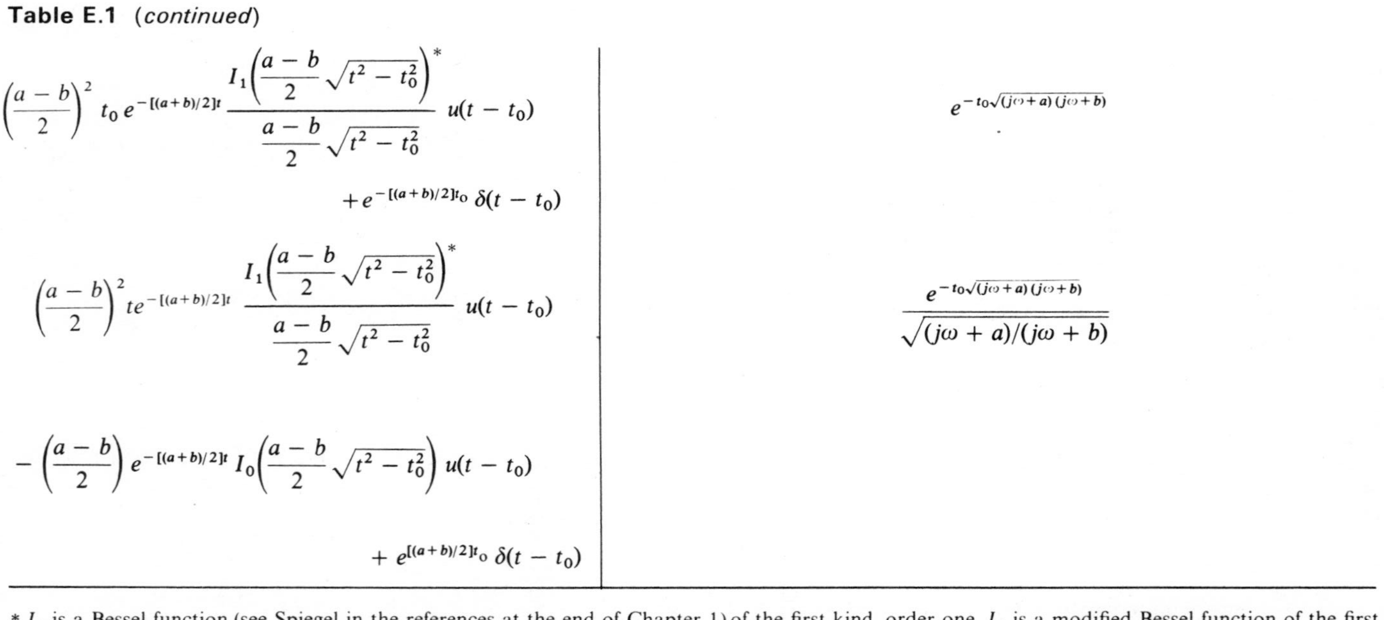

$\left(\frac{a-b}{2}\right)^2 t_0 e^{-[(a+b)/2]t} \frac{I_1\left(\frac{a-b}{2}\sqrt{t^2-t_0^2}\right)^*}{\frac{a-b}{2}\sqrt{t^2-t_0^2}} u(t-t_0) + e^{-[(a+b)/2]t_0}\,\delta(t-t_0)$	$e^{-t_0\sqrt{(j\omega+a)(j\omega+b)}}$
$\left(\frac{a-b}{2}\right)^2 t e^{-[(a+b)/2]t} \frac{I_1\left(\frac{a-b}{2}\sqrt{t^2-t_0^2}\right)^*}{\frac{a-b}{2}\sqrt{t^2-t_0^2}} u(t-t_0) - \left(\frac{a-b}{2}\right) e^{-[(a+b)/2]t} I_0\left(\frac{a-b}{2}\sqrt{t^2-t_0^2}\right) u(t-t_0) + e^{[(a+b)/2]t_0}\,\delta(t-t_0)$	$\frac{e^{-t_0\sqrt{(j\omega+a)(j\omega+b)}}}{\sqrt{(j\omega+a)/(j\omega+b)}}$

* J_1 is a Bessel function (see Spiegel in the references at the end of Chapter 1) of the first kind, order one, I_0 is a modified Bessel function of the first kind, order zero. I_1 is a modified Bessel function of the first kind, order one. These transform pairs are useful for plane wave propagation in lossy media or on lossy two-wire lines. The second pair applies to propagation in a lossless waveguide.

Appendix F
Alternate Formulation of **B**

THE LORENTZ TRANSFORMATION

Let us postulate (as Einstein) that the velocity of light is independent of the motion of the observer and derive the consequences. Consider two inertial systems S and S', where S' is moving with uniform velocity $\mathbf{u} = \mathbf{a}_z u$ with respect to S. The two systems are coincident at $t = 0$. The passage of a pulse of light from 0 to x, y, z in S takes a time t such that

$$x^2 + y^2 + z^2 = c^2t^2 \tag{F.1}$$

where c is the velocity of light. The point x, y, z in S is given by x', y', z' in the S' system. Thus, in the S' system (moving) equation F.1 is

$$(x')^2 + (y')^2 + (z')^2 = c^2(t')^2 \tag{F.2}$$

See Figure F.1.

The transformation equations that link the two systems must be such that an observer in S can predict equation F.2, while an observer in S' can predict equation F.1, since they are observing the *same* event. The relations between primed and unprimed quantities must be *linear*, so that a single event

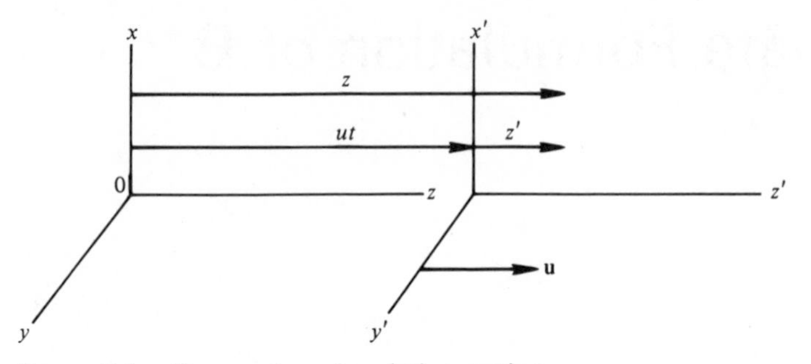

Figure F.1. Two systems in relative motion.

in one system is a single event in the other system. Observers in the two systems will obviously agree about distances in the x and y directions, so,

$$x' = x$$
$$y' = y$$
$$z' = Az + Bt \tag{F.3}$$
$$t' = az + bt$$

Now, $z = ut$ when $z' = 0$, so from equations F.3

$$B = -\frac{Aut}{t} = -Au \tag{F.4}$$

Substituting equations F.3 and F.4 into F.2, and then equating the resultant coefficients with those appearing in equation F.1, we have

$$A^2 - a^2c^2 = 1$$
$$A^2u + abc^2 = 0 \tag{F.5}$$
$$A^2u^2 - b^2c^2 = -c^2$$

Equations F.5 represent three equations and three unknowns, hence,

$$A = b = \frac{1}{\sqrt{1 - u^2/c^2}}$$

$$a = -\frac{u}{c^2}\frac{1}{\sqrt{1 - u^2/c^2}}$$

and

$$z' = \frac{z - ut}{\sqrt{1 - u^2/c^2}}$$
$$t' = \frac{t - uz/c^2}{\sqrt{1 - u^2/c^2}} \tag{F.6}$$
$$x' = x$$
$$y' = y$$

Inversion of equations F.6 gives

$$z = \frac{z' + ut'}{\sqrt{1 - u^2/c^2}}$$

$$t = \frac{t' + uz'/c^2}{\sqrt{1 - u^2/c^2}} \tag{F.7}$$

$$x = x'$$

$$y = y'$$

If $u \ll c$ the Lorentz transformation reduces to the ordinary Galilean transformation (by letting $u/c \to 0$).

Length is different in the two systems. Let $l = z_2 - z_1$ and $l' = z_2' - z_1'$. We desire $l' = f(l)$. From the first of equations F.6,

$$l' = \frac{l}{\sqrt{1 - u^2/c^2}} \tag{F.8}$$

Velocities are different in the two systems also. Suppose an object is moving with velocity u_z' in S' (Figure F.1) and velocity u_z in S. In the S' system at time t' the distance traveled is $z' = u_z' t'$. Thus, from equation F.6,

$$z' = \frac{z - ut}{\sqrt{1 - u^2/c^2}} = u_z' t' = u_z' \frac{t - uz/c^2}{\sqrt{1 - u^2/c^2}}$$

or

$$z - ut = u_z'\left(t - \frac{uz}{c^2}\right)$$

So,

$$\frac{z}{t} = u_z = u + u_z' - \frac{uu_z'}{c^2}\frac{z}{t}$$

or

$$u_z = \frac{u + u_z'}{1 + uu_z'/c^2} \tag{F.9}$$

ELECTRIC FIELD OF A MOVING CHARGE

Consider Figure F.2, where a charge Q at rest in S' moves with velocity $\mathbf{u} = \mathbf{a}_z u$ with respect to S. The charge is at the origin of both systems at $t = t' = 0$. We want to find the electric field as determined by an observer in S and S' due to the moving charge. Electric flux, Ψ_E, and charge Q will be the same in both systems. (Determination of Ψ_E is a process of *counting* flux lines.) The two observers will disagree about $\mathbf{E}$, because they disagree

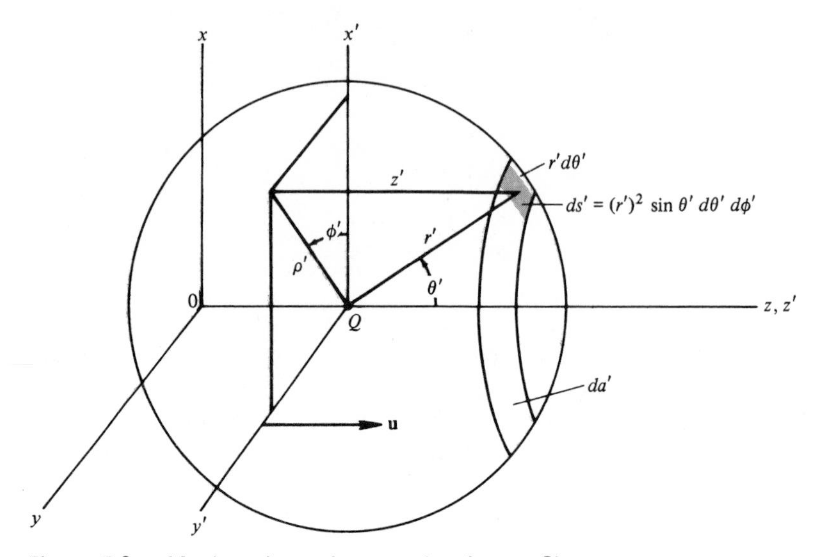

Figure F.2. Moving charge in a moving frame, S'.

about the *area* through which the flux passes and the *distance* from the charge to the point where **E** is measured (**r** or **r**′).

We have

$$\mathbf{D} = \frac{Q}{4\pi r^2}\,\mathbf{a}_r \tag{F.10}$$

and

$$\mathbf{D}' = \frac{Q}{4\pi(r')^2}\,\mathbf{a}_{r'} \tag{F.11}$$

Differential elements of area are (standard spherical coordinates)

$$ds = r^2 \sin\theta \, d\theta \, d\phi$$

and

$$ds' = (r')^2 \sin\theta' \, d\theta' \, d\phi'$$

so,

$$da = r^2 \sin\theta \, d\theta \int_0^{2\pi} d\phi = 2\pi r^2 \sin\theta \, d\theta \tag{F.12}$$

and

$$da' = (r')^2 \sin\theta' \, d\theta' \int_0^{2\pi} d\phi' = 2\pi(r')^2 \sin\theta' \, d\theta' \tag{F.13}$$

Since $d\Psi_E = D_r \, da = D'_r \, da'$, we have $E_r \, da = E'_r \, da'$, or

$$E_r r^2 \sin\theta \, d\theta = E'_r (r')^2 \sin\theta' \, d\theta'$$

From the equations F.6 and A.1 we have ($t = 0$)

$$\rho = \rho' = r \sin \theta = r' \sin \theta'$$

so

$$E_r = E'_r \frac{r' \, d\theta'}{r \, d\theta} = E'_r \frac{\sin \theta \, d\theta'}{\sin \theta' \, d\theta} \tag{F.14}$$

Now,

$$\theta' = \tan^{-1} \frac{\rho'}{z'} = \tan^{-1} \frac{\rho}{z'} = \tan^{-1} \frac{\rho\sqrt{1 - u^2/c^2}}{z}$$

or

$$\theta' = \tan^{-1}\left(\tan \theta \sqrt{1 - \frac{u^2}{c^2}}\right)$$

so

$$\frac{d\theta'}{d\theta} = \frac{\sec^2 \theta \sqrt{1 - u^2/c^2}}{1 + \tan^2 \theta(1 - u^2/c^2)} = \frac{\sqrt{1 - u^2/c^2}}{1 - \sin^2 \theta u^2/c^2} \tag{F.15}$$

Also,

$$\frac{\sin \theta}{\sin \theta'} = \frac{\sin \theta}{\dfrac{\tan \theta \sqrt{1 - u^2/c^2}}{\sqrt{1 + \tan^2 \theta(1 - u^2/c^2)}}} = \frac{\sqrt{1 - \sin^2 \theta u^2/c^2}}{\sqrt{1 - u^2/c^2}} \tag{F.16}$$

With equation F.16, equation F.14 becomes

$$E_r = E'_r \frac{1}{\sqrt{1 - \sin^2 \theta u^2/c^2}} \tag{F.17}$$

Now, $E_z = E_r \cos \theta$ and $E'_z = E'_r \cos \theta'$, so

$$E'_z = E'_r \frac{1}{\sqrt{1 + \tan^2 \theta(1 - u^2/c^2)}} = \frac{E'_r \cos \theta}{\sqrt{1 - \sin^2 \theta u^2/c^2}}$$

or

$$E'_z = E_r \cos \theta$$

from equation F.17. Therefore,

$$E_z = E'_z \tag{F.18}$$

For the ρ components, we have $E_\rho = E_r \sin \theta$ and $E'_\rho = E'_r \sin \theta'$, so substituting equations F.16 and F.17 gives

$$E_\rho = E'_\rho \frac{1}{\sqrt{1 - u^2/c^2}} \tag{F.19}$$

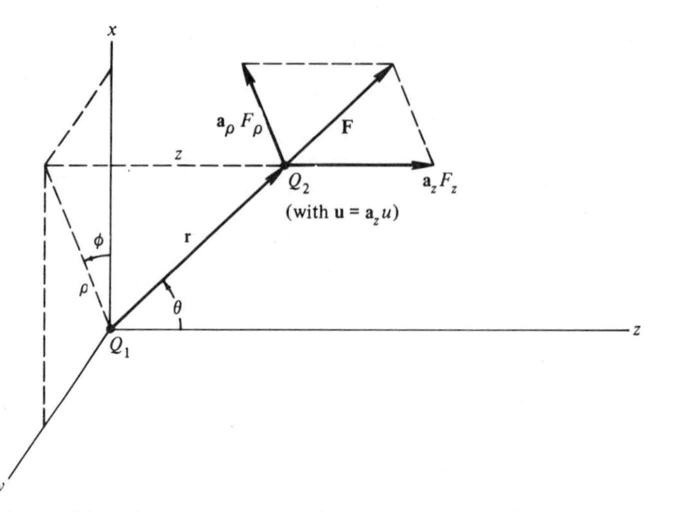

Figure F.3. Force between a fixed and moving charge.

Therefore, the observer in S sees a larger ρ component of electric field (or force) than he would if Q were stationary in S.

For a fixed charge Q_1 and a moving charge Q_2 (with $\mathbf{u} = \mathbf{a}_z u$) as shown in Figure F.3, it is easy to use the results just obtained to show that the force on Q_2 is

$$F_z = \frac{Q_1 Q_2}{4\pi\varepsilon r^2}\cos\theta \tag{F.20}$$

and

$$F_\rho = \frac{Q_1 Q_2}{4\pi\varepsilon r^2}\frac{\sin\theta}{\sqrt{1 - u^2/c^2}} \tag{F.21}$$

INTERACTION BETWEEN TWO MOVING CHARGES

We would now like to extend the results just obtained to include a fixed observer and the interaction between two moving charges. Consider Figure F.4 and the following definitions.

$$\begin{aligned}
\mathbf{u}_1 &= \mathbf{a}_z u_1 = \text{velocity of } Q_1 \text{ in } x, y, z, t\\
\mathbf{u}_2 &= \mathbf{a}_z u_2 = \text{velocity of } Q_2 \text{ in } x, y, z, t\\
\mathbf{u}_{12} &= \mathbf{a}_z u_{12} = \text{velocity of } Q_1 \text{ relative to } Q_2\\
\mathbf{E} &= \text{electric field at } P\text{, due to } Q_1\text{, relative to } 0\\
\mathbf{E}' &= \text{electric field at } P\text{, due to } Q_1\text{, relative to } Q_1\\
\mathbf{E}'' &= \text{electric field at } P\text{, due to } Q_1\text{, relative to } Q_2
\end{aligned}$$

We want ultimately to find the force on Q_2 detected by a fixed (in x, y, z, t) observer. It is immediately obvious from equation F.18 that

$$E_z = E'_z = E''_z$$

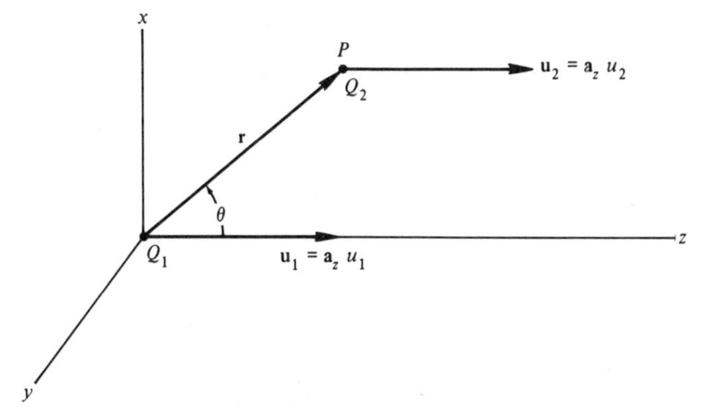

Figure F.4. Two charges with parallel velocities.

From equation F.19 it follows that

$$E_\rho'' = \frac{E_\rho'}{\sqrt{1 - u_{12}^2/c^2}} \tag{F.22}$$

when the notation of Figure F.4 is taken into account. For the same equation we also have

$$E_\rho'' = \frac{E_\rho}{\sqrt{1 - u_2^2/c^2}} \tag{F.23}$$

Combining the last two results gives

$$E_\rho = E_\rho' \frac{\sqrt{1 - u_2^2/c^2}}{\sqrt{1 - u_{12}^2/c^2}} \tag{F.24}$$

where from equation F.9,

$$u_1 = \frac{u_2 + u_{12}}{1 + u_2 u_{12}/c^2}$$

or, solving for u_{12},

$$u_{12} = \frac{u_1 - u_2}{1 - u_1 u_2/c^2} \tag{F.25}$$

Substituting equation F.25 into equation F.24,

$$E_\rho = E_\rho' \frac{\sqrt{1 - u_2^2/c^2}}{\sqrt{1 - u_1^2/c^2}} \frac{\sqrt{1 - u_1^2/c^2}}{\sqrt{1 - \dfrac{1}{c^2}\left(\dfrac{u_1 - u_2}{1 - u_1 u_2/c^2}\right)^2}}$$

which reduces to

$$E_\rho = E_\rho' \frac{1 - u_1 u_2/c^2}{\sqrt{1 - u_1^2/c^2}} \tag{F.26}$$

E'_ρ is the ρ component of the electric field at P due to Q_1, relative to Q_1, so,

$$E'_\rho = \frac{Q_1 \sin\theta}{4\pi\varepsilon r^2}$$

and

$$E_\rho = \frac{Q_1 \sin\theta}{4\pi\varepsilon r^2\sqrt{1 - u_1^2/c^2}} - \frac{Q_1 u_1 u_2 \sin\theta}{4\pi\varepsilon c^2 r^2\sqrt{1 - u_1^2/c^2}}$$

The ρ component of the force on Q_2 is

$$F_\rho = \frac{Q_1 Q_2 \sin\theta}{4\pi\varepsilon r^2\sqrt{1 - u_1^2/c^2}} - \frac{Q_1 Q_2 u_1 u_2 \sin\theta}{4\pi\varepsilon c^2 r^2\sqrt{1 - u_1^2/c^2}}$$

while the z component is

$$F_z = F'_z = \frac{Q_1 Q_2 \cos\theta}{4\pi\varepsilon r^2}$$

Since the coordinate system is standard, it is easy to show that $-u_1 u_2 \sin\theta \mathbf{a}_\rho = \mathbf{u}_2 \times (\mathbf{u}_1 \times \mathbf{a}_r)$. The complete force field is then·

$$\mathbf{F} = \frac{Q_1 Q_2}{4\pi\varepsilon r^2}\left(\mathbf{a}_\rho \frac{\sin\theta}{\sqrt{1 - u_1^2/c^2}} + \mathbf{a}_z \cos\theta\right) + \frac{Q_1 Q_2}{4\pi\varepsilon c^2 r^2}\frac{\mathbf{u}_2 \times (\mathbf{u}_1 \times \mathbf{a}_r)}{\sqrt{1 - u_1^2/c^2}} \tag{F.27}$$

Equation F.27 also holds for nonparallel velocities, $\mathbf{u}_1$, $\mathbf{u}_2$ (not proved). In the usual case, $u_1 \ll c$, so equation F.27 becomes

$$\mathbf{F} = \mathbf{a}_r \frac{Q_1 Q_2}{4\pi\varepsilon r^2} + \frac{Q_1 Q_2}{4\pi\varepsilon c^2 r^2}\mathbf{u}_2 \times (\mathbf{u}_1 \times \mathbf{a}_r) \tag{F.28}$$

If we *define*

$$\mu \equiv \frac{1}{\varepsilon c^2} \tag{F.29}$$

and

$$\mathbf{B} \equiv \mu\mathbf{H} = \mu\mathbf{u}_1 \times \mathbf{D} \tag{F.30}$$

where, of course,

$$\mathbf{D} = \varepsilon\mathbf{E} = \frac{Q_1}{4\pi r^2}\mathbf{a}_r$$

then equation F.28 finally becomes

$$\mathbf{F} = Q_2(\mathbf{E} + \mathbf{u}_2 \times \mathbf{B}) \tag{F.31}$$

Equation F.31 for the Lorentz force was obtained *directly* from Coulomb's law, the Lorentz transformation, and the definition of **B**, equation F.30. It agrees with equation 7.72, the classical result, but equation 7.72

required the *additional* use of the Biot–Savart law. The development of equation F.31 merely gave us a relativistic correction to the electric field, which we decided to call the magnetic field. A complete development of electromagnetic theory can be pursued, beginning with equation F.31, if desired.

Appendix G Proofs of Integral Relations

THE LAPLACIAN OF $1/|\mathbf{r} - \mathbf{r}'| = 1/R$

Many of the proofs involving integration in field theory may be expedited by the interpretation of $\nabla^2(1/|\mathbf{r} - \mathbf{r}'|)$. It is a straightforward matter to show that $\nabla^2(1/|\mathbf{r} - \mathbf{r}'|)$ is zero if $\mathbf{r} \neq \mathbf{r}'$, whereas $\nabla^2(1/|\mathbf{r} - \mathbf{r}'|)$ is undefined if $\mathbf{r} = \mathbf{r}'$. It can also be rigorously established that

$$\iiint_{\text{vol}'} \nabla^2 \frac{1}{|\mathbf{r} - \mathbf{r}'|}\, dx'\, dy'\, dz' = \begin{cases} -4\pi, & \mathbf{r} \text{ inside vol}' \\ 0, & \mathbf{r} \text{ outside vol}' \end{cases}$$

The geometry is shown in Figure G.1. In this case, $\nabla^2(1/|\mathbf{r} - \mathbf{r}'|)$ meets the conditions required for being represented by a Dirac delta function. That is,

$$\nabla^2 \frac{1}{|\mathbf{r} - \mathbf{r}'|} = -4\pi\, \delta(\mathbf{r} - \mathbf{r}') \tag{G.1}$$

where $\delta(\mathbf{r} - \mathbf{r}')$ is the shorthand representation for the three-dimensional Dirac delta function, $\delta(x - x')\,\delta(y - y')\,\delta(z - z')$. It has the dimension $(\text{meters})^{-3}$. The singular nature of equation G.1 should not be surprising

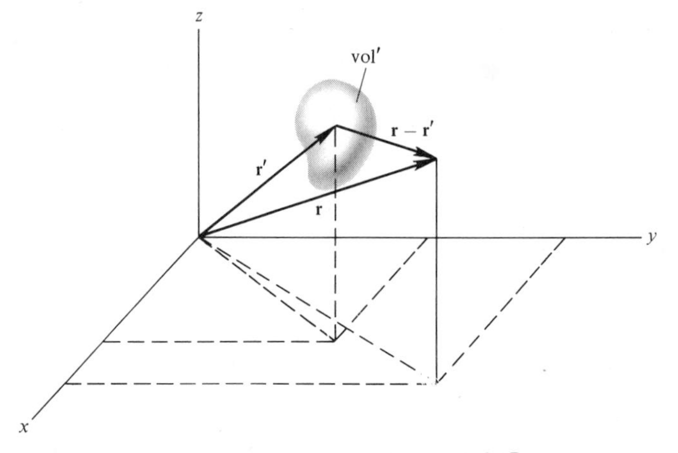

Figure G.1. Geometry for establishing $\nabla^2(1/R)$.

because $1/|\mathbf{r} - \mathbf{r}'|$ is $4\pi\varepsilon$ times the electrostatic potential at $\mathbf{r}$ due to a unit point charge at $\mathbf{r}'$. In other words, Poisson's equation,

$$\nabla^2\Phi(\mathbf{r}) = -\frac{\rho_v(\mathbf{r})}{\varepsilon}$$

becomes

$$\nabla^2\Phi(\mathbf{r}) = -\frac{\delta(x)\,\delta(y)\,\delta(z)}{\varepsilon} = -\frac{\delta(\mathbf{r})}{\varepsilon}$$

for a unit point charge at the origin. The solution to the last equation is

$$\Phi(\mathbf{r}) = h(\mathbf{r}) = \frac{1}{4\pi\varepsilon r}$$

so for a unit point charge at $\mathbf{r}'$ the unit impulse response is

$$h(\mathbf{r} - \mathbf{r}') = \frac{1}{4\pi\varepsilon|\mathbf{r} - \mathbf{r}'|}$$

The last result is sometimes called the Green's function for Poisson's equation.

THE DIVERGENCE OF **E**

Given the superposition integral for $\mathbf{E}$, we have

$$\nabla \cdot \mathbf{E} = \nabla \cdot \left\{\frac{1}{4\pi\varepsilon}\iiint_{\text{vol}'} \frac{\rho_v(\mathbf{r}')\mathbf{a}_R}{R^2}\,dv'\right\}$$

$$= \frac{1}{4\pi\varepsilon}\iiint_{\text{vol}'} \rho_v(\mathbf{r}')\left\{\nabla \cdot \frac{\mathbf{a}_R}{R^2}\right\} dv' \qquad \text{(G.2)}$$

since $\mathbf{\nabla}$ operates on $\mathbf{r}$, not $\mathbf{r}'$. Then,

$$\begin{aligned}\mathbf{\nabla}\cdot\frac{\mathbf{a}_R}{R^2} = \mathbf{\nabla}\cdot\frac{\mathbf{r}-\mathbf{r}'}{|\mathbf{r}-\mathbf{r}'|^3} &= \frac{\partial}{\partial x}\left\{\frac{x-x'}{[(x-x')^2+(y-y')^2+(z-z')^2]^{3/2}}\right\} \\ &+ \frac{\partial}{\partial y}\left\{\frac{y-y'}{[(x-x')^2+(y-y')^2+(z-z')^2]^{3/2}}\right\} \\ &+ \frac{\partial}{\partial z}\left\{\frac{z-z'}{[(x-x')^2+(y-y')^2+(z-z')^2]^{3/2}}\right\} \\ &= -\frac{\partial^2}{\partial x^2}\left\{\frac{1}{[(x-x')^2+(y-y')^2+(z-z')^2]^{1/2}}\right\} \\ &- \frac{\partial^2}{\partial y^2}\left\{\frac{1}{[(x-x')^2+(y-y')^2+(z-z')^2]^{1/2}}\right\} \\ &- \frac{\partial^2}{\partial z^2}\left\{\frac{1}{[(x-x')^2+(y-y')^2+(z-z')^2]^{1/2}}\right\}\end{aligned}$$

According to equation G.1, the last result may be written as

$$\mathbf{\nabla}\cdot\frac{\mathbf{a}_R}{R^2} = -\nabla^2\frac{1}{|\mathbf{r}-\mathbf{r}'|} = -\nabla^2\frac{1}{R} = 4\pi\,\delta(\mathbf{r}-\mathbf{r}')$$

Therefore,

$$\mathbf{\nabla}\cdot\mathbf{E} = \frac{1}{4\pi\varepsilon}\iiint_{\text{vol}'}\rho_v(\mathbf{r}')[4\pi\,\delta(\mathbf{r}-\mathbf{r}')]\,dv'$$

and

$$\mathbf{\nabla}\cdot\mathbf{E} = \begin{cases}\dfrac{\rho_v(\mathbf{r})}{\varepsilon}, & \mathbf{r}\text{ inside vol}' \\ 0, & \mathbf{r}\text{ outside vol}'\end{cases} \tag{G.3}$$

by the sampling property of the impulse function.

THE SCALAR HELMHOLTZ INTEGRAL (STATIC)

The Helmholtz integral, for scalar potential is purported to be a solution to Poisson's equation and Laplace's equation,

$$\nabla^2\Phi(x, y, z) = -\frac{\rho_v(x, y, z)}{\varepsilon}$$

and

$$\nabla^2\Phi(x, y, z) = 0$$

respectively. That this is indeed the case is easy to prove with equation G.1. The potential is given by

$$\Phi(x, y, z) = \frac{1}{4\pi\varepsilon} \iiint_{\text{vol}'} \frac{\rho_v(x', y', z')}{|\mathbf{r} - \mathbf{r}'|} \, dx' \, dy' \, dz'$$

So

$$\nabla^2 \Phi(x, y, z) = \frac{1}{4\pi\varepsilon} \iiint_{\text{vol}'} \rho_v(x', y', z') \nabla^2 \frac{1}{|\mathbf{r} - \mathbf{r}'|} \, dx' \, dy' \, dz'$$

since ∇^2 operates only on x, y, z. With equation G.1

$$\nabla^2 \Phi(x, y, z) = \frac{1}{4\pi\varepsilon} \iiint_{\text{vol}'} \rho_v(x', y', z')[-4\pi \, \delta(\mathbf{r} - \mathbf{r}')] \, dx' \, dy' \, dz'$$

Therefore, utilizing the sampling property of the delta function

$$\nabla^2 \Phi(x, y, z) = \begin{cases} -\dfrac{\rho_v(x, y, z)}{\varepsilon}, & \mathbf{r} \text{ inside vol}' \\ 0, & \mathbf{r} \text{ outside vol}' \end{cases} \tag{G.4}$$

AMPERE'S CIRCUITAL LAW (MAGNETOSTATIC)

It can be shown that Ampere's circuital law follows from the Biot–Savart law. The Biot–Savart law is given by equation 7.5.

$$\mathbf{H}(x, y, z) = \frac{1}{4\pi} \iiint_{\text{vol}'} \mathbf{J}(x', y', z) \times \frac{\mathbf{r} - \mathbf{r}'}{|\mathbf{r} - \mathbf{r}'|^3} \, dx' \, dy' \, dz'$$

It is easy to show that

$$\nabla \frac{1}{|\mathbf{r} - \mathbf{r}'|} = -\nabla' \frac{1}{|\mathbf{r} - \mathbf{r}'|} = -\frac{\mathbf{r} - \mathbf{r}'}{|\mathbf{r} - \mathbf{r}'|^3}$$

so the Biot–Savart law may be written

$$\mathbf{H}(x, y, z) = \frac{1}{4\pi} \iiint_{\text{vol}'} \nabla \frac{1}{|\mathbf{r} - \mathbf{r}'|} \times \mathbf{J}(x', y', z') \, dx' \, dy' \, dz'$$

Now, since by vector identity (see Appendix A)

$$\nabla \frac{1}{\alpha} \times \mathbf{C} \equiv \nabla \times \frac{\mathbf{C}}{\alpha} - \frac{1}{\alpha} \nabla \times \mathbf{C}$$

it follows that

$$\nabla \frac{1}{|\mathbf{r} - \mathbf{r}'|} \times \mathbf{J}(x', y', z') = \nabla \times \left[\frac{\mathbf{J}(x', y', z')}{|\mathbf{r} - \mathbf{r}'|}\right] - \frac{1}{|\mathbf{r} - \mathbf{r}'|} \nabla \times \mathbf{J}(x', y', z')$$

Since ∇ operates only on x, y, z, the last term on the right side of the preceding equation is zero. Therefore,

$$\mathbf{H}(x, y, z) = \frac{1}{4\pi} \iiint_{\text{vol}'} \nabla \times \left[\frac{\mathbf{J}(x', y', z')}{|\mathbf{r} - \mathbf{r}'|}\right] dx'\, dy'\, dz' \tag{G.5}$$

or

$$\mathbf{H}(x, y, z) = \frac{1}{4\pi} \nabla \times \iiint_{\text{vol}'} \frac{\mathbf{J}(x', y', z')}{|\mathbf{r} - \mathbf{r}'|} dx'\, dy'\, dz'$$

So,

$$\mathbf{H}(x, y, z) = \frac{1}{\mu} \nabla \times \mathbf{A}(x, y, z) \tag{G.6}$$

where

$$\mathbf{A}(x, y, z) = \frac{\mu}{4\pi} \iiint_{\text{vol}'} \frac{\mathbf{J}(x', y', z')}{|\mathbf{r} - \mathbf{r}'|} dx'\, dy'\, dz' \tag{G.7}$$

the *vector* Helmholtz integral! Returning to equation G.5, we take the curl of both sides, giving

$$\nabla \times \mathbf{H}(x, y, z) = \frac{1}{4\pi} \iiint_{\text{vol}'} \nabla \times \left[\nabla \times \frac{\mathbf{J}(x', y', z')}{|\mathbf{r} - \mathbf{r}'|}\right] dx'\, dy'\, dz'$$

Another vector identity (see Appendix A)

$$\nabla \times \left(\nabla \times \frac{\mathbf{C}}{\alpha}\right) \equiv \nabla\left(\nabla \cdot \frac{\mathbf{C}}{\alpha}\right) - \nabla^2 \frac{\mathbf{C}}{\alpha}$$
$$= \nabla\left(\frac{1}{\alpha} \nabla \cdot \mathbf{C}\right) + \nabla\left(\mathbf{C} \cdot \nabla \frac{1}{\alpha}\right) - \nabla^2 \frac{\mathbf{C}}{\alpha}$$

gives

$$\nabla \times \left[\nabla \times \frac{\mathbf{J}(x', y', z')}{|\mathbf{r} - \mathbf{r}'|}\right] = \nabla\left[\frac{1}{|\mathbf{r} - \mathbf{r}'|} \nabla \cdot \mathbf{J}(x', y', z')\right] + \nabla\left[\mathbf{J}(x', y', z') \cdot \nabla \frac{1}{|\mathbf{r} - \mathbf{r}'|}\right] - \nabla^2 \frac{\mathbf{J}(x', y', z')}{|\mathbf{r} - \mathbf{r}'|}$$

The first term on the right side of the preceding equation is zero, because ∇ operates on x, y, z. So, $\nabla \cdot \mathbf{J}(x', y', z') = 0$. $\nabla(1/|\mathbf{r} - \mathbf{r}'|)$ can be replaced by $-\nabla'(1/|\mathbf{r} - \mathbf{r}'|)$. Thus,

$$\nabla \times \left[\nabla \times \frac{\mathbf{J}(x', y', z')}{|\mathbf{r} - \mathbf{r}'|}\right] = -\nabla\left[\mathbf{J}(x', y', z') \cdot \nabla' \frac{1}{|\mathbf{r} - \mathbf{r}'|}\right] - \mathbf{J}(x', y', z')\nabla^2 \frac{1}{|\mathbf{r} - \mathbf{r}'|}$$

Another vector identity,

$$\mathbf{C} \cdot \nabla' \frac{1}{\alpha} \equiv \nabla' \cdot \frac{\mathbf{C}}{\alpha} - \frac{1}{\alpha} \nabla' \cdot \mathbf{C}$$

can be used to write

$$\nabla \times \left[\nabla \times \frac{\mathbf{J}(x', y', z')}{|\mathbf{r} - \mathbf{r}'|} \right] = - \nabla \left[\nabla' \cdot \frac{\mathbf{J}(x', y', z')}{|\mathbf{r} - \mathbf{r}'|} \right] + \nabla \left[\frac{1}{|\mathbf{r} - \mathbf{r}'|} \nabla' \cdot \mathbf{J}(x', y', z') \right] - \mathbf{J}(x', y', z') \nabla^2 \frac{1}{|\mathbf{r} - \mathbf{r}'|}$$

The second term on the right side of the preceding equation is zero because $\nabla' \cdot \mathbf{J}(x', y', z') = 0$ by the conservation of charge equation for statics. Thus,

$$\nabla \times \mathbf{H}(x, y, z) = - \frac{1}{4\pi} \nabla \iiint_{\text{vol}'} \nabla' \cdot \frac{\mathbf{J}(x', y', z')}{|\mathbf{r} - \mathbf{r}'|} \, dx' \, dy' \, dz' + \iiint_{\text{vol}'} \mathbf{J}(x', y', z') \, \delta(\mathbf{r} - \mathbf{r}') \, dx' \, dy' \, dz'$$

The first integral can be converted into a surface integral with the aid of the divergence theorem, so

$$\nabla \times \mathbf{H}(x, y, z) = - \frac{1}{4\pi} \nabla \oiint_{s'} \frac{\mathbf{J}(x', y', z')}{|\mathbf{r} - \mathbf{r}'|} \cdot d\mathbf{s}' + \begin{cases} \mathbf{J}(x, y, z), & \mathbf{r} \text{ inside vol}' \\ 0, & \mathbf{r} \text{ outside vol}' \end{cases}$$

The original volume of integration was as large as necessary to enclose all of **J**. It can be larger if so desired. For the term containing the surface integral we make the volume large enough so that none of the current touches the surface. Then this term is zero, and

$$\nabla \times \mathbf{H}(x, y, z) = \begin{cases} \mathbf{J}(x, y, z), & \mathbf{r} \text{ inside vol}' \\ 0, & \mathbf{r} \text{ outside vol}' \end{cases}$$

which is the point form of Ampere's law (statics), or simply Maxwell's equation. Now,

$$\iint_s \nabla \times \mathbf{H} \cdot ds = \iint_s \mathbf{J} \cdot d\mathbf{s} = I$$

so, with Stoke's theorem,

$$I = \iint_s \nabla \times \mathbf{H} \cdot ds = \oint \mathbf{H} \cdot d\mathbf{l}$$

Finally, then

$$\oint \mathbf{H} \cdot d\mathbf{l} = I \tag{G.8}$$

which is Ampere's law as derived from the Biot–Savart law.

THE VECTOR HELMHOLTZ INTEGRAL (STATIC)

Proof that the vector Helmholtz integral for magnetostatics, equation 7.49, ($\mu_0 \to \mu$), or equation G.7,

$$\mathbf{A}(x, y, z) = \frac{\mu}{4\pi} \iiint_{\text{vol}'} \frac{\mathbf{J}(x', y', z')}{|\mathbf{r} - \mathbf{r}'|} \, dx' \, dy' \, dz'$$

satisfies equation 7.48 is now easy to supply. We have

$$\nabla^2 \mathbf{A}(x, y, z) = \frac{\mu}{4\pi} \iiint_{\text{vol}'} \mathbf{J}(x', y', z') \nabla^2 \frac{1}{|\mathbf{r} - \mathbf{r}'|} \, dx' \, dy' \, dz'$$

or

$$\nabla^2 \mathbf{A}(x, y, z) = \frac{\mu}{4\pi} \iiint_{\text{vol}'} \mathbf{J}(x', y', z')[-4\pi \, \delta(\mathbf{r} - \mathbf{r}')] \, dx' \, dy' \, dz'$$

Utilizing the sampling property of the Dirac delta function,

$$\nabla^2 \mathbf{A}(x, y, z) = \begin{cases} -\mu \mathbf{J}(x, y, z) & \mathbf{r} \text{ inside vol}' \\ 0, & \mathbf{r} \text{ outside vol}' \end{cases} \tag{G.9}$$

THE VECTOR HELMHOLTZ INTEGRAL—GENERAL CASE

Equation 9.69,

$$\mathbf{A}(x, y, z, t) = \frac{\mu}{4\pi} \iiint_{\text{vol}'} \frac{\mathbf{J}(x', y', z', t - |\mathbf{r} - \mathbf{r}'| \sqrt{\mu\varepsilon})}{|\mathbf{r} - \mathbf{r}'|} \, dx' \, dy' \, dz'$$

is supposed to be a solution to the inhomogeneous and homogeneous vector wave equations, equations 9.67 and 9.71, respectively. If this can be proved, then the proof will hold for the inhomogeneous and homogeneous *scalar* wave equations also, for certainly the proof will hold for any rectangular scalar component of $\mathbf{A}$. A proof can be obtained for the general case where $\mathbf{J}(x, y, z, t)$ is *any* function of time, but it is involved and rather lengthy. A somewhat simpler proof exists if we assume sinusoidal time variation (phasor notation). This proof can be extended to the general case by means of the Fourier integral, or, in particular, the inverse Fourier transform. This proof can also be used for the static cases (already given) by simply letting k (or ω) become zero.

In *phasor form*, we have

$$\mathbf{A}(x, y, z, \omega) = \frac{\mu}{4\pi} \iiint_{\text{vol}'} \frac{\mathbf{J}(x', y', z', \omega)\, e^{-jk|\mathbf{r}-\mathbf{r}'|}}{|\mathbf{r} - \mathbf{r}'|}\, dx'\, dy'\, dz'$$

from equation 9.76 (with $k = \omega\sqrt{\mu\varepsilon}$), which is supposed to be a solution to equation 9.67 in phasor form. That is, does the integral for **A** satisfy

$$\nabla^2 \mathbf{A}(x, y, z, \omega) + k^2 \mathbf{A}(x, y, z, \omega) = -\mu \mathbf{J}(x, y, z, \omega)$$

Now,

$$\nabla^2 \mathbf{A} + k^2 \mathbf{A} = \frac{\mu}{4\pi} \iiint_{\text{vol}'} \mathbf{J}(x', y', z', \omega)(\nabla^2 + k^2)\left\{\frac{e^{-jk|\mathbf{r}-\mathbf{r}'|}}{|\mathbf{r} - \mathbf{r}'|}\right\} dx'\, dy'\, dz'$$

Applying the operator $\nabla^2 + k^2$ to $\{|\mathbf{r} - \mathbf{r}'|^{-1} e^{-jk|\mathbf{r}-\mathbf{r}'|}\}$ gives

$$\begin{aligned}\nabla^2 \mathbf{A} + k^2 \mathbf{A} = &\frac{\mu}{4\pi} \iiint_{\text{vol}'} \mathbf{J}(x', y', z', \omega)(e^{-jk|\mathbf{r}-\mathbf{r}'|})\nabla^2 \frac{1}{|\mathbf{r} - \mathbf{r}'|}\, dx'\, dy'\, dz' \\ &- \frac{\mu k^2}{4\pi} \iiint_{\text{vol}'} \mathbf{J}(x', y', z', \omega) \frac{e^{-jk|\mathbf{r}-\mathbf{r}'|}}{|\mathbf{r} - \mathbf{r}'|}\, dx'\, dy'\, dz' \\ &+ \frac{\mu k^2}{4\pi} \iiint_{\text{vol}'} \mathbf{J}(x', y', z', \omega) \frac{e^{-jk|\mathbf{r}-\mathbf{r}'|}}{|\mathbf{r} - \mathbf{r}'|}\, dx'\, dy'\, dz'\end{aligned}$$

or

$$\begin{aligned}\nabla^2 \mathbf{A} + k^2 \mathbf{A} = &\frac{\mu}{4\pi} \iiint_{\text{vol}'} \mathbf{J}(x', y', z', \omega)(e^{-jk|\mathbf{r}-\mathbf{r}'|})[-4\pi\, \delta(\mathbf{r} - \mathbf{r}')]\, dx'\, dy'\, dz' \\ &- k^2 \mathbf{A}(x, y, z, \omega) + k^2 \mathbf{A}(x, y, z, \omega)\end{aligned}$$

Utilizing the sampling property of the Dirac delta function, the preceding equation becomes

$$\nabla^2 \mathbf{A}(x, y, z, \omega) + k^2 \mathbf{A}(x, y, z, \omega) = \begin{cases} -\mu \mathbf{J}(x, y, z, \omega), & \mathbf{r} \text{ inside vol}' \\ 0, & \mathbf{r} \text{ outside vol}' \end{cases} \tag{G.10}$$

DIVERGENCE THEOREM

Consider a volume in space bounded by a closed surface s with a well-behaved vector field **F** present. Let dv be a volume element and let $d\mathbf{s}$ be a surface element of the *external* surface s. Now imagine that the entire volume is partitioned into N small volumes $\Delta v_1, \Delta v_2, \ldots, \Delta v_n, \ldots, \Delta v_N$ each having *closed* surfaces $\Delta s_1, \Delta s_2, \ldots, \Delta s_n, \ldots, \Delta s_N$. For each small volume $\mathbf{\nabla} \cdot \mathbf{F}$ is almost uniform and equal to the value it has as Δv approaches zero at a point. According to its definition, equation 2.20, $\mathbf{\nabla} \cdot \mathbf{F}$ is the limit of the flux of the quantity (designated by $\mathbf{F}$) per unit volume as the volume approaches zero. That is,

$$(\mathbf{\nabla} \cdot \mathbf{F})_n = \lim_{\Delta v_n \to 0} \frac{\oint\!\!\oint_{\Delta s_n} \mathbf{F}_n \cdot d\mathbf{s}_n}{\Delta v_n}$$

or

$$(\mathbf{\nabla} \cdot \mathbf{F})_n \approx \frac{\oint\!\!\oint_{\Delta s_n} \mathbf{F}_n \cdot d\mathbf{s}_n}{\Delta v_n}$$

or

$$(\mathbf{\nabla} \cdot \mathbf{F})_n \, \Delta v_n \approx \oint\!\!\oint_{\Delta s_n} \mathbf{F}_n \cdot d\mathbf{s}_n$$

Writing similar expressions for all the small volumes and summing,

$$\sum_{n=1}^{N} (\mathbf{\nabla} \cdot \mathbf{F})_n \, \Delta v_n \approx \sum_{n=1}^{N} \oint\!\!\oint_{\Delta s_n} \mathbf{F}_n \cdot d\mathbf{s}_n$$

The right side of the preceding equation is the flux out of the *external* surface s, that is,

$$\oint\!\!\oint_{s} \mathbf{F}|_s \cdot d\mathbf{s}$$

because the contributions from all the surfaces (or partial surfaces) *internal* to s cancel. They cancel, because at interface surfaces common to adjacent small volumes, $d\mathbf{s}_n$ will be the same in magnitude, but opposite in direction. Thus,

$$\sum_{n=1}^{N} (\mathbf{\nabla} \cdot \mathbf{F})_n \, \Delta v_n \approx \oint\!\!\oint_{s} \mathbf{F}|_s \cdot d\mathbf{s}$$

Now if we let N approach infinity so that Δv_n approaches zero, the sum on the left becomes a volume integral and the approximation becomes exact.

$$\iiint_{\text{vol}} \mathbf{\nabla} \cdot \mathbf{F} \, dv = \oint\!\!\oint_{s} \mathbf{F}|_s \cdot d\mathbf{s} \tag{G.11}$$

Equation G.11 is known as the divergence theorem.

STOKE'S THEOREM

Consider an *open* surface s in space whose periphery is c with a well-behaved vector field $\mathbf{A}$ present. Figure 3.6 demonstrates the geometry and the positive sense for the vector quantities involved. The surface s (not necessarily planar) is subdivided into $\Delta\mathbf{s}_1, \Delta\mathbf{s}_2, \ldots, \Delta\mathbf{s}_n, \ldots, \Delta\mathbf{s}_N$. If each $\Delta\mathbf{s}_n$ is sufficiently small, then $\mathbf{A}_n$ may be assumed to be constant over it. We may write for the circulation

$$\oint_c \mathbf{A} \cdot d\mathbf{l} \approx \sum_{n=1}^{N} \oint_{c_n} \mathbf{A}_n \cdot d\mathbf{l} \tag{G.12}$$

This result occurs because the line integrals over segments of the c_n interior to c are zero since the line integrals along a common boundary for adjacent areas are taken in *opposite* directions. Only those parts of the c_n's which coincide with c contribute to the sum. Letting N approach infinity so that Δs_n approaches zero gives

$$\oint_c \mathbf{A} \cdot d\mathbf{l} = \lim_{N\to\infty} \sum_{n=1}^{N} \oint_c \mathbf{A}_n \cdot d\mathbf{l} \tag{G.13}$$

The component in the $\mathbf{a}_n$ direction of the curl of $\mathbf{A}$, $(\nabla \times \mathbf{A})$, at a point is defined by equation 3.41 as the limit of the circulation per unit area as the area $(\mathbf{a}_n \, \Delta s)$ approaches zero.

$$(\nabla \times \mathbf{A})_n = \lim_{\Delta s\to 0} \frac{1}{\Delta s} \oint_c \mathbf{A} \cdot d\mathbf{l}$$

or

$$\mathbf{a}_n \cdot (\nabla \times \mathbf{A}) = \lim_{\Delta s\to 0} \frac{1}{\Delta s} \oint_c \mathbf{A} \cdot d\mathbf{l}$$

or

$$\mathbf{a}_n \cdot (\nabla \times \mathbf{A}) \approx \frac{1}{\Delta s} \oint_c \mathbf{A} \cdot d\mathbf{l}$$

or, applied to $\Delta\mathbf{s}_n$,

$$\mathbf{a}_n \cdot (\nabla \times \mathbf{A}) \approx \frac{1}{\Delta s_n} \oint_{c_n} \mathbf{A}_n \cdot d\mathbf{l}$$

Therefore,

$$\mathbf{a}_n \cdot (\nabla \times \mathbf{A}) \, \Delta s_n \approx \oint_{c_n} \mathbf{A}_n \cdot d\mathbf{l} \tag{G.14}$$

Substituting equation G.14 into equation G.13

$$\oint_c \mathbf{A} \cdot d\mathbf{l} = \lim_{\substack{N \to \infty \\ \Delta s_n \to 0}} \sum_{n=1}^{N} \mathbf{a}_n \cdot (\nabla \times \mathbf{A}) \, \Delta s_n$$

$$= \lim_{\substack{N \to \infty \\ \Delta s_n \to 0}} \sum_{n=1}^{N} (\nabla \times \mathbf{A}) \cdot \Delta \mathbf{s}_n$$

But the limit is an (open surface) integral in the usual way, so

$$\oint_c \mathbf{A} \cdot d\mathbf{l} = \iint_s (\nabla \times \mathbf{A}) \cdot d\mathbf{s} \tag{G.15}$$

Equation G.15 is known as Stoke's theorem.

Appendix H
Special Functions

RECTANGULAR COORDINATES

Separation of Variables

$$\frac{\partial^2 \alpha}{\partial x^2} + \frac{\partial^2 \alpha}{\partial y^2} + \frac{\partial^2 \alpha}{\partial z^2} + k^2 \alpha = 0 \{\text{scalar Helmholtz (wave) equation} \qquad \text{(H.1)}$$

Assume $\alpha(x, y, z) = X(x)Y(y)Z(z)$ and substitute into equation H.1, then divide by α, giving

$$\frac{1}{X}\frac{d^2 X}{dx^2} + \frac{1}{Y}\frac{d^2 Y}{dx^2} + \frac{1}{Z}\frac{d^2 Z}{dz^2} + k^2 = 0 \qquad \text{(H.2)}$$

Each term is separately a constant, or

$$\frac{1}{X}\frac{d^2 X}{dx^2} = -k_x^2, \qquad \frac{1}{Y}\frac{d^2 Y}{dy^2} = -k_y^2, \qquad \frac{1}{Z}\frac{d^2 Z}{dz^2} = -k_z^2 \qquad \text{(H.3)}$$

so the separation equation is

$$k^2 = k_x^2 + k_y^2 + k_z^2 \qquad \text{(H.4)}$$

The first of equations H.3 has as a solution the sum of any *two* harmonic functions, $h(k_x x)$, where

$$h(k_x x) = \cos(k_x x), \qquad \sin(k_x x), \qquad e^{jk_x x} \quad \text{or} \quad e^{-jk_x x} \tag{H.5}$$

The second and third of equations H.3 have similar solutions, so a general solution to H.1 is

$$\alpha(x, y, z) = h(k_x x)h(k_y y)h(k_z z) \tag{H.6}$$

CYLINDRICAL COORDINATES

Separation of Variables

$$\frac{1}{\rho}\frac{\partial}{\partial \rho}\left(\rho \frac{\partial \alpha}{\partial \rho}\right) + \frac{1}{\rho^2}\frac{\partial^2 \alpha}{\partial \phi^2} + \frac{\partial^2 \alpha}{\partial z^2} + k^2\alpha = 0 \{\text{scalar Helmholtz (wave) equation} \tag{H.7}$$

Assume $\alpha(\rho, \phi, z) = R(\rho)\Phi(\phi)Z(z)$ and substitute into equation H.7, then divide by α, giving

$$\frac{1}{\rho R}\frac{d}{d\rho}\left(\rho \frac{dR}{d\rho}\right) + \frac{1}{\rho^2 \Phi}\frac{d^2\Phi}{d\phi^2} + \frac{1}{Z}\frac{d^2 Z}{dz^2} + k^2 = 0 \tag{H.8}$$

The third term of equation H.8 must be a constant, so

$$\frac{1}{Z}\frac{d^2 Z}{dz^2} = -k_z^2 \tag{H.9}$$

Equation H.9 has a solution (see equation H.5), which is the sum of any two harmonic functions, $h(k_z z)$. Then equation H.8 becomes

$$\frac{\rho}{R}\frac{d}{d\rho}\left(\rho \frac{dR}{d\rho}\right) + \frac{1}{\Phi}\frac{d^2\Phi}{d\phi^2} + (k^2 - k_z^2)\rho^2 = 0 \tag{H.10}$$

The second term of equation H.10 must now be a constant, so

$$\frac{1}{\Phi}\frac{d^2\Phi}{d\phi^2} = -n^2 \tag{H.11}$$

Equation H.11 has a solution which is the sum of any two harmonic functions, $h(n\phi)$. Then, equation H.10 becomes

$$\frac{\rho}{R}\frac{d}{d\rho}\left(\rho \frac{dR}{d\rho}\right) - n^2 + (k^2 - k_z^2)\rho^2 = 0 \tag{H.12}$$

With the separation equation,

$$k^2 = k_\rho^2 + k_z^2 \tag{H.13}$$

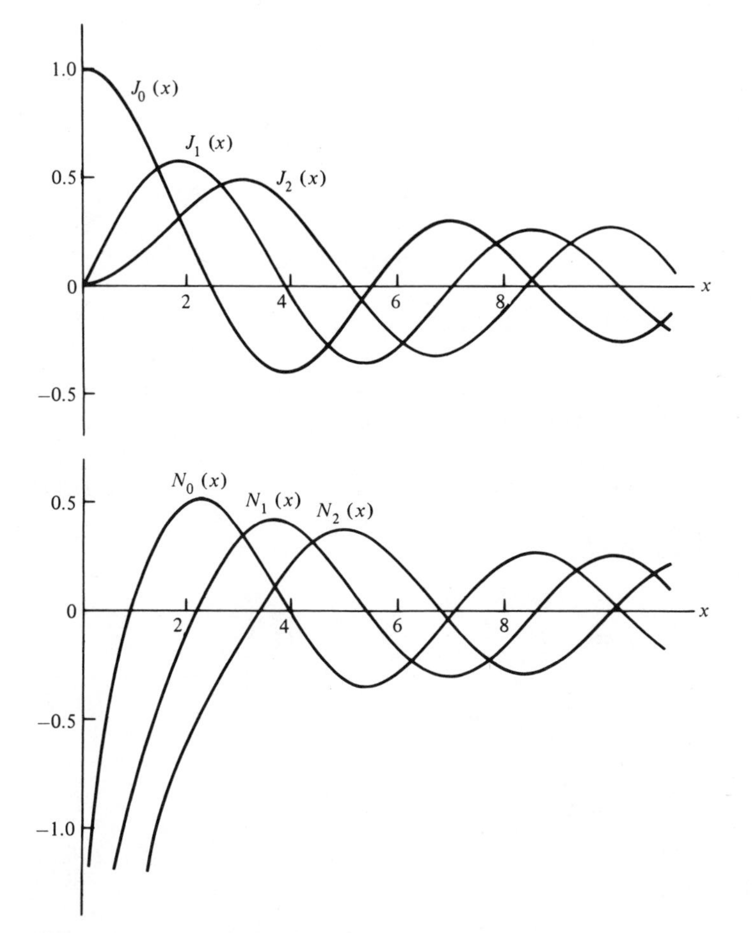

Figure H.1. $J_n(x)$ and $N_n(x)$ versus x, $n = 0, 1, 2$.

equation H.12 can be written

$$\frac{\rho}{R}\frac{d}{d\rho}\left(\rho\frac{dR}{d\rho}\right) - n^2 + k_\rho^2\rho^2 = 0 \tag{H.14}$$

which is Bessel's equation of order n. This equation has as a solution the sum of *Bessel functions*[1] of the first kind, $J_n(k_\rho\rho)$, and Bessel functions of the second kind, $N_n(k_\rho\rho)$ (Neumann functions). Particular linear combinations of J_n and N_n, are called Hankel functions. The first kind is $H_n^{(1)}(k_\rho\rho) = J_n(k_\rho\rho) + jN_n(k_\rho\rho)$, and the second kind is $H_n^{(2)}(k_\rho\rho) = J_n(k_\rho\rho) - jN_n(k_\rho\rho)$. The general solution to equation H.7 is

$$\alpha(\rho, \phi, z) = \begin{Bmatrix} J_n(k_\rho\rho) \\ N_n(k_\rho\rho) \end{Bmatrix} h(n\phi)h(k_z z) \tag{H.15}$$

[1] See Spiegel in the references at the end of Chapter 1.

The recurrence relations are frequently useful. They are ($f_n = J_n$ or N_n)

$$f'_n(x) = f_{n-1}(x) - \frac{n}{x} f_n(x)$$
$$f'_n(x) = -f_{n+1}(x) + \frac{n}{x} f_n(x) \tag{H.16}$$

$J_n(x)$ and $N_n(x)$ for $n = 0$, 1, and 2 are shown in Figure H.1.

CYLINDRICAL WAVEGUIDE (FIGURE H.2)

· *TM to z Mode.* In equation 14.10, let $\mathbf{A} = \mathbf{a}_z A_z$, $A_\rho = A_\phi = 0$ and $\mathbf{F} = 0$. Expand equation 14.10 in circular cylindrical coordinates. We then have

$$\frac{1}{\rho}\frac{\partial}{\partial \rho}\left(\rho \frac{\partial A_z}{\partial \rho}\right) + \frac{1}{\rho^2}\frac{\partial^2 A_z}{\partial \phi^2} + \frac{\partial^2 A_z}{\partial z^2} + k^2 A_z = 0 \tag{H.17}$$

and

$$\begin{aligned} E_\rho &= \frac{1}{j\omega\mu\varepsilon}\frac{\partial^2 A_z}{\partial \rho\, \partial z} & H_\rho &= \frac{1}{\mu\rho}\frac{\partial A_z}{\partial \phi} \\ E_\phi &= \frac{1}{j\omega\mu\varepsilon\rho}\frac{\partial^2 A_z}{\partial \phi\, \partial z} & H_\phi &= -\frac{1}{\mu}\frac{\partial A_z}{\partial \rho} \\ E_z &= \frac{1}{j\omega\mu\varepsilon}\left(k^2 + \frac{\partial^2}{\partial z^2}\right)A_z & H_z &\equiv 0 \end{aligned} \tag{H.18}$$

This, then, is always a *TM to z mode*. A general solution to equation H.17 as given by equation H.15 is

$$\begin{aligned} A_z(\rho, \phi, z) = {} & [AJ_n(k_\rho \rho) + BN_n(k_\rho \rho)] \cdot (C \cos n\phi + D \sin n\phi) \\ & \cdot (E \cos k_z z + F \sin k_z z) \end{aligned} \tag{H.19}$$

where $n = 0, 1, 2, \ldots,$ and the eigenvalues k_z and k_ρ are related by the separation equation, H.13,

$$k_z^2 = k^2 - k_\rho^2$$

No loss of generality occurs *in our problem* if we choose $D = 0$. We also choose $F = -jE$ (as in Chapter 14) to obtain the propagating form, $Ee^{-jk_z z}$. $N_n(k_\rho \rho)$ is a Bessel function of the second kind. These functions all tend toward $(-)$ infinity as ρ approaches zero (see Figure H.1). We must include $\rho = 0$ in our solution, and must therefore exclude $N_n(k_\rho \rho)$. Thus, we set $B = 0$. With $ACE \equiv 1$, we have

$$A_z = J_n(k_\rho \rho) \cos n\phi \; e^{-jk_z z} \tag{H.20}$$

The boundary conditions are (see Figure H.2)

$$E_\phi = E_z = 0, \qquad \rho = a \tag{H.21}$$

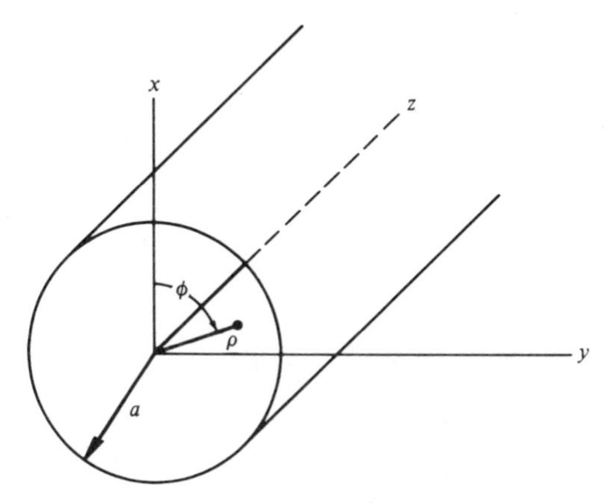

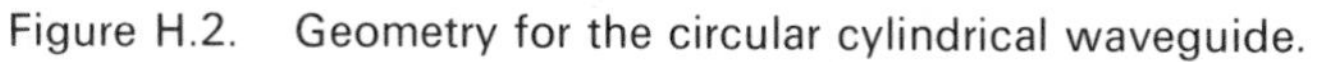

Figure H.2. Geometry for the circular cylindrical waveguide.

Equations H.18 reveal that these conditions are satisfied if A_z is zero itself when $\rho = a$. Therefore,

$$J_n(k_\rho a) = 0 \tag{H.22}$$

Inspection of Figure H.1 reveals that $J_n(k_\rho a)$ has an infinite number of zeros for *each* value of n. We will denote these roots as x_{nl}. That is, x_{nl} is the lth root of $J_n(x) = 0$. The first few roots are shown in Table H.1. We now have

$$k_\rho = \frac{x_{nl}}{a} \begin{cases} n = 0, 1, 2, \ldots \\ l = 1, 2, 3, \ldots \end{cases} \tag{H.23}$$

and

$$k_z^2 = k^2 - \left(\frac{x_{nl}}{a}\right)^2 \tag{H.24}$$

The potential is now given by

$$A_z = J_n\left(x_{nl}\frac{\rho}{a}\right) \cos n\phi e^{-jk_z z} \tag{H.25}$$

The field is obtained by substituting equation H.25 into equation H.18.

Table H.1 x_{nl}, lth ROOT OF $J_n(x) = 0$

l \ n	0	1	2	3
1	2.405	3.832	5.136	6.380
2	5.520	7.016	8.417	9.761
3	8.654	10.173	11.620	13.015

Cutoff occurs when $k_z = 0$, or $k_c = x_{nl}/a$. The cutoff frequency is

$$f_c = \frac{x_{nl}}{2\pi a\sqrt{\mu\varepsilon}} \tag{H.26}$$

so that

$$k_z = \begin{cases} k\sqrt{1 - \left(\dfrac{\omega_c}{\omega}\right)^2} = \beta, & \omega > \omega_c \\ -jk_c\sqrt{1 - \left(\dfrac{\omega}{\omega_c}\right)^2} = -j\alpha, & \omega < \omega_c \end{cases} \tag{H.27}$$

The transverse wave impedance is

$$\eta_{TM}^{+} = \frac{E_\rho}{H_\phi} = -\frac{E_\phi}{H_\rho} = \frac{k_z}{\omega\varepsilon} \tag{H.28}$$

which is the same form as for the rectangular waveguide.

· *TE to z Mode.* In equation 14.17, let $\mathbf{F} = \mathbf{a}_z F_z$, $F_\rho = F_\phi = 0$ and $\mathbf{A} = 0$. Expand equation 14.17 in circular cylindrical coordinates. We then have

$$\frac{1}{\rho}\frac{\partial}{\partial\rho}\left(\rho\frac{\partial F_z}{\partial\rho}\right) + \frac{1}{\rho^2}\frac{\partial^2 F_z}{\partial\phi^2} + \frac{\partial^2 F_z}{\partial z^2} + k^2 F_z = 0 \tag{H.29}$$

$$k_z^2 = k^2 - k_\rho^2 \tag{H.30}$$

and

$$\begin{aligned} E_\rho &= -\frac{1}{\rho}\frac{\partial F_z}{\partial\phi} & H_\rho &= \frac{1}{j\omega\mu}\frac{\partial^2 F_z}{\partial\rho\partial z} \\ E_\phi &= \frac{\partial F_z}{\partial\rho} & H_\phi &= \frac{1}{j\omega\mu\rho}\frac{\partial^2 F_z}{\partial\phi\,\partial z} \\ E_z &\equiv 0 & H_z &= \frac{1}{j\omega\mu}\left(k^2 + \frac{\partial^2}{\partial z^2}\right)F_z \end{aligned} \tag{H.31}$$

This is a TE to z mode. An appropriate solution to equation H.29 is

$$F_z = J_n(k_\rho \rho)\cos n\phi e^{-jk_z z} \tag{H.32}$$

which is the same as equation H.20. The boundary conditions are the same as those given by equation H.21. Equations H.31 reveal that E is identically zero, and the boundary condition on E_ϕ is satisfied if

$$\frac{\partial F_z}{\partial\rho}\Big|_{\rho=a} = 0$$

This condition can be satisfied if

$$\frac{\partial J_n(k_\rho \rho)}{\partial\rho}\Big|_{\rho=a} = 0 \tag{H.33}$$

Table H.2 x'_{nl}, lth ROOT OF $J'_n(x) = 0$

l \ n	0	1	2	3
1	3.832	1.841	3.054	4.201
2	7.016	5.331	6.706	8.015
3	10.173	8.536	9.969	11.346

Inspection of Figure H.1 reveals that the derivative of any order (n) Bessel function has an infinite number of zeros. We denote the roots by x'_{nl}. That is, x'_{nl} is the lth root of $J'_n(x) = 0$. The first few values of x'_{nl} are shown in Table H.2. The eigenvalue k_ρ is

$$k_\rho = \frac{x'_{nl}}{a} \begin{cases} n = 0, 1, 2, \ldots \\ l = 1, 2, 3, \ldots \end{cases} \tag{H.34}$$

The separation equation is then

$$k_z^2 = k^2 - \left(\frac{x'_{nl}}{a}\right)^2 \tag{H.35}$$

and the potential is now given by

$$F_z = J_n\left(x'_{nl}\frac{\rho}{a}\right)\cos n\phi e^{-jk_z z} \tag{H.36}$$

The field is obtained by substituting equation H.36 into equations H.31.

Cutoff, as usual, occurs when $k_z = 0$, or $k_c = x'_{nl}/a$. The cutoff frequency is

$$f_c = \frac{x'_{nl}}{2\pi a\sqrt{\mu\varepsilon}} \tag{H.37}$$

Notice that equation H.37 is not the same as equation H.26. The propagation constant is

$$k_z = \begin{cases} k\sqrt{1 - \left(\dfrac{\omega_c}{\omega}\right)^2} = \beta, & \omega > \omega_c \\ -jk_c\sqrt{1 - \left(\dfrac{\omega}{\omega_c}\right)^2} = -j\alpha, & \omega < \omega_c \end{cases} \tag{H.38}$$

The wave impedance is

$$\eta_{TE}^{+} = \frac{E_\rho}{H_\phi} = -\frac{E_\phi}{H_\rho} = \frac{\omega\mu}{k_z} \tag{H.39}$$

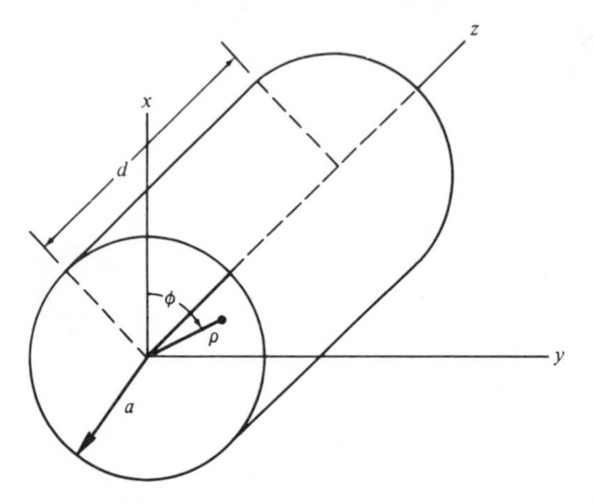

Figure H.3. Geometry for the circular cylindrical cavity.

which is the same form as the wave impedance for rectangular waveguides (TE mode).

Circular cylindrical waveguides have a number of applications besides the normal use as a structure for conveying microwave power from one point to another. Any time mechanical rotation is required (such as for feeding a rotating antenna), a circular geometry is called for. Another application of the circular waveguide which takes advantage of the circular symmetry is found in the "waveguide beyond cutoff" type attenuator. When a waveguide is operated beyond cutoff, the attenuation is exponentially distributed (an evanescent field) along the axial direction from the source toward the receiving end. A pickup probe can be mounted in a barrel which is threaded onto the waveguide. Thus, the probe can be moved along the waveguide by rotating the barrel. The attenuation *in decibels* will be linear with axial distance, and therefore linear with barrel turns.

CYLINDRICAL CAVITY (FIGURE H.3)

· *TM to z Mode.* Based on results for the rectangular cavity,

$$A_z = J_n\left(x_{nl}\frac{\rho}{a}\right)\cos n\phi \cos\frac{q\pi z}{d} \tag{H.40}$$

where $n = 0, 1, 2, 3, \ldots, l = 1, 2, 3, \ldots, q = 0, 1, 2, 3, \ldots$, and the resonant frequency is

$$f_r = \frac{1}{2\pi a\sqrt{\mu\varepsilon}}\sqrt{(x_{nl})^2 + \left(\frac{q\pi}{d}\right)^2} \tag{H.41}$$

The field is given by equation H.18.

· *TE to z Mode.* $F_z = J_n\left(x'_{nl}\frac{\rho}{a}\right)\cos n\phi \sin\frac{q\pi z}{d}$ (H.42)

where $n = 0, 1, 2, 3, \ldots, l = 1, 2, 3, \ldots, q = 1, 2, 3, \ldots$, and

$$f_r = \frac{1}{2\pi a\sqrt{\mu\varepsilon}}\sqrt{(x'_{nl})^2 + \left(\frac{q\pi}{d}\right)^2} \tag{H.43}$$

The field is given by equation H.31.

A circular cylindrical cavity is very useful where a tunable cavity is required. The rotational symmetry is again used to advantage in changing the axial dimension d for such a cavity. When $d \leq 2a$ the dominant mode is the TM_{010} mode.

SPHERICAL COORDINATES

Separation of Variables

$$\frac{1}{r^2}\frac{\partial}{\partial r}\left(r^2\frac{\partial\alpha}{\partial r}\right) + \frac{1}{r^2 \sin\theta}\frac{\partial}{\partial\theta}\left(\sin\theta\frac{\partial\alpha}{\partial\theta}\right) + \frac{1}{r^2\sin^2\theta}\frac{\partial^2\alpha}{\partial\phi^2} + k^2\alpha = 0 \text{\{scalar Helmholtz (wave) equation} \tag{H.44}$$

Assume $\alpha(r, \theta, \phi) = R(r)\Theta(\theta)\Phi(\phi)$ and substitute into equation H.44, then divide by α, giving

$$\frac{\sin^2\theta}{R}\frac{d}{dr}\left(r^2\frac{dR}{dr}\right) + \frac{\sin\theta}{\Theta}\frac{d}{d\theta}\left(\sin\theta\frac{d\Theta}{d\theta}\right) + \frac{1}{\Phi}\frac{d^2\Phi}{d\phi^2} + k^2r^2\sin^2\theta = 0 \tag{H.45}$$

The third term of equation H.45 must be a constant, so

$$\frac{1}{\Phi}\frac{d^2\Phi}{d\phi^2} = -m^2 \tag{H.46}$$

Equation H.46 has as a solution the sum of any two harmonic functions, $h(m\phi)$. Then equation H.45 becomes

$$\frac{1}{R}\frac{d}{dr}\left(r^2\frac{dR}{dr}\right) + \frac{1}{\Theta\sin\theta}\frac{d}{d\theta}\left(\sin\theta\frac{d\Theta}{d\theta}\right) - \frac{m^2}{\sin^2\theta} + k^2r^2 = 0 \tag{H.47}$$

Now,

$$\frac{1}{\Theta\sin\theta}\frac{d}{d\theta}\left(\sin\theta\frac{d\Theta}{d\theta}\right) - \frac{m^2}{\sin^2\theta} = -n(n+1) \quad \text{(constant)} \tag{H.48}$$

which has as a solution the sum of any two *associated Legendre functions*,[2] $P_n^m(\cos\theta)$ (first kind), or $Q_n^m(\cos\theta)$ (second kind). Substitution of equation H.48 into H.47 leaves

$$\frac{d}{dr}\left(r^2\frac{dR}{dr}\right) + [(kr)^2 - n(n+1)]R = 0 \tag{H.49}$$

Solutions to equation H.49 are *spherical Bessel functions*,[2] related to ordinary Bessel functions by

$$\begin{Bmatrix} j_n(kr) \\ n_n(kr) \end{Bmatrix} = \sqrt{\frac{\pi}{2kr}}\begin{Bmatrix} J_{n+1/2}(kr) \\ N_{n+1/2}(kr) \end{Bmatrix} \tag{H.50}$$

The general solution to equation H.44 then is

$$\alpha(r,\theta,\phi) = \begin{Bmatrix} j_n(kr) \\ n_n(kr) \end{Bmatrix} \cdot \begin{Bmatrix} P_n^m(\cos\theta) \\ Q_n^m(\cos\theta) \end{Bmatrix} \cdot h(m\phi) \tag{H.51}$$

Spherical Cavity

Here we desire to find solutions, which are either TM to r ($H_r = 0$) or TE to r ($E_r = 0$), suitable for representing the fields inside a spherical cavity. For the rectangular and cylindrical cavity we were able to let $\alpha(x, y, z)$ or $\alpha(\rho, \phi, z)$ equal A_z(TM to z) or F_z(TE to z) since $\nabla^2 A_z$ (or $\nabla^2 F_z$) $= (\nabla^2\mathbf{A})_z$ [or $(\nabla^2\mathbf{F})_z$]. Now, $\nabla^2 A_r$ (or $\nabla^2 F_r$) $\neq (\nabla^2\mathbf{A})_r$ [or $(\nabla^2\mathbf{F})_r$], so a new formulation is necessary. The starting point for A_r is equation 9.62 written in phasor form.

$$\nabla\times\nabla\times\mathbf{A} - k^2\mathbf{A} = -j\omega\mu\varepsilon\nabla\Phi_a \tag{H.52}$$

where $k^2 = \omega^2\mu\varepsilon$, $\mathbf{J} \equiv 0$ (inside the cavity), and

$$\mathbf{B} = \nabla\times\mathbf{A} \tag{H.53}$$

Expanding equation H.52 in spherical coordinates gives

$$\begin{aligned} \frac{1}{r^2\sin\theta}\frac{\partial}{\partial\theta}\left(\sin\theta\frac{\partial A_r}{\partial\theta}\right) + \frac{1}{r^2\sin^2\theta}\frac{\partial^2 A_r}{\partial\phi^2} + k^2 A_r &= j\omega\mu\varepsilon\frac{\partial\Phi_a}{\partial r} \\ -\frac{\partial^2 A_r}{\partial r\,\partial\theta} &= j\omega\mu\varepsilon\frac{\partial\Phi_a}{\partial\theta} \\ -\frac{\partial^2 A_r}{\partial r\,\partial\phi} &= j\omega\mu\varepsilon\frac{\partial\Phi_a}{\partial\phi} \end{aligned} \tag{H.54}$$

if $A_\theta = A_\phi = 0$. The last two of equations H.54 are satisfied if we choose

$$\frac{\partial A_r}{\partial r} = -j\omega\mu\varepsilon\Phi_a \tag{H.55}$$

[2] See Spiegel in the references at the end of Chapter 1.

which is *not* the Lorentz condition! Substituting equation H.55 into the first of equations H.54 gives

$$\frac{1}{r}\left[\frac{\partial^2 A_r}{\partial r^2} + \frac{1}{r^2 \sin\theta}\frac{\partial}{\partial\theta}\left(\sin\theta\,\frac{\partial A_r}{\partial\theta}\right) + \frac{1}{r^2\sin^2\theta}\frac{\partial^2 A_r}{\partial\phi^2} + k^2 A_r\right] = 0 \tag{H.56}$$

It is a straightforward problem to show that the left side of equation H.56 is identical with

$$\nabla^2\left(\frac{A_r}{r}\right) + k^2\left(\frac{A_r}{r}\right) = 0$$

so A_r/r is a solution to the scalar Helmholtz equation, which, we have already seen, is α in equation H.51. Therefore,

$$A_r = r\alpha$$

or

$$A_r(r, \theta, \phi) = r\begin{Bmatrix} j_n(kr) \\ n_n(kr) \end{Bmatrix}\begin{Bmatrix} P_n^m(\cos\theta) \\ Q_n^m(\cos\theta) \end{Bmatrix} h(m\phi) \tag{H.57}$$

For TE to r modes we start with

$$\nabla \times \nabla \times \mathbf{F} - k^2\mathbf{F} = -j\omega\mu\nabla\Phi_f \tag{H.58}$$

and

$$\mathbf{E} = -\nabla \times \mathbf{F} \tag{H.59}$$

from the equations preceding equations 14.13. If we choose

$$\frac{\partial F_r}{\partial r} = -j\omega\mu\Phi_f \tag{H.60}$$

we have

$$\nabla^2\left(\frac{F_r}{r}\right) + k^2\left(\frac{F_r}{r}\right) = 0$$

and F_r/r satisfies the scalar Helmholtz equation. Therefore,

$$F_r = r\alpha$$

or

$$F_r(r, \theta, \phi) = r\begin{Bmatrix} j_n(kr) \\ n_n(kr) \end{Bmatrix}\begin{Bmatrix} P_n^m(\cos\theta) \\ Q_n^m(\cos\theta) \end{Bmatrix} h(m\phi) \quad . \tag{H.61}$$

The TM to r fields are given by equation H.53 and Maxwell's equation, $\mathbf{E} = (1/j\omega\mu\varepsilon)\nabla \times \mathbf{B}$, as

$$E_r = \frac{1}{j\omega\mu\varepsilon}\left(k^2 + \frac{\partial^2}{\partial r^2}\right)A_r \qquad H_r \equiv 0$$

$$E_\theta = \frac{1}{j\omega\mu\varepsilon}\frac{\partial^2 Ar}{\partial r\,\partial\theta} \qquad H_\theta = \frac{1}{\mu r \sin\theta}\frac{\partial A_r}{\partial\phi} \tag{H.62}$$

$$E_\phi = \frac{1}{j\omega\mu\varepsilon r\sin\theta}\frac{\partial^2 A_r}{\partial r\,\partial\phi} \qquad H_\phi = -\frac{1}{\mu r}\frac{\partial A_r}{\partial\theta}$$

The TE to r fields are given by equation H.59 and Maxwell's equation, $\mathbf{H} = -(1/j\omega\mu)\nabla \times \mathbf{E}$, as

$$E_r \equiv 0 \qquad H_r = \frac{1}{j\omega\mu}\left(k^2 + \frac{\partial^2}{\partial r^2}\right)F_r$$

$$E_\theta = -\frac{1}{r\sin\theta}\frac{\partial F_r}{\partial\phi} \qquad H_\theta = \frac{1}{j\omega\mu r}\frac{\partial^2 F_r}{\partial r\,\partial\theta} \tag{H.63}$$

$$E_\phi = \frac{1}{r}\frac{\partial F_r}{\partial\theta} \qquad H_\phi = \frac{1}{j\omega\mu r\sin\theta}\frac{\partial^2 F_r}{\partial r\,\partial\phi}$$

It is convenient to use Schelkunoff's[3] Bessel functions,

$$\hat{J}_n(kr) = krj_n(kr) = \sqrt{\frac{\pi kr}{2}}\,J_{n+1/2}(kr) \tag{H.64}$$

and

$$\hat{N}_n(kr) = krn_n(kr) = \sqrt{\frac{\pi kr}{2}}\,N_{n+1/2}(kr) \tag{H.65}$$

Then the potentials may be written

$$A_r(r, \theta, \phi) \text{ or } F_r(r, \theta, \phi) = \begin{Bmatrix}\hat{J}_n(kr) \\ \hat{N}_n(kr)\end{Bmatrix}\begin{Bmatrix}P_n^m(\cos\theta) \\ Q_n^m(\cos\theta)\end{Bmatrix}h(m\phi) \tag{H.66}$$

Since $Q_n^m(+1) \to \infty$, a finite field inside a sphere is given by

$$\begin{Bmatrix}A_r(r, \theta, \phi) \\ F_r(r, \theta, \phi)\end{Bmatrix} = \hat{J}_n(kr)P_n^m(\cos\theta)h(m\phi) \tag{H.67}$$

with m and n integers.

[3] S. A. Schelkunoff, *Electromagnetic Fields*, Blaisdell Publishing Company, New York, 1963.

Notice that

$$j_0(kr) = \frac{\sin kr}{kr}, \qquad n_0(kr) = -\frac{\cos kr}{kr}$$

$$j_1(kr) = \frac{\sin kr}{(kr)^2} - \frac{\cos kr}{kr}, \qquad n_1(kr) = -\frac{\cos kr}{(kr)^2} - \frac{\sin kr}{kr}$$

$$P_0^0 = 1$$

$$P_1^0 = u$$

$$P_2^0(u) = \frac{3u^2 - 1}{2}$$

$$P_1^1(u) = -\sqrt{1 - u^2}$$

$$P_2^1(u) = -3\sqrt{1 - u^2}\,u$$

The special functions of this section are tabulated in mathematical references.[4]

[4] *Handbook of Mathematical Functions*, U.S. Department of Commerce, National Bureau of Standards, Applied Mathematics Series 55, 1964.

Answers to Selected Problems

CHAPTER 1

1. $\mathbf{F}_1 = -\mathbf{F}_2 = 3.18 \times 10^{-9} (\mathbf{a}_x - \mathbf{a}_z)$ (N)
2. $Q_2 = -1.30 \times 10^{-9}$ (C)
3. $E_z = \dfrac{\rho_l}{4\pi\varepsilon_0}\left[\dfrac{1}{\sqrt{\rho^2 + (z-h)^2}} - \dfrac{1}{\sqrt{\rho^2 + (z+h)^2}}\right]$ (V/m)
4. $x = 0.32$ (m), $y = 0$ (a line)

9. $\mathbf{E}(0, 0, 1) = -\dfrac{18\pi}{\sqrt{6}}(\mathbf{a}_x - \mathbf{a}_y + 2\mathbf{a}_z)$ (V/m)
10. $\mathbf{E}(1, 0, 0) \approx 54\mathbf{a}_x$ (V/m)

12. $Q = 0.17\text{m}$ (C)

17. (a) $E_z \approx 18\pi$ (V/m) (b) $E_z \approx 1.8$ (V/m)
    (c) $E_z \approx 9 \times 10^{-7}$ (V/m)
18. (c) 1 (e) 2

CHAPTER 2

1. $\mathbf{E}(1, 1, 1) = 9\mathbf{a}_x + 9\mathbf{a}_y + 18\pi\mathbf{a}_z$ (V/m)

6. $E_\rho = \dfrac{\rho_{v0}}{\rho\varepsilon_0}\left[\dfrac{e^{-\alpha\rho}}{\alpha^2}(-\alpha\rho - 1) + \dfrac{1}{\alpha^2}\right]$ (V/m)

7. (a) $\frac{1}{12}$ (C) (b) $\pi/4$ (C) (c) $\frac{2}{3}$ (C)

10. $W = 0$

14. $\mathbf{E}(2, 2, 1000) = 51.19\mathbf{a}_x + 51.19\mathbf{a}_y + 25.6 \times 10^3\mathbf{a}_z$ (V/m)

15. $\Psi_E = \rho_l/6$ (C)

17. (a) $\Psi_E = 0.146Q$ (C) (b) $\Psi_E = 0.354Q$ (C)

19. $\rho_{sb} = -\dfrac{\rho_{sa}a}{b}$ (C/m²)

21. Flux = 0.307

CHAPTER 3

1. $W = 2.08 \times 10^{-18}$ (J)

2. $Q = 2.22 \times 10^{-9}$ (C)

4. $\Phi = -9/2a$ (V)

8. (a) Yes (b) No (c) Yes (d) No

9. (a) $W_E = \dfrac{\pi\varepsilon_0 V_0^2}{\ln(b/a)}$ (J) (b) $C = \dfrac{2\pi\varepsilon_0}{\ln(b/a)}$ (F/m)

10. (a) $F = 19.78 \times 10^{19}$ (N) (b) $\mathbf{F}_m = -9.8m\mathbf{a}_r$ (N)
    (c) $\mathbf{g} = -9.8\mathbf{a}_r$ (m/s²) (d) $\Phi_m = -3.985 \times 10^{14}\ m/r$ (J)
    (e) $u = 11.2 \times 10^3$ (m/s)

12. (a) 0 (b) 0 (c) 0 (d) $\nabla \times \mathbf{F} = 0$, except when $\rho = 0$
    (e) $\nabla \times \mathbf{F} = 0$, except for $z = 0, d$

15. $\Phi = -\rho_s|z|/2\varepsilon_0$ (V)

16. $\mathbf{a} = -\dfrac{20xyz\mathbf{a}_x + 10x^2z\mathbf{a}_y + 10x^2y\mathbf{a}_z}{(400x^2y^2z^2 + 100x^4z^2 + 100x^4y^2)^{1/2}}$

20. $W_E = \dfrac{\rho_l^2}{4\pi\varepsilon_0^2}\ln\dfrac{b}{a}$ (J)

CHAPTER 4

3. $\Phi = V_0$ (V)

4. (a) $\mathbf{a}_n = 0.6\mathbf{a}_x + 0.8\mathbf{a}_y$, away from conductor ($\rho_s > 0$)
   (b) $|\rho_s| = 10^{-9}/72\pi$ (C/m²)

5. $\varepsilon_R = 1.5$

9. $C = 40$ (pF/m) (air), $C = 90$ (pF/m) (poly.)

10. radius = 0.685 (cm)

12. $C = 36.9$ (pF)
13. $F = Q^2/2\varepsilon_0 s$ (N)
18. $C = 115.3$ (pF/m)
22. $E_x(0, 100) = 116.2$ (μV/m)

CHAPTER 5

2. Incomplete set of boundary conditions
3. $\Phi_1 \equiv \Phi_2$
4. $\Phi = V_0 \phi/\phi_0$ (V)
8. $\rho_v = -\dfrac{\varepsilon_0 V_0}{a} e^{-r/a}\left(\dfrac{1}{a} - \dfrac{2}{r}\right)$ (C/m³), $Q = 0$ (net)
14. An infinite number
18. $\rho_s = -2\varepsilon V_0/a \sum_{n=1}^{\infty} [1 - (-1)^n] \text{ctnh}\left(\dfrac{n\pi b}{a}\right)\sin\left(\dfrac{n\pi x}{a}\right)$ (C/m²)
21. $$\Phi(x, y) = \frac{2V_0}{\pi} \sum_{n=1}^{\infty} [1 - (-1)^n] \times \frac{\sinh(n\pi y/a) + \sinh[n\pi(b - y)/a]}{n \sinh(n\pi b/a)} \sin(n\pi x/a) \quad \text{(V)}$$
24. (a) $\Phi_1 = 6.62$ (V), $\Phi_2 = 26.47$ (V), $\Phi_3 = 20.22$ (V), $\Phi_4 = 58.82$ (V), $\Phi_5 = 54.41$ (V), $\Phi_6 = 38.60$ (V)
    (b) $\Phi_1 = 0.15$ (V), $\Phi_2 = 0.59$ (V), $\Phi_3 = 0.43$ (V), $\Phi_4 = 48.31$ (V), $\Phi_5 = 46.33$ (V), $\Phi_6 = 36.58$ (V)

CHAPTER 6

2. (a) 0 (b) 2.18 (A) (c) 0
3. $I = 0.126$ (A)
5. $R = 0.177$ (Ω)
9. (a) transit time = 0.506 (ns) (b) ≈ 200 (MHz)
13. $R \approx \dfrac{1}{2\pi\sigma\delta}\left[\dfrac{1}{a} - \dfrac{1}{b}\right]$ (Ω/m)
16. $C = 35.37$ (pF)
17. $G = 1.06$ (μ℧/m)
18. $t = 0.286 \times 10^{-6}$ (m)
19. $E = 1.06 \times 10^5$ (V/m)
21. $t = 2 \times 10^{-5}$ (m)
23. Nothing happens

CHAPTER 7

3. $2aJ_0$ (A)

6. $\mathbf{F} = -\dfrac{\mu_0 I^2}{2b}\mathbf{a}_y$ (N) (on lower conductor)

8. A circle

12. $W_m = mB_0$ (J)

13. $\mathbf{H} = \dfrac{I}{2\pi}\dfrac{\mathbf{a}_y}{\sqrt{x^2 + y^2}}$ (A/m)

15. (a) $\mathbf{F} = \dfrac{\mu_0 I_1 I_2 \mathbf{a}_y}{4\pi}$ (N) (b) $\mathbf{T} = 0$

17. (a) $\mathbf{T} = \pi\mathbf{a}_z$ (N · m) (b) $P = 329$ (W)

18. $\mathbf{B} = 4.45 \times 10^{-7}$ (Wb/m^2)

CHAPTER 8

3. $\mathscr{R} = 5000$ (H^{-1})

4. $B_{\text{inner}} = 0.625$ (Wb/m^2), $B_{\text{outer}} = 0.3125$ (Wb/m^2)

6. $L = 80$ (mH)

9. $F = 39.8$ (N)

20. (a) $L_{12} = 24.7$ (nH) (b) $L_{12} = 98.8$ (nH)

21. $B = 0.5$ (Wb/m^2)

22. $L_{12} = 61.4$ (nH)

23. (a) $G = 10^{-2}$ (℧) (b) $C = 8.84$ (pF/m)
    (c) $L_{\text{ext}} = 1.26$ (μH/m) (d) $D \approx 22.1$ (cm)

24. At a and c, or at b and d

CHAPTER 9

1. $\oint \mathbf{E} \cdot d\mathbf{l} = 1,\ B_z = -t$ (Wb/m^2)

3. emf $= -\omega B_0 s \cos 2\omega t$ (V)

4. emf $= -B_0 lu$ (V)

7. emf $= -\omega_0 a^2 B_0/2$ (V)

9. $H = 0.127$ (A/m)

14. $A_z = \pm E_0\sqrt{\mu_0 \varepsilon_0}\, x u(t \pm z\sqrt{\mu_0 \varepsilon_0})$ (Wb/m)

16. 0.5×10^6 (m)

20. $\mathbf{A} = \dfrac{\mu I a}{4}\dfrac{\cos \omega(t - \sqrt{a^2 + z^2}\sqrt{\mu\varepsilon})}{\sqrt{a^2 + z^2}}\mathbf{a}_y$ (Wb/m)

CHAPTER 10

6. $\mathbf{E} = -10\eta e^{-j10y}\mathbf{a}_x$ (V/m)
8. (a) $\omega = 300$ (MHz) (b) $\lambda = 2\pi$ (m)
   (c) $\langle \mathbf{S} \rangle = 0.133$ (W/m^2) (d) $u_p = 3 \times 10^8$ (m/s)
   (e) $u_e = 3 \times 10^8$ (m/s)
11. (a) $\alpha = \beta = 47.9 \times 10^3$ (Np/m) (b) $\alpha = \beta = 1513 \times 10^3$ (Np/m)
    (c) $\alpha \approx 1.09 \times 10^{10}$ (Np/m), $\beta \approx 2.09 \times 10^{14}$ (rad/m)
12. (a) $\Gamma = 0.629 \angle 164.1^\circ$ (b) $T = 0.431 \angle 23.9^\circ$ (c) SWR = 4.39
13. $J_s = 0.25$ (mA/m)
14. 86.47%
17. (a) $\theta_i = 83.62^\circ$ (b) $\theta_i = 6.38^\circ$
18. (a) $\theta_i = 6.42^\circ$ (b) $\theta_i = 38.68^\circ$
23. Copper must be 1.031 times as thick.
27. (a) $E_x^{\text{ref}} = -\frac{10}{3}\delta(t - z\sqrt{\mu_0 \varepsilon_0})$ (V/m),
    $E_x^{\text{tran}} = \frac{20}{3}\delta(t + z\sqrt{\mu_0 \varepsilon_0})$ (V/m)

CHAPTER 11

1. $R_0 = 55.56$ (Ω), $L = 0.185$ (μH/m)
2. $\Gamma_l = 0.447 \angle -116.57^\circ$, VSWR = 2.62, $Z_{\text{in}} = 43.5 + j34.2$ (Ω)
3. $Z_l = 38.83 \angle -51.73^\circ$ (Ω)
5. $Z_l = 25 \angle 2.2^\circ$ (Ω)
8. $\langle P_l \rangle = 64$ (W)
9. $z = 5.4$ (cm)
12. $R_0 = 70.71$ (Ω), $\dfrac{|V(\lambda/4)|}{|V(0)|} = 0.706$
17. $\langle P_l \rangle / \langle P_{\text{inc}} \rangle = 4(\text{VSWR})/(\text{VSWR} + 1)^2$
19. $C = 31.8$ (pF)
29. $V_T = \dfrac{V_g}{\cos kl + j(Z_g/R_0)\sin kl}$ (V),

    $Z_T = R_0 \dfrac{Z_g \cos kl + jR_0 \sin kl}{R_0 \cos kl + jZ_g \sin kl}$ (Ω)

CHAPTER 12

3. $\Gamma_l = 0.62 \angle -7.1^\circ$, VSWR = 4.27, $Z_{\text{in}} = 13.44 + j18.51$ (Ω)
4. $Z_l = 20.5 + j7.0$ (Ω)
5. $Z_l = 31.5 - j44.5$ (Ω), $R_0 = 50$ (Ω)
6. $l' = 0.078\lambda$ or $l'' = 0.175\lambda$, $s' = 0.053\lambda$ or $s'' = 0.448\lambda$

9. VSWR = 7.9
10. 1.414 (A) (peak)
12. $Z_l = (2.64 + j1.46)R_0$ (Ω)
16. VSWR = 2.4
18. $R_0' \approx 86.7$ (Ω), $d \approx 0.335\lambda$
19. Bandwidth = 174 (MHz), $\text{VSWR}_{\text{max}} = 4$

CHAPTER 13

3. $\alpha = 4.72 \times 10^{-6}$ (Np/m), $\beta = 22.07 \times 10^{-6}$ (rad/m),
   $Z_0 = 691 \ \angle -11.8^\circ$ (Ω), $u_p = 2.85 \times 10^8$ (m/s)
4. $L_{\text{add}} = 180$ (μH/m), $\alpha = 1.06 \times 10^{-6}$ (Np/m)
6. $Q = 2.34$
8. $Q = 108.7$
11. $C = 20.97$ (pF/m), $G = 5.25$ ($\mu\mho$/m), $L = 1.197$ (μH/m),
    $R = 1.66$ (Ω/m)
17. $\alpha = 12.43 \times 10^{-3}$ (Np/m), $\alpha = 15.43 \times 10^{-3}$ (Np/m)
18. $P_{\text{rec}} = 0.938 \times 10^{-12}$ (W)
20. $\lambda = 2\sqrt{\dfrac{\pi}{f\mu\sigma_c}}$ (m)

CHAPTER 14

3. $Z_l = 570$ (Ω)
4. $b = a/2 = 3.54$ (cm)
5. $\alpha = 177.9$ (Np/m)
7. 1.139×10^{-12} (s)
8. $f_r = 5.303$ (GHz)
9. $a = 2.9$ (cm), $b = 1.45$ (cm)
11. $\langle P_f \rangle = 1.05 \times 10^6$ (W)
14. Simultaneously
17. $\langle \mathbf{S} \rangle = 0$

CHAPTER 15

2. $Q_c = 11{,}100$, 10 (GHz); $Q_c = 12{,}300$, 7 (GHz)
3. $Q_c = 11{,}340$ (silver), $Q_c = 5580$ (brass)
4. $Q_c = 7825$
7. $Q_c = 12{,}340$
9. 0.43 (MHz)
12. $R_e = 0.95 \times 10^6$ (Ω), $L_e = 5.12 \times 10^{-9}$ (H),
    $C_e = 0.176 \times 10^{-12}$ (F) (brass)

CHAPTER 16

4. $E_\phi = \dfrac{k^2\eta}{4\pi}\pi a^2 I \dfrac{e^{-jkr}}{r}\sin\theta$ (V/m), $H_\theta = -\dfrac{E_\phi}{\eta}$ (A/m)

7. (a) $R_r = 40(kh)^2$ (Ω) (b) $R_r = 2.55 \times 10^{-3}$ (Ω)

   (c) $E_\theta = j13.42\dfrac{e^{-jkr}}{r}$ (V/m)

12. $E_\theta(r, t) = \dfrac{60}{r}\left[\delta\left(t - \dfrac{r}{c}\right) + 2\sum_{n=1}^{\infty}(-1)^n\delta\left(t - \dfrac{r}{c} - \dfrac{nh}{c}\right)\right]$ (V/m)

13. (a) VSWR = 1.653 (b) $|E_\theta| = 0.133$ (V/m) (rms)
    (c) $|E_\theta| = 0.137$ (V/m) rms
14. $R_r = 73.08$ (Ω)
18. $\langle P_l \rangle = 1.6 \times 10^{-6}\lambda^2$ (W)
20. $\langle P_l \rangle = 0.631 \times 10^{-6}\lambda^2$ (W)
23. It is scattered (reradiated).
24. $\langle P_l \rangle = 4.1 \times 10^{-6}$ (W)
25. $|E| = 148$ (mV/m) at 1 mile
26. $\theta_B \approx 11.5°$

Index

83 84 85 86   10 9 8 7 6 5 4